Lecture Notes
in Control and Information Sciences

200

Editor: M. Thoma

Photo: Yasaka-no-To (Yasaka Tower), Higashiyama-ku, Kyoto

Design: Yasuyoshi Yokokohji and Masamitsu Kurisu

Tsuneo Yoshikawa and Fumio Miyazaki (Eds.)

Experimental Robotics III

The 3rd International Symposium, Kyoto, Japan,
October 28–30, 1993

Springer-Verlag London Ltd.

Series Advisory Board

Editors

Tsuneo Yoshikawa, PhD
Department of Mechanical Engineering, Faculty of Engineering, Kyoto University,
Yoshidahonmachi, Kyoto 606, Japan

Fumio Miyazaki, PhD
Department of Mechanical Engineering, Faculty of Engineering Science,
Osaka University, Machikaneyama-cho 1-1, Toyonaka, Osaka 560, Japan

ISBN 978-3-540-19905-2   ISBN 978-3-540-39355-9 (eBook)
DOI 10.1007/978-3-540-39355-9

British Library Cataloguing in Publication Data
A catalogue record for this book is available from the British Library

Typesetting: Camera ready by authors

69/3830-543210 Printed on acid-free paper

Preface

Experimental Robotics III – *The Third International Symposium* took place at Kyodai-Kaikan (a conference hall affiliated with Kyoto University) in Kyoto, Japan, in October 1993. It was the third in a series of specialized small-sized international symposia dedicated to presenting and discussing in-depth research results and on-going developments in robotics which have both theoretical foundations and experimental validations. The number of the presentations was kept rather small to guarantee a thorough discussion and friendly interaction among the participants during the three-day, single-session meeting.

The first symposium was held in Montreal, Canada, in June 1989, organized by V. Hayward and O. Khatib, and the second was held in Toulouse, France, in June 1991, organized by R. Chatila and G. Hirzinger. The Proceedings of these two symposia were also published from the Springer-Verlag as *Experimental Robotics* I and II (Lecture Notes on Control and Information Sciences, Vol.139 and Vol.190). The next symposium will be held in Stanford, USA, in 1995 and will be organized by Oussama Khatib and Ken Salisbury. As can be seen from these data, this meeting is being held every two years in a circular fashion around North America, Europe, and Asia.

All of the 43 papers from 10 countries presented at the Symposium are included in this volume. Thanks to the wonderful attendees who are actively working at the forefront of their research fields and made their valuable contributions to the Symposium, we are very pleased to say with confidence that this Proceedings volume represents an essence of the state of the art of the experimental robotics research and gives a good insight into its future directions. The book consists of nine sections: Planning and Arm Control, Force Control, Visual Servoing, Sensing and Learning, Dynamic Skills, Robot Design, Teleoperation, Mobile Robots, and Space Robotics and Flexible Manipulators. Each section is forwarded by a brief introduction of the papers in the section for the convenience of the reader. The keynote lecture "Insect-Model Based Micro-robotics" was given by Prof. Hirofumi Miura from the University of Tokyo, whose experimental research effort for the new AI (Artificial Insects) strongly impressed the audience. In addition to the regular sessions, two video sessions were held and the following researchers presented their video films: K. Yokoi (*MEL*, Japan), J. Trevelyan (*University of Western Australia*, Australia), T. Yoshimi (*Toshiba Corp.*, Japan), M.R. Stein (*University of Pennsylvania*, USA), and A.A. Goldenberg (*University of Toronto*, Canada).

Almost all presentations at the Symposium were accompanied by video films, which is a natural consequence of the theme of the meeting: theoretical work validated by experiments. A compilation of part of these video films, along with those of the video sessions, will be made available soon as Video Proceedings.

The international program committee consists of the following individuals:

T. Yoshikawa	*Kyoto University*, Japan (Co-chair)	
F. Miyazaki	*Osaka University*, Japan (Co-chair)	
R. Chatila	*LAAS/CNRS*, France	
J. Craig	*SILMA*, USA	
P. Dario	*Scuola Superiore S. Anna*, Italy	
V. Hayward	*McGill University*, Canada	
G. Hirzinger	*DLR*, Germany	
O. Khatib	*Stanford University*, USA	
J.-P. Merlet	*INRIA*, France	
K. Salisbury	*MIT*, USA	

We would like to thank the following institutions for their financial support:

Foundation for Promotion of Advanced Automation, Yamanashi, Japan

The Murata Science Foundation, Kyoto, Japan

Fluid Power Technology Promotion Foundation, Osaka, Japan

We are also grateful to Yasuyoshi Yokokohji and the following students at Kyoto University who helped organizing the meeting: Xin-Zhi Zheng, Masamitsu Kurisu, Yong Yu, Syumpei Kawai, Masayuki Kawai, Kensuke Harada, Toshiki Moriguchi, Tamao Okamoto, Masashi Koike, Ryuji Haruna, Norio Hosotani, Mitsumasa Mizuno, Atsushi Matsumoto, Jyun Ueda, Hitoshi Ueda, Takashi Kubota, Keigo Kobayashi, Kenji Tanimoto, and Katsuki Matsudera. Especially, Xin-Zhi Zheng helped a lot in editing this volume also.

Tsuneo Yoshikawa (Kyoto University)
Fumio Miyazaki (Osaka University)

List of Participants

Abdul Aziz, A. A. M.
Mech. Eng. Lab.
1-2, Namiki,Tsukuba, Ibaraki 305
Japan

Alami, R.
LAAS-CNRS
7,Avenue du Colonel Roche
31077 Toulouse Cedex
France

Allotta, Benedetto
Scuola Superiore S. Anna
Via Carducci,40-Pisa 56127
Italy

Andersen, N. A.
Institute of Automatic Control Systems
Technical University of Denmark
Servolaboratoriet, Building 326
DK-2800 Lyngby
Denmark

Anderson, R. J.
Intelligent Systems and Robotics Center
Sandia National Laboratories
Div.1661,Po.Box 5800
Albuquerque, NM 87185
USA

Andreas Baader
Institute of Robotics and System Dynamics
DLR/German Aerospace Research
Post Box 1116d-82230 Wessling
Germany

Buehler, M.
Department of Mechanical Engineering
Centre for Intelligent Machines
McGill University
McConnell 423, 3480 University Street
Montréal, Québec H3A 2A7
Canada

Chatila, R.
LAAS/CNRS
7, Avenue du Colonel-Roche
31077 Toulouse Cedex
France

Chaves, P. R. de S.
Mech. Eng. Lab.
1-2 Namiki, Tsukuba, Ibaraki 305
Japan

Cherif, M.
LIFIA/INRIA Rhône-Alpes
46 Avenue Félix-Viallet
38031 Grenoble Cedex 1
France

Corke, P. I.
CSIRO
Division of Manufacturing Technology
Locked Bag 9
Preston 3072
Australia

Dario, P.
ARTS Lab.
Scuola Superiore S.Anna
Via Carducci,40
56127 Pisa
Italy

Eberman, B. S.
MIT Artificial Intelligence Laboratory
545 Technology Sq Rm831
Cambridge,MA 02139
USA

Ellis, R. E.
Department of Computing and Information Science
Goodwin Hall
Queen's University
Kingston, Ontario K7L 3N6
Canada

Espiau, B.
INRIA Rhône-Alpes
46, avenue Félix Vialet
38031 Grenoble
France

Goldenberg, A.
Robotics and Automation Laboratory
Department of Mechanical Engineering
University of Toronto
5 King's College Rd., Toronto Ontario
M5S 1A4
Canada

Hashino, S.
Mech. Eng. Lab
1-2 Namiki, Tsukuba, Ibaraki 305
Japan

Hayward, V.
McGill University
Center for intelligent Machines
3480 University Street
Montréal, Québec
Canada H3A 2A7

Hirzinger, G.
DLR Oberpfaffenhofen
German Aerospace Research Establishment
Institute for Robotics and System Dynamics
Oberpfaffenhofen, D-82234 Wessling
Germany

Howe, R. D.
Division of Applied Sciences
Harvard University
29 Oxford Street, Cambridge, MA 02138
USA

Ichikawa, J.
Research Center for Advanced Science and Technology
The University of Tokyo
4-6-1 Komaba,Meguro-ku, Tokyo 153
Japan

Ienne, P.
Swiss Federal Institute of Technology
Microcomputing Laboratory
IN-F Ecublens, CH-1015 Lausanne
Switzerland

Kaneko, M.
The Robotics Institute
Department of Industrial and Systems Eng.
Hiroshima University
1-4-1 Kagamiyama,Higashi-Hiroshima
Hirosima 724
Japan

Katsurakawa T.
Kawasaki Heavy Industries
1-1 Kawasaki-cho, Akashi 673
Japan

Khatib, O.
Robotics Laboratory
Department of Computer Science
Stanford University
Stanford, CA 94305
USA

Khosla, P. K.
The Robotics Institute
Carnegie Mellon University
Pittsburgh, PA 15213-3891
USA

Konno, A.
Department of Aeronautics and Space
Engineering
Tohoku University
Aramaki-aza-Aoba, Aoba-ku, Sendai 980
Japan

Krotkov, E.
Robotics Institute
Carnegie Mellon University
5000 Forbes Avenue
Pittsburgh, PA 15213
USA

Laugier, C.
LIFIA/INRIA Rhône-Alpes
46 Avenue Félix Viallet 38031
Grenoble Cedex 1
France

Lenarčič, J.
"Jozef Stefan" Institute
University of Ljubljana
Jamova 39, 61111 Ljubljana
Slovenia

Mazer, E. F.
LIFIA/INRIA Rhône-Alpes
46, Avenue Felix Viallet
38031 Grenoble cedex
France

Merlet, J.-P.
INRIA Sophia-Antipolis
BP 93
06902 Sophia-Antipolis Cedex
France

Miura, H.
The Univ. of Tokyo
7-3-1 Hongo, Bunkyo-ku
Tokyo 113
Japan

Miyazaki, F.
Department of Mechanical Engineering
Faculty of Engineering Science
Osaka University
Machikaneyama-tyo 1-1, Toyonaka
Osaka 560
Japan

Morel, G.
Laboratoire de Robotique de Paris
Centre Univ. de Technologie
10-12, av. de l'Europe
78140 Vélizy
France

Moritz, W.
FB 10 - Automatisierungstechnik
University of Paderborn
Pohlweg 55
D-33098 Paderborn
Germany

Nakashima, K.
Kawasaki Heavy Industries
1-1, Kawasaki-cho, Akashi 673
Japan

Nagai, K.
Depertment of Computer Science and
Systems Engineering
Faculty of Science and Engineering
Ritsumeikan University
Kitamachi, Kita-ku
Kyoto 603
Japan

Nishikawa, A.
Osaka University
Machikaneyama-tyo 1-1,Toyonaka
Osaka 560
Japan

Noborio, H.
Osaka E. C. Univ.
Hatsu-cho 18-8, Neyagawa
Osaka 572
Japan

Papegay, Y. A.
Electrotechnical Lab.
1-1-4 Umezono, Tsukuba, Ibaraki 305
Japan

Pardo, Gerardo
Aerospace Robotics Laboratory
Durand 017
Stanford University
Stanford, CA 94305
USA

Pin, F. G.
Oak Ridge National Laboratory
P.O. Box 2008
Oak Ridge, TN 37831-6364
USA

Rives, P.
INRIA-centre de Sophia Antipolis
2004 Route des Lucioles
BP93 06902 Sophia Antipolis
France

Rizzi, A.
Artificial Intelligence Laboratory
Department of Electrical Engineering and
Computer Science
University of Michigan
1101 Beal Avenue
Ann Arbor, MI 48109-2110
USA

Salcudean, T.
Department of Electrical Engineering
University of British Columbia
Vancouver, B.C.2356, Main Mall V6T
1Z4
Canada

Sano, A.
Department of Mechanical and Control
Engineering
University of Electro-Communications
1-5-1 Chofu, Tokyo 182
Japan

Sato, T.
Research Center for Advanced Science
and Technology
The University of Tokyo
4-6-1 Komaba,Meguro-ku, Tokyo 153
Japan

Schütte, H.
University of Paderborn, Pohlweg 55
D-33098 Paderborn
Germany

Siciliano, B.
Dipartimento di Informatica e Sis-
temistica
Università degli Studi di Napoli Federico
II
DIS, Via Claudio 21, 80125 Napoli
Italy

Sikka, Pavan
Department of Computing Science
University of Alberta
615 General Services Building
Edmonton, Alberta T6G 2H1
Canada

Stein, M. R.
University of Pennsylvania
3401 Walnut St. #330C, Philadelphia
PA 19104
USA

Trevelyan, J.
Department of Mechanical Engineering
University of Western Australia
Nedlands, 6009
Australia

Uchiyama, M.
Department of Aeronautics and Space
Engineering
Tohoku University
Aramaki-aza-Aoba, Aoba-ku, Sendai 980
Japan

Ueyama, K.
SANYO Electric
1-18-13 Hashiridani Hirakata
Osaka 573
Japan

Ulivi, Giovanni
Dipartimento di Informatica e Sistemistica
Università degli Studi di Roma
"La Sapienza"
Via Eudossiana, 18-00184 Roma
Italy

Whitcomb, L. L.
Department of Mathematical Engineering and Information Physics
University of Tokyo
7-3-1 Hongo, Bunkyo
Tokyo 113
Japan

Yang, B.-H.
Center for Information-Driven Mechanical Systems
Department of Mechanical Engineering
Massachusetts Institute of Technology
77 Massachusetts Ave. Rm 3-351 Cambridge, MA 02139
USA

Yokoi, K.
Robotics Department Mechanical Engineering Laboratory
Namiki 1-2, Tsukuba, Ibaraki 305
Japan

Yokokohji, Y.
Department of Mechanical Engineering
Faculty of Engineering
Kyoto University
Yoshidahonmachi, Kyoto 606
Japan

Yoshida, K.
Dept. of Mechano-Aerospace Engineering
Tokyo Institute of Technology
2-12-1, O-okayama, Meguro
Tokyo 152
Japan

Yoshikawa, T.
Department of Mechanical Engineering
Faculty of Engineering
Kyoto University
Yoshidahonmachi, Kyoto 606
Japan

Yoshimi, T.
Toshiba R&D Center
Toshiba Corporation
4-1 Ukishima-cho,Kawasaki-ku,Kawasaki
Kanagawa 210
Japan

Yuta, S.
Intelligent Robotics Laboratory
University of Tsukuba
Tsukuba 305
Japan

Zelinsky, A.
Intelligent Machine Behaviour Section
Intelligent Systems Division
Electrotechnical Laboratory
1-1-4 Umezono, Tsukuba Science City
Ibaraki 305
Japan

Contents

Section 9: Space Robotics and Flexible Manipulators . 529

Author Index
— (section) page —

Section 1
Planning and Arm Control

Motion planning and control of robot manipulators have been one of the major research fields in robotics and a vast number of works have been reported. This field still continues to be so, one of the reasons for this being the importance and wide applicability in practice.

Avoiding singularities is a fundamental problem that any motion control algorithm should face. Chiaverini, Siciliano, and Egeland demonstrate the effectiveness of their weighted damped-least square inverse kinematics solution with a feedback correction error term using the six-joint ABB IRb 2000 manipulator.

Williams and Khatib describe an object-level control methodology for manipulating objects by multiple arms while controlling the internal forces to the desired values. This method is based on their two concepts: The virtual linkage that provides a physical interpretation of the internal forces and the augmented object that consists of the arms and the object whose dynamics are all described in an operational space.

Kaneko and Hayashi propose a control law that overcomes a problem of deadlock in the task of contact point detection by using the self-posture changing motion of a robot finger in contact with an object. This motion is useful in detecting approximately the contact point and shape of an unknown object in contact.

The paper by Lenarčič and Žlajpah studies a problem of minimum joint torque motion for redundant manipulators, especially for a planar 2-R manipulator which lifts a load from the initial straight downward arm configuration to the final straight upward one. A quadratic cost function of the joint torques is used to find a locally optimum Cartesian path and two control schemes are given which include singularity-avoidance function too.

Pardo-Castellote, Li, Koga, Cannon, Latombe, and Schneider present an overall system architecture design for achieving task-level operation on a two-armed robotic assembly workcell. Their system consists of four components: (1) graphical user interface which receives high-level task specification from the user, (2) planner which decomposes the specified task into subtasks and finds path of the arms for each subtask, (3) the dual-arm robot control and sensor system, and (4) on-line simulator which masquerade as the robot for fast prototyping and testing of the rest of the system.

Mazer, Ahuactzin, Talbi, Bessiere, and Chatroux describe an implementation of a path planner for a six degrees of freedom arm on a parallel machine with 128 transputers. Although this implementation achieves a very small planning time (0.5 to 3 seconds for the case with 10 obstacles), the authors are not satisfied yet because their original idea was to demonstrate the possibility of including a global path planner into an incremental trajectory generator.

Experimental Results on Controlling a 6-DOF Robot Manipulator in the Neighborhood of Kinematic Singularities

Stefano Chiaverini   Bruno Siciliano

Dipartimento di Informatica e Sistemistica
Università degli Studi di Napoli Federico II
80125 Napoli, Italy

Olav Egeland

Institutt for teknisk kybernetikk
Norges tekniske høgskole
7034 Trondheim, Norway

Abstract— This work reports experimental results on kinematic control of a 6-dof industrial robot manipulator in the neighborhood of kinematic singularities. A weighted damped-least squares inverse kinematics solution with varying damping and weighting factors is devised; a feedback correction error term is added. The performance of the solutions is investigated in two case studies of critical trajectories passing by the shoulder and wrist singularities of the manipulator.

1. Introduction

Close to a kinematic singularity, the usual inverse differential kinematics solutions based on Jacobian (pseudo-)inverse become ill-conditioned, and this is experienced in the form of very high joint velocities and large control deviations. In real-time and sensory control of robotic manipulators, the reference trajectory is not known a priori and some remedies must be taken in order to counteract the unexpected occurrence of singularities. The same kind of problem is encountered in joy-stick control of a robot if the operator attempts to lead the robot through —or nearby— a singularity using end-effector motion increments.

Several methods and/or algorithms aimed at computing well-behaved or robust inverse kinematics solutions have been proposed in the literature. Most of them are based on a modification of the exact inverse differential kinematics mapping by resorting to approximate mappings that offer robustness to singularities at the expense of reduced tracking accuracy.

In the framework of our joint research project, following the earlier satisfactory experimental results obtained on the five-joint ABB Trallfa TR400 manipulator [1], in this work we investigate the performance of a number of schemes for control in singular configurations on the six-joint ABB IRb 2000 manipulator.

The basic inverse kinematics solution is derived using a weighted damped least-squares inverse [2],[3] of the end-effector Jacobian where the use of proper weighting allows shaping of the solution along given end-effector space directions. A recently proposed estimation algorithm of the two smallest singular values is utilized to compute varying damping and weighting factors [4]. Further, the introduction of a feedback correction term to avoid numerical drift instabilities [5] is proposed and its effects on solution accuracy and feasibility are discussed.

A description of the laboratory set-up is provided. Experimental case studies are illustrated with the manipulator passing nearby either a single (wrist) or a double (wrist and shoulder) singularity. The influence of the weighting factor and feedback correction term on solution accuracy is extensively tested out, and the performance of the various solutions is compared.

2. Kinematics

The ABB IRb 2000 is a six-revolute-joint robot manipulator manufactured by ABB Robotics. The manipulator is shown in Fig. 1. Its Craig-Denavit-Hartenberg parameters, joint working ranges and maximum speeds are reported in the tables below.

Link	ϑ [rad]	d [m]	a [m]	α [rad]
1	$\pi/2$	0	0	0
2	$\pi/2$	0	0	$\pi/2$
3	$\pi/2$	0	0.710	0
4	0	0.850	0.125	$\pi/2$
5	0	0	0	$\pi/2$
6	0	0.100	0	$\pi/2$

Joint	Working range [rad]	Max speed [rad/s]
1	$-0.99 \div +0.99$	2.01
2	$-0.85 \div +0.85$	2.01
3	$-2.72 \div -0.49$	2.01
4	$-3.43 \div +3.43$	4.89
5	$-2.00 \div +2.00$	5.24
6	$-3.14 \div +3.14$	5.24

The inner three joints are in the same arrangement as in an elbow manipulator, while the outer three joints constitute the spherical wrist commonly used in industrial robots.

Let q denote the (6×1) joint vector. The (6×6) Jacobian matrix J relates the joint velocity vector $\dot{q}$ to the (6×1) end-effector velocity vector ν through the mapping

$$\nu = \begin{bmatrix} \dot{p} \\ \omega \end{bmatrix} = J(q)\dot{q} \tag{1}$$

where $\dot{p}$ and ω represent end-effector linear and angular velocities, respectively.

In configurations where J has full rank, any end-effector velocity can be attained. When J is rank deficient, i.e. rank(J) < 6, constraints on the feasible end-effector velocity occur and the manipulator is said to be at a singular configuration or at a *singularity*.

The ABB IRb 2000 manipulator has a simple kinematic structure and its singularities are well-understood. We have:

- If $a_3 \cos \vartheta_3 + d_4 \sin \vartheta_3 = 0$, the elbow is stretched out and the manipulator is in the so-called *elbow* singularity; this does *not* correspond to a reachable configuration of the manipulator, due to the mechanical joint range for ϑ_3, and then is of no interest.

- If the wrist point lies on the axis of joint 1, its position cannot be changed by a rotation of ϑ_1 and the manipulator is in the so-called *shoulder* singularity; from simple geometry, it can be found that the shoulder singularity occurs when $a_2 \sin \vartheta_2 + a_3 \cos(\vartheta_2 + \vartheta_3) + d_4 \sin(\vartheta_2 + \vartheta_3) = 0$.

- If $\vartheta_5 = 0$ the two roll axes of the wrist are aligned and the manipulator is in the so-called *wrist* singularity.

3. Kinematic Control

It is well-known that the control system of a robotic manipulator operates in the joint space generating the driving torques at the joint actuators. The reference trajectory for the joint control servos is to be generated via kinematic inversion of the given end-effector trajectory; this is known as the *kinematic control* problem. When the arm is at —or close to— a singularity, large joint velocities may occur or degenerate directions may exist where end-effector velocity is not feasible. Therefore, the control system of a robotic manipulator should be provided with the capability of handling singularities.

A well-known effective strategy that allows motion control of manipulators in the neighborhood of kinematic singularities is the *damped least-squares* technique originally proposed in [2],[6]. The joint velocity solution can be formally written as

$$\dot{q} = \left[J^{\mathrm{T}}(q) J(q) + \lambda^2 I \right]^{-1} J^{\mathrm{T}}(q) \nu \tag{2}$$

where $\lambda \geq 0$ is a damping factor. Small values of λ give accurate solutions but low robustness to occurrence of singular and near-singular configurations. Large values of λ result in low tracking accuracy even when a feasible and accurate solution would be possible.

The damping factor λ determines the degree of approximation introduced with respect to the pure least-squares solution; then, using a constant value for λ may turn out to be inadequate for obtaining good performance over the entire manipulator workspace. An effective choice is to adjust λ as a function of some measure of closeness to the singularity at the current configuration of the manipulator.

A singular region can be defined on the basis of the estimate of the smallest singular value of J; outside the region the exact solution is used, while inside the region a configuration-varying damping factor is introduced to obtain the desired approximate solution. The factor must be chosen so that continuity of joint velocity $\dot{q}$ is ensured in the transition at the border of the singular region. We have selected the damping factor according to the following law [7]:

$$\lambda^2 = \begin{cases} 0 & \hat{\sigma}_6 \geq \varepsilon \\ \left(1 - \left(\frac{\hat{\sigma}_6}{\varepsilon}\right)^2\right) \lambda^2_{max} & \hat{\sigma}_6 < \varepsilon \end{cases} \tag{3}$$

where $\hat{\sigma}_6$ is the estimate of the smallest singular value, and ε defines the size of the singular region; the value of λ_{max} is at user's disposal to suitably shape the solution in the neighborhood of a singularity.

We have computed an estimate of the smallest singular value by the recursive algorithm recently proposed in [4] which is an extension of the algorithm in [8]; this algorithm allows estimating not only the smallest but also the second smallest singular value and then is congenial to handle closeness to a double singularity, whenever crossing of smallest singular values occurs.

The above damped least-squares method achieves a compromise between accuracy and robustness of the solution. This is performed without specific regard to the components of the particular task assigned to the manipulator's end effector. The *user-defined accuracy* strategy introduced in [7] based on the weighted damped least-squares method in [2] allows discriminating between directions in the end-effector space where higher accuracy is desired and directions where lower accuracy can be tolerated.

Let a weighted end-effector velocity vector be defined as

$$\tilde{\nu} = W\nu \tag{4}$$

where W is the (6×6) task-dependent weighting matrix taking into account the anisotropy of the task requirements. Substituting (4) into (1) gives

$$\tilde{\nu} = \tilde{J}(q)\dot{q} \tag{5}$$

where $\tilde{J} = WJ$. It is worth noticing that if W is full-rank, solving (1) is equivalent to solving (5), but with different conditioning of the system of equations to solve. This suggests selecting only the strictly necessary weighting action in order to avoid undesired ill-conditioning of $\tilde{J}$. The damped least-squares solution to (5) is

$$\dot{q} = \left[\tilde{J}^T(q)\tilde{J}(q) + \lambda^2 I\right]^{-1} \tilde{J}^T(q)\tilde{\nu} \tag{6}$$

For the typical elbow geometry with spherical wrist, it is worthwhile to devise a special handling of the wrist singularity which is difficult to predict at the planning level in the end-effector space. It can be recognized that, at the wrist singularity, there are only two components of the angular velocity vector that can be generated by the wrist itself. The remaining component might be generated by the inner joints, at the expense of loss of accuracy along some other end-effector space directions though. For this reason, lower weight should be put on the angular velocity component that is infeasible to the wrist. For the ABB IRb 2000, this is easily expressed in the frame attached to link 4; let R_4 denote the rotation matrix describing orientation of this frame with respect to the base frame, so that the infeasible component is aligned with the x-axis. We propose then to choose the weighting matrix as

$$W = \begin{bmatrix} I & O \\ O & R_4 \text{diag}\{w, 1, 1\} R_4^T \end{bmatrix} \tag{7}$$

Similarly to the choice of the damping factor as in (3), the weighting factor w is selected according to the following expression:

$$(1 - w)^2 = \begin{cases} 0 & \hat{\sigma}_6 \geq \varepsilon \\ \left(1 - \left(\frac{\hat{\sigma}_6}{\varepsilon}\right)^2\right)(1 - w_{min})^2 & \hat{\sigma}_6 < \varepsilon \end{cases} \tag{8}$$

where $w_{min} > 0$ is a design parameter [7],[3].

The above inverse kinematics solutions are expected to suffer from typical numerical drift, when implemented in discrete time. In order to avoid this drawback, a *feedback correction* term [5] can be keenly introduced by replacing the end-effector velocity ν by

$$\nu_d + Ke \tag{9}$$

where the subscript "d" denotes the desired reference end-effector velocity, K is a positive definite —usually diagonal— (6×6) matrix, and e expresses the error between the desired and actual end-effector location. The error e is computed as

$$e = \begin{bmatrix} e_t \\ e_o \end{bmatrix} = \begin{bmatrix} p_d - p \\ \frac{1}{2}(n \times n_d + s \times s_d + a \times a_d) \end{bmatrix} \tag{10}$$

where the translation error is given by the (3×1) vector e_t and the orientation error is given by the (3×1) vector e_o. The end-effector position is expressed by the (3×1) position vector p while its orientation by the (3×3) rotation matrix $R = [n \quad s \quad a]$, with n, s, a being the unit vectors of the end-effector frame.

It is important to notice that, in the neighborhood of a singularity, end-effector errors typically increase along the near-degenerate components of the given end-effector velocity and convergence is slowed down [9]. Therefore, we propose to shape the action of the feedback correction term around the singularities using $K = \varrho K_0$, where K_0 is a constant matrix and ϱ is a varying factor to be properly adjusted.

We have found that it is important to have $\varrho = 0$ inside the singular region defined by $\sigma_6 \leq \varepsilon$. Indeed, if a velocity is assigned along a near-degenerate direction and a nonzero gain is used for the feedback correction term, the error e will eventually approach zero; however, the resulting joint velocities may cause the manipulator to reach the joint limits. Outside the singular region, interpolation is used to achieve a smooth solution and full value to the gains ($\varrho = 1$) is set when far enough from the singularity. In our experience, interpolation had to be performed when $\varepsilon < \sigma_6 < 4\varepsilon$ using a quadratic type function, i.e.

$$\varrho = \begin{cases} 0 & \sigma_6 \leq \varepsilon \\ \dfrac{(\sigma_6 - \varepsilon)^2}{(3\varepsilon)^2} & \varepsilon < \sigma_6 < 4\varepsilon \\ 1 & \sigma_6 \geq 4\varepsilon \end{cases} \tag{11}$$

4. Experimental results

The experiments were run on an ABB IRb 2000 robot manipulator. The original joint servos of the S3 industrial control system were used, and an interface was established at the joint increment level. This allowed implementation of a kinematic control strategy, that is an inverse kinematics module based on the foregoing damped least-squares solution providing the reference inputs to the manipulator joint servos. This was done in cooperation with ABB Robotics who slightly modified the S3 control system by adding an interface board with a dual-port RAM, as well as software modules to establish the required communication protocol. The resulting communication facilitated the transfer of joint variables and increments at a sampling time of 12 [ms], and in addition it allowed for remote initialization and activation of the ABB S3 system.

The inverse kinematics were computed on a 25 [MHz] Motorola 68040 VME board which communicated with the ABB S3 control system through the dual-port RAM. The software was developed on a SUN 350 workstation, and the executable code was downloaded to the 68040 board and executed using Vx-Works. A block diagram showing the interconnection between major components of the laboratory set-up is sketched in Fig. 2.

The program system which was executed on the Motorola 68040 was written in C, and consisted of two activities; an interactive activity containing initialization and user interface, and the real-time controller activity. The real-time controller activity had the following steps: input joint angles, calculate joint increments, output joint increments, store data for logging.

The interactive user interface contained initialization of the S3 control system and of the communication between the ABB controller and the real-time activity. Further it contained initialization of kinematic parameters and a menu-driven system for adjusting kinematic parameters, specifying trajectory data, and selecting the algorithm for the inverse kinematics solution. In addition, a function for transferring logged data to the SUN system was included. The logged data was subsequently written to a file in MATLAB format which allowed for postprocessing and plotting in the SUN UNIX environment.

A simple solution to the problem of communication between the real-time and interactive activities was achieved by including both activities in the same C program. The interactive activity was the main program, while the real-time activity was called from a function which was started by a timer interrupt every 12 [ms]. The data transfer between the two activities was performed through global data structures containing parameters for the inverse kinematics algorithms, trajectory data, logged data and logical variables for algorithm selection.

In the following we present two case studies to demonstrate the application of the preceding schemes to real-time kinematic control of the ABB IRb 2000 robot manipulator.

Trajectory 1: A reference trajectory through the wrist singularity was studied. The initial configuration was $q = [0 \quad \pi/12 \quad -\pi/2 \quad 0 \quad 0.15 \quad 0]^T$ [rad]. An end-effector increment $\Delta p = [0.18 \quad 0.45 \quad -0.45]^T$ [m] was interpolated using linear segments with parabolic blends. The blend time was 0.2 [s], and the total time of the trajectory was 1.5 [s]. The resulting cruise velocity between 0.2 [s] and 1.3 [s] was approximately 0.5 [m/s]. The wrist singularity was encountered after approximately 0.6 [s].

The damping factor was computed from (3) with $\varepsilon = 0.04$ and $\lambda_{max} = 0.04$. The estimate $\hat{\sigma}_6$ of the smallest singular value was computed using the algorithm in [4]

The basic damped least-squares scheme (2) was used first; then, for comparison, the same trajectory was tested with the weighted damped least-squares scheme based on (6). The weighting matrix was chosen as in (7),(8) with $w_{min} = 0.1$. The results are shown in Fig. 3 for both the damped least-squares and weighted damped least-squares schemes, without feedback correction term. In both cases the joint velocities were feasible with peak values of approximately 2 [rad/s]. Without weighting, the norm of the translation error at final time t_f was $\|e_t(t_f)\| = 0.055$ [m], while the orientation error norm was $\|e_o(t_f)\| = 0.06$ [rad]. With weighting, the errors were $\|e_t(t_f)\| = 0.0025$ [m] and $\|e_o(t_f)\| = 0.12$ [rad].

This result clearly demonstrates the advantage of using weighting since our main concern was to achieve accuracy in translation. In fact, weighting resulted in a reduction of the translation error by a factor of approximately 20, while the orientation error was increased only by a factor of two. The effect of weighting on the smallest singular values is seen from Fig. 3; a significant reduction of $\tilde{\sigma}_6$ around the singularity is obtained compared to the corresponding σ_6 in the non-weighted case.

The same experiments were repeated with a feedback correction term according to (9),(10) with $K = \text{diag}\{12\dots 12\}$. The results are shown in Fig. 4. In this case the joint velocities were higher. In fact, with the damped least-squares solution, joint velocities 4 and 6 saturated between 0.6 and 0.8 [s] at 5 [rad/s] and -5 [rad/s], respectively. With the introduction of weighting, the joint velocities were feasible with peak values less than 5 [rad/s]. Thanks to the feedback correction term, the end-effector error e converged to zero after leaving the singular region (see Fig. 4). This resulted in a reorientation of joints 4 and 6 by $\pm\pi$ in the non-weighted case, which reflects the fact that large increments in joint angles may result from small end-effector increments when the manipulator is close to a singularity. Remarkably, Figure 3 reveals also that outside the singular region $\tilde{\sigma}_6$ tends to σ_6; obviously, this was not the case in the previous experiments without feedback correction (see Fig. 3).

Trajectory 2: A reference trajectory involving both the shoulder singularity and the wrist singularity was studied. The initial configuration was $q = [0 \quad 0.7893 \quad -\pi/2 \quad \pi/2 \quad -0.05 \quad 0]^T$ [rad]. An end-effector increment $\Delta p = [0.1 \quad 0.1 \quad 0]^T$ [m] was interpolated using linear segments with parabolic blends. The blend time was 0.15 [s], and the total time of the trajectory was 1.0 [s]. The resulting cruise velocity between 0.15 [s] and 0.85 [s] was approximately 0.166 [m/s]. The wrist singularity was encountered after approximately 0.22 [s].

Also for this trajectory, the damping factor was computed from (3) with $\varepsilon = 0.04$ and $\lambda_{max} = 0.04$. The basic damped least-squares solution (2) was used. The estimate $\hat{\sigma}_6$ of the smallest singular value was found: a) by computing both $\hat{\sigma}_5$ and $\hat{\sigma}_6$ as in [4], and b) by computing $\hat{\sigma}_6$ as in [8].

The results without feedback correction are shown in Figs. 5,6. The damped least-squares scheme performs well also in this case. The norm of the translation error at final time t_f was $\|e_t(t_f)\| = 0.03$ [m], while the norm of the orientation error was $\|e_o(t_f)\| = 0.015$ [rad].

In case a) the crossing of the two smallest singular values associated with the wrist and shoulder singularity was successfully detected at 0.15 [s] and 0.37 [s], and an accurate estimate $\hat{\sigma}_6$ of the smallest singular value was found. This gave satisfactory damping around the wrist singularity. The resulting joint velocities had peak values less than 1.2 [rad/s]. In case b) the crossing of the two smallest singular values caused the estimate $\hat{\sigma}_6$ to track σ_5, and the wrist singularity appearing at 0.22 [s] was not detected. This resulted into a low damping factor around the wrist singularity, and high joint velocities were experienced; in particular, the velocity of joint 1 saturated at -2 [rad/s]. Finally, note that the final errors were a little smaller in case b) since the incorrect estimate of σ_6 produced lower damping on the solution throughout the singular region.

5. Conclusions

The weighted damped least-squares solution with varying damping factor, varying weighting factor and feedback correction term has been implemented on an industrially available hardware indicating that enhanced real-time kinematic control of robot manipulators through singularities is definitely possible.

Acknowledgments

The research work described in this paper was supported partly by *Consiglio Nazionale delle Ricerche* under contract 93.00905.PF67, and partly by *ABB Robotics*.

References

[1] S. Chiaverini, B. Siciliano, and O. Egeland, "Robot control in singular configurations — Analysis and experimental results," *2nd International Symposium on Experimental Robotics*, Toulouse, F, June 1991, in *Experimental Robotics II*, Lecture Notes in Control and Information Sciences 190, R. Chatila and G. Hirzinger (Eds.), Springer-Verlag, Berlin, D, pp. 25–34, 1993.

[2] Y. Nakamura and H. Hanafusa, "Inverse kinematic solution with singularity robustness for robot manipulator control," *ASME J. of Dynamic Systems, Measurements, and Control*, vol. 108, pp. 163–171, 1986.

[3] S. Chiaverini, O. Egeland, R.K. Kanestrøm, "Weighted damped least-squares in kinematic control of robotic manipulators," *Advanced Robotics*, vol. 7, pp. 201–218, 1993.

[4] S. Chiaverini, "Estimate of the two smallest singular values of the Jacobian matrix: Application to damped least-squares inverse kinematics," *J. of Robotic Systems*, vol. 10, no. 8, 1993.

[5] L. Sciavicco and B. Siciliano, "Coordinate transformation: A solution algorithm for one class of robots," *IEEE Trans. on Systems, Man, and Cybernetics*, vol. 16, pp. 550–559, 1986.

[6] C.W. Wampler, "Manipulator inverse kinematic solutions based on vector formulations and damped least-squares methods," *IEEE Trans. on Systems, Man, and Cybernetics*, vol. 16, pp. 93–101, 1986.

[7] S. Chiaverini, O. Egeland, and R.K. Kanestrøm, "Achieving user-defined accuracy with damped least-squares inverse kinematics," *Proc. 5th Int. Conf. on Advanced Robotics*, Pisa, I, pp. 672–677, June 1991.

[8] A.A. Maciejewski and C.A. Klein, "Numerical filtering for the operation of robotic manipulators through kinematically singular configurations," *J. of Robotic Systems*, vol. 5, pp. 527–552, 1988.

[9] C.W. Wampler and L.J. Leifer, "Applications of damped least-squares methods to resolved-rate and resolved-acceleration control of manipulators," *J. of Dynamic Systems, Measurement, and Control*, vol. 110, pp. 31–38, 1988.

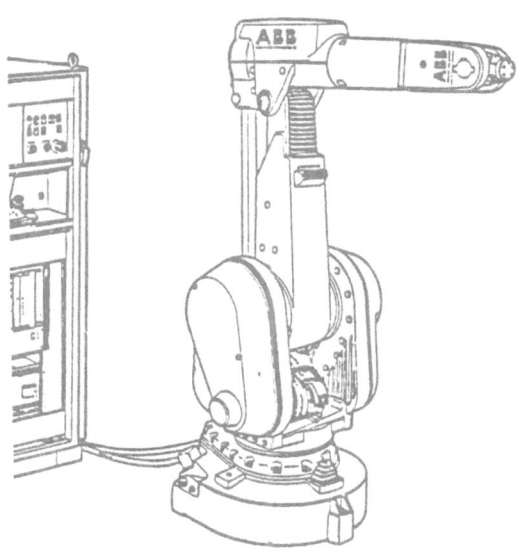

Fig. 1 — The ABB IRb 2000 industrial robot manipulator.

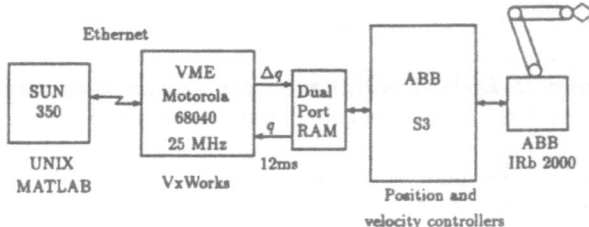

Fig. 2 — Block diagram of laboratory set-up.

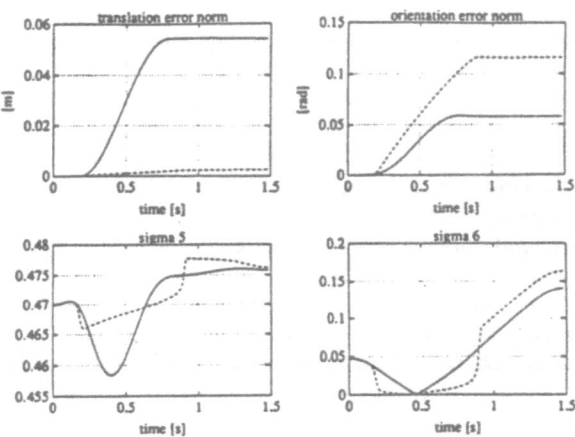

Fig. 3 — Translation and orientation error norms and estimates of the two smallest singular values for Trajectory 1 without feedback correction; damped least-squares scheme (solid), weighted damped least-squares scheme (dashed).

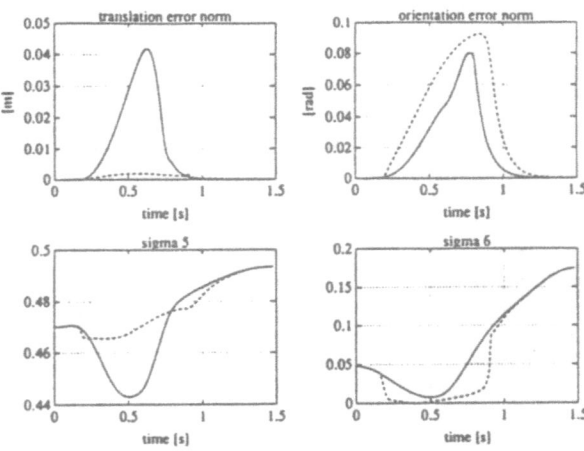

Fig. 4 — Translation and orientation error norms and estimates of the two smallest singular values for Trajectory 1 with feedback correction; damped least-squares scheme (solid), weighted damped least-squares scheme (dashed).

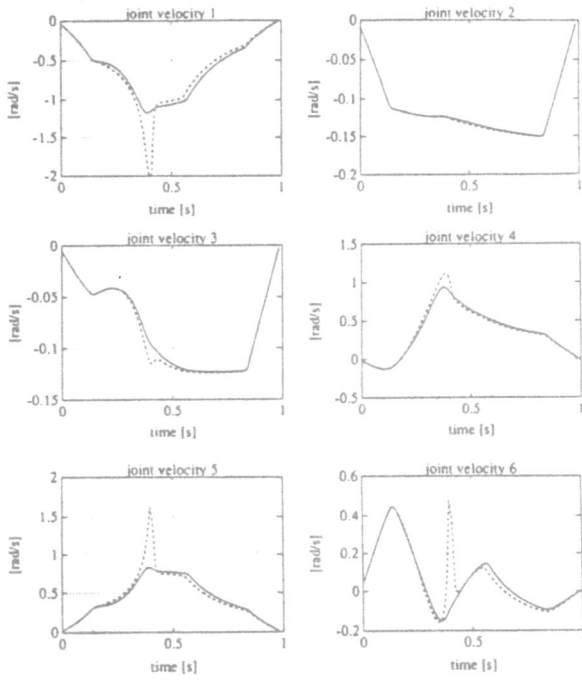

Fig. 5 — Joint velocities for Trajectory 2; case **a)** (solid), case **b)** (dashed).

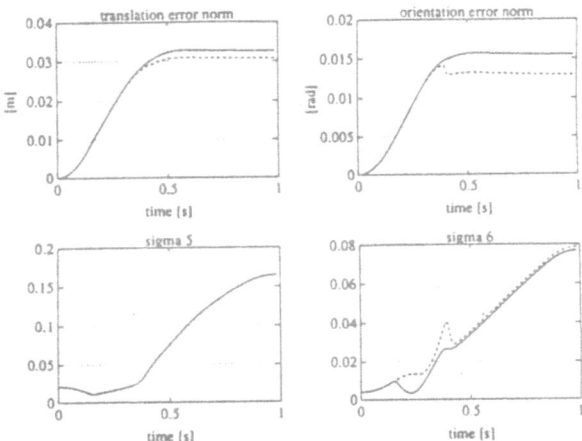

Fig. 6 — Translation and orientation error norms and estimates of the two smallest singular values for Trajectory 2; case **a)** (solid), case **b)** (dashed).

Experiments in Multi-Grasp Manipulation

David Williams and Oussama Khatib

Robotics Laboratory
Computer Science Department
Stanford University

Abstract— In this paper, we present results in dextrous, multi-grasp manipulation based on the integration of two basic concepts: the *virtual linkage* and the *augmented object*. The virtual linkage provides a physical representation of internal forces and moments during multi-grasp manipulation. It does this by representing the manipulated object with an equivalent closed-chain mechanism. To control object motion, we describe the operational-space, rigid-body dynamics for the multi-manipulator/object system using the *augmented object*. With this concept, motion of the manipulated object and the forces it applies to the environment are controlled by operational forces at some selected point on the object. This control force is then partitioned among the arms in such a way as to minimize internal (strain-causing) forces. These forces are characterized by the virtual linkage and controlled independently. This approach enables a multi-arm robot system to manipulate objects while performing accurate control of internal forces. Simulation results are presented, as well as experimental results for a multi-arm system of three PUMA 560 manipulators.

1. Introduction

Research dealing with the characterization of internal forces and moments has focused essentially on two issues; minimizing internal forces during manipulation [10, 11], or determining a set of grasp forces that satisfy frictional and stability constraints for point contact without applying excessive force to the manipulated object [3, 4, 9]. Typically these approaches establish the relationship between the forces applied to an object and the resultant of these forces at some reference point. They then proceed to solve this relationship using a generalized inverse. The basis for the nullspace associated with this generalized inverse is used to characterize the internal forces.

While such methods allow closed-loop minimization of internal forces, they are not as useful for maintaining a specific set of non-zero internal forces. This is because their characterizations of these forces are not based on any physical model. The virtual linkage model described below is designed to address precisely this problem, providing a realistic characterization of internal forces.

Once internal forces have been characterized, a controller must be developed for the multiple-arm/object system. In early work, control of multi-arm robot systems was often treated as motion coordination problem. One of the first methods for controlling a two-arm system involved organizing the controller in a master/slave fashion, and using a motion coordination procedure to minimize errors in the relative position of the two manipulators [2]. In another study

[14], one manipulator was treated as a "leader" and the other as a "follower." Control of the follower was based on the constraint relationships between the two manipulators.

Neither of these methods generalized easily to systems with more than two arms. In addition, they usually required very accurate position sensing and/or force-sensing to avoid excessive internal forces. This was because kinematic, joint-space constraints were used to achieve coordination in these methods. A control methodology was needed that operated directly on the generalized coordinates of the manipulated object (i.e. object-level control) and accounted for the complete closed-chain dynamics of the system.

The *augmented object* method [6] was developed to do exactly that. It controls a multi-arm system by using its closed-chain dynamics to decouple the system in operational space. Using this method, the system can be treated as a single *augmented object* which represents the total system dynamics as perceived at the operational point. Dynamic behavior of this system is described by a relationship between the vector of total forces to be applied at the operational point (i.e. the command vector) and the dynamic characteristics of the augmented object. In the physical system, the command vector results from the collective action of the end-effectors. Therefore, a command force can be applied to the system by partitioning it into a set of forces to be applied by the various end effectors. This force partitioning can be based on any criteria such as minimization of total joint actuator effort or internal forces.

Recently, there has been an increased interest in closed-loop minimization of internal forces in multi-arm systems. In [13], a controller was developed for a pair of one-dimensional arms with unilateral constraints. This work was extended in [8] to provide dynamic control of contact conditions. Simulations of a planar two-arm system with rolling contacts were presented.

In [1], a multi-arm control system was proposed to close a force-control loop around a resolved-rate controller for each arm. The forces to be applied by each arm were calculated using a cost function that minimized a weighted combination of the applied forces while maintaining the desired resultant. Each arm was then controlled to produce the requested force. Experimental results were presented for two PUMA 600 robots manipulating an instrumented beam.

A controller was presented in [7] which minimized internal forces by requiring each arm to apply an equal force at some reference point on the object. A model following technique was used to provide a reference trajectory for each arm based on the desired state and the actual position/forces. This reference trajectory was followed using a non-linear controller which decoupled each arm in the absolute coordinate system of its end-effector. Experimental results for a pair of planar, three-degree-of-freedom arms were presented.

Each of these methods is able to control the trajectory of an object while minimizing internal forces and moments. However, none provide the ability to specify non-zero internal forces, which can be quite useful for pre-loading in assembly tasks. The importance of this feature was demonstrated in [15], where a trajectory planner was developed to perform the specific task of pre-loading a flexible beam using two coordinating PUMA 560 manipulators.

Although a great deal of progress has been made in multi-grasp manipulation, none of the work described above considers internal force dynamics. Below, we review our model of internal forces and derive the dynamic model for a closed-chain multi-grasp manipulation system. We then present a controller which dynamically decouples the system to allow independent control of object motion and internal forces.

2. The Virtual Linkage

When an object is being manipulated by a set of robot arms, we often want to control its internal stress state during motion. However, these stresses are a complicated function of both the object's geometry and its motion. To characterize them for control, it is important to define a model that is both a reasonable approximation of the actual object and is easy to compute in real time.

Consider the quasi-static case where object velocities and accelerations do not affect the internal stress state. In this case, we can characterize object stresses by considering internal forces in an equivalent *virtual linkage* [12], defined as follows:

A virtual linkage associated with an n-grasp manipulation task is a $6(n-1)$-degree-of-freedom mechanism whose actuated joints characterize the object's internal forces and moments.

In other words, this method characterizes the object's internal stress state by replacing it with an equivalent "virtual linkage" which has much simpler internal force relationships, but is able to resist the same applied forces as the original object. Forces and moments applied at the grasp points of a virtual linkage cause joint forces and torques at its actuators. When these actuated joints are subjected to the opposing forces and torques, the virtual linkage becomes a statically determinate locked structure. The internal forces and moments in the object are then characterized by these forces and torques. There is a one-to-one correspondence between the amount of actuator redundancy and the number of internal forces and moments. This ensures that each internal force can be controlled independently.

2.1. Kinematic Structure

To clarify these ideas, consider a rigid body manipulation task being performed by three manipulators, as illustrated in Figure 1. If we assume that each manipulator grasps the object rigidly, and that none is in a singular configuration, the resulting system has eighteen actuator degrees of freedom. Since specifying the resultant forces and moments requires six degrees of freedom, twelve degrees of actuator redundancy exist. These twelve degrees of freedom can be used to produce twelve independent internal forces and moments.

The kinematic structure of the virtual linkage we propose for this manipulation task is one in which three actuated prismatic joints are connected by passive revolute joints to form a closed-chain, three-degree-of-freedom mechanism. A spherical joint with three actuators is then connected at each grasp

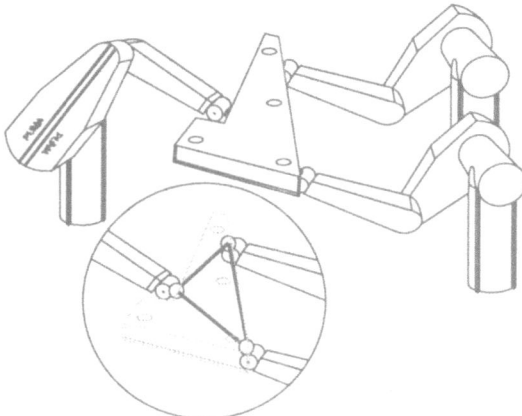

Figure 1. A Multi-Arm Manipulation Task and an Equivalent Virtual Linkage

point to resist internal moments. An equivalent virtual linkage for the three-arm manipulation task is shown in the inset of Figure 1.

The selection of this kinematic structure is motivated as follows: Since forces applied by the arms act to produce stresses throughout the object, they are best represented as a set of interactions between arms. This is the justification for actuated members in the virtual linkage. Applied moments, however, are local phenomena, and are best modeled by introducing actuated spherical joints at each grasp point.

The virtual linkage described above constitutes the basic structure for virtual linkages with additional grasps. Indeed, the virtual linkage model can be extended to any number of grasp points. Each additional grasp point results in six new actuator degrees of freedom which must be characterized by the virtual linkage.

Specifically, each new grasp introduces three internal forces and three internal moments. These are accounted for by introducing three more prismatic actuators and one more spherical joint to the linkage. This is accomplished by connecting new grasp points to existing grasp points through three actuated virtual members and one spherical joint. For instance, in the virtual linkage presented above, a fourth grasp is added as shown in Figure 2.

In general, any virtual linkage associated with an n-grasp manipulation task requires $3(n-2)$ actuated members. We can independently specify forces in these $3(n-2)$ members, along with n internal moments at the grasp points (corresponding to $3n$ actuator degrees of freedom).

2.2. The Grasp Description Matrix

It was shown in [12] that the relationship between the forces and moments applied by each arm at the grasp points ($\mathbf{f}_i$), the resultant force and moment at the operational point ($\mathbf{F}_R$), and the internal forces and moments ($\mathbf{F}_I$), can be written as

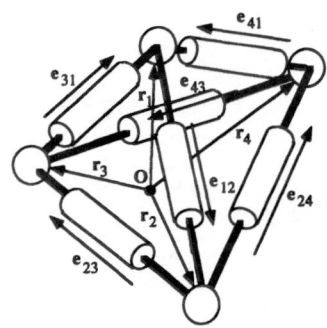

Figure 2. A Four Grasp Virtual Linkage

$$
\left[\begin{array}{c} \mathbf{F}_R \\ \mathbf{F}_I \end{array} \right] = \mathbf{G}_{grasp} \left[\begin{array}{c} \mathbf{f}_1 \\ \vdots \\ \mathbf{f}_n \end{array} \right] ; \tag{1}
$$

where $\mathbf{G}_{grasp}$ is the $6n \times 6n$ *grasp description matrix*. The explicit formulas for $\mathbf{G}_{grasp}$ and $\mathbf{G}_{grasp}^{-1}$ are given in [12]. An equivalent matrix, $\mathbf{G}$, can be defined which relates the forces applied by each arm at the operational point to the resultant and internal forces. $\mathbf{G}$ and $\mathbf{G}^{-1}$ are easily computed from the corresponding matrices developed in [12]. For some grasp configurations, the rank of the grasp description matrix $\mathbf{G}$ might be deficient. In such a situation, the virtual linkage is at a kinematic singularity; two or more adjacent members are aligned. This implies that the system has lost the ability to independently specify all internal forces and moments. Since these singularities are a function only of grasp geometry, not of the object motion, they are easily avoided.

3. Object-Level Control

To effectively control internal forces, the virtual linkage method must be integrated with a controller that accounts for the closed-chain dynamics of the multi-arm system. Such a controller was developed in [6], and introduced the *augmented object concept* to characterize the system's rigid-body dynamics. Below, we review the augmented object model for a system of manipulators rigidly connected to the manipulated object. We then extend it to consider internal force dynamics and manipulation of objects with simultaneous control of internal forces.

3.1. The Augmented Object Model

The problem of manipulating an object with multiple manipulators is illustrated in Figure 1. The end effectors of these manipulators are each assumed to have the same number of degrees of freedom and to be rigidly connected to the manipulated object. We choose one point on the object to be controlled and call this point the *operational point*. Because we have assumed a rigid grasp, this point is fixed with respect to each of the effectors. In [6], we showed how the dynamics of the individual manipulators and the object could be combined, resulting in a new *augmented object* which models the dynamics at the operational point for the multi-arm robot system.

This multi-grasp system can be controlled by the generalized operational force which is the sum of all end-effector forces expressed at the operational point. These end-effector forces are in turn generated by the corresponding manipulator's actuators. The generalized joint force vector corresponding to each end-effector force is found using the Jacobian matrix associated with that manipulator, computed with respect to the operational point. Since this will be important later, we consider the augmented object equations of motion in more detail.

3.1.1. System Equations of Motion

As discussed above, consider a multi-manipulator system that results from rigidly connecting the manipulated object to the effectors of N n-degree-of-freedom manipulators. In the case of non-redundant manipulators, the number of degrees of freedom, $n_\oplus$ associated with the combined system can be shown [5] to be equal to 3 in the planar case and to 6 in the spatial case. In other words, the number of operational coordinates, m, is equal to the number of degrees of freedom, $n_\oplus$, of the mechanism formed by the combined object/manipulator system. Therefore, these coordinates form a set of generalized coordinates for the system.

Let $\Lambda_\mathcal{L}(\mathbf{x})$ represent the kinetic energy matrix, with respect to the operational point, of the manipulated object in this coordinate system. When the object is held by N effectors, these inertial properties are modified. To characterize these changes, the N-effector/object system can be viewed as an *augmented object* representing the total inertias perceived at the operational point. If $\Lambda_i(\mathbf{x})$ represents the kinetic energy matrix associated with the i th effector at this point, the operational-space inertial characteristics of the augmented object can be shown to satisfy

$$\Lambda_\oplus(\mathbf{x}) = \Lambda_\mathcal{L}(\mathbf{x}) + \sum_{i=1}^{N} \Lambda_i(\mathbf{x}). \qquad (2)$$

It was shown in [6] that the augmented object equations of motion are

$$\Lambda_\oplus(\mathbf{x})\ddot{\mathbf{x}} + \mu_\oplus(\mathbf{x}, \dot{\mathbf{x}}) + \mathbf{p}_\oplus(\mathbf{x}) = \mathbf{F}_\oplus; \qquad (3)$$

where the vector $\mu_\oplus(\mathbf{x}, \dot{\mathbf{x}})$, of centrifugal and Coriolis forces also possesses the additive property of the inertia matrices as given in equation 2:

$$\mu_\oplus(\mathbf{x}) = \mu_\mathcal{L}(\mathbf{x}) + \sum_{i=1}^{N} \mu_i(\mathbf{x}). \qquad (4)$$

The gravity vector is also a linear combination:

$$\mathbf{p}_\oplus(\mathbf{x}) = \mathbf{p}_\mathcal{L}(\mathbf{x}) + \sum_{i=1}^{N} \mathbf{p}_i(\mathbf{x}); \tag{5}$$

where $\mathbf{p}_\mathcal{L}(\mathbf{x})$ and $\mathbf{p}_i(\mathbf{x})$ are the operational-space gravity vectors associated with the object and the ith effector.

If the object is not in contact with the environment, the generalized operational force, $\mathbf{F}_\oplus$, is the resultant of the forces produced by each of the N end-effectors at the operational point.

$$\mathbf{F}_\oplus = \sum_{i=1}^{N} \mathbf{F}_i. \tag{6}$$

These operational forces, $\mathbf{F}_i$, are generated by the corresponding manipulator actuators. The generalized joint force vector $\boldsymbol{\Gamma}_i$ corresponding to $\mathbf{F}_i$ is given by

$$\boldsymbol{\Gamma}_i = J_i^T(\mathbf{q}_i)\,\mathbf{F}_i;$$

where $\mathbf{q}_i$ and $J_i^T(\mathbf{q}_i)$ are, respectively, the vector of joint coordinates and the Jacobian matrix computed with respect to $\mathbf{x}_o$ and associated with the ith manipulator.

3.1.2. Multi-Arm System Control Structure

The dynamic decoupling and motion control of the augmented object in operational space is achieved by selecting the following control structure

$$\mathbf{F}_\oplus = \widehat{\Lambda}_\oplus(\mathbf{x})\mathbf{F}^\star + \widehat{\mu}_\oplus(\mathbf{x}, \dot{\mathbf{x}}) + \widehat{\mathbf{p}}_\oplus(\mathbf{x}); \tag{7}$$

where $\mathbf{F}^\star$ is the control force corresponding to the dynamically decoupled system. Typically, this force is calculated with a proportional/derivative control law or a trajectory tracking law. The feedback gains for these laws should be calculated based on control of the unit-mass, decoupled system.

The control structure in equation 7 provides the net force $\mathbf{F}_\oplus$ to be applied to the augmented object at the operational point. Any control structure developed for the multi-arm/object system must apply a total force of $\mathbf{F}_\oplus$ in order to properly decouple the system. In the next section, we discuss the dynamics of internal forces, and show how they modify the results presented here. We then describe a control structure that decouples and controls object motion as well as internal forces.

3.2. Internal Force Dynamics

To analyze the dynamics of internal forces, we start with the operational-space equations of motion for one arm, i, in the multiple-arm/object system

$$\Lambda_i(\mathbf{q}_i)\ddot{\mathbf{x}}_i + \mu_i(\mathbf{q}_i, \dot{\mathbf{q}}_i) + \mathbf{p}_i(\mathbf{q}_i) = \mathbf{F}_i - \mathbf{F}_{si}; \tag{8}$$

where $\mathbf{F}_i$ is the control force applied to arm i at the operational point; $\mathbf{F}_{si}$ is the force that it exerts on the object at that point, and $\mathbf{x}_i$ is the operational-space position of the arm. The equations of motion for the manipulated object can be written as

$$\Lambda_{obj}(\mathbf{x}_{obj})\ddot{\mathbf{x}}_{obj} + \mu_{obj}(\mathbf{x}_{obj}, \dot{\mathbf{x}}_{obj}) + \mathbf{p}_{obj}(\mathbf{x}_{obj}) = \sum_{i=1}^{N} \mathbf{F}_{si}; \tag{9}$$

where $\mathbf{x}_{obj}$ is the position of the manipulated object. The sensed forces for each arm, $\mathbf{F}_{si}$, are given by

$$\mathbf{F}_{si} = \mathbf{K}_i(\mathbf{x}_i - \mathbf{x}_{obj}); \tag{10}$$

where $\mathbf{K}_i$ is grasp stiffness of grasp i, represented at the operational point of that arm. Since we ultimately plan to control forces in the object frame, we will determine how the forces change in that frame. Taking two derivatives of equation 10 yields

$$\ddot{\mathbf{F}}_{si} = \mathbf{K}_i(\ddot{\mathbf{x}}_i - \ddot{\mathbf{x}}_{obj}). \tag{11}$$

Substituting from equations 8 and 9 gives an equation for the sensor dynamics of each arm in the object frame,

$$\ddot{\mathbf{F}}_{si} = \mathbf{K}_i[\Lambda_i^{-1}(\mathbf{F}_i - \mathbf{F}_{si} - \mu_i - \mathbf{p}_i) - \Lambda_{obj}^{-1}(\sum_{i=1}^{N} \mathbf{F}_{si} - \mu_{obj} - \mathbf{p}_{obj})]. \tag{12}$$

Next, define the $6n \times 6$ coupling matrix

$$\mathcal{I} = \begin{bmatrix} \mathbf{I} \\ \vdots \\ \mathbf{I} \end{bmatrix}; \tag{13}$$

Using this equation, equation 12 can be rewritten as

$$\mathbf{\Psi}\ddot{\mathbf{F}}_t + \mathbf{\Phi}\mathbf{F}_t + \mathbf{\Upsilon} = \mathcal{F}. \tag{14}$$

where

$$\begin{aligned} \mathbf{\Psi} &= \Lambda_{sys}\mathbf{K}_{sys}^{-1}\mathbf{G}^{-1} \\ \mathbf{\Phi} &= (\mathbf{I} + \Lambda_{sys}\mathcal{I}\Lambda_{obj}^{-1}\mathcal{I}^T)\mathbf{G}^{-1} \\ \mathbf{\Upsilon} &= \mu_{sys} + \mathbf{p}_{sys} - \Lambda_{sys}\mathcal{I}\Lambda_{obj}^{-1}(\mu_{obj} + \mathbf{p}_{obj}) \end{aligned}$$

In these equations, $\Lambda_{sys} = diag(\Lambda_i)$ and $\mathbf{K}_{sys} = diag(\mathbf{K}_i)$. μ_{sys}, $\mathbf{p}_{sys}$, $\mathcal{F}$, and $\mathbf{F}_s$ are the vectors whose components are the sets of vectors μ_i, $\mathbf{p}_i$, $\mathbf{F}_i$, and $\mathbf{F}_{s,i}$, respectively. Equation 14 describes the internal force dynamics of the multiple-grasp system.

3.3. An Extended Object-Level Controller

To decouple and control the multi-arm/object system described above, we select the control structure

$$\mathbf{\Psi}\ddot{\mathbf{F}}_t + \mathbf{\Phi}\mathbf{F}_t + \mathbf{\Upsilon} = \widehat{\mathbf{\Psi}}\mathbf{F}_t^\star + \widehat{\mathbf{\Phi}}\mathbf{F}_{t(d)} + \widehat{\mathbf{\Upsilon}} \tag{15}$$

where $\mathbf{F}_t^\star$ is the vector of resultant and internal control forces

$$\mathbf{F}_t^\star = -\mathbf{K}_v\dot{\mathbf{F}}_t - \mathbf{K}_p(\mathbf{F}_t - \mathbf{F}_{t(d)}); \tag{16}$$

and $\mathbf{F}_{t(d)}$ is the vector of desired resultant and internal forces

$$\mathbf{F}_{t(d)} = \begin{bmatrix} \widehat{\Lambda}_{obj}\mathbf{F}^\star + \widehat{\mu}_{obj} + \widehat{\mathbf{p}}_{obj} \\ \mathbf{F}_{I(d)} \end{bmatrix}. \tag{17}$$

In the above equations, the hats denote estimated values. If we assume perfect modeling, this control structure gives a closed-loop system behavior of

$$\ddot{\mathbf{F}}_t + \mathbf{K}_v\dot{\mathbf{F}}_t + (\mathbf{K}_p + \mathbf{\Psi}^{-1}\mathbf{\Phi})(\mathbf{F}_t - \mathbf{F}_{t(d)}) = 0; \tag{18}$$

Using this control structure, it can be shown that the individual arm control forces, $\mathbf{F}_i$, satisfy

$$\mathbf{F}_i = \widehat{\Lambda}_i\mathbf{F}^\star + \widehat{\mu}_i + \widehat{\mathbf{p}}_i + \mathbf{F}_{int,i}; \tag{19}$$

with

$$\mathbf{F}_{int,i} = \widehat{\Lambda}_i\widehat{\mathbf{K}}_i^{-1}\mathbf{F}_{s,i}^\star + \mathbf{F}_{s(d),i}; \tag{20}$$

where $\mathbf{F}_{s(d),i}$ is the ith component of the feedforward force at the operational point

$$\mathbf{F}_{s(d)} \triangleq \begin{bmatrix} \mathbf{F}_{s(d),1} \\ \vdots \\ \mathbf{F}_{s(d),N} \end{bmatrix} = \mathbf{G}^{-1}\mathbf{F}_{t(d)}; \tag{21}$$

and $\mathbf{F}_{s,i}^\star$ is the ith component of the control force vector corresponding to the internal and resultant control forces

$$\mathbf{F}_s^\star \triangleq \begin{bmatrix} \mathbf{F}_{s,1}^\star \\ \vdots \\ \mathbf{F}_{s,N}^\star \end{bmatrix} = \mathbf{G}^{-1}\mathbf{F}_t^\star. \tag{22}$$

Notice that, when $\mathbf{K}$ approaches infinity, equation 19 reduces to

$$\mathbf{F}_i = \widehat{\Lambda}_i\mathbf{F}^\star + \widehat{\mu}_i + \widehat{\mathbf{p}}_i + \mathbf{F}_{s(d),i}.$$

Adding these equations for all arms, and using equation 9 gives

$$\mathbf{F}_\oplus = \widehat{\Lambda}_\oplus\mathbf{F}^\star + \widehat{\mu}_\oplus + \widehat{\mathbf{p}}_\oplus.$$

This shows how, in the case of rigid body motion, our controller reduces to an augmented-object controller. Equation 19 shows that each arm in the multi-arm system handles it's own dynamics while minimizing internal forces. An additional term, $\mathbf{F}_{int,i}$ is then added to control the internal force dynamics.

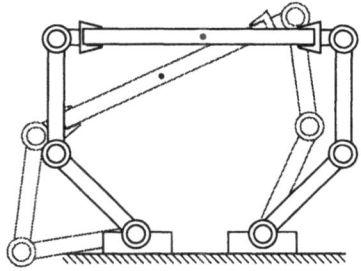

Figure 3. A Two-Arm Planar Manipulation

Item	Length	Mass	Inertia
Link 1	1.0	1.0	0.08333
Link 2	1.0	1.0	0.08333
Link 3	0.2	0.2	0.00067
Object	1.0	1.0	1.0

Table 1. Kinematic and Inertial Simulation Parameters

4. Results

To verify the performance of the controller presented above, a simulation was performed using two planar, three-degree-of-freedom arms connected to a rigid object with flexible grasps, as shown in Figure 3. The kinematic and inertial parameters used in the model are summarized in Table 1, while the grasp stiffnesses and damping are given in Table 2. The controller was implemented digitally with a servo rate of 1000 Hz.

Equation 18 predicts that the error dynamics of internal forces are coupled to those of the resultant forces applied to the object. Because of this, both the resultant force on the object and the internal forces and moments must be controlled. This coupling also has implications regarding the type of motion control used for the system. If the commanded resultant changes too quickly because of sudden changes in desired object acceleration, the controller may not be able to track the change. This has a noticeable effect on the internal forces. An example of this phenomenon is shown in Figure 4. The commanded motion is a 10° counter-clockwise rotation coupled with a motion down and to the left. In this figure, and those which follow, solid lines denote the system

Grasp	K_x	K_y	K_θ	β_x	β_y	β_θ
1	800.0	800.0	20.0	40.0	40.0	1.0
2	800.0	800.0	20.0	40.0	40.0	1.0

Table 2. Grasp Simulation Parameters

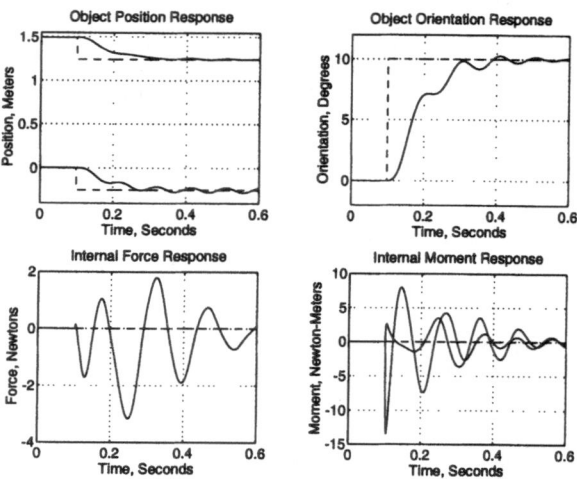

Figure 4. Open-Loop Internal Force Response During High Acceleration

response, while dashed lines denote desired values.

Here, the task was performed using a goal-position controller with no control of internal (or resultant) forces. The result is a rather large oscillation of the object relative to the arms. The oscillation could be improved by increasing the amount of damping at the grasp points, but the coupling remains. Even with closed-loop control of internal forces, it is difficult to eliminate the effect of large accelerations on internal forces, due to the limited bandwidth of any internal/resultant force controller, as illustrated in Figure 5. If the grasps were rigid, the internal forces and moments would be identically zero. In reality, the finite grasp stiffness and damping characteristics cause deviations from this ideal situation. This is the cause of the initial overshoot in internal forces.

To ensure that the desired acceleration changes smoothly, a fifth-order polynomial trajectory can be used with zero velocity and acceleration constraints specified at each endpoint. Figure 6 shows the system response to such a trajectory with no control of internal forces and moments. The final position of the object and the time to reach the goal are approximately the same as for the goal-position controller mentioned above.

The internal force response can be improved dramatically by adding feedback control of internal forces and moments as characterized by the virtual linkage. When closed-loop control of internal and resultant forces is added, the system responds to the above trajectory as shown in Figure 7. Internal forces and moments are reduced by approximately an order of magnitude. Trajectory tracking is also improved.

The main advantage of the virtual linkage is it's ability to characterize internal forces in a manner which is physically significant. This allows non-zero internal forces to have a predictable effect on the object. Using the same motion as before, we now command the system to maintain a 10 newton tension in the object, and internal moments of plus and minus one newton-meter, respectively,

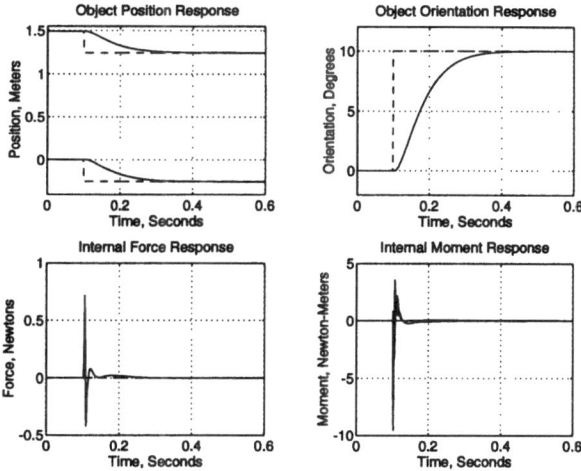

Figure 5. Closed-Loop Internal Force Response During High Acceleration

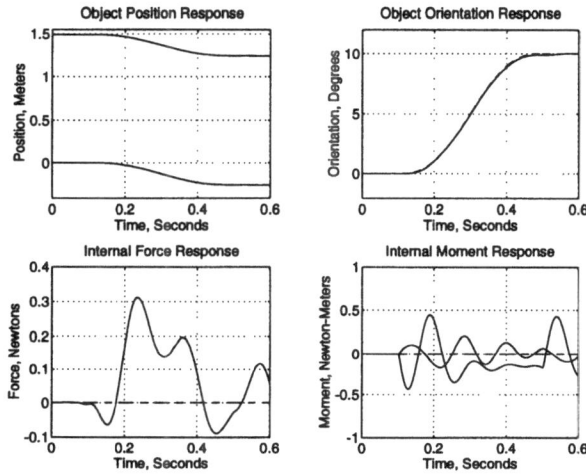

Figure 6. Open-Loop Internal Force Response During Motion

at each grasp. This type of operation might be used in preloading the object for assembly. The system response is shown in Figure 8.

The results of applying this controller to a system of three cooperating PUMA 560 manipulators holding a piece of Plexiglas are shown below. The controller is implemented on a multiprocessor system with a message-passing operating system kernel. The servo rate is 170 Hz.

Figure 9 shows the system's open-loop response to a set of internal force step inputs with no motion. The first plot shows the object position, the next three plots show the internal force in each of the three virtual links, as shown in Figure 1. Two points are immediately obvious from this plot. First, there is a great deal of friction in the system. Second, the grasps are lightly damped,

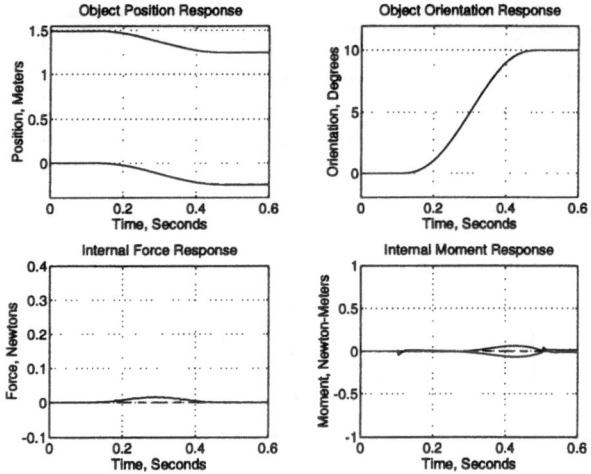

Figure 7. Closed-Loop Internal Force Response During Motion

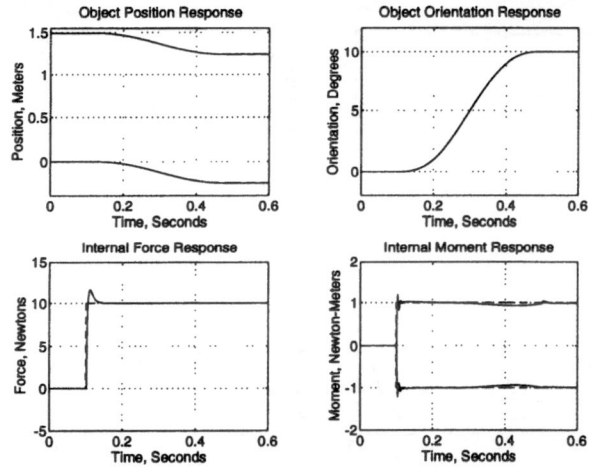

Figure 8. Non-Zero Internal Force Control During Motion

as evidenced by the 14 Hz vibrations.

The corresponding closed-loop response is shown in Figure 10. Internal force response is significantly improved.

5. Conclusions

In this paper, we describe an object-level methodology for performing dextrous, multi-grasp manipulation with simultaneous control of internal forces. This approach is based on the integration of two basic concepts: the *virtual linkage* and the *augmented object*. Compared to other methods of characterizing internal forces, the virtual linkage has the advantage of providing a physical represen-

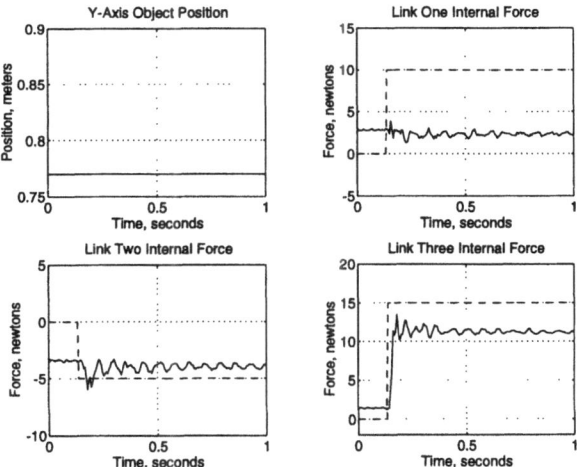

Figure 9. Static Open-Loop Internal Forces for a Three Arm System

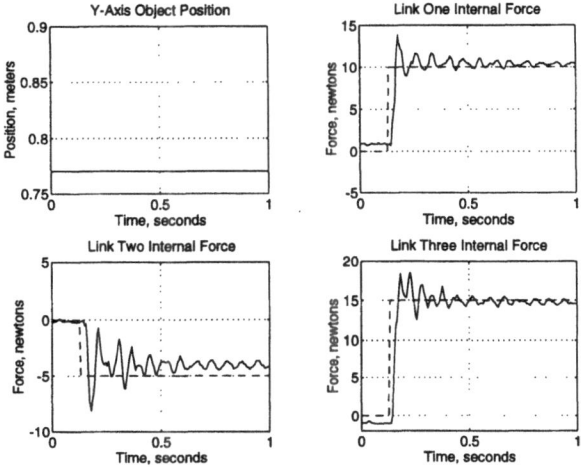

Figure 10. Static Closed-Loop Internal Forces for a Three Arm System

tation of internal forces and moments. This allows control of non-zero internal forces to have a physically meaningful effect on the manipulated object. As with the *augmented object* concept, motion of the manipulated object and control of the forces it applies to the environment are accomplished by specifying operational forces at some selected point on the object.

Simulation results are presented for a pair of three-degree-of-freedom manipulators grasping an object. These simulations show the effectiveness of the method, as well as the effect of object motion on internal forces. Experimental results are also presented for a system of three cooperating PUMA 560 manipulators.

28

References

[1] T.E. Alberts and D.I. Soloway. "Force Control of a Multi-Arm Robot System," *Proceedings of the 1988 IEEE International Conference on Robotics and Automation*, pages 1490-1496. Philadelphia, PA. April 1988.

[2] C.O. Alford and S.M. Belyeu. "Coordinated Control of Two Robot Arms," *Proc. IEEE Robotics and Automation*, pages 468-473, Atlanta, GA. 1984.

[3] Y. Chen, I. Walker, and J. Cheatham, "A New Approach to Force Distribution and Planning for Multifingered Grasps of Solid Objects," *Proceedings of the 1991 IEEE Conference on Robotics and Automation*, pages 890-896, Sacramento, CA. April 1991.

[4] G. Guo and W.A. Gruver, "Fingertip Force Planning for Multifingered Robot Hands," *Proceedings of the 1991 IEEE Conference on Robotics and Automation*, pages 665-672, Sacramento, CA. April 1991.

[5] O. Khatib, "A Unified Approach to Motion and Force Control of Robot Manipulators: The Operational Space Formulation," IEEE Journal on Robotics and Automation, Vol. 3, no. 1, 1987

[6] O. Khatib, " Object Manipulation in a Multi-Effector Robot System," The International Symposium of Robotics Research, Santa Cruz, CA. August 1987.

[7] M. Koga, K. Kosuge, K. Furuta, and K. Nosaki, "Coordinated Motion Control of Robot Arms Based on the Virtual Internal Model," *IEEE Transactions on Robotics and Automation*. Vol. 8, No. 1, pages 77-85, February 1992.

[8] V. Kumar, X. Yun, E. Paljug, and N. Sarkar, "Control of Contact Conditions for Manipulation with Multiple Robotic Systems," *Proceedings of the 1991 IEEE Conference on Robotics and Automation*, pages 170-175, Sacramento, CA. April 1991.

[9] V. Kumar and K.J. Waldron, "Suboptimal Algorithms for Force Distribution in Multifingered Grippers," *IEEE Transactions on Robotics and Automation*. Vol 5. No. 4, pages 491-498, August 1989.

[10] Y. Nakamura, "Minimizing Object Strain Energy for Coordination of Multiple Robotic Mechanisms," *Proc. 1988 American Control Conference*, pages 499-504.

[11] Y. Nakamura, K. Nagai, and T. Yoshikawa, "Mechanics of Coordinative Manipulation by Multiple Robotic Mechanisms," *Proceedings of the 1987 IEEE Conference on Robotics and Automation*, pages 991-998.

[12] D. Williams and O. Khatib, "The Virtual Linkage: A Model for Internal Forces in Multi-Grasp Manipulation," *Proceedings of the 1993 IEEE Conference on Robotics and Automation*, pages 1025-1030.

[13] X. Yun. 1991, "Coordination of Two-Arm Pushing," *Proceedings of the 1991 IEEE Conference on Robotics and Automation*, pages 182-187, Sacramento, CA. April 1991.

[14] Y.F. Zheng and J.Y.S Luh, "Joint Torques for Control of Two Coordinated Moving Robots," *Proceedings of the 1986 IEEE International Conference on Robotics and Automation*, pages 1375-1380, San Francisco, CA. April 1986.

[15] Y.F. Zheng and M.Z. Chen, "Trajectory Planning for Two Manipulators to Deform Flexible Beams," *Proceedings of the 1993 IEEE Conference on Robotics and Automation*, pages 1019-1024.

A Control Scheme Avoiding Singular Behavior of Self-Posture Changing Motion

Makoto Kaneko* and Tatsunori Hayashi**

* Faculty of Engineering
Hiroshima University
1-4-1 Kagamiyama, Higashi-Hiroshima, Hiroshima 724 JAPAN
** Mitsubishi Heavy Industry
Araimachi, Takasago, Hyogo 767 JAPAN

Abstract ---This paper proposes a new control scheme avoiding singular behaviors during Self-Posture changing Motion (SPCM) of a link system. With a proper combination of a compliant joint and a position-controlled joint, a link system has the capability of changing its posture while maintaining contact between inner link and environment over an angular displacement at the position-controlled joint. One undesirable behavior of SPCM is that with a sudden increase of contact force, a SPCM completely stops after the particular link posture depending on the frictional coefficient at the point of contact. To avoid this singular behavior, the commanded signal for the compliant joint is actively changed according to the contact force, so that the link system may continuously slip and change its posture. Experimental works are also shown

1. Introduction

With a proper combination of a compliant joint and a position-controlled joint, a link system has the capability of changing its posture while maintaining contact between inner link and environment over an angular displacement at the position-controlled joint. The series of these motions is called *Self-Posture Changing Motion (SPCM)*, which can be conveniently used for detecting an approximate contact point between inner link and unknown object [1], [2]. Furthermore, a series of *SPCM* contributes a simple way to both contact point detection for *Whole-Arm Manipulation (WAM)* [3] and object-shape-sensing for unknown object. The basic motions of *SPCM* and related experiments have been included in the IEEE '92 ICRA video proceedings [4].

A series of SPCM continuously makes progress irrespective of frictional coefficient at the point of contact, when the link posture is close to straight-line. However, contact force shoots up after a particular link configuration and any further *SPCM* perfectly stops after the link configuration [3]. This characteristic is called the *standing-up characteristic of contact force*. We have shown in [3] that the link posture leading to the *standing-up characteristic* strongly depends on the frictional coefficient at the point of contact. Providing a new control strategy to avoid such singular behavior is especially required when applying a series of *SPCM* for object-shape-sensing. Because the link posture is no more close to straight-line and the link system may result in the *standing-up characteristic of contact force* without carefully planning the motion of position-controlled joint.

The goal of this paper is to remove a distinct *standing-up characteristic of contact force* during a *SPCM* by introducing a new control scheme. We begin by explaining a general discussion of *SPCM* and then address a question why such a singular behavior appears. Through the analysis of experimental results, we show that when a contact force lying on the boundary of the friction cone passes through the compliant joint, the *standing-up characteristic of contact force* appears. The basic idea in this work is that we actively increase the tangential force component at the point of contact according to the increase of the normal force component, so that we can break the contact force entering the friction cone. This idea can be easily realized by simply changing the command signal for the compliant joint according to the normal force component at the point of contact. The proposed control scheme works especially when the link posture is close to the singular link posture, while it does not when the posture is close to straight-line. The effectiveness of the proposed control scheme is verified through experiments.

2. General Discussions of *SPCM*

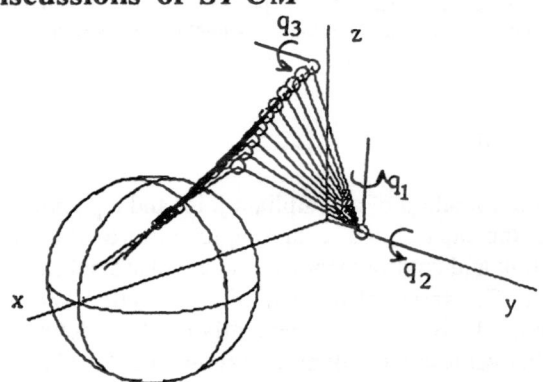

(a) Case 1 (joint 1: locked, joint 2: compliant, joint 3: position-controlled)

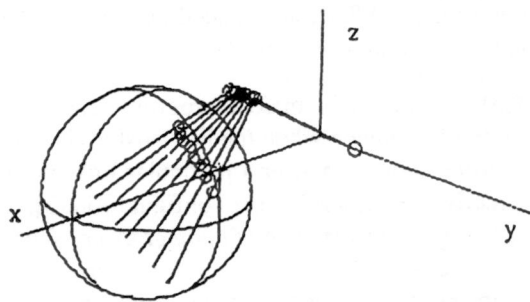

(b) Case 2 (joint 1: compliant, joint 2:locked, joint 3:position-controled)
Fig.1 Examples of SPCM for a 3D link system

A simple 3D link system as shown in Fig.1 is used to explain the *self-posture changing motion*. The link system has three degrees-of-freedom and the object is assumed to be fixed to the environment. Now, assume that for the position-controlled joint we impart an arbitrary angular displacement in the direction shown by the arrow of q3 in Fig.1 for the following two cases: Case 1 in which joint 1, joint 2, and joint 3 are assigned as a locked joint, a compliant joint and a position-controlled joint, respectively; and Case 2 in which joint 1, joint 2, and joint 3 are assigned as a compliant joint, a locked joint and a position-controlled joint, respectively. As a result, the link posture will automatically change its posture according to the angular displacement given to the position-controlled joint as shown in Figs.1(a) and (b), while maintaining contact between link and object. These characteristics constitute *self-posture changeability (SPC)* and the series of motions that bring about *SPC* is termed *self-posture changing motion (SPCM)*. When the axis of a position-controlled joint is selected to be parallel to the compliant joint, the motion of link is restricted in a 2D plane as shown in Fig.1(a), while that in Fig.1(b) is not. This particular joint assignment can be conveniently used when finding either an approximate contact point between finger link and an unknown object or the local shape of object.

Before addressing the mathematical definition of *SPCM*, we set the following assumptions to simplify the discussion.

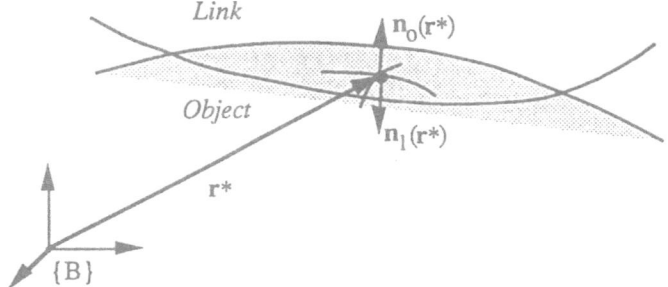

Fig. 2 Relationship between link and object in contact with each other

Assumption A (general)

Assumption 1:   Serial link system is assumed.

Assumption 2:   Object does not move during a *SPCM*.

Assumption 3:   Both object compliance and link compliance are negligibly small.

Assumption 4:   System compliance derives only from joint compliance control.

Assumption 5:   Mass of each link element is negligibly small.

We now define *self-posture changeability* with the following mathematical expressions.

Definition 1:   Assume a 3-D link system and that the h-th link is in contact with an object. For a set of angular displacement vector $dq_p=[dq_{p1}, ..., dq_{pn}]^t$ ($1<p_i<h$), *Self-posture changeability (SPC)* is guaranteed if vectors r^* and $r^\#$ satisfying the following equations exists continuously during a change of link posture:

$$G_0(r^*)=0 \tag{1}$$
$$G_1(r^*, q)=0 \tag{2}$$
$$G_0(r^\#)\geq0 \tag{3}$$
$$G_1(r^\#, q)=0 \tag{4}$$

where q and p_i are the joint angle vector $q=[q_1, ..., q_n]^t$ and a position-controlled joint number. Functions $G_0(r)$ and $G_1(r, q)$ are defined to possess the following characteristics:

$G_0(r)>0$ for a point outside of object
$G_0(r)=0$ for a point over object
$G_0(r)<0$ for a point inside of object
$G_1(r, q)>0$ for a point outside of link system
$G_1(r, q)=0$ for a point over link system
$G_1(r, q)<0$ for a point inside of link system

The series of motions brought about *SPC* is defined *Self-Posture Changing Motion (SPCM)*.

(1) and (2) mean that the point indicated by r^* exists over not only the surface of object but also the surface of link. Mathematically, these two conditions still allow for the hypothetical condition that a part of link system is inside the object. With additional (3) and (4), we can remove such a geometrical relationship between link and object. (3) means that a point over link system is never inside the object. In other words, it is either a point outside of object or a point on the surface of the object.

3. Utilization of *SPCM*

For practical applications, we set several additional assumptions.

Assumption B (specific)
Assumption 6 :   Link motion is limited to a 2D plane.
Assumption 7 :   Using the notation V_l for the workspace of the link and V_0 for the space occupied by object :

$$V_l \cap V_0 \neq \phi \tag{5}$$

Assumption 8 :   Each joint has a torque sensor and a joint position sensor.
Assumption 9 :   The link element is assumed to have the crosssection of a
very thin rectangular

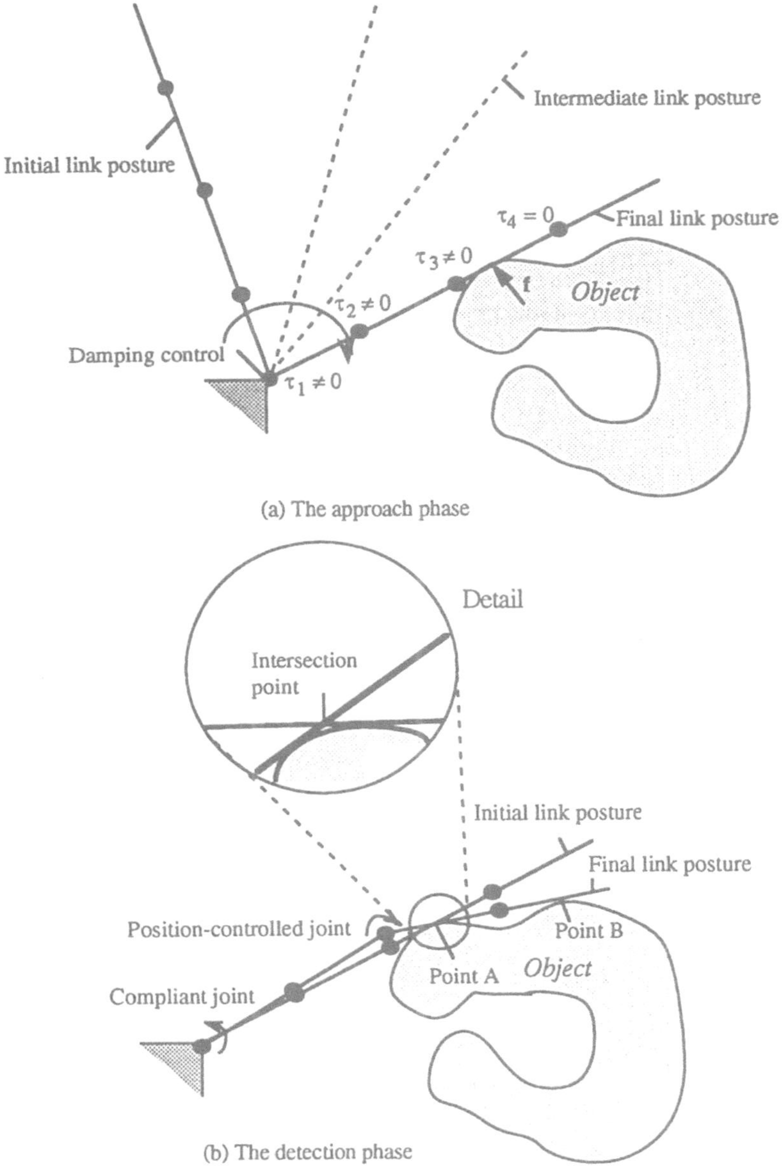

(a) The approach phase

(b) The detection phase

Fig.3 An approach to detect contact point

3.1 Contact point detection

While a number of approaches [5]-[8] have been reported for detecting contact location between robot and environment, we briefly explain the approach using *SPCM*. The sensing scheme is composed of two parts, one is the approach phase, where the link approaches until a part of link makes contact with an object as shown in Fig.3(a), and the other is the detection phase, where a compliance controller is used for the first joint and a position controller for the joint connecting the contact link. By imparting a small angular displacement to the position-controlled joint, the link posture changes while maintaining contact between the link and the object as shown in Fig.3(b). To maintain contact between link and object, the destination of the contact point of the link is set inside the object. This condition is given by

$$dr_d{}^t n_0 < 0 \qquad\qquad (6)$$

where n_0 is the unit normal vector of object and dr_d is a command displacement vector of link at the contact point by imparting an angular displacement to a position-controlled joint. Therefore, the motion planning for realizing a series of *SPCM* is so simple and easy. Fig.3 (b) shows the contact link postures before and after *SPCM*. We can regard the intersection point as the contact point between link and object.

3.2 Contact point detection for *WAM*

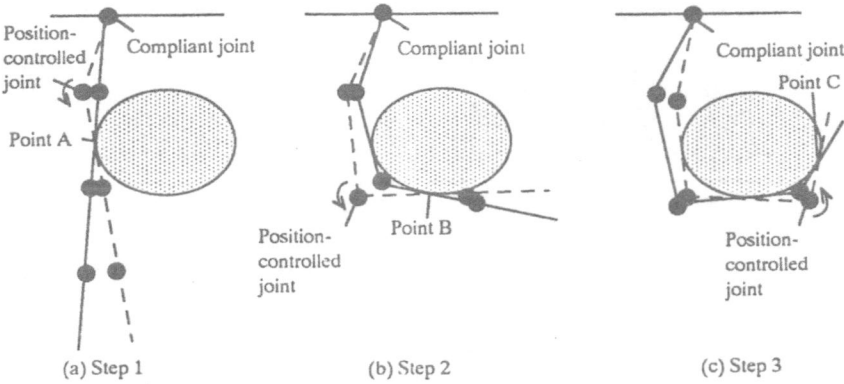

(a) Step 1   (b) Step 2   (c) Step 3

Fig.4 A procedure detecting contact points for Whole-Arm Manipulation using SPCM

The basic idea can also be applied to contact point detection for *WAM*. Fig.4 shows a procedure to detect contact point for *WAM* using *SPCM*, where (a), (b) and (c) are explaining each step to detect approximate contact points *A*, *B*, and *C*, respectively. In step 1, the link system approaches the object until part of link contacts with the object. Then, a small angular displacement is imparted to the position-controlled joint as shown in Fig.4(a). Through this motion, the robot can compute an approximate contact point *A* by applying the scheme explained in Fig.3. In step 2, an angular displacement is imparted to the third joint until another part of the link again contacts with the object as shown in the real line in Fig.4(b). At the end of this particular motion, the link contacts with the object at both points A and B. When a further small angular displacement is imparted at the same joint, two point contacts can not be maintained any more, and as s result, the second link will be away from the object as shown in the dotted line in Fig.4(b). Through this motion, the robot can detect an approximate contact point B. By applying a similar procedure, the robot can eventually detect an approximate contact point C as shown in Fig.4(c).

3.3 Object shape sensing

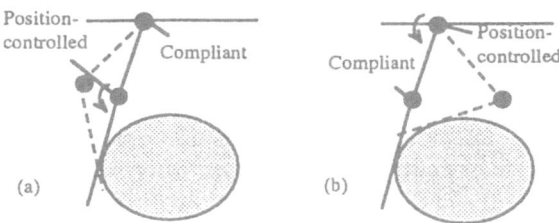

Fig.5 Object-shape-sensing using SPCM

Fig.5 explains an approach detecting a possible object shape by using *SPCM*. In Fig.5(a), a compliant and a position controllers are used for the first and the second joints, respectively and vice versa in Fig.5(b). Note that when the last link (the second link in Fig.5) is in contact with the object, there are two possible contact states, one at the link tip and the other at the inner link, where the possible contact point in the inner link can be computed by using two slightly different link postures as explained in Section 3.1. Without assuming a further tactile sensor, it is difficult to judge which contact state happens actually. Therefore, when the last link is in contact with the object, both points should be candidates for actual contact points. During a series of *SPCM*, temporarily two points are stored in the memory in such a case. Since the link can not penetrate the object, we have to remove such points that the link passes during a series of *SPCM*. Finally, we can obtain a possible object shape by connecting points remained.

4.Standing-up behavior of contact force

For the wire driven robot having two d.o.f [9], we design a joint servo system in the following manner:

(a) A damping control system with a torque feedback loop is implemented for joint *1*. This control law is given by

$$\tau_{1r} = k_v(\dot{q}_{r1} - \dot{q}_1) \tag{7}$$

where k_v is damping coefficient and $\dot{q}_l$ and $\dot{q}_{lr}$ are actual and commanded joint angular velocities.

(b) A position control system is implemented for the second joint

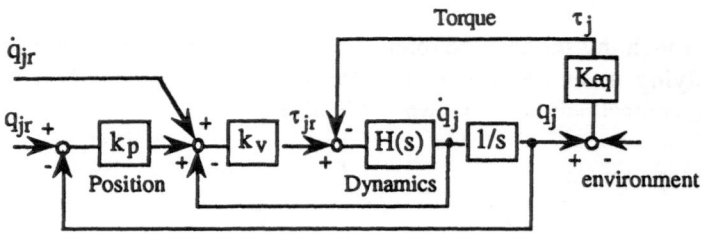

Fig. 6 Joint servo system with friction feedback loop

This control system is shown in Fig.6, where $H(s)$, k_p, and K_{eq} are dynamic characteristic of the actuator [rad/Nms], position forward gain [1/s], and equivalent environment stiffness [Nm/rad], respectively. The damping control is realized by setting $k_p=0$ and the position control by setting $\dot{q}_{2r}=0$ and $k_p\neq0$. When the link makes contact with an object, the torque about joint 1 is given by $\tau_1 = k_v \dot{q}_{lr}$. Since joint 1 can rotate freely for a torque that slightly exceeds the torque given by $\tau_1 = k_v \dot{q}_{lr}$, we can continuously use this system as an equivalent compliance controller during a series of *SPCM*. This is very convenient for practical applications, because when a link contacts an object, it automatically stops with a proper contact force.

Through experimental approaches using the wire driven robot, it has been shown that the contact force shoots up after a particular link configuration and any further *SPCM* is perfectly blocked out after the link configuration. We called this characteristic the standing-up characteristic of contact force during *SPCM* and the characteristic has been addressed as a kind of singular behavior of *SPCM* [3]. Let us now consider why such a singular behavior appears, assuming that each actuator is mounted on the base and the power is transmitted from each actuator to individual joint through pulleys and tendons. This transmission system is often introduced into robotic finger driving systems.

Fig.7(a) shows a conventional control scheme, where damping control and position control systems are implemented for joint 1 and 2, respectively and f is the contact force. Assume that the direction of contact force exactly coincides with the normal direction of object. Since the torque around joint 1 is given by $\tau_1 = rxf$ for such a tendon-driven robot, f does not generate any torque around joint 1 when directions of both vectors r and f coincide with each other, where r is the vector directing from joint 1 to joint 2. As a result, *SPCM* stops with the standing-up characteristic of contact force as shown in Fig.7 (b), where q2 is the angular displacement of joint 2. As it is seen from Fig.7(b), *SPCM* completely stops with standing-up behavior of contact force when the angular displacement q2 becomes approximately 0.7. Through careful observations of direction of contact force, we found that contact force shoots up along the boundary of the friction cone during *standing-up-phase*. Therefore, we can determine the standing-up-angle q_{2s} with the following equation (8).

$$q_{2s} = \frac{\pi}{2} - \alpha \tag{8}$$

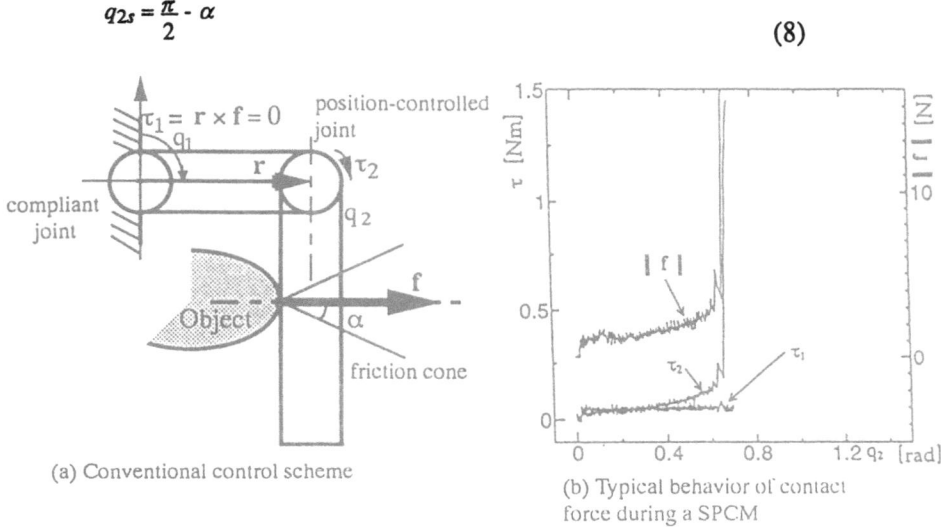

(a) Conventional control scheme

(b) Typical behavior of contact force during a SPCM

Fig.7 Conventional control scheme and contact force during a SPCM

Since any *SPCM* stops after standing-up characteristic, q_{2s} provides the maximum angle leading to *SPCM*. Under a frictionless contact, $q_{2s}=\pi/2$.

5. A new control scheme avoiding singular behavior

The goal of this paper is to remove such a distinct standing-up characteristic of contact force during SPCM by introducing a new control scheme. Fig.7(a) shows the proposed control scheme where a damping control is used for joint *1* and its reference value is adaptively changed according to the torque τ_2 sensed at joint *2*. The control law for the first joint is given by

$$\tau_{1r} = k_v\,(\dot{q}_{1r} - \dot{q}_1) - \beta\tau_2 \tag{9}$$

where β is torque coefficient as shown in Fig.8(b). This control law becomes active especially when the link posture is close to the singular one as shown in Fig.8(a). Because both contact force and τ_2 increases in such a particular link posture, while they are small for a link posture close to the straight-line. Generally, when $f_t > \mu f_n$, a slip occurs between link and object, where f_t and f_n are tangential and normal directional forces, respectively, and μ is the frictional coefficient at the point of contact. When $q_2=\pi/2$, this condition is replaced by

$$\left|\frac{\tau_1 b}{\tau_2 l_1}\right| > \mu \tag{10}$$

where b and l_1 are contact distance at link 2 and length of link 1, respectively. Since $\tau_1 = \tau_{1r} - \beta\tau_2$,

$$\left|\frac{\tau_{1r}}{\tau_2} - \beta\right| > \mu\frac{l_1}{b} \tag{11}$$

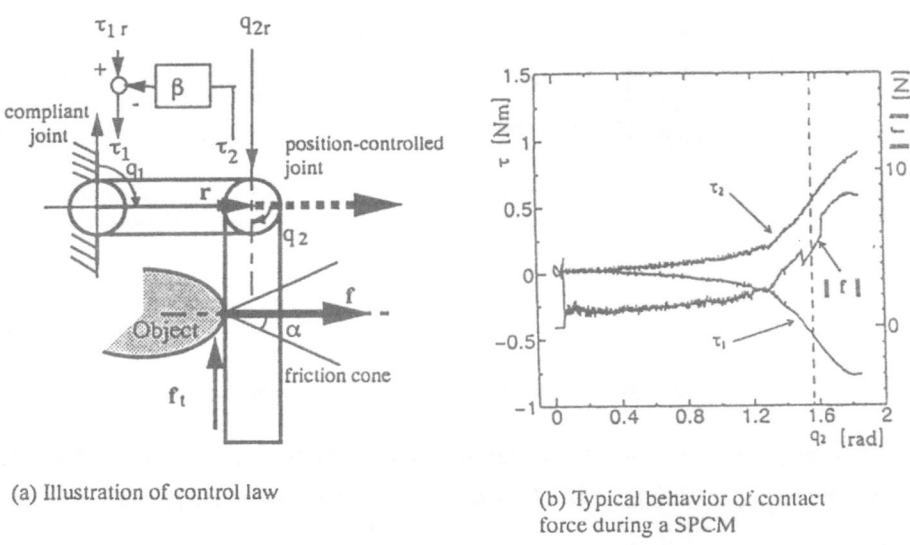

(a) Illustration of control law

(b) Typical behavior of contact force during a SPCM

Fig.8 The proposed control scheme and cotact force during a SPCM

When the link posture is close to the singular one, f_n and τ_2 increases while τ_{1r} keeps constant. Since τ_{1r}/τ_2 becomes small under the singular posture, we finally obtain the following relationship

$$|\beta| > \mu\frac{l_1}{b} \tag{12}$$

By selecting a relatively large gain β, this condition is always guaranteed regardless μ, b and l_1. With this condition, a large tangential force f_t appears at the point of contact which surely breaks the static frictional condition. As a result, the second link will slip in the direction shown by the arrow. Thus, we can expand the link posture capable of realizing *SPCM*. The idea taken here is surprisingly simple but it works quite effective. Fig.8(b) shows an experimental result, where no distinct standing-up characteristic of contact force is observed and *SPCM* is successfully realized even under $q_2 = \pi/2$ corresponding to the singular posture in the conventional control scheme. Note that the total system stability is guaranteed if each joint controller is originally designed to be stable. Because the command value of the first joint is simply changed according to the second joint torque. The control signal unilaterally flows from joint 2 to joint l.

6. Conclusion

Main results obtained through this work are in the following:
(1) We discussed several applications of *SPCM*.
(2) We explained qualitatively how and when the *standing-up characteristic of contact force* appears during a series of *SPCM*.
(3) We proposed a new control scheme avoiding the standing-up behavior of contact force by actively changing the command value of the compliant joint according to the contact force.

Finally, we would like to express their sincere thanks to Mr. Imamura and Mr. Yoshinari, research staff of Kobe Steel Co Ltd., for their cooperation with the development of the wire driven robot used in the experiments.

References:
[1]M. Kaneko and K. Tanie: Contact Point Detection for Grasping an Unknown Object Using Self-Posture Changeability, Proc. of the IEEE Int. Conf. on Robotics and Automation, p864, 1990.
[2]M. Kaneko and K. Tanie: Self-Posture Changeability (SPC) for 3-D Link System, Proc. of the IEEE Int. Conf. on Robotics and Automation, p1635, 1992.
[3]M. Kaneko, T. Hayashi, H. Maekawa, K. Tanie: Contact Point Detection between Robotic Fingers and Object Using Joint Compliance, 1992 IEEE Video Proceedings, 1992.
[4]M. Kaneko and T. Hayashi: Standing-up Characteristic of Contact Force during Self-Posture Changing Motions, Proc. of the IEEE Int. Conf. on Robotics and Automation, p202, 1993
[5]J. K. Salisbury: Interpretation of contact geometries from force measurements, Proc. of the 1st Int. Symp. on Robotics Research, 1984.
[6]T. Tsujimura and T. Yabuta: Object detection by tactile sensing method employing force/moment information, IEEE Trans. on Robotics and Automation, vol.5, no.4, p444, 1988.
[7]A. Bicchi, J. K. Salisbury, and D. L. Brock: Contact sensing from force measurements, MIT A.I. Memo, No. 1262, p4, 1990.
[8]S. J. Gordon and W. T. Townsend: Integration of tactile-force and joint-torque information in a Whole-Arm manipulator, Proc. of the IEEE Int. Conf. on Robotics and Automation, p464, 1989.
[9]M. Kaneko and N. Imamura: Development of a Tendon-Driven Finger with Single Pulley-type TDT Sensors, Proc. of the IEEE Int. Workshop on Intelligent Robots and Systems, p752, 1991.

Control Considerations on Minimum Joint Torque Motion

Jadran Lenarčič and Leon Žlajpah

The Robotics Laboratory
"J. Stefan" Institute, University of Ljubljana
Jamova 39, 61111 Ljubljana, Slovenia

Abstract - The problem of minimum joint torque motion is considered in the low-level control. The optimum motion is obtained by a local optimisation that uses the calculated derivatives of the cost function represented as a square norm of the joint torques. The experiments are carried out with planar mechanisms lifting loads. It is shown that standard control schemes which include simple singularity-avoidance algorithms and limited joint velocities can be used for this purpose.

1. Introduction

Control of redundant manipulators can be treated as an optimisation problem in which the end effector executes a prescribed path corresponding to the task, while the rest of the mechanism moves in an optimum way with respect to the given set of criteria, such as minimum energy or minimum joint rates [1,2]. The minimum joint torque motion is requested especially in applications of lifting, pushing, and manipulation of heavy objects.

The majority of the existing robots is designed with the objective to provide an abstraction of a disembodied hand whose motion can arbitrarily be programmed to fit a task [3]. Singularities are, therefore, removed from the central region of the workspace. Yet, in singularity configurations the mechanism can support infinite loads in the direction of impossible velocity. Humans, for example, use singularities of their limbs to gain mechanical advantage in lifting or standing. Similar benefits can be obtained by mechanical manipulators. Weak actuation can be compensated by more sophisticated control that exploits singularities.

A minimisation of joint torques requires the mechanism to move in singularity configurations in order to gain full mechanical advantage. This may cause control instabilities. Algorithms with singularity handling capabilities included in various control schemes were reported [4,5], but they introduce errors in timing or position accuracy. A positive aspect is that a path near kinematic singularities already provides a good approximation of the minimum joint torque motion. The error between the optimum and the near-optimum trajectory appears when the joint torques are very small and can be neglected. More critical, as shown in the present paper, are the discontinuities that arise from the local optimisation.

This paper presents observations based on experiments and numerical simulations of the minimum joint torque motion. Two basic control algorithms are analysed. The first possesses an off-line generation of local optimum paths for minimum joint torque motion as reference trajectories with a classical controller. The other includes the minimisation of joint torque's on-line in each time instant.

2. Theory and assumptions

Since they are relatively simple, the local optimisation methods are time-effective and can be implemented in real-time control utilising computed and sensory information for the feedback loops [1]. The main drawback of these methods is that they do not guarantee to completely satisfy a given task. On the contrary, the global optimisation techniques [6] search for the global optimum but they are computationally very expensive and are far from real-time implementation. Moreover, these methods depend on the mathematical model of the system. The parameters of a real robot can unexpectedly change and the implemented model may become inaccurate.

We speculate that the human motion can be represented by a local optimisation in the lower level of control, interpreted as a minimisation of the fatigue. The global optimum is provided in a higher level, in motion planning that is executed well in advance and which is based on experience, intelligence, and intuition. Hence, we are convinced that robots, similarly to humans, should globally optimise their movements by using artificial intelligence, motion planing algorithms, and self-learning techniques, while simple local optimisation methods should be implemented in the lower control.

In this paper, the minimum joint torque motion is considered in the case of lifting loads by a planar n-degrees-of-freedom robot manipulator. If we neglect the gravity of links and their dynamics, the joint torques can be computed by

$$\tau = J^T F \ , \tag{1}$$

where F is the force applied to the end effector, and J is the Jacobian matrix. The end effector position x_1 and y_1 and its first-order derivatives are

$$x_i = x_{i+1} + d_i \cos(\sum_{k=1}^{i} \theta_k) \ ,$$

$$y_i = y_{i+1} + d_i \sin(\sum_{k=1}^{i} \theta_k) \ ,$$

$$\frac{\partial x_1}{\partial \theta_i} = -y_i, \ \frac{\partial y_1}{\partial \theta_i} = x_i \ , \tag{2}$$

$i = n, n-1,..., 1$, θ_i are the joint angles, and d_i are the link lengths. Accordingly, the joint torques caused by the vertical unity force $F = 1$ are

$$\tau_i = x_i \ , i = 1, 2, ..., n. \tag{3}$$

42

One way to specify the lifting problem is to minimise the quadratic cost function

$$f = \frac{c_1}{2}(y_1 - y_0)^2 + \frac{c_2}{2}\sum_{i=1}^{n} k_i x_i^2 \ . \tag{4}$$

Here, the first term is related to the motion in the direction of y, y_0 is the desired value of y_1, the second term is related to the joint torques where the desired values are zero, and k_i are the corresponding weighting factors, as well as the constants c_1 and c_2. The derivatives of f can be expressed as follows

$$\frac{\partial f}{\partial \theta_i} = c_1(y_1 - y_0)x_i - c_2(y_i \sum_{j=1}^{i} k_j x_j + \sum_{j=i+1}^{n} y_j k_j x_j) \ . \tag{5}$$

Once the derivatives of the cost function are established, we can choose among a variety of numerical procedures, including the pseudo inverse approach which is widely used in redundancy resolution methods [7]. In this work, the minimisation of the given cost function (4) is based on the steepest descent method as reported in [8] and is specified by the following iteration

$$\theta_i^{(r+1)} = \theta_i^{(r)} + \alpha^{(r)} \frac{\partial f^{(r)}}{\partial \theta_i^{(r)}} \ , \tag{6}$$

where $r = 1,2,...$ is the number of iteration, $\alpha^{(r)}$ is the iteration step size, and $\theta^{(0)}$ is the initial estimation of joint angles. In order to achieve better convergence, the iteration step size can be computed according to the proposed in [8].

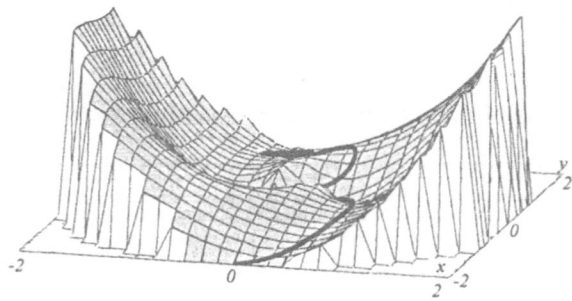

Figure 1. Square norm of joint torques in x-y space

Figure 1 shows the function of joint torques $(\tau_1)^2 + (\tau_2)^2$ in Cartesian space for the 2-R planar robot manipulator with the links $d_1 = 1.1$, $d_2 = 0.9$. In Figure 2, the Cartesian path of minimum joint torques is presented, where the initial configuration is straight downward and the final straight upward.

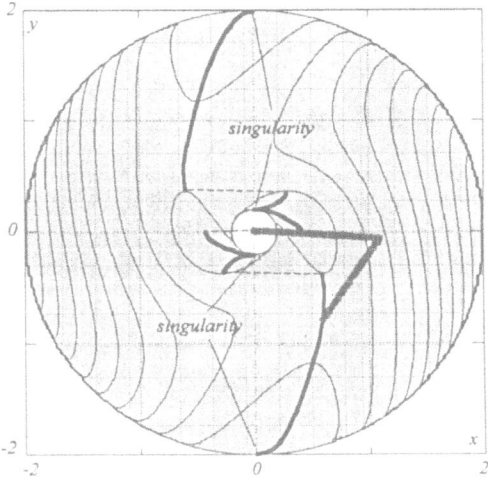

Figure 2. Cartesian path of minimum joint torques

The discontinuities appear near and in the kinematic singularities at the inner border of the workspace. They require the mechanism to instantaneously change the values of joint angles with infinite joint rates. The joint angles as functions of y are shown in Figure 3.

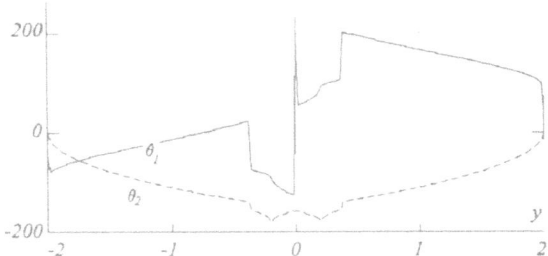

Figure 3. Theoretically optimum joint angles

Clearly, these trajectories cannot be realised in practise. Even if the mechanism was capable to execute extremely high joint velocities, the intermediate configurations, which serve for continuous motion from one optimum position to another, would pass through joint torques that are away from the minimum. A more feasible solution, but still theoretical, is obtained by optimising the path according to the iterative procedure proposed in (6) in which the optimisation is performed following a y_0 that changes from the minimum value -2.0 to the maximum 2.0. The obtained Cartesian path and the trajectories of joint angles as functions of y are shown in Figures 4 and 5, respectively.

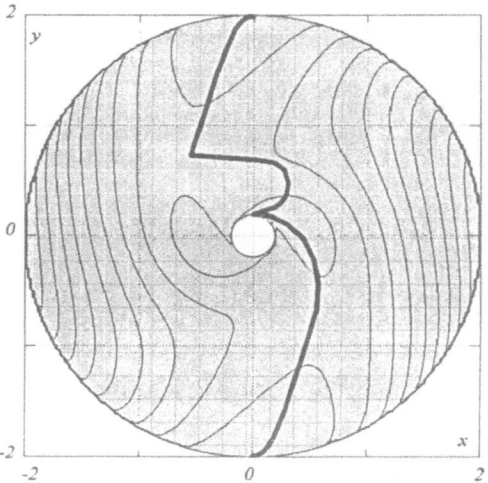

Figure 4. Optimum Cartesian path with respect to (6)

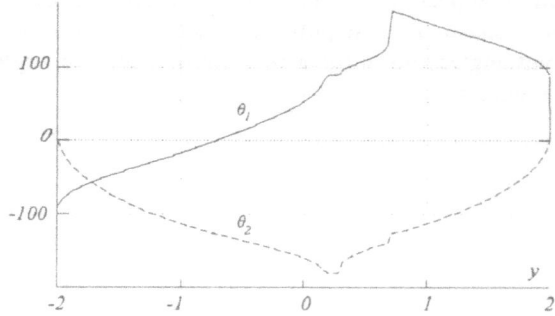

Figure 5. Optimum joint angles with respect to (6)

It can be observed that the procedure avoids the first inner singularity and moves directly to the next. In this kinematic singularity, the mechanism either changes the configuration - elbow up to elbow down - or stays in the same and continues its climbing. In order to limit the joint rates, one of the links stops and then inverts the direction of motion. The same path, specified in terms of joint torques, is illustrated in Figure 1. A critical point can be noticed relatively far from the second inner singularity when the mechanism suddenly changes its position and falls in the globally optimum path. Namely, the algorithm of local optimisation follows a less steep direction until it reaches a point from where it instantaneously turns into the valley of minimum joint torques on the other side of the hill. To avoid discontinuities of this kind, global optimisation schemes should be utilised. Other types or forms of cost functions can also serve to minimise this problem.

3. Experiments and simulations

The following examples show the simulation of the 2-R robot manipulator lifting
a load attached to the end effector from the straight downward to the straight
upward configuration. In the real-time simulation the dynamical model of the
mechanism is given as follows

$$\tau = H(\theta)\,\ddot{\theta} + h(\theta,\dot{\theta}) + B\,\dot{\theta} + g(\theta) \ . \tag{7}$$

Here, H is the inertia matrix, h are the Coriolis and centripetal terms, B is the
diagonal matrix associated with the viscose friction, g is the vector that contains
the gravity of the links of the mechanism and of the load, and θ is the vector of
joint angles.

The utilised control scheme is the usual feed-back loop specified by the
following equation

$$u = k_p(\theta_r - \theta) + k_d(\dot{\theta}_r - \dot{\theta}) + \tau \ . \tag{8}$$

It minimises the error between the reference and the actual joint angles and their
velocities, as well as the actual joint torques τ. The vector u denotes the controlled
torques of the direct-drive motors, k_p are the position feedback constants (in the
treated examples are zero), and $k_d = 400, 100$ are the velocity feedback constants
for u_1 and u_2, respectively.

In the first example, the velocities of the joint angles are assumed to be equal
to the gradient of the cost function (5) in which $c_1 = 0$

$$\vartheta_i = \frac{\partial f}{\partial \theta_i} \ , i = 1,2. \tag{9}$$

These are transformed in the Cartesian space velocities by multiplication with the
corresponding Jacobian

$$\dot{p} = J\,\dot{\vartheta} \ . \tag{10}$$

The reference Cartesian velocities are chosen as

$$\dot{p}_x = \dot{p}_x$$

$$\dot{p}_y = max(\dot{p}_y, v) \ , \tag{11}$$

where v is a desired vertical velocity of the end effector. If the Jacobian is not
singular, the reference joint velocities are obtained by

$$\dot{\theta}_r = J^{-1}\,\dot{p} \ . \tag{12}$$

The obtained Cartesian path of the end effector carrying the load is presented in
Figure 6. Because of the included dynamics of the mechanism and of the load, the
corresponding joint trajectories are continuos. They are shown in Figure 7.

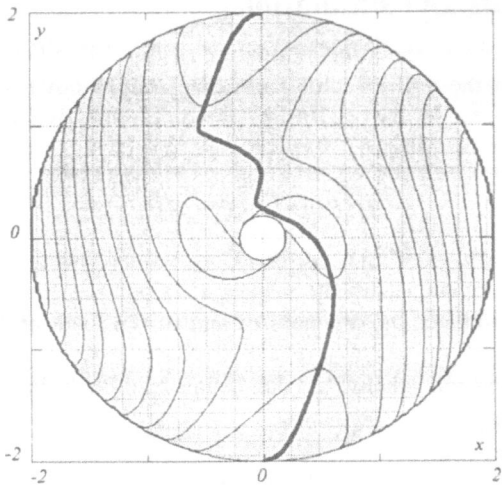

Figure 6. Cartesian path produced by the first controller

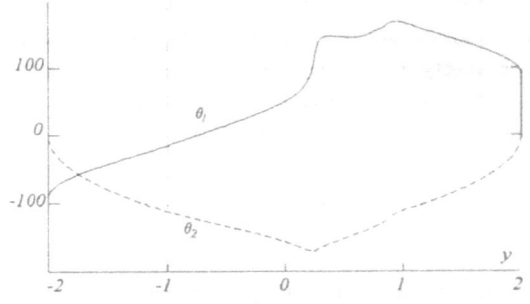

Figure 7. Joint angles produced by the first controller

In the second numerical example, we simulated the lifting task by calculating the reference joint velocities as follows

$$\theta_{r_i} = \frac{\partial f}{\partial \theta_i} \ , i = 1,2,$$

(13)

for $c_1 = 400$, while the desired vertical position of the end effector y_0 is increased in each time instant according to

$$y_0 = y_{min} + vt \ ,$$

(14)

where v is the desired vertical velocity of the end effector. The resulting Cartesian path of the end effector is shown in Figure 8, and the corresponding joint angles in Figure 9.

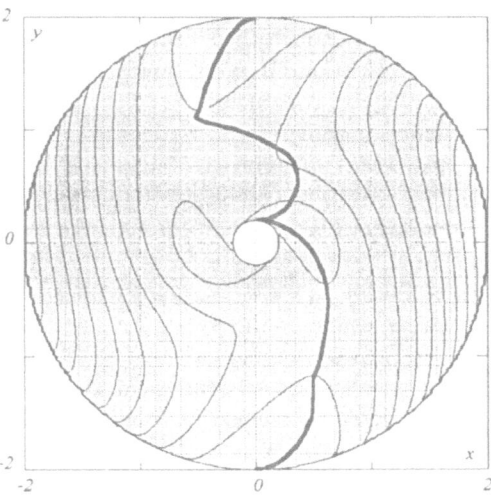

Figure 8. Cartesian path produced by the second controller

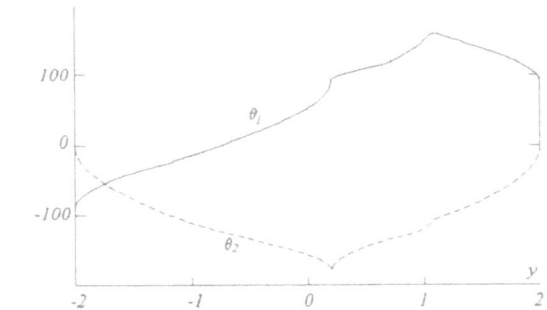

Figure 9. Joint angles produced by the second controller

Note that in both cases the joint rates are limited by multiplication with a variable that is 1 when the velocities are less than the attainable maximum, otherwise

$$c = max(1, \frac{\delta\theta_{r_i}}{\delta\theta_{max_i}}) \ , \ i = 1,2. \tag{15}$$

Both controllers include a singularity avoidance algorithm which becomes effective only very close to the singularity configurations.

4. Discussion and conclusions

The motion with minimum joint torques is considered in this paper. It is synthesised based on the local optimisation that uses a square norm of joint torques as a cost function. The experiments are carried out with a planar 2-R mechanism lifting a load that is attached to the end effector. The objective is to find an optimum path of the end effector from the initial straight downward to the final straight upward configuration. It is assumed in these examples that the mechanism contains relatively weak actuators.

The utilised cost function is shown to be well suited for the real-time control, especially because we can obtain its derivatives in an analytical form that does not require expensive computations. It can be observed that the second link of the 2-R manipulator in the given task always passes through the maximum joint torque without regard to the path of the end effector between its initial and final position. This is valid for the last link of any n-R planar mechanism. There is a freedom, however, to decide at which instant on the path - or values of other joint torques - the last joint torque reaches its maximum.

In order to gain full mechanical advantage in the minimum joint torque motion, the manipulator moves in singularity configurations. This may cause instability problems. Standard control schemes fail in kinematic singularities and must, therefore, be improved by using singularity avoidance algorithms. This introduces additional position and timing error between the obtained and the optimum solution. Nevertheless, if the kinematic singularities are avoided, the obtained solution remains very practical, since it produces an error in joint torques when these are very small or zero. Hence, this error can easily be neglected. More critical may be the error in timing of the task execution. Realistically, if the minimum joint torque is requested, timing should never be defined strictly.

The main problem in the treated examples in this work is associated with the local optimisation of joint torques. Following the less steep direction - minimum derivatives of the cost function - the local optimisation procedure can produce a path that is away from the global optimum which in a concrete example may not be realisable with the given motors. This critical point can be noticed relatively far from the second inner singularity when the mechanism suddenly changes its position and falls in the globally optimum path. Since the algorithm follows the minimum gradients, it continues climbing until it reaches a point from where it instantaneously turns into the valley of minimum joint torques on the other side of the hill. A more optimum solution from the global point of view would be to follow a more steep direction and find a saddle point to cross the hill which can be realised with considerably less joint torques. One possibility to overcome these difficulties is to apply other forms of the cost function dedicated to a specific task. We trust, however, that there is no general and cheap solution to this problem at the low-level control. Off-line motion planning and also artificial intelligence methods must be used in a higher level of control that search for global optima.

Figure 10 represents the quadratic function of joint torques $(\tau_1)^2+(\tau_2)^2$ for: (a) the theoretical optimum in Figure 2, (b) the local optimum path in Figure 4, (c) the first controller in Figure 6, and (d) the second controller in Figure 8. The second controller produces a path which better fits the local optimum. It can be noticed, however, that the first controller obtains a smaller maximum along the path than the second-one and also smaller than the local optimum (b). Near the singularity, it chooses a more steep direction (Figure 1) which, by chance, turns out to be more optimum in a global sense.

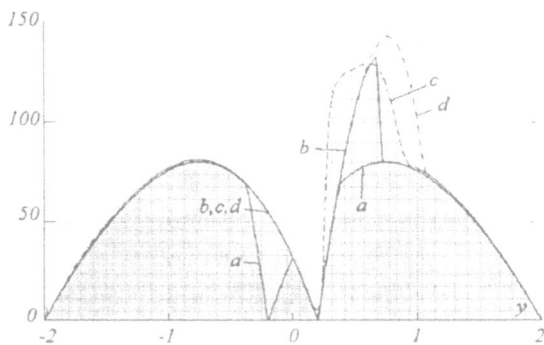

Figure 10. Joint torques depending on y

In this work, we consider the problem of static lifting. It is clear that the obtained results can be ameliorated by including dynamical effects. By exploiting the inertia of the mechanism's links and the load attached to the end effector it would be possible to solve the task of lifting with very small joint torques. We presume, however, that the main problem in using such a strategy would be related to the stability and it would request very sophisticated and robust control schemes. This is still to be explored.

References

[1]   D.N. Nenchev. Redundancy Resolution through Local Optimisation: Review. *J.Robotic Systems*, 6(6):769-789, 1989

[2]   A. Nedungadi and K. Kazerounian. A Local Solution with Global Characteristics for the Joint Torque Optimisation of a Redundant Manipulator. *J. Robotic Systems*, 6(5): 631-654, 1989.

[3]   J. Kieffer and J. Lenarčič. On the Exploitation of Mechanical Advantage Near Robot Singularities. *Proc. 3rd Workshop Advances in Robot Kinematics* (V. Parenti-Castelli and J. Lenarčič, Eds.), Ferrara, pp. 65-72, 1992

[4]   S. Chiaverini. Estimate of the Two Smallest Singular Values of the Jacobian Matrix: Application to Damped Least-Squares Inverse Kinematics. *Proc. 3rd Workshop Advances in Robot Kinematics* (V. Parenti-Castelli and J. Lenarčič, Eds.), Ferrara, pp. 80-87, 1992

[5]   Y. Nakamura and H. Hanafusa. Inverse Kinematic Solutions with Singularity Robustness for Robot Manipulator. *ASME J. Dynamic Systems, Measurement, and Control*, 109: 163-171, 1986

[6]   M. Vukobratović and M. Kirćanski. A dynamic Approach to Nominal Trajectory Synthesis for Redundant Manipulators. *IEEE Trans. Systems, Man, and Cybernetics*, SMC-14: 580-586, 1984

[7]   J. Lenarčič. Computational Considerations on Redundancy Resolution of Multi Link Manipulators, Computational Kinematics, (J. Angeles, P. Kovacs, and ?? Hommell, Eds.), *Kluwer Academic. Publishers Ltd.*, pp. 75-85, Dordrecht 1993

[8]   C.A. Klein and C.H. Huang. Review of Pseudoinverse Control for Use with KInematically Redundant Manipulators. *IEEE Trans. Systems, Man, and Cybernetics*, SMC-13: 245-250, 1983

Experimental Integration of Planning in a Distributed Control System[*]

Gerardo Pardo-Castellote[†], Tsai-Yen Li[‡],
Yoshihito Koga[§], Robert H. Cannon, Jr.[¶],
Jean-Claude Latombe[‖], Stanley A. Schneider[**]
Stanford University
Stanford, California 94305

Abstract— This paper describes a complete system architecture integrating planning into a two-armed robotic workcell. The system is comprised of four major components: user interface, planner, the dual-arm robot control and sensor system, and an on-line simulator.

The graphical user interface provides high-level user direction. The motion planner generates complete on-line plans to carry out these directives, specifying both single and dual-armed motion and manipulation. Combined with the robot control and real-time vision, the system is capable of performing object acquisition from a moving conveyor belt as well as reacting to environmental changes on-line.

The modules communicate through a novel subscription-based network data sharing system called the *Network Data Delivery Service* (NDDS). NDDS allows the different modules to transparently share data, and thus be distributed across different computer systems. Its stateless protocol naturally supports multiple anonymous data consumers and producers, arbitrary data types, on-line reconfiguration and error recovery.

The control software is integrated within the *ControlShell* framework. *ControlShell* provides an object-oriented generic software framework for combining reusable software components into a working, complex system.

This paper presents an overview of the individual system components, as well as a summary of the architecture developed to integrate the system. Much of the paper focuses on the interfaces between components.

1. Introduction

Automation and ease-of-operation are two goals of robotic systems. Ideally, one would specify a high-level task such as an assembly and have it executed automatically. To achieve these goals, sophisticated software modules such as planners, user interfaces, controllers etc. are being developed. However, the complexity of these modules and the fact that they are often developed

---

[*]This work was supported in part by ARPA/Navy Contract No. N00014-92-J-1809.

[†]Department of Electrical Engineering.

[‡]Department of Mechanical Engineering.

[§]Department of Mechanical Engineering.

[¶]Department of Aeronautics and Astronautics.

[‖]Department of Computer Science.

[**]Real Time Innovations Inc.

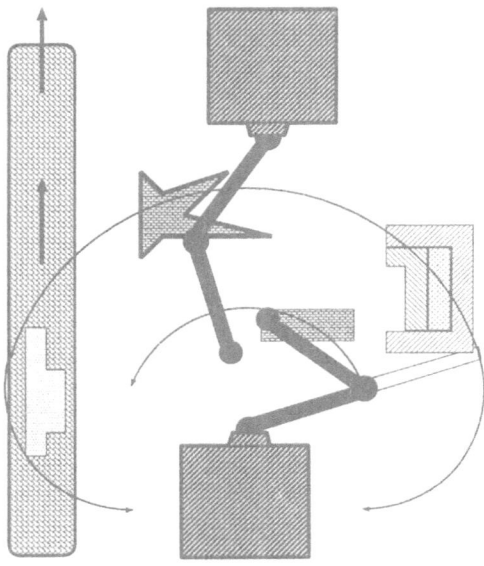

Figure 1. **Experimental Demonstration**

Experimental demonstration consisting of a robotic assembly in the presence of moving objects. The robot has two 4-DOF arms. The parts are delivered by a conveyor.

at different times by different groups of people make system integration and testing very time consuming and often problem-specific.

In a joint effort, the Computer Science Robotics Laboratory and the Aerospace Robotics Laboratory at Stanford University have developed a flexible experimental test-bed to explore these issues. Our goal is to achieve task-level operation on a distributed robotic system in a dynamic environment.

The experimental demonstration is illustrated in Figure 1. Two 4-DOF arms manipulate parts in a dynamic environment containing both *static* and *moving* obstacles and parts. The parts are supplied by a conveyor. A vision system identifies and tracks the moving parts which are picked-up by the robot *while in motion*. Several efficient path planning modules are also implemented to find trajectories to deliver and assemble parts while avoiding the obstacles in the workspace. Due to their size and weight, some of the parts require cooperative manipulation and regrasping by the two arms whereas others are manipulated by a single arm. The user monitors and issues task-level commands using a graphical user interface.

2. System Architecture

Our experimental test-bed is composed of four modules as illustrated in

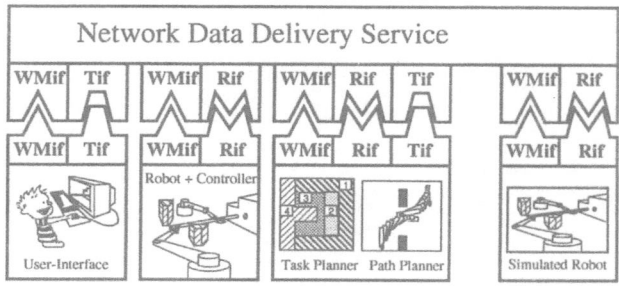

Figure 2. **System Architecture**

The overall system showing its four main modules. Each module communicates using one or more of the three interfaces: The World Model Interface (WMif), the Robot Interface (Rif) and the Task Interface (Tif). These modules are physically distributed. The Network Data Delivery Service plays the role of a bus providing the necessary interconnections.

Figure 2. The user interface receives state information from the robot and sensor systems and provides a graphical representation of the scene to the user. During operation, the high-level task is specified from the graphical user interface. The task planner/path planner receive continuous updates from the robot and sensors. The planners produce robot-commands primitives which are executed by the robot controller.

The simulator and the robot have the same interface to the other modules. This allows the simulator to masquerade as the robot for fast prototyping and testing of the rest of the system.

A novel *network-transparent, subscription-based* data-sharing scheme—the *Network Data Delivery Service* (NDDS)—facilitates communication among the different modules. It allows them to be distributed across different computer systems (with different processor architectures and operating systems[1]) and provides the necessary arbitration between data updates, enabling multiple users to operate and monitor the system concurrently.

The NDDS system [9] builds on the model of information producers (sources) and consumers (sinks). Producers register a set of data instances that they will produce, unaware of prospective consumers and "produce" the data at their own discretion. Consumers "subscribe" to updates of any data instances they require without concern for who is producing them. In this sense the NDDS is a "subscription-based" model. NDDS provides stateless (and hence robust) mechanisms to resolve multiple-producer conflicts and supports multiple-rate consumers.

The use of subscriptions to drastically reduces the update overhead over a classical client-server architecture. Occasional subscription requests, at low

---

[1]The current demonstration involves DEC workstations, Sun Workstations and several VME-based real-time processors

bandwidth, replace numerous high-bandwidth client requests. Latency is also reduced, as the outgoing request message time is eliminated.

NDDS differs from other distributed data-sharing schemes [17, 8, 3, 14] in its transparent support for multiple anonymous data-producers and consumers as well as in its fully-distributed, symmetrical implementation (which contains no privileged nodes).

All modules in the system communicate using one of the following three interfaces (built on top of NDDS): The *World Model* interface, the *Robot* interface and the *Task* interface. The functionality of two of these interfaces is summarized in tables 1 and 2.

This arrangement is analogous to a hardware bus as illustrated in Figure 2. NDDS plays the role of the physical interconnections. The three interfaces being similar to bus-access protocols. This approach is key to reducing system integration time and produces generic, reusable modules.

3. Controller

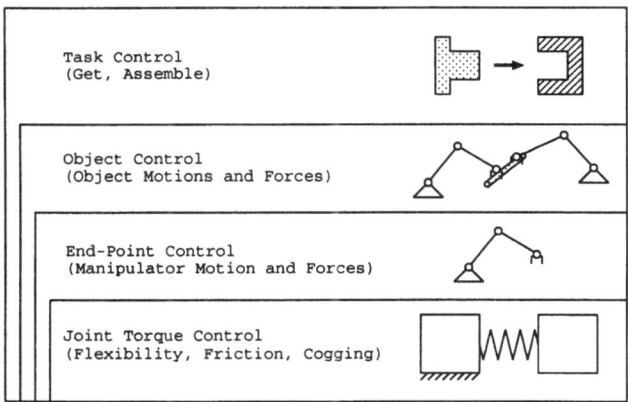

Figure 3. **Four-level control hierarchy for two-armed robot.**

We use a four layer hierarchy to control the two-armed robot. At the lower joint level, we use joint-torque sensors to compensate for the non-idealities of the motor (cogging, non-linearity) and the joint dynamics induced by the joint flexibility. The next layer the arm level control can now assume ideal actuation (i.e. the motors deliver the desired torque to the link itself) and use a Computed Torque approach to compensate for the non-linear arm dynamics. The third object layer, is concerned with object behavior and assumes that the arms are virtual multi-dimensional actuators that apply torques to the object. The top layer implements elementary tasks such as object acquisition and release, insertions etc.

The robot system consists of two four degree-of-freedom SCARA manipulators equipped with joint torque sensors, joint encoders and an end-point 6-DOF force sensor. An overhead vision system provides global sensing of the position

54

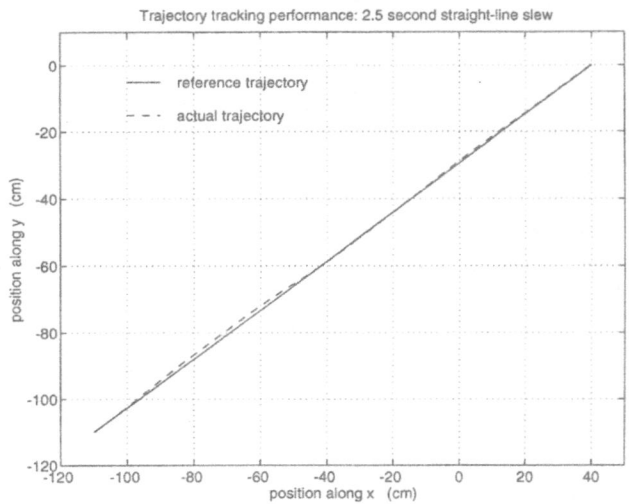

Figure 4. **Experimental tracking performance for right arm.**

Illustration of the tracking response of right arm. The reference is a fifth order polynomial trajectory for the arm endpoint commanding it to follow a 1.75 m straight line path in 2.5 sec. This trajectory requires accelerations of up to 4.3 m/s² (close to 1/2 g). The maximum tracking error is 1.4 cm.

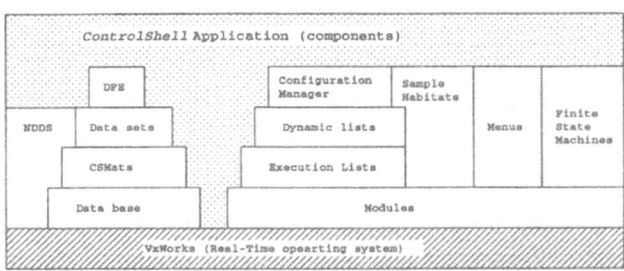

Figure 5. *ControlShell* **Structure.**

The right side of the diagram denotes the "execution" hierarchy; the left side is the "data" hierarchy. The application, consisting partially of a set of reusable components, has access to all facilities at every level. ControlShell provides a layer on top of the real-time operating system VxWorks.

of both the robot and objects. The robot is controlled from a VME-based real-time computer system.

The control software is organized in a four-layer hierarchy originally presented in [10]. The highest level of the control hierarchy illustrated in Figure 3 uses Finite State Machines (FSMs) to coordinate the actions of the two arms and react to both external and internal events. The current implementation uses three FSMs: A global FSM and individual ones for each of the two arms. The global FSM receives commands from the task-planner and may initiate a cooperative two-arm action and/or send stimuli to the individual arm FSMs. Capture of moving objects from the conveyor is achieved using FSM subchains. Subchains are FSM subprograms analogous to subroutines in conventional programming. For example, the subchain in the FSM illustrated in figure 6 is used to perform single-arm object acquisitions from a conveyor.

The next level of the hierarchy (object control) commands the arms to achieve the desired object behavior. The controller used enforces the Virtual Object Impedance Control policy [12]. The third layer uses a Computed Torque approach to achieve dynamic arm control and, at the lowest (joint) level, higher bandwidth joint-torque control-loops are used to control the joint flexibility and compensate for the non-idealities of the motors. Figure 4 illustrates the trajectory-tracking performance of a single arm following a straight line.

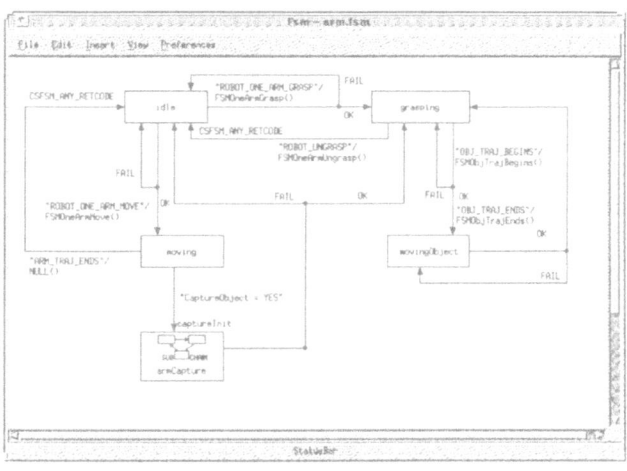

Figure 6. **Finite-State Machine (FSM) used to control an individual arm.**

Stimuli sent to the FSM (shown in quotes) cause the associated transition routines to execute. The return code of these transition routines determines the next state. Subchains are FSM "subprograms".

With the increased complexity demanded from distributed control-systems, the use of CASE tools that facilitate the programming task is becoming essential [2, 15, 13]. All the real-time software in this project has been implemented using the *ControlShell* framework. *ControlShell* [5] is a CASE tool that enables modular design and implementation of real-time software. It contains an object-oriented tool-set which provide a series of execution and data exchange mechanisms that capture both data and *temporal* dependencies. This allows a unique *component-based* approach to real-time software generation and management. *ControlShell* defines temporal events, and provides mechanisms for attaching routines to those events. It provides data structure specifications, and mechanisms for binding data and routines while resolving data dependencies. *ControlShell's* event-driven finite state machine provides the means to weave asynchronous events into a sequential execution stream. *ControlShell* also includes numerous code-generation and maintenance tools such as a graphical component editor, a finite state machine editor, a data-flow editor, an interactive menu facility etc. The structure of *ControlShell* is summarized in figure 5.

Object information from World Model	
location	object location in the global reference frame.
grasps	positions within the object where a grasp is possible
properties	Mass and inertia of a object, whether it can be moved by the robots etc.
shape	object shape for collision avoidance purposes.
Robot information from World Model	
location	robot location in the global reference frame.
joint values	value of each of the joint coordinates
joint limits	limits on the joint coordinates
Denavit-Hartenberg Parameters	Description of robot kinematics
state	Whether the robot is moving, grasping an object etc.

Table 1. Information Available using the World-Model interface

4. Planner

The automatic generation of robot paths for accomplishing a user-specified task is one of the key elements in our automated robot system. Since our goal is to build such an interactive system, the on-line motion planning capability is crucial. Indeed, it is unacceptable for a user to specify a task and then have to wait a few minutes for the motion of the robots to be planned and executed. Due to the high computational cost of robot motion planning [1, 16], we focus our effort on developing effective strategies and efficient path planning algorithms to achieve on-line performance.

command	meaning
move object	Move an object that is being grasped. This command will provide a via-point collision-free path for the object. The robot is controlled using object impedance control.
move arms (operational space)	Move the arms. This command assumes the arms are not grasping an object. The command provides a via-point collision-free path for the arm-endpoints.
move arms (joint space)	Move the arms. This command assumes the arms are not grasping an object. The command provides a via-point collision-free path for the arms joints (This is provided to resolve kinematic ambiguities).
grasp	Grasp an object. This command specifies the object to be grasped.
un-grasp	Un-grasp an object.

Table 2. Robot-Command primitives available through the Robot interface.

Alami, Siméon, and Laumond [11] have proposed and implemented an algorithm for the case of one robot and several movable objects. They assume both a finite number of possible grasps and a finite number of object placements within the workspace. The robot grasps one object at a time and the remaining (currently non-grasped) objects are treated as static obstacles. Their algorithm first considers all the possible arrangements of the robot grasping the object and then decomposes into cells the collision-free subset of these arrangements. These cells are subsequently connected by links which correspond to regrasping operations of the robot. The resulting graph, called the *Manipulation Graph*, is then searched for a path.

In [4], the authors consider a similar scenario dealing with two robots and multiple moving objects on a conveyor belt. The authors present an A^* algorithm to search for a locally-optimal motion sequence for a given assembly represented as a AND-OR graph. In this work, they assume that only the last links of the arms may collide with each other and that the parts are fed slowly enough that the robots can always pick up objects from locations that do not deviate too much from their nominal positions.

Our research differs from these related work in that we consider a more general problem with no prior knowledge about the speed, the feeding rate, the picking position, the picking arm, or the sequence of parts.

The planning system consists of two modules: the *Task Planner* and the *Path Planner*. The task planner is responsible for determining how to utilize the resources (in our case, two manipulator arms) to accomplish the user specified task. The result is a decomposition of the problem into subtasks, which are then sent to the path planner in order to find the necessary motion of the manipulators.

4.1. Task Planner

The role of the task planner is first determine how to solve the user-specified task (e.g. put object A at location P), then make use of the path planning algorithms to actually find the sequence of manipulator motions to complete the task and, finally to send the result to the robots in terms of the robot-command primitives listed in table 2.

The user-specified task may involve moving multiple objects to a set of specified goals. Because of this, the task-planner will break the task down into subtasks and, among other things, determine which object to manipulate first. The criteria for choosing the next object to manipulate is priority-based. The priority of an object is based on the expected time that the (moving) object will take before it leaves the work cell. For example, static objects have lower priorities than objects on the conveyer belt. Once the next object to be manipulated is identified, a manipulator arm(s) is selected to pick-up and deliver the object. The criteria used to select the arm(s) takes into account whether the object/task requires two-arm manipulation, which arm(s) is free (i.e. not currently involved in another task), which arm is closer to the object, the expected time it will take for the object to leave the workspace of each arm, etc. At this point, rather than solving the complete manipulation path, that is moving the arm to grasp the object, grasping it, carrying it to the goal, and then ungrasping it, the task planner further divides the problem into two subtasks. The first subtask is moving the arm to grasp the object, while the second subtask is to deliver the object to its goal.

The task planner runs in a loop selecting the highest priority subtask to plan according to the current state of the world (this state includes both user-specified tasks and subtasks that are already "in-progress"). For example, if there are objects that need to be moved within the workcell, the task planner will detect this at the start of the loop, select the object with highest priority, and invoke the path planner to find a trajectory for one (or both) of the arms to grasp the object. Each time around the loop only the subtask with the highest priority is sent to the path planner. In the event that the planner fails to solve the highest priority subtask, the subtask with the next highest priority is attempted. Once a plan is found, it is broken down into individual robot-command primitives and sent to the robot for execution. Once the motion of the robot (as a response to the command) is detected, the subtask becomes "in-progress" and the task planner returns to the top of the loop. In the event that the execution of this subtask become questionable (e.g. the obstacles have moved or the goal position has been changed) the task planner will first determine if the plan is still valid and if not it will replan accordingly.

4.2. Path Planner

The subtasks that are requested by the task planner to be solved are the following. Note that a free arm refers to a robot arm that is not committed to any subtask.

- Move one arm to grasp a static object while the other arm is free.

- Move one arm to grasp a static object while the other arm is moving.

- Move one arm to catch a moving object while the other arm is free.

- Move one arm to catch a moving object while the other arm is moving.

- Deliver the object that is grasped to its goal location while the other arm is free.

- Deliver the object that is grasped to its goal location while the other arm is holding another object.

- Move two arms to grasp a static object.

- Move two arms to catch a moving object.

Notice that we do not consider the case where both arms deliver two independent objects at the same time or the case where one arm is moving and the other arm delivers its object to the goal. For these cases we decouple the problem and plan the motion of the arms sequentially. Our reasoning is that though the *robot execution time* for this decoupled approach may be slightly longer than if the arms moved simultaneously, the *planning time* to find the sequential motion will be significantly shorter than the time to find the simultaneous motion of both arms. Indeed since efficiency is the key issue in on-line motion planning, we make several assumptions to simplify the path planning problem associated to each subtask and build a library of efficient primitives to solve them.

In our scenario, reasonable simplifications can be made to reduce the size of the search spaces implied by each subtask, and thus reducing the time required to solve them. Due to the fact that the first two links of the SCARA-type arms move in a plane and our assembly task only involves pick-and-place operations, we simplify the motion planning problem in the three dimensional workspace into a problem of two dimensions. This assumes that when the end-effectors of the arms are as high up as they can go, the arms can move in an unrestricted manner above the obstacles in the workspace. This reduces the search space of each arm from three dimensions to two[2]. Once the arm(s) are grasping an object, the size of the object is considered to be such that it can collide with the obstacles in the workspace—that is, the arms are unable to lift the object above the obstacles. The result is a three dimensional search space which fortunately still presents little difficulty for fast computation.

---

[2]the complexity of motion planning grows exponentially with the dimension of the search space

For each of the aforementioned search spaces, we build a path planning primitive to find a collision-free path connecting two configurations in the particular search space. These primitives are extremely efficient and can find paths in a fraction of a second. Some of them are based on existing algorithms [7, 6], while the others are completely new. We then associate a strategy with each subtask which utilizes these primitives to solve it. For example, the subtask of moving one arm to catch a moving object while the other arm is free has the following strategy. First the path of the moving arm to grasp the object is found while ignoring the presence of the free arm. The second step is to have the free arm comply with the motion of this moving arm. If the subtask is solved, the corresponding motion of the robots is returned to the task planner. Otherwise, the task planner is notified that the particular subtask could not be solved.

Our experiments demonstrate that this planning approach yields on-line performance.

5. Discussion and Conclusions

This paper has presented the overall design of an experimental test bed developed to explore the integration of planning in distributed control systems.

We have concentrated on several key aspects of this problem. First, we have developed a powerful distributed network communication architecture (NDDS), and implemented it within a generic real-time programming framework (ControlShell). We have defined clear interface specifications, and implemented simple-yet-powerful robotic system modules within that architecture.

In particular, we have developed and implemented a graphical user interface, a four-level dynamic control hierarchy, an on-line simulator, and an extremely fast on-line motion and task planning capability. The unique aspects of these modules are described briefly above.

This paper describes work in progress by the authors. Our experiments with *non-moving* objects have shown the power of our approach. Numerous tests in *simulation* have shown the system to be able to operate in the presence of moving objects. We will experimentally demonstrate assembly with moving parts in the immediate future.

References

[1] J.F. Canny. *The Complexity of Robot Motion Planning*. MIT Press, Cambridge, MA, 1988.

[2] A. Joubert D. Simon. The orcad: Towards an open robot controller computer aided design system. Research Report 1396, INRIA, February 1991.

[3] Chris Fedor. TCX task communications. School of computer science / robotics institute report, Carnegie Mellon University, 1993.

[4] Li-Chen Fu and Yung-Jen Hsu. Fully automated two-robot flexible assembly cell. In *IEEE International Conference on Robotics and Automation*, pages 332–338, Atlanta, Georgia, USA, May 1993.

[5] Real-Time Innovations Inc. *ControlShell: Object Oriented Framework for Real-Time Software User's Manual.* 954 Aster, Sunnyvale, California 94086, 4.2 edition, August 1993.

[6] Y. Koga and J.C. Latombe. Experiments in dual-arm manipulation planning. In *IEEE International Conference on Robotics and Automation*, pages 2238–2245, Nice, France, May 1992.

[7] J.C. Latombe. *Robot Motion Planning.* Kluger Academic Publishers, Boston, MA, 1991.

[8] MBARI (Monterrey Bay Aquarium Research Institute. Data manager user's guide. Internal Documentation, 1991.

[9] G. Pardo-Castellote and S. Schneider. The network data delivery service: A real-time data connectivity system. ARL Memo 99-1993, Stanford Aerospace Robotics Laboratory, September 1993.

[10] Lawrence E. Pfeffer. *The Design and Control of a Two-Armed, Cooperating, Flexible-Drivetrain Robot System.* PhD thesis, Stanford University, Stanford, CA 94305, (December) 1993. To be published.

[11] T. Siméon R. Alami and J.P. Laumond. A geometrical approach to planning manipulation tasks: The case of discrete placements and grasps. In H. Miura and S. Arimoto, editors, *Robotics Research 5*, pages 453–459, Cambridge, MA, 1990. The MIT Press.

[12] S. Schneider and R. H. Cannon. Object impedance control for cooperative manipulation: Theory and experimental results. *IEEE Journal of Robotics and Automation*, 8(3), June 1992. Paper number B90145.

[13] Reid Simmons and Chris Fedor. Task control architecture programmer's guide. School of computer science / robotics institute report, Carnegie Mellon University, 1992.

[14] Sparta, Inc., 7926 Jones Branch Drive, McLean, VA 22102. *ARTSE product literature.*

[15] D. B. Stewart, D. E. Schmitz, and P.K. Khosla. Chimera ii: A real-time multiprocessing environment for sensor-based robot control. In *Proceedings of the IEEE International Symposium on Intelligent Control*, Albany, NY, September 1989.

[16] G. Wilfong. Motion planning in the presence of movable obstacles. In *4th ACM Symp. of Computational Geometry*, pages 279–288, Urbana-Champaign, Illinois, June 1988.

[17] J. D. Wise and Larry Ciscon. *TelRIP Distributed Applications Environment Operating Manual.* Rice University, Houston Texas, 1992. Technical Report 9103.

Parallel Motion Planning with the Ariadne's Clew Algorithm

E. Mazer, J.M. Ahuactzin, E. Talbi, P. Bessiere and T. Chatroux

Laboratoire d'Informatique Fondamentale et d'Intelligence
Artificielle
46 Avenue Felix Viallet
38000 Grenoble, France

Abstract— We describe an implementation of a real time path planner for a robot arm with six degrees of freedom moving among dynamical obstacles. The planner is based on a novel technique called the Ariadne's Clew Algorithm. A brief description of this algorithm and parallel implementation of it are presented. Finally we analyze experiments made with this planner.

1. Introduction

Our experimental testbed includes two robot's arms each having six degrees of freedom. Our goal was to plan the motions of one of these robots (the Controled Robot) with a path planner fast enought to "re-plan" in line new trajectories when the other arm (the Dynamical Obstacle Robot) was placed on its way towards the goal. The idea was to demonstrate the possibility of including a global path planner into an incremental trajectory generator. This idea was first successfully experimented with a 2D version of our planner. A simulated mobile robot uses this planner to plan its trajectories among obstacles. Using the simulator the user was able to change the positions of the obstacles while the mobile robot was moving towards the goal, in that case, the planner reacts immediately by re-computing a new trajectory from its current position to the goal. For example if the door represented figure 1 was closed before the robot reached it, the planner was quickly (in less than 0.05s) able to devise a new path . As a result the robot was continuously progressing towards the goal despite the tentatives of the operator to move obstacles on its way. This report describes an attempt to reach the same level of reactivity with a six degree of freedom arm moving in a three dimensional environment.

2. Previous Work

A recent survey of robot motion planning techniques can be found in [2]. A detailed presentation of the field exists also in [3]. Many attempts have been made to obtain "real-time" path planners. For example, in the proceedings of the last conference on Robotic and Automation one can find: a method to implement the planner of Latombe and Baranquant on a parallel machine [4], a method which uses an optically computed potential fields [5] and three "fast

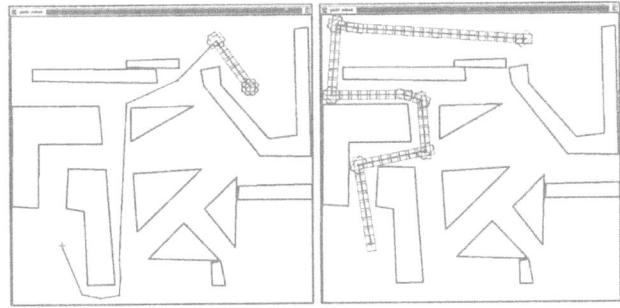

Figure 1. Re - planning when a door is closed

path planners" [7, 8, 6]. However, while we try to reach the same goal, the work described in this paper is more related to the approach taken by Overmars in [9] and to the Sandros [10] motion planner. Both methods use a set of landmarks to represent the free space and a local planner which is used to connect the landmarks as well as the initial and final positions. In the work described by Overmars, the landmarks are placed randomly in the search space, when a landmark is place into a C-Obstacle it is moved to a close free location. In the Sandros motion planner the landmarks are placed in "slices" of the configuration space. The completeness of boths method have been proved, and fast response time have been reported. The main advantage of our planner is it's ability to place landmarks more efficiently than the previous methods and, thanks to an implementation on a massively parallel machine, it is also much faster.

3. The Ariadne's clew algorithm

A detailed description of the Ariadne's clew algorithm can be found in [11, 13], here we only sketch its principle. The ultimate goal of a path planner is to find a path in the configuration space from the initial position to the target. However, while searching for this path, an interesting sub-goal may be to try to collect information about the free space and about the possible paths to go about that space. The ARIADNE'S CLEW algorithm tries to do both at the same time and it is made of two sub-algorithms: SEARCH and EXPLORE. The EXPLORE algorithm collects information about the free space with an increasingly fine resolution, while, in parallel, the SEARCH algorithm opportunistically checks if the target can be reached. The EXPLORE algorithm works by placing landmarks in the search space in such a way that a path from the initial position to any landmark is known. In order to learn as much as possible about the free space the EXPLORE algorithm tries to spread the landmarks all over the space. To do so, it tries to put the landmarks as far as possible from one another. For each new landmark produced by the EXPLORE algorithm the SEARCH algorithm checks with a local method if the target may be reached from that landmark. Both the EXPLORE and the SEARCH algorithms may be seen as solving optimization problems on a special

64

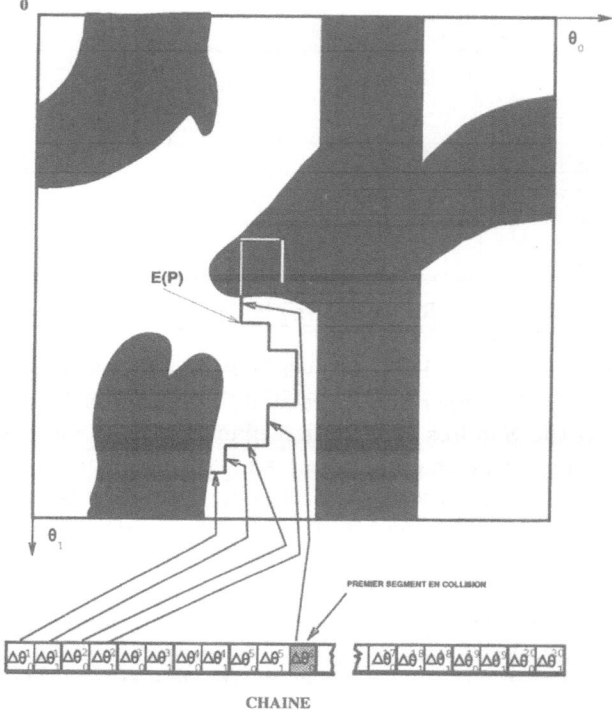

CHAINE

Figure 2. A Manhattan path in the configuration space

set: namely the set of **Manhattan paths** of fixed length.

3.1. The Manhattan paths

Let $(\theta_1, \ldots, \theta_n)$ denotes the configuration of a system with n degrees of freedom. We define a Manhattan path of length 1 as the path consisting of moving each degree of freedom once. We denote such a path as:

$$p^1 = \{\Delta\theta_1^1, \Delta\theta_2^1, \ldots, \Delta\theta_i^1, \ldots, \Delta\theta_n^1\}$$

More generally we define a Manhattan path of length k as the concatenation of k Manhattan paths of length 1.

$$p \in \Re^{k \times n} = \{\Delta\theta_1^1, \ldots, \Delta\theta_i^j, \ldots, \Delta\theta_n^k\}$$

Given a Manhattan path p of length k we denote by $E(p)$ the point of the configuration space corresponding to the extremity of the last collision free segment of p (see figure 2). Basic geometric computations [1] permit to compute $E(p)$.

3.2. SEARCH

Given a goal configuration Θ^g we define a function $F : \Re^{k \times n} \xrightarrow{F} \Re^+$ as :

$$F(p) = \| E(p) - \Theta^g \|$$

If a Manhattan path p_0 towards the goal exists then $F(p_0) = 0$. As a result, finding a Manhattan path p of length k can be seen as a minimization problem:

$$\min_{p \in \Re^{k \times n}} F(p)$$

While we do not have any analytical expression for F we can easily compute its value at any point of $\Re^{k \times n}$. Many methods exist to minimize such a function, we use a genetic algorithm. Even in the presence of local minima, we can implicitly define a region of $\Re^{k \times n}$ called the "back-projection" of Θ^g on which the minimization method find a global minimum for F. The goal of EXPLORE is to place a reachable point in that region.

3.3. EXPLORE

One can visualize the EXPLORE algorithm by imagining a robot placing landmarks in the free space starting from its initial position. Each time it places a new landmark it tries to place it as far as possible from landmarks previously placed. Each of the landmarks is connected by a free path to at least one of the other landmarks. We denote by EL_q the set of existing landmarks at step q. The explore-algorithm begins with $EL_1 = \{L_1\}$ were L_1 is the initial location then, in next step, EL is incremented with the new landmark. If the search space is bounded, then the robot will fill the free space connected to its initial position with landmarks. These landmarks will become closer and closer as the search time is becoming larger. At a given point in time the last generated landmark will necessarily fall into the back-projection region of the goal and a solution will be found !

We denote by PL the set of all Manhattan paths starting from the landmarks $L_i \in EL_q$, PL is indexed by $[0, 1, \ldots q] \times \Re^{k \times n}$. We denote $E(p)$ the extremity of the last non-colliding segment of a particular path $p \in PL$.

Let p_2 the path of PL which maximize

$$p_2 : \max_{p \in PL} \| L_1 - E(p) \|$$

According to our definitions, the point $L_2 = E(p_2)$ is the most further location of the search space reachable from the initial location L_1, we choose it as the second landmark. Now, we have have $EL_2 = \{L_1, L_2\}$. Given a path p starting either from L_1 or L_2 we consider the minimum value between $\| E(p) - L_1 \|$ and $\| E(p) - L_2 \|$ and we try to maximize this value over EL_2 in order to find a new reachable landmark which is as far as possible from L_1 and from L_2. In other word:

$$p_3 : \max_{p \in PL} \min\{\| L_2 - E(p) \|, \| L_1 - E(p) \|\}$$

More generally if we have already n landmarks we can find the $n+1$ landmark by maximizing the following expression:

$$p_{n+1} : \max_{p \in PL} \min_{j=1,n} \| L_j - E(p) \| \tag{1}$$

By taking $L_{n+1} = E(p_{n+1})$ we get our new $n+1^{th}$ landmark. Lets consider the function:

$$\forall n \geq 2 : V(n) = \max_{p \in PL} \min_{j=1,n-1} \| L_j - E(p) \|$$

If the search space is bounded then:

$$\lim_{n \to \infty} V(n) = 0$$

or

$$\forall \varepsilon \exists n : \forall j > n \quad V(j) < \varepsilon$$

Then, if G is a point of the accessible free space (ie: it exists a path from L_1 to G) then we have:

$$\forall \varepsilon : \exists n \| L_n - G \| < \varepsilon \tag{2}$$

We call this property : *epsilon-reachability*. The epsilon-reachability has a strong consequence for planning a path in a continuous space : if one can find ε such that there is a function which solve the path planning problem in any ball of diameter ε (the search function), then by combining it with the explore algorithm we get a deterministic method to plan a path between any two points of the configuration space.

3.4. A Parallel Implementation

We have implemented this algorithm on a parallel machine with 128 transputers. The figure 3 represents the configuration we choose to implement the Ariadne's clew algorithm. The numbers inside the rectangles indicate the physical number of each processor, the edges correspond to the physical links between two transputers. It is possible to use a programmable switch board to configure the machine with this particular topology. In this architecture one has to consider three levels of parallelism.

1. **The parallel execution of "Search" and "Explore".** At the first level the EXPLORE algorithm, which uses the processor 1 to 60, runs in parallel with the SEARCH algorithm (processors 61 to 120). The processor 128 is used by EXPLORE to communicate the landmarks to SEARCH.

2. **The parallel execution of the genetic algorithm.** Both EXPLORE and SEARCH uses a genetic algorithm as an optimization technique. A description of parallel genetic algorithms can be found in [12]. In our case the population used by the genetic algorithm is reduced to six

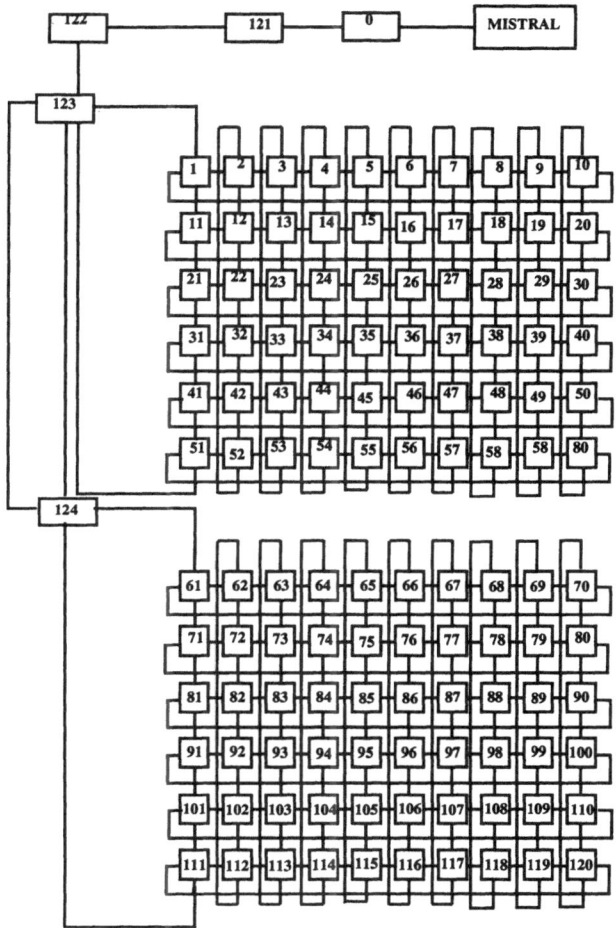

Figure 3. Actual implemetation on 128 Transputers

individual. Each individual evaluates in parallel the cost function with its own "farm" of processors. For examples the "individuals" of SEARCH are located on processors 1-11-21-31-41-51 and are organized in a single ring (see figure 4).

3. **The parallel evaluation of the cost function.** In both cases, the evaluation of the cost function boiled down to many serial computations of the legal range of motion for a single link. In turn, the computation of each legal range can be split into three types of elementary ra·ge computation denote as A,B and C. For each individual the computational load is spread over its farm of processors by having each processor of the farm responsible for a given elementary computation. For example the processor 21 uses the processors 22-23-24 to perform the computation of type A, the processors 25-26-27 for the computation of type B and the

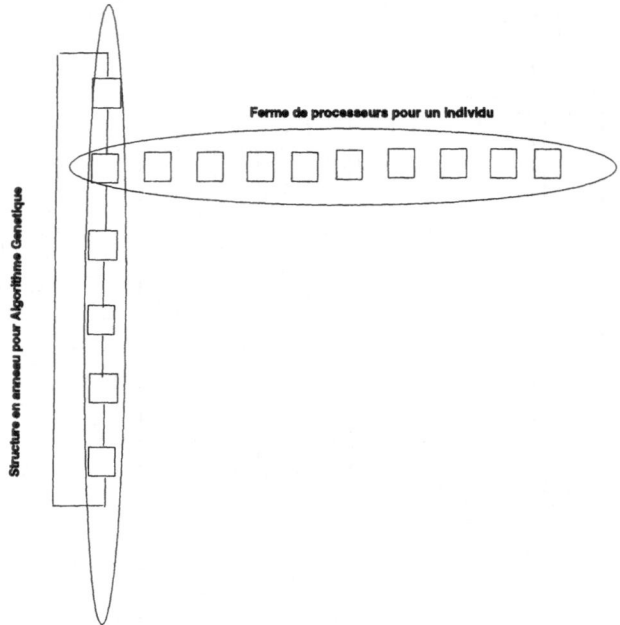

Figure 4. Ring structure and processor farm

processors 28-29-30 for the type C. .

4. Experiments

The figure 5 represents the architecture of our experiment setup.

The robot I is under the control of the Mega-Node (via Kali) running a parallel implementation of the Ariadne's Clew algorithm. The robot II is used as a dynamical obstacle: it is controlled (via Kali) by a random motion generator.

First we use our robot simulation package ACT to describe the scene with the two robots. We compile this representation into a special geometrical representation which enclosed each obstacle and each link of the robots into a box to minimize the number of geometric computation.

This model is downloaded to all the processors as well as the goal position for robot I, then the following algorithm is executed :

1. **Step 1 :** The curent position of the two robots is diffused to all the processors.

2. **Step 2 :** The Mega-node produces a plan which assumes robot II is standing still.

3. **Step 3 :** Kali executes only the first part of the plan (a manhattan pass of length one).

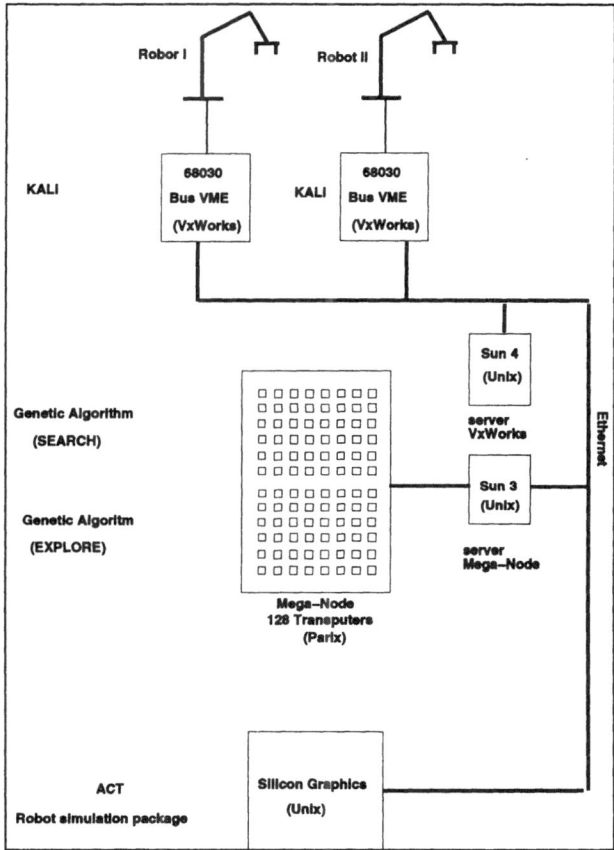

Figure 5. A parallel Architecture for Robot Motion Planning

4. **Step 4 :** The random motion planner is called and produces a random motion which is executed by robot II (with robot I standing still).

5. **Go to Step 1**

5. Experimental Results

To speed up the computation of the evaluation function we only use the enclosing boxes of the obstacles and of the links. The controlled robot is made of 6 boxes and there are 10 boxes used as obstacles including the moving obstacles (see figure 6).

In that particular environment, a Manhattan path of length one requires $\frac{6^2+6}{2} * 10 = 210$ computations for the rotational ranges. Each computation of the range requires 240 elementary contact computations. So, to evaluate a path of length 5, 252,000 elementary contact computations are necessary. To perform the evaluation of a single generation of the genetic algorithm with 6 individuals we reach a total $1,512,000$ contact computations. It would not be possible

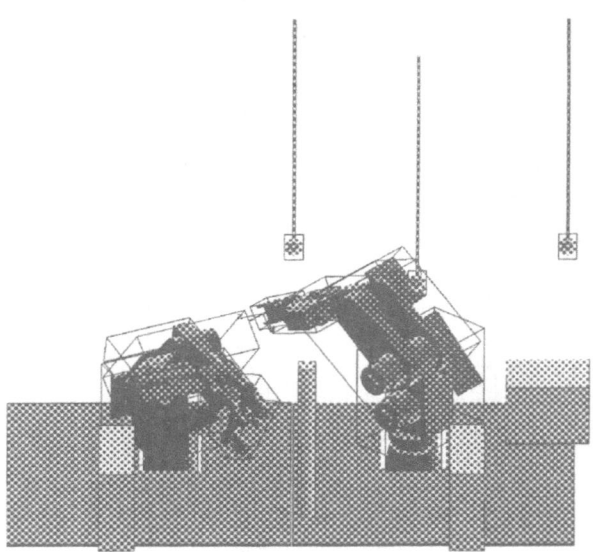

Figure 6. The experimental scene with enclosing boxes

to handle such a large amount of computation without the use of geometric filters which permit to discard quickly the link-obstacles pairs which cannot interact. In practice the use of geometric filters permit to reduce the amount of "real" geometric computation by a factor 10. However the computational load remains very heavy, even for a simple scene such as the one we use in our example. Only the use of a massively parallel machine permits to keep the planning time reasonably small : between 0.5 to 3 seconds for a six degrees of freedom arm with 10 obstacles. It is also important to note that the number of generations used by the genetic algorithm to produce a solution is very small : 5. As a consequence we believe that very little optimization is made with the genetic algorithm, it seems that the genetic algorithm is only useful to find the most promising landmark which is "closer" from the free space not already explored. Nevertheless we have successfully tried the Ariadne's clew algorithm on very complex scene necessitating non-trivial motion to reach the goal. This suggest that it may be possible to randomly find a path in a scene considered as "difficult" (note that placing landmarks is still a very useful process). We conjecture that many planners found in the literature are tested on scene where many randomly generated paths will lead to a point in the back-projection of the goal.

For an industrial application of path planning the number of potentially colliding (link-obstacle) pairs is several order of magnitude bigger than the one used in our experiment (210). In industry the models of the robot and the models of the obstacles are very detailed and contains hundreds of geometrical entities. In that case, the evaluation of a single Manhattan path will be impossible even on massively parallel machine.

6. Conclusion

We have presented a new motion planner for a six degrees of freedom arm. The planner achieves fast response time compare to existing planner but remains slow and cannot be used as part of a trajectory generator. Given a fixed number of obstacles the planning time is proportional to the difficulty of the task. The planning time is also proportional to the number of link-obstacles pairs found in the scene. We are currently working on a new version of the ARIADNE'S CLEW algorithm which will use a classical optimization algorithm as well as a hierarchy of geometrical models for the environment.

References

[1] T. Lozano-Pérez : *A simple motion-planning algorithm for general robot manipulators.*, IEEE Int. Jour. on Robotics and Automation, RA-3, June 1987.

[2] Yong K. Hwang and Narendra Ahuja : *Gross Motion Planning - A Survey.*, ACM computing Surveys, Vol 24, No 3, pages 119-289, September 92.

[3] Jean-Claude Latombe: *Robot Motion Planning*, Ed. Kluwer Academic Publisher, 1991.

[4] D.J Challou, M. Gini, and V. Kumar : *Parallel Search Algorithms for robot Motion Planning* IEEE Int. Conf. On Robotics an Automation, Atlanta, May 93.

[5] M.B Reid : *Path planning Using Optically Computed Potential Fields* IEEE Int. Conf. On Robotics an Automation, Atlanta, May 93.

[6] C.W Warren *Fast path Planning Using Modified A* Method* IEEE Int. Conf. On Robotics an Automation, Atlanta, May 93.

[7] P.K Pal and K. Jayarajan : *Fast Path planning for Robot Manipulators Using Spatial Relations in the Configuration Space* IEEE Int. Conf. On Robotics an Automation, Atlanta, May 93.

[8] P.Fiorini and Z.Shiller : *Motion Planning in Dynamic Environments Using the Relative Velocity Paradigm* IEEE Int. Conf. On Robotics an Automation, Atlanta, May 93.

[9] M.H. Overmars : *A Random approach to motion planning* Spring School on Robot motion planning, Rodez(France), March 93.

[10] P.C Chen and Y.K Hwang : *SANDROS: A motion planner with performance proportional to task difficulty* IEEE Int. Conf. On Robotics an Automation, Nice, May 92.

[11] J. Ahuactzin, G. Talbi, P. Bessière and E.Mazer : *"Using Genetic Algorithm for robot motion planning".* ECAI, 92 Vienne 1992.

[12] G. Talbi, J. Ahuactzin,P. Bessière and E.Mazer : *"A parallel implementation of a robot motion planner".* CONPAR92, Lyon, 1992.

[13] E.Mazer, G. Talbi, J. Ahuactzin, P. Bessière : *"The Ariadne's Clew Algorithm"*, SAB92 Honolulu 1992.

Section 2
Force Control

Force control is recognized to be one of the fundamental techniques that should find many applications, especially in various unstructured environments. It seems, however, that this technique must be made more robust and easier to handle before we see its wide practical application.

Sano, Furusho, Hattori, and Nishina utilize the H^∞ control theory to design a robust force controller for a two-link planar manipulator with flexibility due to harmonic drives at the two joints and due to a force sensor between the second link and the end-effector. The superiority of the proposed robust force controller over a constant gain feedback controller has been demonstrated by experiment.

ElMaraghy and Massoud present a dynamic hybrid position/force control algorithm for manipulators with flexible joints interacting with rigid environment, and its extension to an adaptive hybrid position/force control algorithm to treat cases where the robot parameters are uncertain. The latter algorithm has the same basic structure as the former, but in addition it has a robust sliding observer and a parameter identification law. Experimental results which show the validity of the former algorithm are given.

Morel and Bidaud propose a two-level controller of manipulators for assembly tasks which involve contacts with unknown and varying environment. The lower level controller is a simplified impedance controller. Simplification is done regarding the dynamics compensation and force feedback in the Hogan's impedance control algorithm. The higher level controller is a fuzzy logic supervisor for on-line modification of the desired impedance and reference trajectory. Preliminary experimental results are reported.

Manipulation involves recognizing and controlling changes in contact conditions among objects. Eberman and Salisbury describe a procedure for representing and segmenting the temporal contact signals from force sensors using an auto-regressive model. This procedure produces segments of approximately constant spectrum, and is shown experimentally to be useful in detecting abrupt changes in noisy signals without calibration.

Trevelyan describes how a type of open-loop force control algorithm has been used successfully in a sheep shearing robot. The developed robot has an end-effector (cutter and comb sliding on sheep skin) which is mounted on a manipulator arm through a hydraulic piston/cylinder actuator with its oil pressure being kept at a desired value by a servo-valve. Based on this experience he suggests that robots which need to exert controlled forces should be equipped with special end-effectors, if possible, rather than new arm or controller designs.

Merlet, Mouly, Borrelly, and Itey describe an experimental robotic cell they have developed for deburring polygonal objects. The cell consists of three manipulators which do tasks of picking up a polygonal object arriving in a random position, placing it on a special plate, probing its vertices to determine the shape, and finally deburring it. Various force measurements are extensively used to perform these tasks.

Whitcomb, Arimoto, Naniwa, and Ozaki present their recent effort to develop an experimental robot system for performing an exhaustive comparative performance evaluation of different model-based force control algorithms. Among them is an adaptive force control algorithm they have proposed for the simultaneous position and force trajectory tracking control of a robot arm in contact with a smooth surface. A new overload-protected force sensing wrist is also introduced. Preliminary experimental results are shown.

Robust Force Control of Manipulators based on H^∞ Control Theory

Akihito SANO, Junji FURUSHO, Koji HATTORI,
Masahiro NISHINA

Department of Mechanical and Control Engineering,
The University of Electro-Communications,
Chofu, Tokyo 182, JAPAN
phone:0424-83-2161, fax:0424-84-3327, E-mail:sano@mce.uec.ac.jp

Abstract— In order to realize a sophisticated control, it is necessary to model a controlled system properly. However, it is impossible to obtain a complete model which can represent a physical system perfectly. Therefore, a robust control system has to be designed by considering an uncertainty of the model.

The so-called H^∞ control problem has drawn much attention in the area of the control system design as a post-modern control theory. By using this theory, the loop shaping procedure in the frequency-domain and the modern control theory based on the state space are fused, and the uncertainty in the model can be treated quantitatively and systematically.

The present study applies this robust control theory for a force control which is important for the robots of next generation. In particular, the force control of the manipulator with a gear system is examined and the H^∞ controller which guarantees low sensitivity as well as robustness against the modeling error and the variation of the system is designed.

1. Introduction

In most electric-powered manipulators, each joint is driven through a reduction gear unit such as a harmonic drive. Due to the elasticity and the friction of the driving system including the uncertainties, large problems have occured in many manipulators [1],[2].

Recently, the H^∞ control is applied for the flight control [3], the flexible structures [4] and the electromagnetic suspension [5] etc. from the robustness point of view. In the early 1980s, the H^∞ theory requires advanced mathematics and the design procedure is complicated. However, in recent years, the procedure for solving the H^∞ control problem has been considerably simplified [6],[7]. Now, it is well known that we can obtained the H^∞ suboptimal controller in the state space setting only by solving two algebraic Riccati equations.

In this study, the robust force control of the manipulator that the end-effector is constrained on the environment is discussed. In Section 2, 2-link horizontal manipulator with the flexibility in its driving system is modeled; in Section 3, the hierarchical force control based on controlling the angular velocity at each joint is discussed; in Section 4, the H^∞ controller which guarantees both the low sensitivity and the robust stability against the modeling error and the variation of the system caused by changing the configulation is designed; in Section 5, the effectiveness of the proposed robust control strategy is confirmed by the experiment using the robot arm.

2. Modeling

2.1. 2-link Arm and Equations of Motion

In this section, the modeling of 2-link horizontal manipulator with the flexibility in its driving system is discussed [8]. Figure 1 shows a model of the manipulator arm. In this model, there are an end-effector at the end of the 2nd link and a

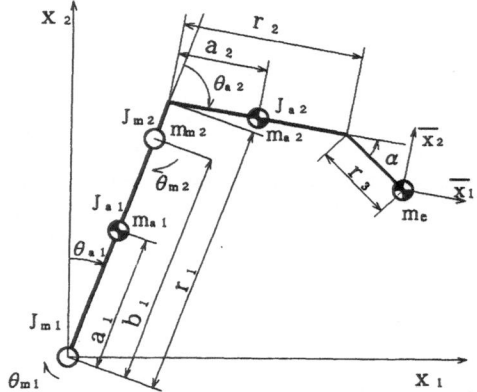

Figure 1. Model of Manipulator

force sensor between the 2nd link and end-effector. Because of the flexibility of the driving system with harmonic gear, the joint angle and the motor angle are represented as θ_{aj} and θ_{mj}, respectively. The position of the end-effector is expressed as $\bar{x}$. And, $\Phi = [\theta_a^T, \theta_m^T, \bar{x}^T]^T$ is adopted as a generalized coordinate.

From Lagrange's equation, the equations of motion for the manipulator are

$$A(\Phi)\ddot{\Phi} + b(\Phi, \dot{\Phi}) = \begin{bmatrix} M_a \\ M_m \\ F_e \end{bmatrix} \tag{1}$$

where $A(\Phi)$ is the 6×6 inertia matrix related to the motion of the arm and includes the translational motion of the center of gravity of the motor, $b(\Phi, \dot{\Phi})$ is the centrifugal force term. Moreover, M_a, M_m and F_e are the generalized force vectors corresponding to the generalized coordinate vector θ_a, θ_m and $\bar{x}$, respectively. Concretely, the force vector F_e acting on the end-effector is given as follows:

$$F_e = -H_e\bar{x} \tag{2}$$

H_e: spring constant matrix of force sensor

The respective elements of M_a and M_m are the following torque.

M_{aj} : the output torque of the reduction gear unit namely, the jth joint-driving torque.

M_{mj} : the sum of the torque electromagnetically generated by the armature current and the reaction torque from the input shaft of the reduction gear unit; namely, the sum of the torques acting on the jth motor.

2.2. Model of Driving System

In this section, by using the model proposed by Good et al. [9] for the robot arms employing harmonic drives, the model for the driving system of the jth joint is given. When the twisting angle of the driving system of the jth joint is denoted by ε_j, the relationship between θ_{aj} and θ_{mj} is expressed as follows:

$$\theta_{aj} = n_j \theta_{mj} + \varepsilon_j \qquad j = 1, 2 \tag{3}$$

where n_j is the reduction ratio of the driving system of the jth joint. According to the model of Good et al., the respective elements of M_a and M_m are.

$$\begin{cases} M_{aj} = & -c_{aj}\dot{\theta}_{aj} - c_{bj}(\dot{\theta}_{aj} - n_j\dot{\theta}_{mj}) \\ & -k_j(\theta_{aj} - n_j\theta_{mj}) - P_{aj}\mathrm{sign}(\dot{\theta}_{aj}) \\ M_{mj} = & -c_{mj}\dot{\theta}_{mj} + n_j c_{bj}(\dot{\theta}_{aj} - n_j\dot{\theta}_{mj}) \\ & +n_j k_j(\theta_{aj} - n_j\theta_{mj}) - P_{mj}\mathrm{sign}(\dot{\theta}_{mj}) + T_{mj} \\ & \qquad\qquad\qquad\qquad j = 1, 2 \end{cases} \tag{4}$$

where

c_{aj}, c_{bj}, c_{mj} : coefficients of viscous friction [N·m·s/rad]

$k_j(\cdot)$: function of nonlinear spring [N·m]

P_{aj}, P_{mj} : coefficients of Coulomb friction [N·m]

In this study, the input to the servomotor is a voltage input. Namely, torque T_{mj} is expressed as follows:

$$\begin{cases} T_{mj} & = & c_{Tj}i_j \\ v_j & = & R_j i_j + c_{vj}\dot{\theta}_{mj} \qquad j = 1, 2 \end{cases} \tag{5}$$

where
c_{Tj} : torque constant [N·m/A]
i_j : armature current [A]
v_j : input voltage [V]
R_j : armature resistance [Ω]
c_{vj} : back emf voltage constant [V·s/rad]

2.3. Constraint on the Environment

When the end-effector is constrained on the environment (the equations of constraint: $q(\Phi) = 0$), the dynamic equations of motion are obtained by introducing Lagrange's undetermined multipliers Γ as follows [8]:

$$A(\Phi)\ddot{\Phi} + d(\Phi, \dot{\Phi}) = S(\Phi)\Gamma + Dv \tag{6}$$

$$S(\boldsymbol{\Phi}) = \left[\frac{dq}{d\boldsymbol{\Phi}}\right]^T \tag{7}$$

where

$$d(\boldsymbol{\Phi}, \dot{\boldsymbol{\Phi}}) = b(\boldsymbol{\Phi}, \dot{\boldsymbol{\Phi}}) + C\dot{\boldsymbol{\Phi}} + e(\boldsymbol{\Phi}) + p$$
$$D = [0_{2\times 2}, \quad \mathrm{diag}\{c_{T1}/R_1, c_{T2}/R_2\}, \quad 0_{2\times 2}]^T$$
$$v = [v_1, v_2]^T$$

where, C, $e(\boldsymbol{\Phi})$ and p are viscous matrix, nonlinear spring term and Coulomb friction term, respectively. $\boldsymbol{\Gamma}$ that satisfies the equations of constraint is given as follows:

$$\boldsymbol{\Gamma} = -(S^T A^{-1} S)^{-1}\{S^T A^{-1}(-d(\boldsymbol{\Phi}, \dot{\boldsymbol{\Phi}}) + Dv) + \dot{S}\dot{\boldsymbol{\Phi}}\} \tag{8}$$

Next, the linear model for a whole arm around the equilibrium state ($\boldsymbol{\Phi} = \boldsymbol{\Phi}_s, \dot{\boldsymbol{\Phi}} = 0, v = v_s, \boldsymbol{\Gamma} = \boldsymbol{\Gamma}_s$) is determined. By neglecting Coulomb friction and $b(\boldsymbol{\Phi}, \dot{\boldsymbol{\Phi}})$ term higher than the second order, and replacing the function of the nonlinear spring with $\bar{k}_1, \bar{k}_2$ approximately, the linear model is obtained as follows:

$$A_s \Delta\ddot{\boldsymbol{\Phi}} + C\Delta\dot{\boldsymbol{\Phi}} + (E - W)\Delta\boldsymbol{\Phi} = D\Delta v + S_s \Delta\boldsymbol{\Gamma} \tag{9}$$

$$\Delta\boldsymbol{\Gamma} = (S_s^T A_s^{-1} S_s)^{-1} S_s^T A_s^{-1}\{C\Delta\dot{\boldsymbol{\Phi}} + (E - W)\Delta\boldsymbol{\Phi} - D\Delta v\} \tag{10}$$

where $A_s = A(\boldsymbol{\Phi})$, $S_s = S(\boldsymbol{\Phi})$. E and W are the following matrices.

$$E = \begin{bmatrix} E_{11} & E_{12} & E_{13} \\ E_{21} & E_{22} & E_{23} \\ E_{31} & E_{32} & E_{33} \end{bmatrix}$$

$$\begin{cases} E_{11} = \mathrm{diag}\{\bar{k}_1, \ \bar{k}_2\} \\ E_{12} = E_{21} = \mathrm{diag}\{-n_1\bar{k}_1, \ -n_2\bar{k}_2\} \\ E_{22} = \mathrm{diag}\{n_1^2\bar{k}_1, \ n_2^2\bar{k}_2\} \\ E_{13} = E_{23} = E_{31} = E_{32} = 0_{2\times 2}, \ E_{33} = H_e \end{cases}$$

$$W = \left\{\left[\frac{\partial}{\partial\Phi_1}\left[\frac{\partial q}{\partial\boldsymbol{\Phi}}\right]\right]^T_{\boldsymbol{\Phi}=\boldsymbol{\Phi}_s}\boldsymbol{\Gamma}_s, \cdots, \left[\frac{\partial}{\partial\Phi_6}\left[\frac{\partial q}{\partial\boldsymbol{\Phi}}\right]\right]^T_{\boldsymbol{\Phi}=\boldsymbol{\Phi}_s}\boldsymbol{\Gamma}_s\right\}$$

3. Hierarchical Force Control

3.1. Control Law

Generally, with respect to the force control, there are two kinds of the control method, namely a direct force control and indirect (hierarchical) force control. At the latter method, the high gain local feedback is carried out at each joint. By the high gain feedback, the system becomes the robust against the Coulomb friction and the variation of torque of the driving system. And, the reference

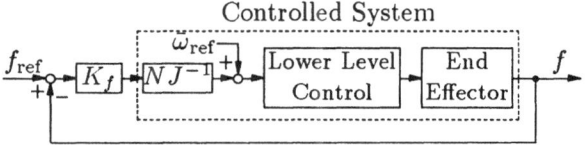

Figure 2. Block Diagram of Force Control

signal to the local feedback controller is modified by utilizing the force signal from the force sensor mounted on the end-effector. In this study, the following hierarchical control structure is adopted (see Fig.2) [8].

Lower Level   In the lower level, the angular velocity at each joint is feedback, the proportional plus integral control is the basis, and it can be expressed as follows:

$$\begin{cases} v_j = K_{Pj}(\omega_{\text{ref}j} - \dot{\theta}_{mj}) + K_{Ij}z_j \\ \dot{z}_j = \omega_{\text{ref}j} - \dot{\theta}_{mj} \qquad j = 1,2 \end{cases} \tag{11}$$

where

$\omega_{\text{ref}j}$: reference angular velocity of the motor driving jth joint

K_{Pj} : proportional feedback coefficient of the controller for the jth joint

K_{Ij} : integral feedback coefficient of the controller for the jth joint

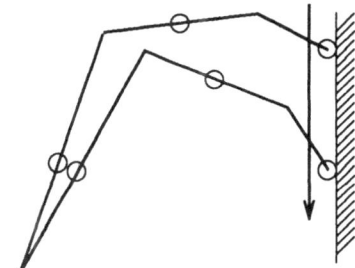

Figure 3. Movement along The Wall

Upper Level   In the upper level, the reference signal to the local controller is modified by the force signal obtained from the force sensor. The reference angular velocity $\omega_{\text{ref}j}$ is generated as follows:

$$\omega_{\text{ref}j} = \bar{\omega}_{\text{ref}j} + \Delta\omega_{\text{ref}j} \tag{12}$$

where $\bar{\omega}_{\text{ref}j}$ is the angular velocity which needs to make the end-effector move at the desired speed along the wall (see Fig.3). $\Delta\omega_{\text{ref}j}$ is the modified value of the reference velocity which is calculated from the deviation of the force $(f_{\text{ref}} - f)$, and it can be expressed as follows:

$$\begin{bmatrix} \Delta\omega_{\text{ref}1} \\ \Delta\omega_{\text{ref}2} \end{bmatrix} = NJ^{-1}K_f(f_{\text{ref}} - f) \tag{13}$$

where

f_{ref} : reference force signal
N : reduction ratio matrix
J : Jacobian matrix (constant)
K_f : proportional feedback gain

3.2. Unstability caused by Variation of System

One view of the 2-link manipulator used in this experiment is shown in Fig.4. In this section, the simple constant gain feedback is examined.

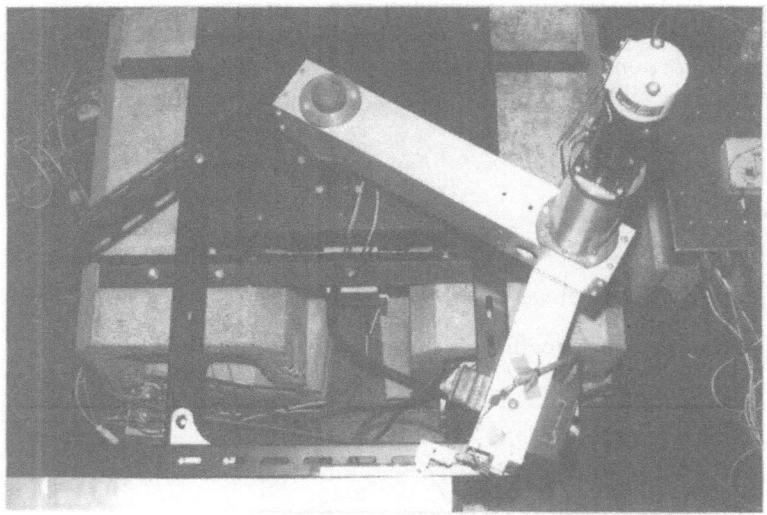

Figure 4. 2-link manipulator

Figure 5 shows the experimental results for the step response. Figure 5(a) represents the force data when the manipulator does not move along the wall (stays at the center position). Figure 5(b) represents the force data under the movement along the wall. In this experiment, the proportional feedback gain K_f is 3.36×10^{-3} which is decided at the center position experimentally.

As seen from the figures, in the case of the simple constant gain feedback, the desired quick response can not be accomplished. Especially, in case of the movement along the wall, the system becomes unstable. As one of the reasons, the variation of the system caused by changing the configulation of manipulator is considered.

4. Design of H^∞ Controller

4.1. H^∞ Control Theory

The H^∞ control theory in which the H^∞ norm of the closed loop transfer function is less than a given number γ is proposed by Doyle et al. [6],[7]. Figure 6 shows the basic block diagram where $G(s)$ stands for a generalized plant and contains what is usually called the plant plus all weighting functions., $C(s)$ the

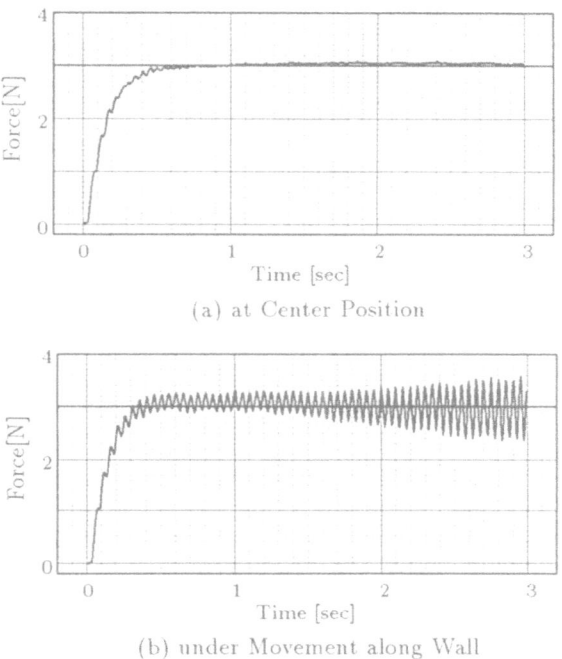

(a) at Center Position

(b) under Movement along Wall

Figure 5. Step Response ($K_f = 3.36 \times 10^{-3}$)

Table 1. Parameters of Model

	Link-1	Link-2	Effector
a	$0.1422m$	0.1405	—
b_1	$0.31m$	—	—
r	$0.31m$	0.19	0.130767
m_a	$4.96kg$	5.08	0.15
J_a	$0.1072416 kg\, m^2$	0.0367925	0.0
m_{m2}	—	$0.25kg$	—
J_m	$8.9475 \times 10^{-5} kg\, m^2$	4.298×10^{-5}	—
k	$5199.06 N\, m/rad$	1229.06	—
c_V	$0.194V\, s/rad$	0.17536	—
c_T	$0.18718N\, m/A$	0.175364	—
R	18.1Ω	12.1	—
n	0.01	0.01	—
P_a	$0.0N\, m$	0.0	—
P_m	$0.0680N\, m$	0.0147	—
c_a	$1.89N\, m\, s/rad$	1.0	—
c_b	$0.3N\, m\, s/rad$	0.1	—
c_m	$0.00068N\, m\, s/rad$	0.00017	—
α	—	—	$36.587deg$

controller, u the control input, y the observation output, z the error signals, and w the input disturbance. Now, let us consider the following feedback system.

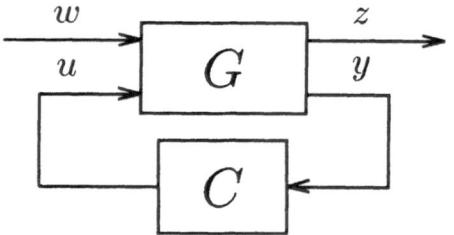

Figure 6. Generalized Plant

$$\left[\begin{array}{c} z \\ y \end{array}\right] = G(s) \left[\begin{array}{c} w \\ u \end{array}\right] \quad , \quad u = C(s)y \tag{14}$$

The H^∞ control problem is to choose the controller, C, connecting the observation vector y to u, that internally stabilizes the closed-loop system and such that the closed-loop transfer function, T_{zw}, is denoted as follows:

$$z = T_{zw}w, \quad T_{zw} = G_{11} + G_{12}C(I - G_{22}C)^{-1}G_{21} \tag{15}$$

For a given number $\gamma > 0$, the H^∞ control theory finds all controllers such that the H^∞ norm of T_{zw} is less than γ.

$$\parallel T_{zw} \parallel_\infty < \gamma \tag{16}$$

where $\parallel \cdot \parallel_\infty = sup_w \bar{\sigma}(\cdot)$ ($\bar{\sigma}(\cdot)$ is the maximum singular value).

In this study, a mixed sensitivity problem for single-input and single-output (SISO) system is discussed. In the mixed sensitivity problem, both a sensitivity reduction at the low frequency and the robust stabilization at the high frequency are accomplished. In practical, the trade-off is considered by the suitable weighting functions $W_S(s), W_T(s)$, and the controller $C(s)$ which satisfies the Eq.(17) is designed.

$$\left\parallel \left[\begin{array}{c} W_S S \\ W_T T \end{array}\right] \right\parallel_\infty < 1 \tag{17}$$

$$T := PC(1 + PC)^{-1}$$

$$S := (1 + PC)^{-1}$$

where $S(s)$ is a sensitivity function, $T(s)$ is a complementary sensitivity function.

4.2. Nominal Model

In this section, the nominal model P_0 of the controlled system is derived. The manipulator can be expressed approximately by the mathematical model of order 14. However, it is not suitable to design the controller by using this model, because the order of the obtained controller becomes high.

The frequency response of the system obtained from the experiment is shown in Fig.7. With respect to the configuration of the manipulator, θ_{a1} and θ_{a2} are chosen as 0.524 and 1.396 [rad], respectively. From this frequency

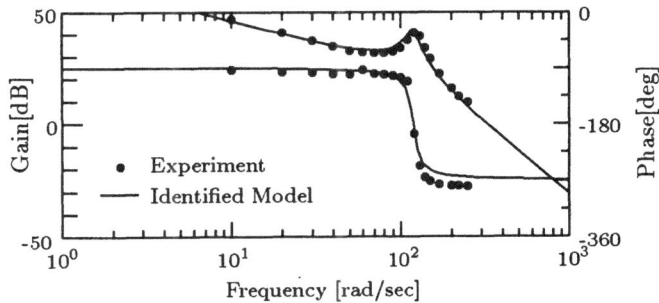

Figure 7. Frequency Response of Controlled System

response, the controlled system is identified as the model of lower order, the model of order 3 which is included the poles (complex conjugate roots) at 120 [rad/sec] and the pole at the origin related to the force control method based on manipulating the angular velocity at each joint mentioned in Section 3. The nominal model is derived as follows:

$$P_0 = \frac{28673000}{s(s + 7.85 + 119j)(s + 7.85 - 119j)} \tag{18}$$

4.3. Perturbation and Weighting Function

As regarded with the perturbation, the modeling error caused by the model reduction and the variation of the system caused by changing the configuration of the manipulator are considered. In this study, all perturbation of the system are assumed as the following multiplicative perturbation $\Delta_m(s)$.

$$P(s) = P_0(s)\{1 + \Delta_m(s)\} \tag{19}$$

The amplitude of the perturbation at the each frequency, $|\Delta_m(j\omega)|$ is derived from the several experiments. Figure 8 shows the obtained perturbation. For the reference, the perturbation which is calculated from the mathematical model is plotted together.

One the other hand, the weighting function W_T for the robust stabilization is chosen so as to suppress this perturbation (See Fig.8).

$$|\Delta_m(j\omega)| < |W_T(j\omega)| \qquad \forall\omega \tag{20}$$

We set

$$W_T = \frac{(1/40\,s + 1)^3}{6} \tag{21}$$

The weighting function W_S serves to satisfy the control specification for the sensitivity characteristic and the response characteristic. We set W_S as ρ/s. ρ is the free parameter.

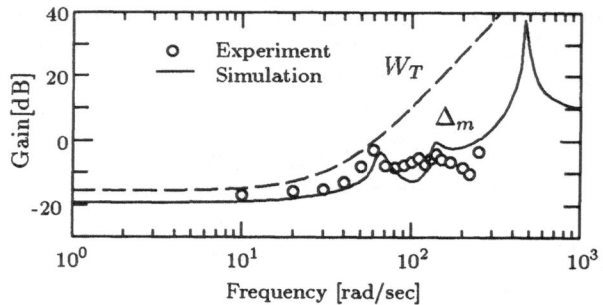

Figure 8. Perturbation Δ_m and Weighting Function W_T

4.4. H^∞ Controller

We may make use of CAD MATLAB due to derive the H^∞ controller in case of the above weighting functions ($\rho = 39$). Since the nominal model P_0 (Eq.(18)) includes the pole at the origin, the generalized controlled function $G(s)$ has to be transformed in order to apply the usual algorithm given in CAD. The controller C is derived as follows:

$$C = 36.4 \frac{(s + a)(s + \bar{a})}{(s + 2721)(s + b)(s + \bar{b})} \tag{22}$$
$$a = 7.85 + 119j \quad , \quad b = 86.4 + 68.9j$$

Figure 9 shows the frequency characteristic of the gain and the phase of the controller.

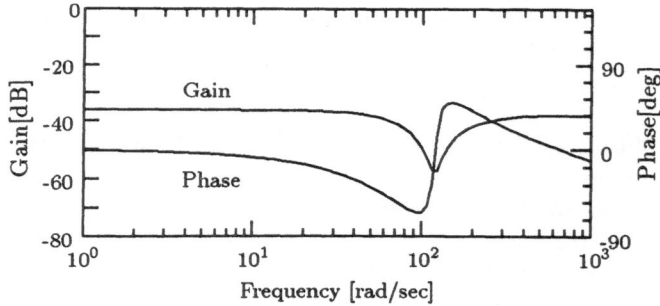

Figure 9. Controller

5. Experimental Results and Conclusions

Figure 10 shows the frequency response of the closed loop system. The bandwidth in case of the constant gain feedback mentioned in Section 3 is merely

10 [rad/sec]. As seen from this figure, the bandwidth is improved up to 40 [rad/sec].

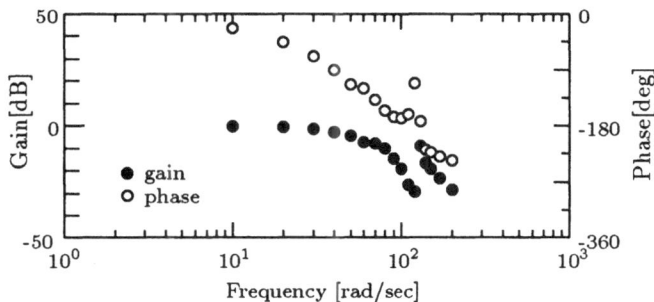

Figure 10. Frequency Response of Closed Loop System

Figure 11 shows the step response under the movement of the end-point along the wall. As seen from this figure, the rise time is very short as compared with the case of the constant gain feedback, and the robustness of the H^∞ controller can be confirmed.

In this study, the H^∞ control theory was applied from the robustness point of view. The uncertainty in the model was treated quantitatively as the perterbation and the controller was derived computationally. The effectiveness of the proposed robust force control was confirmed by the above-mentioned experiments.

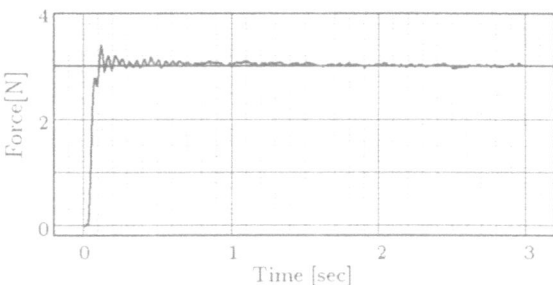

Figure 11. Step Response under Movement along The Wall

References

[1]   Furusho, C.K. Low, A. Sano, T. Sahashi and H. Nagao: Vibration Absorption Control of Robot Arm with Flexible Drive System (The Case of One Link Arm Using Harmonic Drive), Trans. Jpn. Soc. Mech. Eng., (in Japanese), Vol.55, No.517,

C pp.2376-2383, 1989.

[2]   Furusho, H. Nagao and M. Naruse: Multivariable Root Loci of Control Systems of Robot Manipulators with Flexible Driving Systems (Distortion Feedback), JSME Int. Journal, Vol.35, No.1, Series III pp.65-73, 1992.

[3]   Isaac, P.P. Khargonekar and G. Robel: Design of Localized Capture and Track Modes for A Lateral Autopilot Using H^{∞} Synthesis, IEEE Control Systems Magazine, Jun. pp.13-21, 1990.

[4]   Kimura, T. Oike, K. Akai and T. Kida: Robust Stability-Degree Assignment and Its Applications to The Control of Flexible Structures, Int. J. of Robust and Nonlinear Control, 1, pp.153-170, 1991.

[5]   Fujita, T. Namerikawa, F. Matsumura and K. Uchida: $\mu-$ Synthesis of An Electromagnetic Suspension System, Proc. of the 31th Conf. on Decis. and Control, pp.2574-2579, 1992.

[6]   Glover and J.C. Doyle: State-Space Formulae for All Stabilizing Controllers that Satisfy an H_{∞}-norm Bound and Relations to Risk Sensitivity, Systems & Control Letters, Vol.11-2, pp.167-172, 1988.

[7]   C. Doyle, K. Glover, P.P. Khargonekar and B.A. Francis: State-Space Solutions Standard H_2 and H_{∞} Control Problems, IEEE Trans. on AC-34-8, pp.831-847, 1989.

[8]   Furusho and T. Ohwaki and A. Sano: Force Control of Robot Arms Based on Joint Velocity Manipulation (The Case of Flexible Driving Systems), Trans. Jpn. Soc. Mech. Eng., (in Japanese), Vol.58, No.550, C pp.1860-1867, 1992.

[9]   C. Good, L.M. Sweet and K.L. Storobel: Dynamic Models for Control System Design of Integrated Robot and Drive Systems, Trans. ASME, J. Dyn. Syst Meas. Control., Vol.107, No.1 pp.53, 1985.

Adaptive Dynamic Hybrid Position and Force Control of Flexible Joint Robot Manipulators

Hoda A. ElMaraghy
Professor and Director

Atef T. Massoud
Ph. D. Candidate

Flexible Manufacturing Centre, Mechanical Engineering Department, McMaster University, Hamilton, Ontario, Canada L8S 4L7

Abstract— An adaptive control approach is presented to control the position and force of flexible joint robot manipulators interacting with rigid environment when the robot parameters are uncertain and an observer is used to estimate the state vector used in the feedback.

A feedback linearizable fourth order model of the flexible joint robot is constructed. Also, a constraint frame is derived which enables decoupling of the system into a position subsystem and a force subsystem.

A model–based and an adaptive control approaches are presented. In the model–based algorithm, the robot parameters are assumed to be known and full state feedback is available using the robot dynamic model. It is shown that a nonlinear feedback control law can be designed which linearizes the system, uses the constraint frame to decouple the system into the position and the force subsystems, and imposes desired closed loop characteristics in each subsystem independently. In the adaptive control algorithm, the robot parameter are uncertain and only link position is available. This control algorithm consists of a nonlinear control law which has the same structure as the one derived for the model–based algorithm except that it uses the estimated parameters and observed state, an adaptation law and, a sliding mode observer. The sufficient conditions under which the closed loop system composed of the flexible joint robot interacting with a rigid environment, the observer, the controller, is stable are presented.

An experimental two link flexible joints robot manipulator constrained by a straight rigid wall is used to evaluate the two control algorithms.

1. Introduction

The problem of joint flexibility exists in many robots due to the flexibilities in the drive systems, such as gears, shafts, belts, chains, harmonic drives, etc. [21,26]. As a result, lightly damped oscillations occur when the resonant frequencies due to joint flexibility are excited. This usually happens because these resonant frequencies are very low and often exist inside the control bandwidth. Also, joint flexibility reduces the control bandwidth and performance and can cause control instability. Joint flexibility must be included in the dynamic model and control law in order to achieve better performance.

In many industrial applications, the robot end effector comes into contact with the environment. Therefore, there is a need to control both the position and force at the end effector in order to execute contact tasks efficiently. Much work has been presented to control the position and force of rigid robot manipulators. When the robot motion is constrained by perfectly rigid surfaces, there are kinematic constraints on the motion of the end effector. McClamroch and Wang [16], Mills and Goldenberg [18], and Yoshikawa [28,29] considered these constraints in the dynamic model and in the derivation of the control law. On the other hand, in force control approaches by Khatib [12] and DeLuca et al [6] the constraints haven't been considered for decoupling, but the dynamic interactions were considered.

There is a need to consider joint flexibility in modelling and control [21,26]; also, there is a need to control both position and force of the robot end point. However, the combined problem of controlling position and force of flexible joints robot manipulators has received little study. It is more important to account for joint flexibility when dealing with force control than with pure position control [4,7]. Krishnan and McClamroch [14] applied a linear quadratic optimal control for the linearized constrained model. Mills [17] applied the idea of composite control to a singular perturbation model. Spong [25] applied the idea of integral manifold and corrective control of singularly perturbed nonlinear systems to extend the force control of rigid robots to the case of flexible joints. Jankowski and ElMaraghy [11] proposed two methods with different approaches for coordinate reduction, which ensure an exact feedback linearization for hybrid position–force control.

Some researchers proposed some adaptive control strategies to control the position of flexible joint robot manipulators. Ghorbel et al [9] and Hung [10] applied the adaptive control algorithm developed by Slotine and Li [23] to flexible joint robots. They simply modified the control law derived for rigid robots by adding a corrective term to damp out the elastic oscillations at the joints. The results were applied to an experimental single link flexible joint robot. The variable structure control approach was adopted by Sira–Ramizer et al [22] in the design of the outer loop control law for the linearized system. The system was first linearized by a nonlinear feedback and coordinate transformation. Mrad and Ahmad [19] derived an adaptive control law based on an energy motivated Lyapunov function and a linear parametrization model of the links system only.

Some advances have been made to include the effect of using a nonlinear observer to estimate the velocities used in the feedback. An approach for controlling rigid robot in the joint space using only joints position measurement was presented in [2]. The velocity used in the feedback is estimated using a sliding mode observer. A variable structure control algorithm is presented in [30] . Also, for controlling rigid robot in the joint space using a model based observer for velocity estimate. Only [1] presented an adaptive control approach for joint tracking when the robot parameters are uncertain and a sliding observer is used. The parameter adaptation and state observation were performed simultaneously.

The problem tackled in this paper is tracking of position and force for flexible joint robots considering parameter uncertainty and the estimation of the state for feedback. First, a model–based controller is presented. It assumes that the robot parameters are known and the state used in the feedback is obtained using the robot dynamic model. The controller serves as a base for designing the nonlinear control law in the adaptive algorithm. The adaptive control algorithms consists of a nonlinear feedback control law similar to that of the model–based control law except that it uses the estimated states and parameters, a parameter update law, and a sliding mode observer. Lyapunov stability criterion is used to investigate the stability due to interaction between the robot, the observer, the controller, and the adaptation law. The conditions under which the closed loop system, composed of the robot, observer, controller, and adaptation law, is stable are obtained.

2. Fourth Order Model

In this section, a fourth order model is derived. This model is used in the development of the nonlinear feedback control law and the sliding mode observer. The equations of motion of an n degrees of freedom flexible joints robot manipulator in the presence of contact forces at the tip of the end effector are given by [24]:

$$D(q) \ \ddot{q} + C(q, \dot{q}) + B \ \dot{q} - K \ (q_m - q) = - J^T F \tag{1}$$
$$I_m \ \ddot{q}_m + B_m \ \dot{q}_m + K \ (q_m - q) = \tau \tag{2}$$

where q is the nx1 link angular position vector, q_m is the nx1 motor angular position vector, $D(q)$ is the nxn manipulator inertia matrix, $C(q, \dot{q})$ is the nx1 Coriolos and centrifugal forces vector, K is the nxn diagonal matrix with entries equal to the joint stiffness, J^T is the nxn transpose of the manipulator Jacobian, F is the nx1 forces vector at the end effector expressed in the reference frame, I_m is the nxn diagonal matrix with entries equal to the the the rotors inertia, B_m is the nxn diagonal matrix with entries equal to the coefficients of viscous damping at the motors, and τ is the nx1 applied motor torque vector. By inspecting the above equations, it is clear that the input torques are not related directly to the link position vector and its derivatives nor to the external forces at the end effector. Thus, it is difficult to design the control inputs so that the end effector tracks specified position and force trajectories. Also, it is not a straight forward procedure to design the control inputs to linearize the system. For these reasons, the system is transformed into a new coordinate system. In the new coordinates, the control inputs are related directly to the link position vector, the contact forces and their derivatives. By solving equation (1) for the motor position vector differentiating twice and substituting into equation (2), the following fourth order model results:

$$M(q) \ \overset{4}{q} + h_p + h_f + N(q) \ J^T \ \ddot{F} = \tau \tag{3}$$

where $\overset{n}{x}$ is the nth derivative of x, $M(q)$ is nxn new inertia matrix, h_p nx1 vector, each element is a nonlinear function of $q, \dot{q}, \ddot{q}, \overset{3}{q}, h_f$ is nx1 vector, each element is a nonlinear function of $q, \dot{q}, \ddot{q}, F, \dot{F}$, and $N(q)$ is nxn new matrix the elements of which are function of the inertial parameters of the links and motors and joint stiffness. The idea of constructing the fourth order model was introduced in [8,27] where it was used to implement the inverse dynamics position control for two degrees of freedom flexible joint robots. Also, Jankowski and ElMaraghy [11] constructed a forth order model and used for position–force control. The system given by equation (3) can be transformed to the following state space form

$$\dot{x} = f(x) + g(x_1) \tau \qquad\qquad (4)$$

where $\quad x = [\ x_1\ x_2\ x_3\ x_4\ ]^T = [\ q\ \dot{q}\ \ddot{q}\ \overset{3}{q}\ ]^T$

This model given by equation (4) is the final form that will be used in the control and observation presented next. This model is identical to that obtained by Spong [24] and Sira–Ramerez et al [22], and has been investigated by many other researches who proved that the this model is feedback linearizable.

3. Constraint Frame

In this section the required transformation to decouple the system into a position subsystem and a force subsystem is presented. The development of the constraint frame is adopted from the work by Yoshikawa [28]. Assuming that there are m $(m \leq 6)$ independent frictionless constraint surfaces, each is a holonomic constraint given by the following algebraic equation in the reference frame

$$\phi_i(x) = 0 \qquad\qquad i = 1, 2, ..., m \qquad\qquad (5)$$

where x is the position vector in the reference frame. It can be seen that [28] the following rotation matrix can be obtained

$$R = \begin{bmatrix} R_p \\ R_f \end{bmatrix} = [e_1, e_2, ..., e_6]^T \qquad\qquad (6)$$

where $\quad R_p = [e_1, e_2, ..., e_{6-m}]^T \quad$ and $\quad R_f = [e_{m+1}, e_{m+2} ... ., e_6]^T$

The matrix R_f represents the coordinate axes normal to the constraint surface, while matrix R_p represents the coordinate axes which complement R_f. The coordinate system with its origin at the present end effector position x and with the unit bases $\{\ e_1, e_2, ...e_6\ \}$, defines the constraint frame. The rotation matrix R, composed of R_p and R_f, is the transformation matrix from the reference frame into the constraint frame.

It can be seen that the transformation of the link position vector fourth derivative into the constraint frame is given by the following equation

$$\overset{4}{q} = J^{-1}\ [\ R^{-1}\ (\ \overset{4}{x_t}\ -\ a_x\ )\ -\ a_q] \qquad\qquad (7)$$

where $a_q = 3\ (\ \dot{J}\ \overset{3}{q}\ +\ \ddot{J}\ \ddot{q}\ )\ +\ \overset{3}{J}\ \dot{q}$ and $a_x = 3\ (\ \dot{R}\ \overset{3}{x}\ +\ \ddot{R}\ \ddot{x}\ )\ +\ \overset{3}{R}\ \dot{x}$

where x_t is the end effector position vector in the constraint frame. Additionally, the transformation of the contact force second derivative from the reference frame into the constraint frame is given by

$$\ddot{F} = L\ \ddot{f_f}\ +\ a_f \qquad\qquad (8)$$

where $\quad a_f = 2\ \dot{L}\ \dot{f_f}\ +\ \ddot{L}\ f_f \quad$ and $\quad L = (\ R_f\ T^{-1})^T$

T is 6x6 matrix which transforms the end effector velocity $\dot{x}$ expressed in the reference frame to a 6 dimensional vector consisting of translational velocity along each axis and rotational velocity about each axis. $f_f \in R^m$ represents the contact force F in terms of the unit force vector L which is normal to the constraint surface. It can be shown that the force normal to the constraint surface is given by:

$$f_f = (\ R_f\ J\ D^{-1}\ J^T\ L\ )^{-1}\ [\ R_f\ J\ D^{-1}\ h_1\ +\ R_f\ \dot{J}\ \dot{q}\ +\ \dot{R_f}\ J\ \dot{q}\ ] \qquad (9)$$

4. Model Based Nonlinear Controller

Assuming that the robot parameters are known, then the feedback linearization can be used to linearize the robot model given by equation (3). Based on the same assumption, the state vector required for the feedback is obtained from the dynamic model. Utilizing the transformations of the link position vector and the contact force, the following nonlinear feedback control law is designed

$$\tau = M J^{-1} [R^{-1} (\overset{4}{x}^{*}_{t} - a_x) - a_q] + h_p + h_f + N J^T (L \overset{2}{f}^{*} + a_f)(10)$$

where $\overset{4}{x}^{*}_{t} = \begin{bmatrix} \overset{4}{P}^{*} \\ 0 \end{bmatrix}$

$\overset{4}{P}^{*}$ is 6–m x 1 vector of new input to the position subsystem and $\overset{2}{f}^{*}$ is m x 1 vector of new inputs to the force subsystem. The control law given by equation (10), when applied to the flexible joint robot model given by equation (3), will result in the following closed loop system

$$\overset{4}{P} = \overset{4}{P}^{*} \tag{11}$$

$$\ddot{f}_f = \overset{2}{f}^{*} \tag{12}$$

The nonlinear control law linearizes the system by cancelling the nonlinear terms, uses the constraint frame transformation to decouple the system to the position subsystem and the force subsystem, and finally imposes a desired closed loop characteristic in each subsystem.

Outer Loop Design

Assume that the desired position trajectory is continuous and smooth up to the fourth derivative and the desired force trajectory is also continuous and smooth up to the second derivative. Then, a characteristic frequency and damping can be achieved for each subsystem as follows. By choosing the input $\overset{4}{P}^{*} \in R^{6-m}$ as in the following equation

$$\overset{4}{P}^{*} = \overset{4}{P^d} + K_{p3} (\overset{3}{P^d} - \overset{3}{P}) + K_{p2} (\ddot{P^d} - \ddot{P}) + K_{p1} (\dot{P^d} - \dot{P})$$
$$+ K_{p0} (P^d - P) \tag{13}$$

where K_{p3}, K_{p3}, K_{p3}, and K_{p3} are positive diagonal matrices and $P^d, \dot{P^d}, \ddot{P^d}, \overset{3}{P^d}$ and $\overset{4}{P^d}$ are the desired position and its first four derivatives respectively. Then, the following error equation results:

$$\overset{4}{e}_p + K_{p3}\overset{3}{e}_p + K_{p2}\ddot{e}_p + K_1\dot{e}_p + K_{p0}e_p = 0 \tag{14}$$

where $e_p = P^d - P$ is the position error vector. The same procedures are followed for the force subsystem, the force input $\overset{2}{f}^{*} \in R^m$ is designed as:

$$\overset{2}{f}^{*} = \ddot{f^d} + K_{f1}(\dot{f^d} - \dot{f}) + K_{f0}(f^d - f) \tag{15}$$

where K_{f0} and K_{f1} are positive diagonal matrices and $f^d, \dot{f^d}$ and $\ddot{f^d}$ are the desired normal force and its first two derivatives, respectively. Then the following error equation results

$$\ddot{e}_f + K_{f1}\dot{e}_{f1} + K_{f0}e_f = 0 \tag{16}$$

where $e_f = f^d - f$ is the force error vector. The closed loop poles for each subsystem can be allocated as desired with the choice of the gain matrices. For the position and force subsystem the gains are designed as follows $K_{p0} = \omega_p^4$, $K_{p1} = 4\zeta_p\omega_p^3$, $K_{p2} = (4\zeta_p^2 + 2)\omega_p^2$ and $K_{p3} = 4\zeta_p\omega_p$, also $K_{f0} = \omega_f^2$ and $K_{f1} = 2\zeta_f\omega_f$. where

ω_p, ζ_p, ω_f, and ζ_f are the natural frequency and damping ratio of the position and force closed loops, respectively.

5. Adaptive Controller

The adaptive control system maintains the structure of the model based controller, but in addition has an adaptation law. The basic structure of this adaptive control system is similar to that developed by Craig [5] for position control of rigid robot. The basic idea of this scheme is to apply the control law similar to the model–based approach using the estimated parameters and adding an adaptation law. If the actual states were available for the control law, then the mismatch in the closed loop would have been only due to parameters errors and the adaptation law can be designed to ensure the stability of the closed loop.

The approach presented in this paper is illustrated in figure (1). It is different from Craig's approach [5] in that the state used in the feedback is estimated. As a result the mismatch in the closed loop is due to both parameter and observation errors. The adaptation law has the same structure as that if the actual state were available, but it uses the observed state. The condition under which the closed loop is stable is presented.

5.1 Robust Sliding Observer

In the model–based control, the links angular velocity are obtained using 6 points numerical differentiation, the links acceleration and jerk are computed from the dynamic model. When the robot parameters are uncertain , a sliding mode observer is proposed [3]. The sliding observer is based on the concept of the attractive manifold, which consists of defining a hyper surface (as a function of the observation errors) and determining the conditions such that the manifold is attractive. Once the error trajectories reach this hyper surface, a switching error based action makes the observation errors approaches zero.

The system described by equation (4) can be decomposed and written in the following form:

$$\left.\begin{aligned} \dot{x}_1 &= C x_2 \\ \dot{x}_2 &= A_2\, x_2 + f_2(x_1, x_2) \\ y_1 &= x_1 \end{aligned}\right\} \tag{17}$$

where $x_1 = q_1$, $x_2 = [\, \dot{q}\ \ddot{q}\ \overset{3}{q}\,]^T$

$$A_2 = \begin{bmatrix} 0 & I & 0 \\ 0 & 0 & I \\ 0 & 0 & 0 \end{bmatrix}, \quad C = [\, I\ \ 0\ \ 0\,] \quad \text{and}$$

$$f_2(x_1, x_2) = M(q)^{-1}\,(\, \tau - h_p - h_f - N(q)\, J^T\, \ddot{F}\,)$$

where the fact that the force and its derivatives are functions of the state $x = [x_1, x_2]^T$ has been used from equation (9). $f_2(x_1, x_2)$ is continuous in x_2 and has a Lipschitz constant defined by the following equation:

$$\| f_2(x_1, x_2) - f_2(x_1, \hat{x}_2) \| \ < \mu_x \, \| x_2 - \hat{x}_2 \| \tag{18}$$

The observer is given by:

$$\left.\begin{aligned} \dot{\hat{x}}_1 &= C\hat{x}_2 - \Gamma_1\, \tilde{x}_1 - K_1\, sgn\ (\tilde{x}_1) \\ \dot{\hat{x}}_2 &= A_2\, \hat{x}_2 + f_2(\hat{x}_1, \hat{x}_2) - \Gamma_2\, \tilde{x}_1 - K_2\, sgn\ (\tilde{x}_1) \end{aligned}\right\} \tag{19}$$

where $\tilde{x}_1 = \hat{x}_1 - x_1$, $\tilde{x}_2 = \hat{x}_2 - x_2$, and Γ_1, Γ_2, K_1, and K_2 are positive definite constant diagonal matrices, $\Gamma_1 = diag\ \{\ \gamma^i_{\ 1}\}$, $\Gamma_2 = diag\ \{\ \gamma^i_{\ 2}\}$, and $K_1 = diag\ \{\ \lambda^i_{\ 2}\}$. This is different from the design in [3] in that the switching term gains are constant. The error equation can be obtained by subtracting the system given by equation (17) from the observer (19) and is given by:

$$\left.\begin{aligned}
\dot{\tilde{x}}_1 &= C\tilde{x}_2 - \Gamma_1\ \tilde{x}_1 - K_1\ sgn\ (\tilde{x}_1) \\
\dot{\tilde{x}}_2 &= A_2\ \tilde{x}_2 + f_2(x_1\hat{x}_2) - f_2(x_1,x_2) - \Gamma_2\ \tilde{x}_1 - K_2\ sgn\ (\tilde{x}_1)
\end{aligned}\right\} \quad (20)$$

The hyper plane $\tilde{x}_1$ is attractive within the following set:

$$\tilde{x}_2^{\ i} - \gamma_1^{\ i}\ \hat{x}_{1_i}^{\ i} - \lambda_1^{\ i} < 0\ ;\ \tilde{x}_1^{\ i} > 0\ \forall\ i = 1,2\ ..\ n$$
$$\tilde{x}_2^{\ i} - \gamma_1^{\ i}\ \hat{x}_1^{\ i} - \lambda_1^{\ i} > 0\ ;\ \tilde{x}_1^{\ i} < 0\ \forall\ i = 1,2\ ..\ n$$

Note that one can always start with $\hat{x}_1(0) = x_1(0)$, because the initial position is directly available form the encoder. Therefore at $t=0$, we have $\tilde{x}_1(0) = 0$ which when introduced in (20) gives sufficient conditions for the invariance of $\tilde{x}_1$ that is $\|\ \tilde{x}_2^{\ i}\| < \lambda^i_{\ 1}$ for all t > 0. Thus, the equivalent dynamics on the reduced–order manifold is given by:

$$\dot{\tilde{x}}_2 = (A_2 - K_2\ K_1^{\ -1}\ C)\ \tilde{x}_2 + f_2(x_1\hat{x}_2) - f_2(x_1,x_2) \quad (21)$$

Choose the following Lyapunov function candidate: $V_2 = \tilde{x}_2^{\ T}\ P_2\ \tilde{x}_2$, where P_2 is positive definite symmetric matrix satisfying the Lyapunov equation:

$$(A_2 - K_2\ K_1^{\ -1}\ C)^T\ P_2 + P_2\ (A_2 - K_2\ K_1^{\ -1}\ C) = -\ Q_2 \quad (22)$$

and Q_2 is a positive definite symmetric matrix. The derivative of V_2 along the trajectories of (21) is given by:

$$\dot{V}_2 = \tilde{x}_2^{\ T}\ [(A_2 - K_2\ K_1^{\ -1}\ C)^T\ P_2 + P_2\ (A_2 - K_2\ K_1^{\ -1}\ C)]\ \tilde{x}_2\ +$$
$$\tilde{x}_2^{\ T}\ P_2\ [f_2(x_1\hat{x}_2) - f_2(x_1,x_2)] \quad (23)$$

It can be seen using (18) that:

$$\dot{V}_2 \le -\ \lambda_{min}(\ Q_2\ )\ \|\ \tilde{x}_2\ \|^2 + 2\ \mu_x\ \lambda_{max}\ (\ P_2\ )\ \|\ \tilde{x}_2\ \|^2 \quad (24)$$

where $\lambda_{min}(Q_2)$ and $\lambda_{max}(P_2)$ are the minimum and maximum eigenvalues of Q_2 and P_2, respectively. If $\lambda_{min}(Q_2) > 2\ \mu\ \lambda_{max}\ (\ P_2\ )$, the $\dot{V}_2$ is negative definite and hence $\tilde{x}_2 \Rightarrow 0$ as $t \Rightarrow \infty$. Now, it is clear that the gain matrix K_2 can be designed to ensure the stability of the reduced–order manifold. The procedure developed by Raghavan and Hedrick [20] can be used to design K_2 to ensure the stability of the reduced–order manifold.

5.2 Linear Parametrization Model

It is possible to construct a linear parametrization model for the links subsystem [13], also, a linear parametrization for the motor subsystem can constructed, the models are given by the following two equations:

$$\Psi_l\ \Theta_l + K(q - q_m) + J^T F = 0 \quad (25)$$
$$\Psi_m\ \Theta_m - K(q - q_m) = \tau \quad (26)$$

where Θ_l, Θ_m, Ψ_l, and Ψ_m are the links parameters vector, the motor parameters vector, a matrix which its elements are functions of the link position, velocity, and acceleration, and a matrix which its elements are functions of the motor velocity and acceleration. By inspecting the equations of motion, (1) and (2), or (25) and (26), one can see that each of the links and motors subsystems are coupled by the stiffness force due to joint flexibility. Recognizing that this coupling force is internal, then the two subsystems can be combined by solving any of the two equations for the stiffness force vector and substituting into the other. Using this one can augment the two linear parametrization models together to yield one model that describes the flexible joint robot. This model is given by the following equation:

$$\Psi \Theta + J^T F = \tau \tag{27}$$

where Θ is a vector of the links and motors parameters and Ψ is a regression matrix, which is a function of the links position, velocity, and accelerations, and the motors velocity, and acceleration. The link velocity and acceleration are obtained from the observer, while the motor velocity and acceleration can be obtain by numerical differentiation as proposed and tested by [13,27].

5.3 Adaptive Controller

The control law given by equation (10) when using the estimated parameters and observed state is described by the following equation:

$$\tau = \hat{M} J^{-1} [R^{-1} (\overset{4}{x}^*_t - a_x) - a_q] + \hat{h}_p + \hat{h}_f + \hat{N} J^T (L \ddot{f}^* + a_f) \tag{28}$$

where $\hat{M}, \hat{N}, \hat{h}_p$, and $\hat{h}_f$ are the estimated values of M, N, h_p, and h_f, respectively. The new inputs $\overset{4}{P}^*$ and $\ddot{f}^*$ are designed as given by equations (13) and (15) using the estimated states. Applying the above control law to the model given by equation (3), after some algebraic operations and the use of the linear parametrization model given by equation (27), the following closed loop results:

$$\hat{M} J^{-1} [R^{-1} (\overset{4}{e}_p + K_{p3}\overset{3}{e}_p + K_{p2}\ddot{e}_p + K_{p1}\dot{e}_p + K_{p0}e_p)]$$
$$+ \hat{N} J^T (L (\ddot{e}_f + K_{f1}\dot{e}_f + K_{f0}e_f) = \hat{\Psi} \tilde{\Theta} + \tilde{\Psi} \bar{\Theta} \tag{29}$$

where $\hat{e}_p = P^d - \hat{P}$ and $\hat{e}_f = f^d - \hat{f}$, $\hat{P}$ is the estimated position in the tangential direction of the constraint and $\hat{f}$ is the estimated force normal to the constraint. $\tilde{\Theta} = \Theta - \hat{\Theta}$, $\hat{\Theta}$ is the estiamted parameters vector. $\tilde{\Psi} = \Psi - \hat{\Psi}$, $\hat{\Psi}$ is the estiamted regression matrix. Inspecting the closed loop system given by equation (29), one can see that the right hand side represents the error due to the mismatch in both the parameters and state. Specifically, $\hat{\Psi} \tilde{\Theta}$ is the error in the closed loop due to parameters uncertainty, and $\tilde{\Psi} \Theta$ is the error due to observation of the state. The constraint frame can be used to combine the position vector in the tangential direction and the force vector in the normal direction into a new vector Z. The closed loop of equation (29) can be written as:

$$S \begin{bmatrix} \overset{4}{e}_p + K_{p3}\overset{3}{e}_p + K_{p2}\ddot{e}_p + K_{p1}\dot{e}_p + K_{p0}e_p \\ \ddot{e}_f + K_{f1}\dot{e}_f + K_{f0}e_f \end{bmatrix} = \hat{\Psi} \tilde{\Theta} + \tilde{\Psi} \Theta \tag{30}$$

where S is a new matrix $6x6$, the first $6-m$ columns are given by the first $6-m$ columns of the matrix $\hat{M}J^{-1}R^{-1}$ and the other m columns are given by the matrix $\hat{N}J^{T}L$. The above system can be transformed to the following state space form:

$$\dot{Z} = A_1 \hat{Z} + B_1 \hat{\Phi} \tilde{\Theta} + B_1 \tilde{\Phi} \Theta \tag{31}$$

where

$$Z = [z_1 \ z_2 \ z_3 \ z_4 \ z_5 \ z_6]^T \equiv \left[e_p \ \dot{e}_p \ \ddot{e}_p \ \overset{3}{e}_p \ e_f \ \dot{e}_f \right]^T$$

$$\hat{Z} = \left[\hat{z}_1 \ \hat{z}_2 \ z_3 \ \hat{z}_4 \ \hat{z}_5 \ \hat{z}_6 \right]^T \equiv \left[\hat{e}_p \ \dot{\hat{e}}_p \ \ddot{\hat{e}}_p \ \overset{3}{\hat{e}}_p \ \hat{e}_f \ \dot{\hat{e}}_f \right]^T, \quad B_1 = [0 \ 0 \ 0 \ I \ 0 \ I]^T,$$

$$\Phi = S^{-1} \ \Psi \text{ and } A_1 = \begin{bmatrix} 0 & I & 0 & 0 & 0 & 0 \\ 0 & 0 & I & 0 & 0 & 0 \\ 0 & 0 & 0 & I & 0 & 0 \\ -K_{p0} & -K_{p1} & -K_{p2} & -K_{p3} & 0 & 0 \\ 0 & 0 & 0 & 0 & 0 & I \\ 0 & 0 & 0 & 0 & -K_{f0} & -K_{f1} \end{bmatrix}$$

The system above can be written in the following form:

$$\dot{Z} = A_1 Z + B_1 \hat{\Phi} \tilde{\Theta} + B_1 \tilde{\Phi} \Theta + B_3 \ g_1(x_1,x_2,\hat{x}_2)\tilde{x}_2 \tag{32}$$

where $B_3 \ g_1(x_1,x_2,\hat{x}_2)\tilde{x}_2$ represents the transformation of the $\hat{Z} - Z$ as a function of the error in the joint space. Choose the following Lyapunov function candidate:

$V_1 = Z^T P_1 Z + \tilde{\Theta}^T \Gamma^{-1} \tilde{\Theta}$, where P_1 is positive definite symmetric matrix satisfying the Lyapunov equation:

$$A_1^T P_1 + P_1 A_1 = -Q_1 \tag{33}$$

and Q_1 is a positive definite symmetric matrix and Γ is a symmetric positive definite gain matrix. Choosing the following adaptation law:

$$\dot{\hat{\Theta}} = -\Gamma \ \Phi^T B_1^T P_1 \hat{Z} \tag{34}$$

Noting that all the parameters are constant in time, $\dot{\tilde{\Theta}} = \dot{\hat{\Theta}}$, then, the derivative of V_1 along the trajectories of (29) is given by:

$$\dot{V}_1 = -Z^T Q_1 Z + 2\tilde{\Theta}^T \hat{\Phi}^T B_1^T P_1 g_1(x_1,x_2,\hat{x}_2)\tilde{x}_2 + 2[\Theta^T \tilde{\Phi}^T B_1^T +$$

$$\tilde{x}_2^T \ g_1^T(x_1,x_2,\hat{x}_2)A_1^T B_3^T]P_1 Z \tag{35}$$

Now, the stability of the overall closed loop can be analyzed using the introduced Lyapunov functions V_1 and V_2.

Lemma: Consider the control law given by equation (28), the sliding observer given by equation (19) and initialized such that $\tilde{x}_1(t) = 0$ remains attractive, and the adaptation law given by equation (34), the augmented state vector defined by $s(t) = [Z \ \tilde{x}]^T$ converges to zero as t approaches infinity.

Proof: the closed-loop analysis is performed on the basis of the reduced-order manifold dynamics (21) and the tracking error dynamics (32), i.e.

$$\dot{Z} = A_1 Z + B_1 \hat{\Phi} \tilde{\Theta} + B_1 \tilde{\Phi} \Theta + B_3 \ g_1(x_1,x_2,\hat{x}_2)\tilde{x}_2 \tag{36}$$

$$\dot{\tilde{x}}_2 = (A_2 - K_2 K_1^{-1} C) \tilde{x}_2 + f_2(x_1,\hat{x}_2) - f_2(x_1,x_2) \tag{37}$$

Introducing the following Lyapunov function

$$V = V_1 + V_2 = s^T \begin{bmatrix} P_1 & 0 \\ 0 & P_2 \end{bmatrix} s + \tilde{\Theta}^T \Gamma^{-1} \tilde{\Theta} \tag{38}$$

the time derivative of V is given by:

$$\dot{V} = - s^T Q s + 2 [\tilde{\Theta}^T \hat{\Phi}^T B_1^T P_1 g_1(x_1 x_2 \hat{x}_2) + [f_2(x_1 \hat{x}_2) - f_2(x_1 x_2)] \; P_2]^T \tilde{x}_2$$

$$+ 2[\Theta^T \tilde{\Phi}^T B_1^T + \tilde{x}_2^T g_1^T(x_1, x_2, \hat{x}_2) A_1^T B_3^T]P_1 Z \tag{39}$$

where $\quad Q = \begin{bmatrix} Q_1 & 0 \\ 0 & Q_2 \end{bmatrix}$

Let q_{10} and q_{20} be the minimum eigenvalues of Q_1 and Q_2 , and p_{10} and p_{20} be the maximum eigenvalues of P_1 and P_2, respectively. Introducing the following properties:

1- $\| \Theta^T \tilde{\Phi}^T B_1^T \| \le \delta_1 \| \tilde{x}_2 \|$

2- $\| \tilde{x}_2^T g_1^T(x_1, x_2, \hat{x}_2) A_1^T B_3^T \| \le \delta_2 \| \tilde{x}_2 \|^2$

3- $\| \tilde{\Theta}^T \hat{\Phi}^T B_1^T P_1 g_1(x_1 x_2 \hat{x}_2) \| \le \delta_3 \| \tilde{x}_2 \|$

4- $\| f_2(x_1 x_2) - f_2(x_1 \hat{x}_2) \| < \mu_x \| \tilde{x}_2 \|$

then equation (39) can be written as:

$$\dot{V} \le - [\| Z \| \; \| \tilde{x}_2 \|] \; [H] \begin{bmatrix} \| z \| \\ \| \tilde{x}_2 \| \end{bmatrix} \tag{40}$$

where $\quad H = \begin{bmatrix} q_{10} & - (\delta_1 + \delta_2) \, p_{10} \\ - (\delta_1 + \delta_2) \, p_{10} & q_{20} - \mu_x \, p_{20} + \delta_3 \, p_{10} \end{bmatrix}$

Finally, we get that

$$\dot{V} \le - \lambda_{\min}(H) \; \| s \| \tag{41}$$

where $\lambda_{\min}(H)$ is the smaller eigenvalue of H. It is necessary to have $\lambda_{\min}(H) > 0$ for all $t > 0$ to ensure the closed loop stability, . To satisfy this condition the following two conditions are required:

$$q_{20} - \mu_x \, p_{20} + \delta_3 \, p_{10} > 0$$

$$q_{10} \, (q_{20} - \mu_x \, p_{20} + \delta_3 \, p_{10}) - (\delta_1 + \delta_2)^2 \, p_{10}^2 > 0$$

Assuming that the observer and controller gains can be chosen to satisfy the above two conditions, then the closed loop system will be stable and the state vector s will approach zero as the time approaches infinity.

6. Experimental and Simulation Results

The experimental setup used to implement the above controllers consists of a mechanical arm and a real time controller. The mechanical arm is a two link modular direct drive manipulator which has been designed and built to truly simulate the effect of joint flexibility. Figure 2 shows a solid model of the mechanical arm. The joints flexibility is designed–in using helical torsional springs which can be replaced to change the flexibility characteristics. The current configuration has two resonant frequencies of 5.2 and 7.5 Hz. The robot is equipped with four encoders to read the positions of the motors and the links, and a force/torque sensor at the end effector. The end effector has a roller located after the force sensor to satisfy the assumption that the constraint surface is frictionless. The real time controller was designed and built. It consists of a digital signal processing card based on the TMS320C30 chip which is interfaced through a bus to a number of input output devices for communication with robot sensors and motors. Details of the experimental robot can be found in [15].

The dynamic model of the experimental robot is given by equation (1) and (2) when $n=2$. The $D(q)$, $C(q,\dot{q})$, K, B, I_m, and B_m are given by:

$$D(q) = \begin{bmatrix} d_1 + 2d_2\cos(q_2) & d_3 + d_2\cos(q_2) \\ d_3 + d_2\cos(q_2) & d_3 \end{bmatrix}, I_m = \begin{bmatrix} I_{m1} & 0 \\ 0 & I_{m2} \end{bmatrix}, B_m = \begin{bmatrix} b_{m1} & 0 \\ 0 & b_{m2} \end{bmatrix}$$

$$C(q,\dot{q}) = \begin{bmatrix} -\dot{q}_2(2\dot{q}_1 + \dot{q}_2)\sin(q_2) \\ d_2 \dot{q}_1{}^2 \sin(q_2) \end{bmatrix}, \quad K = \begin{bmatrix} k_1 & 0 \\ 0 & k_2 \end{bmatrix}, \quad B = \begin{bmatrix} b_1 & 0 \\ 0 & b_2 \end{bmatrix}$$

where $d_1 = I_1 + I_2 + m_1 a_1{}^2 + m_{r2} l_1{}^2 + m_2(l_1{}^2 + a_2{}^2)$, $d_2 = 2 m_2 l_1 a_2$, and $d_3 = I_2 + m_2 a_2{}^2$. Where I_1, I_2, m_1, m_2, a_1, a_2, l_1, l_2, and m_{r2} are the moment of inertia about an axis parallel to the axis of rotation passing through the center of mass, the mass, the distance from the center of rotation to the center of mass, length of the first and second link, respectively, and the mass of the second rotor. Table 1 shows two lists of the robot parameters using I–DEAS solid modeler and sine sweep identification.

Figure 3 shows the robot constrained by a straight wall. The task used in the experiment is to move 0.5 m (from point A to point B in figure 3) in the y direction according to a desired position trajectory, while applying a desired normal force to the wall. The position and force controller parameters (ω_p, ζ_p, ω_f, and ζ_f) are 25,0.9,50,1.1, respectively. Figures 4 to 6, present the experimental results for the model based controller. Figure 4, shows plots of the desired and actual trajectories tangential to the constraining wall. The maximum error is less than 7%, which can be seen from figure 5. This figure also shows that the error is approaching zero and a stable response is obtained. Figure 6, shows plots of the desired and actual measured normal force. The error in the force tracking is attributed to the Coulomb friction in the second motor. This friction force affects the force trajectory directly. Attempts to feedforward the friction force at the second motor were tried but caused instability in the force control by loosing contact with the constraint. For the experiment shown, the tangential force amplitude is of about 0.6 N. During experimentation, it was found that, the force subsystem should be designed to be faster than the position subsystem to avoid loosing contact with the environment and thus to avoid instability. This requirement can be met by choosing $\omega_f > \omega_p$.

The results of simulation for the adaptive control algorithm are presented in figure 7 to 9. The task is to move 0.25 m (from point A to point C in figure 2) in the y direction. The link inertia parameters have 25% uncertainty. The position and force controller parameters (ω_p, ζ_p, ω_f, and ζ_f) are 15,0.4,60,1.1, respectively. For the observer $\lambda^1{}_1 = \lambda^2{}_1 = 0.04$. Figure 6 and 7 show plots of the position and force tracking errors. The observation error $\tilde{x}_1$ remained less than 0.00015 during the motion time, which indicates that the sliding observer is working within the manifold $\tilde{x}_1 = 0$ and as a result the error in the link velocities remained less than 0.04 as desired. Figure 9 shows plots of the true and estimated inertia parameters. It can be seen that the estimation parameters reach their corresponding values slowly with some overshoot, however, the controller for both the position and force remained stable.

Conclusion

A model–based and an adaptive control approaches were presented to control the position and force for flexible joint robot constrained by rigid environment. The first approach deals with the problem when the robot parameters are known and serves as a basis for the adaptive control case. A nonlinear feedback control law was designed which uses a constraint frame to decouple the system into a position and a force subsystems, linearize the system, and impose desired closed loop characteristics in each subsystem. The experimental results obtained using a two link direct drive flexible joint robot show the validity of the approach. Improvements can be made by considering the Coulomb friction in the motors.

In the adaptive control approach, the robot parameters are uncertain and the state used in the feedback is estimated. The algorithm consists of a control law similar to the model based control law but uses the estimated parameters and observed state, a robust sliding mode observer, and an adaptation law. The sufficient conditions to ensure the stability of the closed loop system composed of the constrained flexible joint robot, the controller, the observer, and the adaptation law were derived. Simulation results of the adaptive control approach are presented as a preliminary step before testing the algorithm on the experimental robot.

References

[1] Canudas de Wit, C. and Fixot, N., "Adaptive Control of Robot Manipulators via Estimated Velocity Feedback", Proc. of the Int. Workshop on Nonlinear and Adaptive Control: Issues in Robotics, France, pp. 69–82, November 1990.

[2] Canudas de Wit, C., Fixot, N., Astrom, K. J. " Trajectory Tracking in Robot Manipulators via Nonlinear Estimated State Feedback", IEEE J. of Robotics and Automation, Vol. 8, No. 1, pp. 138– 144, February 1992.

[3] Canudas de Wit, C. and Slotine, J.–J. E. " Sliding Observers for Robot Manipulators", Automatica, Vol. 27, No. 5, pp. 859–864, 1991.

[4] Chian, B. C. and Shahinpoor, M.," The Effects of Joint and Link Flexibilities on the Dynamic Stability of Force Controlled Robot Manipulators", Proc. of the IEEE Conf. on Robotics and Automation, Scottsdale, Az, pp. 398–403, May 1989.

[5] Craig, J. " Adaptive Control of Mechanical Manipulators " Addison–Wesely Publishing Company, Inc., 1988.

[6] De Luca, A., Manes, C., and Nicolo, F.," A Task Space Decoupling Approach to Hybrid Control of Manipulators", 2nd IFAC Symposium on Robot Control (SYRCO'88), Karlsruhe, Germany, pp. 54.1–54.6, October 1988.

[7] Eppinger, S. D. and Seering, W. P.," Three Dynamic Problems in Robot Force Control", Proc. of the IEEE Conf. on Robotics and Automation, Scottsdale, AZ, pp. 392–397, May 1989.

[8] Forrest–Barlach, M. G. and Babcock, S. M.," Inverse Dynamics Position Control of Complaint Manipulators", IEEE J. of Robotics and Automation, Vol. RA–3, No. 1, pp. 75–83, February 1987.

[9] Ghorble, F., Hung, J. Y., and Spong, M. W. " Adaptive Control of Flexible Joint Manipulators " IEEE Control System Magazine, pp. 9–13, December, 1989.

[10] Hung J. Y. " Robust Control Design of Flexible Joint Robot Manipulator " Ph. D. thesis, Department of Electrical and Computer Engineering, University of Illinois at Urbana– Champain, 1989.

[11] Jankowski, K. P. and ElMaraghy, H. A. " Dynamic decoupling for Hybrid Control of Rigid–Flexible–Joint Robots " IEEE Trans. on Robotics and Automation, Vol. 8, No. 5, pp. 519–533, Oct., 1992.

[12] Khatib , O.," A Unified Approach for Motion and Force Control of Robot Manipulators: The Operational Space Formulation", IEEE J. of Robotics and Automation, Vol. RA–3, No. 1, pp. 43–53, February 1987.

[13] Khosla, P. " Estimation of the Robot Dynamics Parameters: Theory and Application" Inter. J. of Robotics and Automation, Vol. 3, No. 1, pp. 35–41, 1988.

[14] Krishnan, H. and McClamroch, N. H.," A new Approach to Position and Contact Force Regulation in Constrained Robot Systems", Proc. of the IEEE Conf. on Robotics and Automation, Cincinnati, OH, pp. 1344–1349, May 1990.

[15] Massoud, Atef T. and ElMaraghy, Hoda A. " Design, Dynamics, and Identification of a Flexible Joint Robot Manipulator ", The IASTED Inter. Conf. on Robotics and Manufacturing, Oxford, England, Sept., 1993.

[16] McClamroch, N. H. and Wang, D.," Feedback Stabilization and Tracking of Constrained Robots", IEEE Trans. on Automatic Control, Vol. 33, No. 5, pp. 419–426, May 1988.

[17] Mills, J. K. ," Control of Robot Manipulators with Flexible Joints During Constrained Motion Task Execution", Proc. of the 28th Conf. on Decision and Control, Tampa, FLA, pp. 1676–1681, December 1989.

[18] Mills, J. K. and Goldenberg, A. A.," Force and Position Control of Manipulators During Constraint Motion Tasks", IEEE Trans. on Robotics and Automation, Vol. 5, No. 1, pp. 30 46, February 1989.

[19] Mrad, F. T. and Ahmad, S. " Adaptive Control of Flexible Joint Robot Using Position and Velocity Feedback " Inter. J. of Control, 1992.

[20] Raghavan, S. and Hedrick, J. K. "Observers for a Class of Nonlinear Systems", Int. J. of Control, 1993.

[21] Riven, E. " Mechanical design of Robots", New York: McGraw–Hill Book Co., 1988.

[22] Sira–Ramrez, Hebertt and Spong, M. " Variable Structure Control of Flexible Joint manipulators", IEEE Int. J. of Robotics and Automation, Vol. 3,No. 2, pp. 57–64, 1988.

[23] Slotine, J. –J. E. and Li, W. " On the Adaptive Control of Robot Manipulator " Robotics: Theory and Applications, The winter Annual Meeting of the ASME, DSV 3, pp. 51–56, Anaheim, CA, December, 1986.

[24] Spong, M.," Modeling and Control of Elastic Joint Robots", ASME J. of Dyn. Syst., Meas. Cont., vol. 109, No. 4, pp. 310–319, December 1987.

[25] Spong, M.," On the Force Control Problem of Flexible Joints Manipulators",

IEEE Trans. on Automatic Control, Vol. 34, No. 1, pp. 107–111, January 1989.

[26] Sweet L. M. and Good M. C. " Re–Definition of the Robot Motion Control Problem: Effects of Plant Dynamics, Drive System, Constraints and User Requirements " Proc. of 23rd Conf. on Decision and Control, Las Vegas, NV, pp. 724–732, Dec. 1984.

[27] Uhlik C. R. " Experiments in High–Performance Nonlinear and Adaptive Control of a Two–Link Flexible–Drive–Train Manipulator " Ph. D. thesis, Depart ment of Aeronautics and Astronautics, Stanford University, Stanford CA 94305, May 1990.

[28] Yoshikawa, T.," Dynamic Hybrid position Force Control of Robot Manipulators– Description of Hand Constraints and Calculation of Joint Driving Forces", IEEE J. of Robotics and Automation, Vol. RA–3, No. 5, pp. 386–392, October 1987.

[29] Yoshikawa, T.," Dynamic Hybrid position Force Control of Robot Manipulators– Controller Design and Experiment", IEEE J. of Robotics and Automation, Vol. 4, No. 6, pp. 699–705, December 1988.

[30] Zhu, W.–h, Chen, H.–t., and Zhang, Z.–j. "A Variable structure Robot Control Algorithm with an Observer", IEEE Trans. on Robotics and Automation, Vol. 8, No. 4, pp. 486–492, August 1992.

Table 1. Experimental Robot parameters from design
and Sine Sweep Identification

	I–DEAS	Sine Sweep
l_1 (m)	0.400	
l_2 (m)	0.350	
d_1 (Kg.m^2)	2.110	2.087
d_2 (Kg.m^2)	0.223	0.216
d_3 (Kg.m^2)	0.085	0.084
b_1 (N.m.s/rad)		2.041
b_2 (N.m.s/rad)		0.242
b_{m1} (N.m.s/rad)		1.254
b_{m2} (N.m.s/rad)		0.119
k_1 (N.m/rad)	198.49	125.56
k_2 (N.m/rad)	51.11	31.27
I_{m1} (Kg.m^2)	0.1226	0.1224
I_{m2} (Kg.m^2)	0.017	0.0168

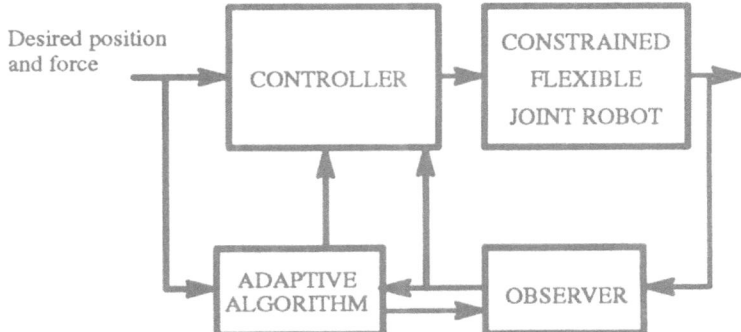

Fig. 1 Layout of the Adaptive Control Approach

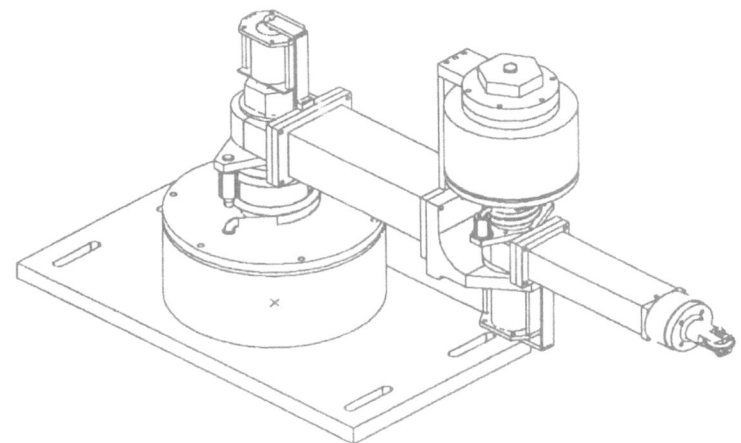

Fig. 2 Solid model of the experimental robot

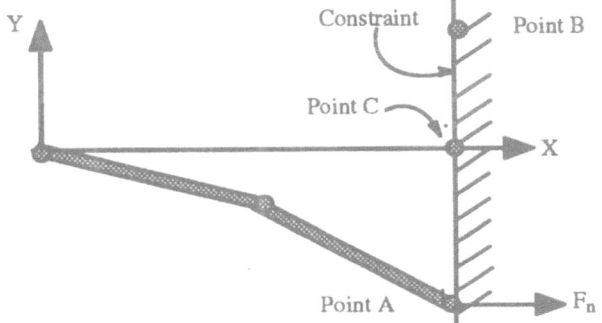

Fig. 3 Layout of the experimental robot

102

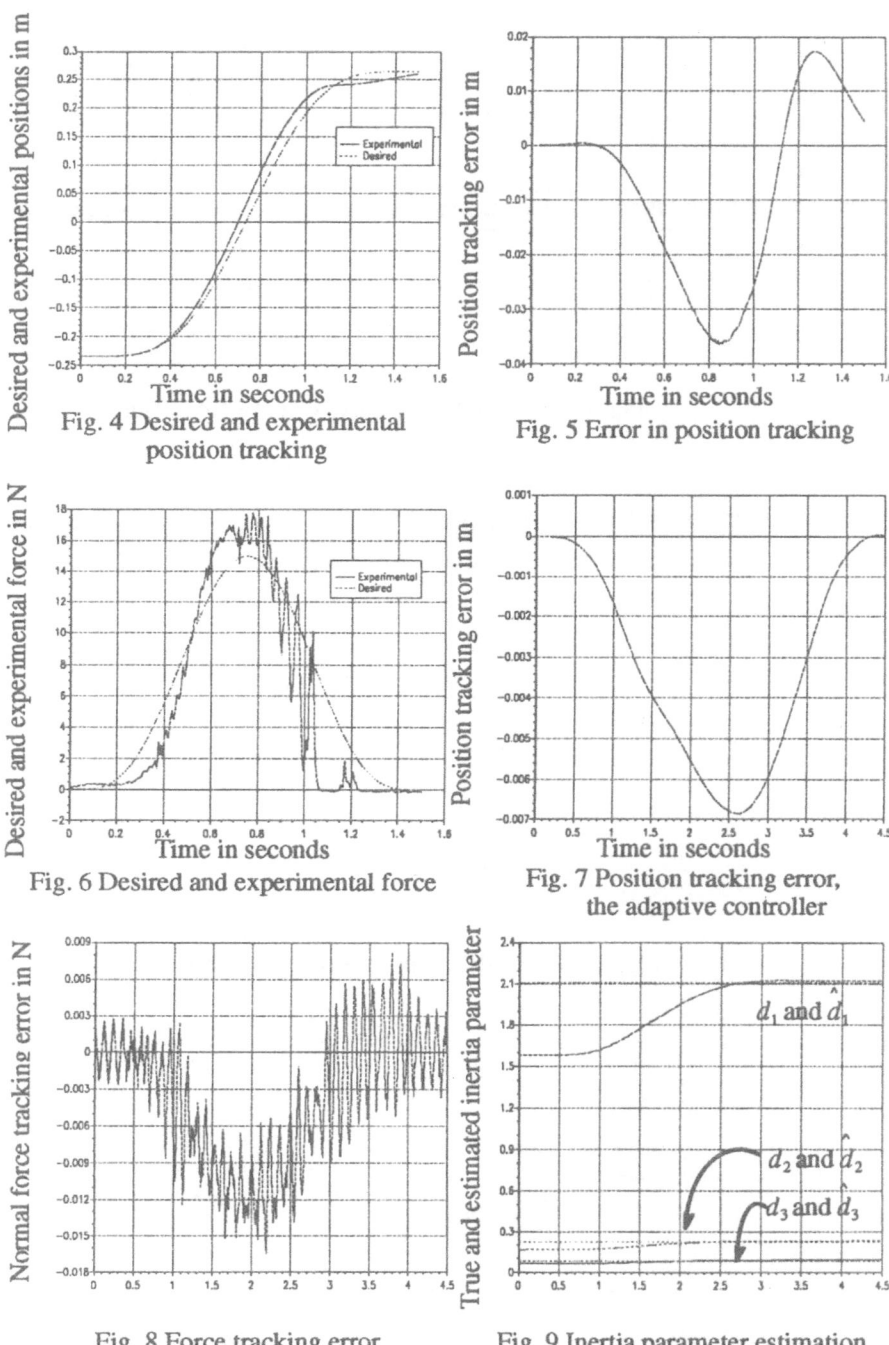

Fig. 4 Desired and experimental
position tracking

Fig. 5 Error in position tracking

Fig. 6 Desired and experimental force

Fig. 7 Position tracking error,
the adaptive controller

Fig. 8 Force tracking error,
the adaptive controller

Fig. 9 Inertia parameter estimation,
the adaptive controller

Experiments on Impedance Control to Derive Adaptive Strategies

Guillaume Morel and Philippe Bidaud

Laboratoire de Robotique de Paris
Centre Universitaire de Technologie
10-12 av. de l'Europe - 78140 Vélizy - FRANCE.
e-mail : morel@robo.jussieu.fr

Abstract— This paper presents some experimental results on impedance control applied to an industrial manipulator, while considering the constraints and perturbations that occur during a typical assembly task (unknown and varying environment parameters, impacts, friction, noises,...). An experimental and theorical analysis is developed first to derive a simplified scheme from the original Hogan's controller. The simplifications concern the dynamic compensation stage and the force feedback, which has been suppressed to deal with experimental conditions. To increase the performances of the resulting controller, a higher level is added to modify on-line the desired impedance and/or the reference trajectory. This supervisor has been developed using fuzzy logic. Experimental results on an industrial IBM Scara illustrate the ability of the system to absorb large external impacts and perturbations due to variations in the behavior of the manipulator and its environment.

1. Introduction

In recent years, assembly robotics has been the object of a large number of research progams and industrial developments. These applications, involving contacts between end effector and the external environment, require the control of the interaction wrench. To cope with this problem, two methods have been proposed : impedance control [1] and hybrid force/position (F/P) control [2]. Both are the object of numerous implementations [3],[4],[5]. Their use for complex contact tasks, involving contacts with a partially unknown, nonlinear, and dynamically variable environment, as those encountered for instance in a multi-component assembly system or two hands assembly structures, implies to increase their efficiency.

More precisely, the following difficulties have to be mastered :

- variations of the mechanical behavior of the different manipulated components and the manipulation structure, changes in the nature of constraints during fine motion phases,

- impact dynamics : when the contact occurs between parts to be mated, large impulses can be produced. The dynamical characteristics change (the inertia properties are drastically modified when the manipulator comes in contact), and a simple switch between motion control and force control creates unstabilities in the controlled system [6],

- friction which can induce jamming, stick-slip or wedging and cause part damages or task failures,

- uncertainties on contact model and measurement noises.

The developed controller is based on an impedance control scheme. Impedance control provides a unified framework for controlling a manipulator both in free motions and contact motions. Also, recent developments have corrected the inherent difficulties in impedance control for task specification : Lasky and Hsia [7] proposed to add a trajectory modifying controller in the outer-loop for force tracking; a similar scheme with adaptive properties has been proposed by Seraji and Colbaugh in [8]; the selection matrices of the hybrid F/P scheme, which allow the user to easily describe task constraints, have been introduced within an impedance scheme by Anderson and Spong [9]. However, very few works experimenting this technique for complex tasks have been presented yet. In the next section, we analyse the impedance controller structure and propose some simplifications. Experimental results are given for both free motion control and force control for elementary tasks. Considering more difficult conditions, such as uncertainties on environment behavior or high speed impacts, we develop in section 3, basic principles of a supervisor for on-line adaptation of the controller parameters.

2. Experimental analysis of impedance control

2.1. Hogan's Controller

The basic principle of impedance control is to program the dynamic relationship between end-effector velocity $\dot{x}$ and interaction force F_a:

$$- F_a = M_d.\ddot{x} + B_d.(\dot{x} - \dot{x}_r) + K_d.(x - x_r) \qquad (1)$$

where x_r is the reference trajectory and M_d, K_d and B_d are respectively the desired inertia, stiffness and damping matrices.

Then, since the desired behavior is naturally choosen to be uncoupled (and isotropic for free motions), the dynamics of the manipulator has to be compensated in the cartesian space. The cartesian space dynamics can be expressed from the joint space dynamics :

$$\tau_c - J^{-t}(q).F_a = H(q).\ddot{q} + b(q, \dot{q}) + g(q) \qquad (2)$$

where q is the joint configuration, $J(q)$ is the Jacobian matrix, τ_c is the command torque vector, $H(q)$ is the joint space kinetic energy matrix, $b(q, \dot{q})$ is the vector of Coriolis and centrifugal torques, and $g(q)$ the gravity torque vector. Using the transformation relationships : $F_c = J^{-t}(q).\tau_c$ and $\dot{x} = J(q).\dot{q}$, one obtains the *operational space* dynamics :

$$F_c - F_a = \Lambda(q)\ddot{x} + \mu(q, \dot{q}) + p(q) \qquad (3)$$

where $\Lambda(q) = J^{-t}(q).H(q).J^{-1}(q)$ is the operational space kinetic energy matrix, $\mu(q, \dot{q}) = J^{-t}(q).b(q, \dot{q}) - \Lambda(q).\dot{J}(q).\dot{q}$ is the vector of Coriolis and centrifugal forces, and $p(q) = J^{-t}(q).g(q)$ is the gravity force. To achieve the desired behavior described by eq. (1), the control law developed by Hogan is :

$$F_c = \tilde{\Lambda}.M_d^{-1}.\{B_d.(\dot{x}_r - \dot{x}) + K_d.(x_r - x)\} + \tilde{\mu} + \tilde{p} + (1 - \tilde{\Lambda}.M_d^{-1})F_a \qquad (4)$$

where ~ designates estimated values. This control law realizes a dynamic decoupling (the three first terms) plus a force feedback term. For free motions, when $F_a = 0$, the controller is equivalent to the operational space motion controller developed by Khatib [10].

2.2. Simplifiation of dynamics decoupling

The complexity of the control law (4) is often an obstacle for its implementation. In order to reduce it, a usual way is to not consider the dynamic compensation stage. So, a simple Jacobian transpose scheme is sometimes suggested to realize an impedance controller :

$$\tau_c = J^t.\{B_d.(\dot{x}_r - \dot{x}) + K_d.(x_r - x)\} \qquad (5)$$

However, An & Hollerbach [11] have shown clearly the role of the dynamic compensation in cartesian force control. They demonstrated *not only that using a dynamic model leads to more accurate control, but also that not using this model can in certain cases make force control unstable*. To evaluate more precisely the role of dynamics in the control law, we have proceeded in two steps :

- first, we experimentally compared the free trajectory tracking performances of the Hogan's impedance controller (4) (without force feedback for free motions) to a J^t scheme (5),

- secondly, we analysed the influence of the different terms of the dynamic model to evaluate those we can neglect.

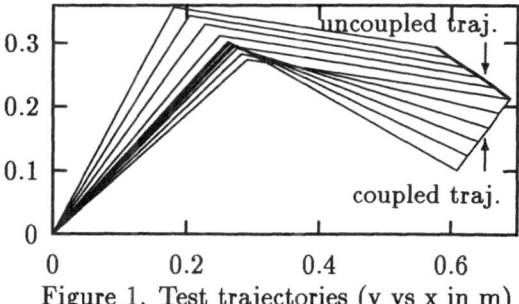

Figure 1. Test trajectories (y vs x in m)

We have tested the performances of different controllers for two different trajectories (in the x-y plane) on a 7576 IBM Scara (fig. 1). The first one is an *uncoupled trajectory*, i.e. the non-diagonal terms of $\Lambda(q)$ are closed to zero. In addition the variation of the diagonal terms during the motion is quite small (about $+/-$ 15%). For the second trajectory, the non-diagonal terms of the kinetic energy matrix vary from zero to quite large values so that the system becomes highly coupled. The two trajectories used here are perpendicular straight lines. The maximum velocity is 0.15 m/s and the acceleration is 0.5 m/s^2. The two controllers have been designed for those experiments to provide a 5 Hz bandwidht and a 0.8 damping coefficient. The sampling frequency is 200 Hz for both of them. We give in appendix A the dynamic model that we used to design the impedance controller and to determine the trajectories.

Results are given in fig. 2-a for uncoupled trajectory. Since the coupling terms are negligible, and the gains of the J^t scheme have been tuned from the local values of $\Lambda(q)$, the results of the two controllers are very similar. Joint friction is not considered here, and there is no integration in the correctors; that makes a small error remains. However, results for coupled trajectory (fig. 2-b)

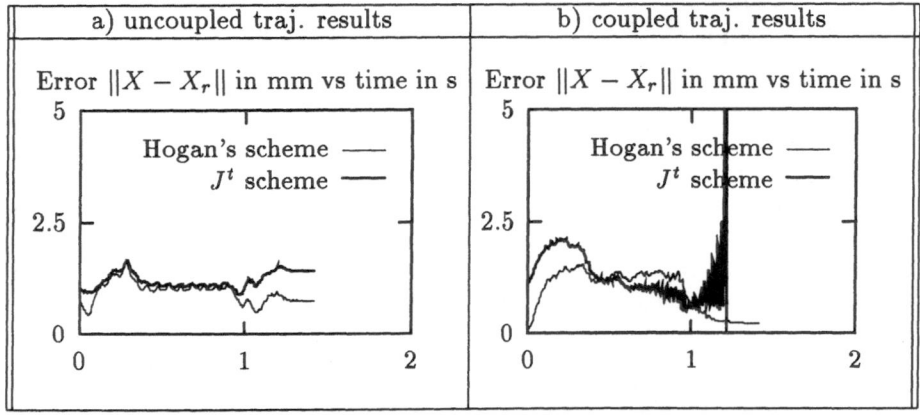

Figure 2. Results for free trajectory tracking

show clearly that the J^t scheme is unable to control the end-effector motion for coupled configurations (it becomes unstable). Vibrations can be suppressed by reducing the stiffness and overdamping the control law, but then, the trajectory tracking performances are degraded.

These experimental results can be interpreted by the observation of $\Lambda(q)$ terms which are strongly variable in the workspace. In addition, in certain configurations, non diagonal terms can be plainly greater than the diagonal terms. To improve this analysis, we have ploted in fig. 3, the first term of both $H(q)$ and $\Lambda(q)$, for the same range of variations of q. One can see that the ratio between maximum and minimum value is less than 2 for $H_{11}(q)$ and greater than 120 for $\Lambda_{11}(q)$. This explain the poor performances of the J^t law compared to a regular joint P.D. controller, and justify the necessity of a dynamic model in cartesian force control for this type of kinematics.

Nevertheless, the dynamic compensation stage can be simplified on the basis of the two following points : (i) usually, velocities during an assembly task are small enough (rarely much more than few cm/s) so that we can neglect Cori-

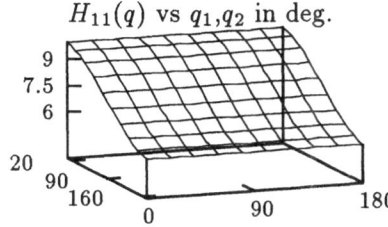
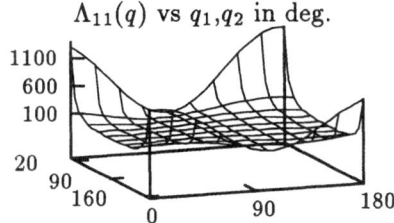

Figure 3. IBM 7576 dynamics plot

olis and centrifugal forces; (ii) furthermore, the previous evaluation of inertia matrices suggests that the nonlinear variations of $\Lambda(q)$ essentially come from inverse jacobian terms, so that we propose to use as an estimation :

$$\tilde{\Lambda}(q) = J^{-t}(q).H^0.J^{-1}(q) \qquad (6)$$

where $H^0 = \mathrm{diag}(H(q^0))$ is a constant diagonal matrix computed for $q_2 = 90°$ which is the maximum manipulability configuration. Fig. 4 represents

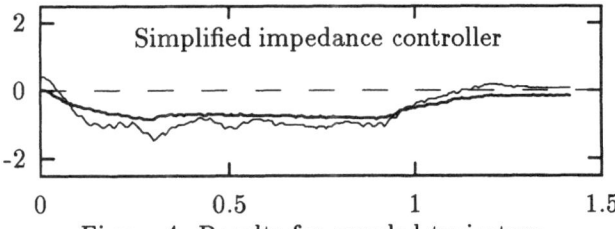

Figure 4. Results for coupled trajectory

the performances obtained for the coupled test trajectory and shows that the simplified impedance controller is sufficient for free motions control. Notice that a similar technique with a diagonal mass matrix has been successfully experimented recently [12].

2.3. The effect of force feedback

When the end-effector becomes in contact with the environment, the force feedback is active. Hogan discussed on the effect of this force feedback in [5]. He indicated that the role of this term is to change apparent inertia when the manipulator is in contact, and experimentally proved on an apparatus Scara its efficiency. However, we won't show here any results on the impedance control with the force feedback because we didn't succeed in making it stable. The reason of this check can be given using a 1 d.o.f. example. Let's consider a mass M in contact with its environment. In this case the control law is :

$$F_c = \frac{M}{M_d}(B_d(\dot{x}_r - \dot{x}) + K_d(x_r - x)) + (1 - \frac{M}{M_d}).F_s$$

$$\text{or}: F_c = \frac{M}{M_d}(B_d(\dot{x}_r - \dot{x}) + K_d(x_r - x) - F_s) + F_s$$

where F_s is the measured force. A scheme of this impedance controller is represented in fig. 5, where the principle of the force feedback action appears

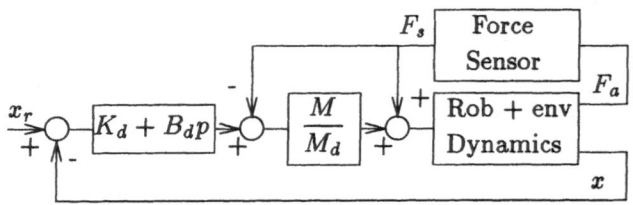

Figure 5. Impedance control of a 1 d.o.f. robot

clearly. The force measurement is used to directly compensate the force exerted by the environment to the end-effector and to be reintroduced at the output of the position corrector. In fact, the output of the position corrector can be considered as a force reference and $\frac{M}{M_d}$ is a dimensionless proportionnal gain which multiply the force error. Then, the relative values of M and M_d have a critical role in the closed loop behavior :

- if $M = M_d$, the force feedback is inoperative (i.e. if we don't want to change the apparent inertia, we don't have to use the force feedback)

- if $M \gg M_d$, the system is unstable

- if $M \ll M_d$, the system is uncontrolled. Indeed, the command force can be written as :

$$F_c = \frac{M}{M_d}.\varepsilon_F + F_s \tag{7}$$

where ε_F is the force error. Since $\frac{M}{M_d}$ is very small, one can approximate $F_c \sim F_s$ i.e. the major part of the command is the direct force compensation. In addition, force measurement is usually noised, so that the contoller's very unlikely to succeed, and will provide unstability.

In conclusion, the choice of M_d depends on the value of M (we have to set M_d to a few times M). Considering that, for a multi d.o.f. system, $\Lambda(q)$ is highly variable in the workspace, the M_d matrix can not respect everwhere the previous condition to keep the system stable. This is the reason why we did not use the force feedback in our impedance controller. So, there is no particular condition on the choice of M_d, which has been arbitrarily set to unity.
Finally, the implemented control law is :

$$F_c = \tilde{\Lambda}(q)\{B_d.(\dot{x}_r - \dot{x}) + K_d.(x_r - x)\}$$
$$\text{or}: \quad \tau_c = H^0 J^{-1}\{B_d.(\dot{x}_r - \dot{x}) + K_d.(x_r - x)\} \tag{8}$$

This control law does not provide the desired behavior described by eq. (1). Indeed, if we neglect the Coriolis and centrifugal forces, and assuming that $\tilde{\Lambda}(q) \sim \Lambda(q)$, the closed loop dynamics of the manipulator becomes:

$$-\Lambda^{-1}(q).F_a = \ddot{x} + B_d.(\dot{x} - \dot{x}_r) + K_d.(x - x_r) \tag{9}$$

This means that the environment behavior looks like coupled from the end-effector point of view. In addition, this *virtual* environment behavior depends on joint configuration. A solution to this problem will be developed in section 3 by implementing adaptive strategies to modify B_d and K_d.

2.4. Influence of environment behavior

Applying the control law (9) to the previous one-dimensional problem, the closed loop behavior can be characterized by the natural frequency ω_c and the damping coefficient ζ :

$$\omega_c = \sqrt{k_d + \frac{k_e}{M}} \tag{10}$$

$$\zeta = \frac{B_d}{2} \cdot \sqrt{\frac{M}{M.K_d + k_e}} \tag{11}$$

where the environment has been modelized as a pure stiffness k_e. Since the interaction dynamics is drastically modified between free and constrained motions (k_e turns from 0 to a quite large value), desired impedance has to be modified. Practically, for free motions control, the desired stiffness has to be maximized while respecting the stability constraints to minimize position errors, when for constrained motions, the desired stiffness must be decreased in the normal direction to keep stability. To cope with this, for a three dimensional case, the desired impedance must be decomposed in two subspaces with respect to the task-frame [9]. The resulting impedance parameters in a reference frame K_d and B_d are given by :

$$\begin{aligned} B_d &= b_{d_n}.\Omega + b_{d_t}.\check{\Omega} \\ K_d &= k_{d_n}.\Omega + k_{d_t}.\check{\Omega} \\ \Omega &= ({}_0^t R)^t .S.{}_0^t R \\ \check{\Omega} &= ({}_0^t R)^t .(1 - S).{}_0^t R \end{aligned} \tag{12}$$

where S is the selection matrix used in the hybrid force/position control theory [2], ${}_0^t R$ is the rotation matrix between fixed frame and task frame, b_{d_n} and k_{d_n} are scalar parameters in constrained directions (i.e. which have to be choosen according to the environment stiffness) and b_{d_t} and k_{d_t} are also scalars which can be chosen to take friction into account if necessary. Ω and $\check{\Omega}$ are the *generalized task specification matrices* previously defined by Khatib [13].

2.5. Contact experiments

Lasky *et al.* [7], and more recently Seraji *et al.* [8] have developed a method to provide the force tracking ability to an impedance controller. The principle is to add an external loop when the manipulator is in contact with its environmnent, which modify the reference trajectory, along the normal, depending on a force error signal :

$$\dot{x}_{r_n} = f(\varepsilon_{F_n}) \tag{13}$$

The more simple way to provide this kind of behavior is to implement a proportionnal corrector. Since force measurement signal is noised, and to avoid delays introduced by force filtering, we use fuzzy logic to implement this corrector. Force error is normalized with respect to the desired force F_d. The fuzzy rules are given simply by :

$$\text{If } \hat{\varepsilon}_{F_n} \text{ is } \mathcal{X}, \text{ Then } \hat{\dot{x}}_{r_n} \text{ is } \mathcal{X} \tag{14}$$

where ^indicates the fuzzy value, and $\mathcal{X}$ is Negative Large (NL), Negative Small (NS), Zero (Z), Positive Small (PS) or Positive Large (PL). Fuzzy members of

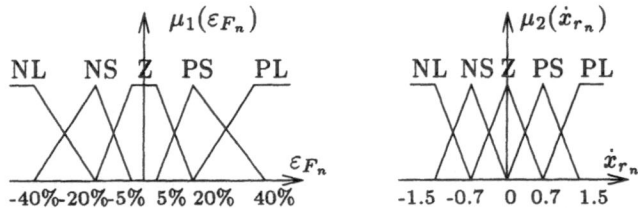

Figure 6. Fuzzy members for the external force loop

the force error signal $\varepsilon_{F_n} = (F_d - F_n)/F_d$ and the velocity output are given in fig. 6.

A set of experiments have been done to improve the force tracking performances. The manipulator contacts its environment with a normal trajectory. The environment stiffness has been measured to $k_e \sim 15000 N/m$. When the contact is detected (for a normal force greater than 3 N to deal with noises), normal gains b_{d_n} and k_{d_n} are switched from their initial values to the contact values which have been determined from the evaluation of k_e and experimentally refined. In addition, the external fuzzy force loop is switched on. The controller have been successfully experimented when the approach velocity is small enough (fig. 7 [1]), even if a force error due to the membership profile

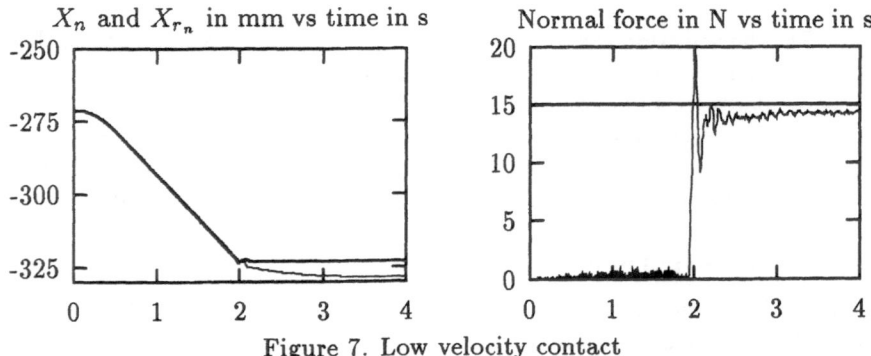

Figure 7. Low velocity contact

$\mu_1(\varepsilon_{F_n})$ (an error less than 5% is fuzzified as purely Z) remains.
However, when the end-effector becomes in contact with its environment, the

---

[1]See appendix B for experimental data

produced impulses depend on the end-effector velocity : in the normal direction, the impulse P^n is equal to the linear momentum component jump along the normal. Denoting the equivalent end-effector mass as $m_{eq} = n^t.\Lambda(q).n$, we can write :

$$P^n = \int_{t_1}^{t_2} F_a^n(t)dt = m_{eq}.\dot{x}_{r_n}(t_1) - m_{eq}.\dot{x}_{r_n}(t_2) \tag{15}$$

where t_1 and t_2 are the time of the begin and the end of impact. Clearly, if the velocity is too large, or for a configuration such that m_{eq} is big, the impulse will be large and make naturally the end-effector to rebound. Furthermore, the effect of the external loop is disastrous, as illustrated in fig. 8 [2]: during the first impact, desired force is overpassed; then the external loop reacts with a back motion, and since the time response is significative with regard to impact duration, this amplify the restitution phenomena and increase the risk of rebound. In fact, impact impulses are very brief, so that the conventional way

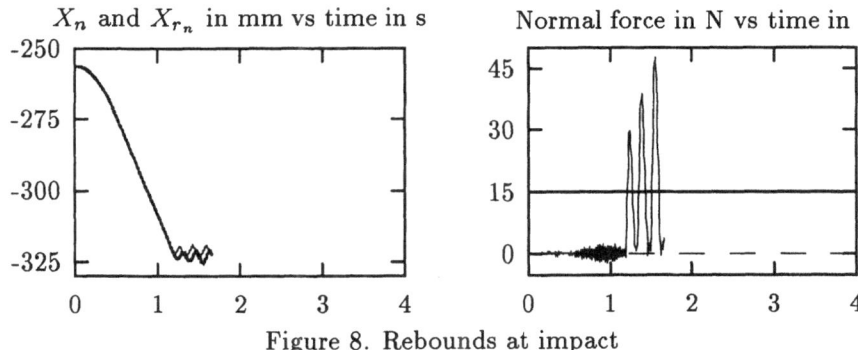

Figure 8. Rebounds at impact

we have used to control the task is not sufficient. A specific strategy has to be developed. Note that a PD controller rather than a simple P controller for the external loop would allow to increase the maximum velocity for which the contact is stable. But the obtained performances do not consist a significant progress for the impact control.

3. Implementation of a fuzzy supervisor

3.1. Motivations and principle

In the previous section, we have shown that both the impedance and the reference velocity have to be adapted with respect to the environment dynamics (eq. (10-11)), manipulator's dynamics (eq. (9)) and perturbations (fig. 8). It seems natural to propose a global scheme for the control of forces and motions, by adding a supervisor as illustrated in fig. 9. The supervisor is used on-line to tune the gains and compute the reference velocity, from informations on state variables and task specification (x_d, F_d). Desired impedance is computed in the task frame and transformed using eq. (12). For the experiments presented in

---

[2]see Appendix B for experimental data

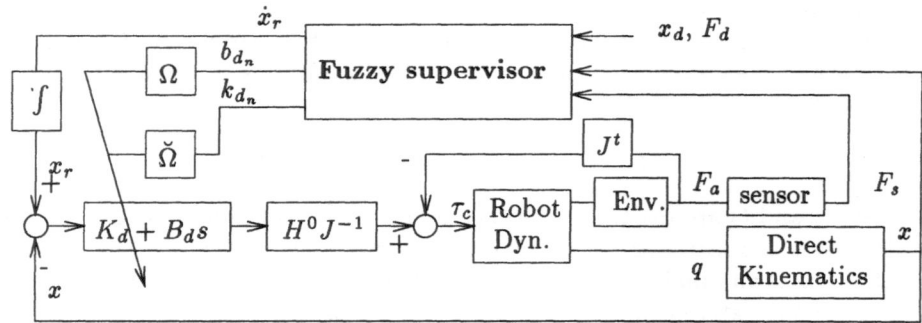

Figure 9. Structure of the adaptive impedance controller

this paper, which concern normal impacts control, friction won't be considered. Then, in the tangential directions, we set fixed values for : $k_{d_t} = k_d$, $b_{d_t} = b_d$, and $\dot{x}_{r_t} = \dot{x}_d - n^t.\dot{x}_d$, where n is the unit normal oriented from end-effector to environment. In the normal direction, k_{d_n}, b_{d_n}, and $\dot{x}_{r_n}$ are continously adjusted.

If we consider realistic experimental conditions, we have to take account of:

- Real time constraints : during impact, the reaction has to be very fast, so that the time computation of the supervisor must be as short as possible.

- Measurement noises : it is well known that informations from a force sensor are strongly noised.

- Uncertainties on environment behavior : the environment stiffness is generally not exactly known, and often variable in the space.

Since a complete analytic computation for the supervisor is not reasonable, we have prefered to explore the used of a reactive technique. The supervisor must deal with some very various situations during an assembly sequence, a fuzzy algorithm was prefered to the neuromimetic solution, which is generally used to learn a repetitive task. To compute the desired impedance, we implemented a multi-level fuzzy algorithm, using two kinds of rules :

- **Switching rules** which are used to deal with the problems due to transitions between free and constrained motions. This stage includes contact detection, impact detection and two groups of fuzzy rules : the first is used to switch between a set of nominal impedance parameters for contact mastering $(k_{d_n}^0, b_{d_n}^0)$ and the contactless values (k_d, b_d) and the second activates the set of rules to reference velocity.

- **Tuning rules** to on-line modify the nominal impedance parameters, which are initially off-line determined from an estimation of environment stiffness and manipulator's dynamics. This level also includes the fuzzy external force loop presented in section 2.5

3.2. Transition control

The objective of this section is to clarify the fuzzy rules used to switch the control parameters between free and constrained motions. To design the switching rules, we suppose to know two set of parameters : (k_d, b_d) which are the desired gains when the system is out of constraints and $(k_{d_n}^0, b_{d_n}^0)$ which are the desired values when the system is constrained. Results in fig. 7 show that if the approach velocity is small enough, a simple switch is sufficient. The only problem we encountered in this case is due to the contact detection : because of noises and inertial effects on the masses between the contact point and the force sensor, the measured force for free motion is not zero. To deal with that,

Contact Detection

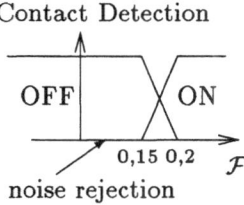

noise rejection

C D	α_k	α_b	α_x
OFF	1	1	1
ON	$\dfrac{k_{d_n}^0}{k_d}$	$\dfrac{b_{d_n}^0}{b_d}$	0

Figure 10. Detection rules

we construct a fuzzy contact detection (CD) signal (figure 10), from $\mathcal{F} = \frac{F_n}{F_d}$. CD is used to compute α_k, α_b and α_x from fuzzy table represented in figure 10, and the output of the supervisor are computed by :

$$
\begin{aligned}
k_{d_n}^{act} &= \alpha_k . k_d \\
b_{d_n}^{act} &= \alpha_b . b_d \\
\dot{x}_{r_n} &= \alpha_x \dot{x}_{d_n} + (1 - \alpha_x) f(\varepsilon_{F_n})
\end{aligned}
\tag{16}
$$

where $\dot{x}_{d_n}$ is the desired velocity for free motions and $f(\varepsilon_{F_n})$ is the external force loop.

Considering now problems due to "high velocity" collisions, a specific strategy for impact control must be implemented, which suppose first to detect impacts. An impact detection (ID) signal is constructed using the following rule :

> if $\quad \hat{dF}$ is PL
>
> or $\quad \hat{ID}$ is ON and ($\hat{\varepsilon}_{F_n}$ is PL or $\hat{dF}$ is not Z)
>
> then $\quad \hat{ID}$ is ON
>
> else $\quad \hat{ID}$ is OFF

where $\hat{dF}$ is the fuzzy value of the variation of force, computed from $dF = F_n(kT) - F_n((k-1)T)$, where T is the sampling period, using the membership represented in fig. 11. The first line is the detection of the impact begining, and the second line is used to detect the end of the impact.

Many strategies have been implemented to control impacts. Khatib and Burdick [10] experimented a method based on maximal damping during impact impulse. Volpe and Kholsa have discussed the role of the desired mass for the impact process control [14] and experimented a method equivalent to an impedance scheme with a large target mass and using the Hogan's force feed-

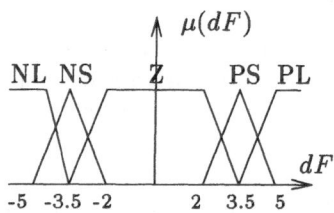

Figure 11. Fuzzy membership of force variation

back as a feedforward term. The developed method is inspired by those previous works. Both gains and reference velocity are adjusted. To maximize damping, we use directly the ID signal :

$$k_{d_n}^{act} = (1 - 0.8ID)\alpha_k.k_d \qquad (17)$$

$$b_{d_n}^{act} = (1 + ID)\alpha_b.b_d \qquad (18)$$

such that the stiffness is deacreased and damping coefficient is increased during impact. To compute the reference velocity, we exploit the fuzzy table 1. Since, impact impulse is very short and can not be avoided, the goal of this table is to prepare the end of impact rather than to track the desired force. Thus, during

$\hat{\dot{x}}_{r_n}$		$\hat{dF}$				
		NL	NS	Z	PS	PL
	NL	PS	Z	Z	Z	NS
	NS	PS	PS	PS	PS	Z
$\hat{\varepsilon}_{F_n}$	Z	PL	PS	PS	PS	PS
	PS	PL	PL	PL	PS	PS
	PL	PL	PL	PL	PL	PS

Table 1. Fuzzy rules for the reference velocity computation during impact

compression, desired penetration is augmented to prepare the restitution phase. We experimented such a strategy for different configurations and environments. The maximum impact velocity without rebounds is obviously increased. Fig. 12 shows the efficiency of the impact controller [3], for a velocity which is twice the one of the experiment which previously failed (fig. 8), and a maximum impulse force which is four times the previous one. After the impact phase, when $ID = 0$, the external force loop is switched on to achieve the desired force F_d.

3.3. On-line tuning

3.3.1. Providing a desired behavior

In the previous section, we have supposed that the set of contact gains ($k_{d_n}^0$, $b_{d_n}^0$) is *a priori* determined. However, those values have to be tuned with

---

[3]See appendix B for experimental data

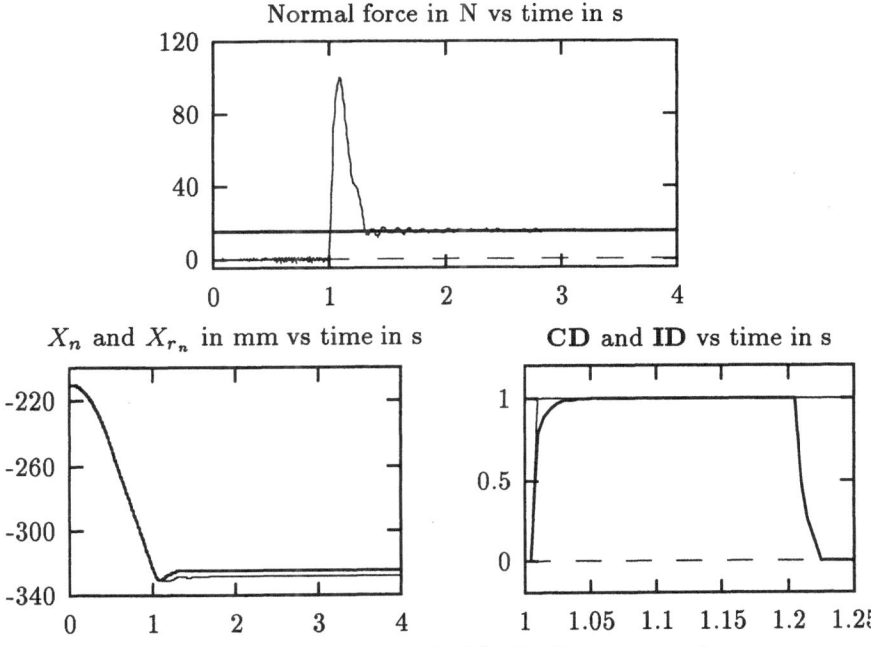

Figure 12. Impact control with the fuzzy zupervisor

respect to the environment behavior and manipulator's dynamics. More precisely, rewritting eq. (9) for the static equilibrium of a one-dimensional sample, we have :

$$M.k_{d_n}(x_{r_n} - x_n) = F_n \qquad (19)$$

This means that the actual stiffness of the end-effector is not k_{d_n} but $M.k_{d_n}$. In the same way, the actual damping is $M.b_{dn}$ rather than b_{dn}. So, k_{d_n} and b_{dn} have to be changed taking account on the manipulator's dynamics. Since $\Lambda(q)$ and the normal of the contact n are not exactly known, we have prefered to estimate the actual stiffness and damping rather than to relie on a model to compute it. k_a and b_a are estimated assuming a static equilibrium :

$$k_a = \frac{F_n}{x_{r_n} - x_n} \qquad (20)$$

$$b_a = \frac{k_a}{k_{d_n}}.b_{dn} \qquad (21)$$

To provide a desired behavior (k_{a_d}, b_{a_d}), we modify the gains $(k^0_{d_n}, b^0_{d_n})$ from the construction of stiffness and damping errors :

$$\varepsilon_k = \frac{k_a - k_{a_d}}{k_{a_d}} \qquad (22)$$

$$\varepsilon_b = \frac{b_a - b_{a_d}}{b_{a_d}} \qquad (23)$$

116

Two corrective signals δ_1 and δ_2 are obtained by fuzzy proportional laws similar to those used for the external force loop (see sec 2.5). Fuzzy membership of ε and δ are given in fig. 13, where ε represents ε_k or ε_b and δ is δ_1 or δ_2. The

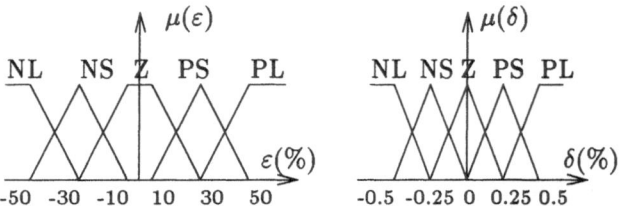

Figure 13. Fuzzy membership for the control of the actual stiffness and damping

clarified values of δ_1 and δ_2 are used to adapt the gains of the controller with :

$$k^0_{d_n}(kT) \quad = \quad (1+\delta_1).k^0_{d_n}((k-1)T) \qquad (24)$$
$$b^0_{d_n}(kT) \quad = \quad (1+\delta_2).b^0_{d_n}((k-1)T) \qquad (25)$$

3.3.2. Experimental results

To illustrate the adaptive properties of the controller, we experimented a wall following task : the objective is to follow the surface of a flexible beam embed-

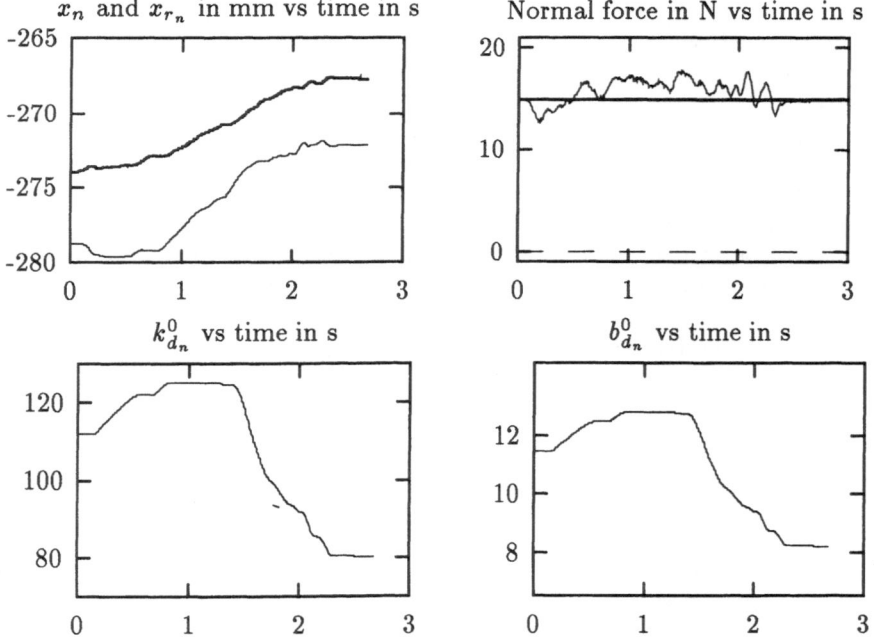

Figure 14. Adaptation results for a wall following task

ded at its two extremities, while the manipulator's dynamics changes during the motion. In the experiment represented in fig. 14, the range of variation of the diagonal terms of $\Lambda(q)$ is about 20%. The maximum tangential velocity is

0.07 m/s.

Results show that while the gains are modified during the motion, the actual stiffness (i.e. the ratio between F_n and $(x_{x_n} - x_n)$ is approximatively constant and closed to the desired value $k_{a_d} = 3000 N/m$. However, force tracking performances are affected during the motion. This is due to the variation of the stiffness of the environment (k_e turns from 4000 to 10000 N/m), which modify the closed loop behavior of the controller.

In fact, the choice of the desired stiffness and damping k_{a_d} and b_{a_d} has to be on-line modified to take into account the variations of the environment behavior. A solution to this problem, based on the principle given in the next section, is currently developed.

3.3.3. Tuning a desired behavior

We have seen in section 2.4 that the choice of the desired stiffness and damping depends on the environement behavior, such that a solution to tune the desired behavior can be based on an estimated value of the environment stiffness. However, this requires the exact knowledge of the nominal (unconstrained) environment position. Since k_a and b_a governs the stability properties of the controller, and the force tracking performances, we prefer to tune the gains

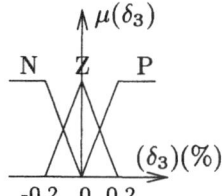

$\hat{\delta}_3$		$\hat{dF}$	
	N	Z	P
$\hat{\varepsilon}_{F_n}$ N	Z	N	N
Z	Z	Z	Z
P	P	P	Z

Figure 15. Rules to tune the actual stiffness

from an estimation of the performances of the controller rather than from an estimation of environment behavior.

The adaptation rules use the force error and the force variation as inputs. The fuzzy rules are inspired by a similar work done for a PI controller developed by Suh *et al.* [15]. Two corrective signals are fuzzy computed, δ_3 (fig. 15),

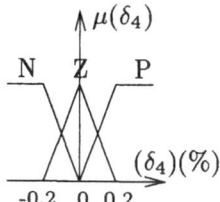

$\hat{\delta}_4$		$\hat{dF}$	
	N	Z	P
$\hat{\varepsilon}_{F_n}$ N	Z	N	P
Z	P	Z	P
P	P	N	Z

Figure 16. Rules to tune the actual damping

and δ_4 (fig. 16), which are respectively used to modify the desired stiffness and

damping coefficient, i.e. the ratio between k_{a_d} and b_{a_d}.
Finally :

$$k_{a_d}(kT) = (1 + \delta_3)(1 - \delta_4).k_{a_d}((k - 1)T) \tag{26}$$

$$b_{a_d}(kT) = (1 + \delta_3)(1 + \delta_4).b_{a_d}((k - 1)T) \tag{27}$$

4. Conclusion

In this paper, we have developed a global control scheme based on impedance control, and a fuzzy supervisor.

Preliminary experiments have shown the practical limitations of the original impedance controller. From those results, we have proposed a simplified impedance controller and determined the rules for the supervisor. Results on impact control and gains adaptation to compensate the manipulator's dynamic perturbations have been shown.

Future works concern the implementation of the adaptative algorithms presented in section 3.3.3 to deal with the variations of environment behavior. When they will be effective, we will have an efficient control tool to experiment complex assembly tasks on a two arms system which is currently developed in our lab.

Appendix A

A part of the dynamic model of the IBM 7576 Scara has been identified in [16]. We computed the diagonal terms from those results taking account on the force sensor and end-effector mass. We identified the motor inertia of the second axis to compute the non diagonal terms. Finally, we use for the kinetic energy matrix :

$$H(q) = \begin{pmatrix} 7.76 + 2.25\cos(q_2) & 1.07 + 0.94\cos(q_2) \\ 1.07 + 0.94\cos(q_2) & 1.91 \end{pmatrix}$$

$$J(q) = \begin{pmatrix} -0.4s_1 - 0.4s_{12} & -0.4s_{12} \\ 0.4c_1 + 0.4c_{12} & 0.4c_{12} \end{pmatrix}$$

$$\Lambda(q) = J^{-t}(q)H(q)J^{-1}(q)$$

where $s_1 = \sin(q_1)$, $s_{12} = \sin(q_1 + q_2)$, $c_1 = \cos(q_1)$ and $c_{12} = \cos(q_1 + q_2)$.

Appendix B

The following table gives the values of significative parameters for experiments reported in fig 9,10 and 13.

References

[1] N. Hogan. Impedance control : A new approach to manipulation. *ASME J. of Dyn. Sys. Meas. and Control*, 107:1–24, 1985.

[2] M.H. Raibert and J.J. Craig. Hybrid position/force control of manipulators. *ASME J. of Dyn. Sys. Meas. and Control*, 102:126–133, 1981.

Contact velocity	fig 8 : 0.03 m.s^{-1}
	fig 9 : 0.08 m.s^{-1}
	fig 13 : 0.15 m.s^{-1}
Contact stiffness	$\sim$15000 N.m^{-1}
Desired Force	15 N
Normal P gain $k_{d_n}^0$	79 N.m^{-1}
Normal D gain $b_{d_n}^0$	12.5 N.s.m^{-1}
Projected mass $(n^t.\Lambda.n)$	$\sim$ 46 kg

Table 2. Impact tasks parameters

[3] X. Delebarre. *Commandes en Position et Force de Deux Bras Manipulateurs pour l'Exploration Planétaire*. PhD thesis, Université de Montpellier II, 1992.

[4] J.E. Colgate. *The Control of Dynamically Interacting Systems*. PhD thesis, M.I.T., Dept. Mech. Ing., 1988.

[5] N. Hogan. Stable execution of contact tasks using impedance control. In *Int. Conf. on Rob. and Automation*, pages 1047–1054. IEEE, 1987.

[6] J.K. Mills. A generalized lyapunov approach to robotic manipulation stability during transition to and from contact tasks. In *Japan-USA Conf. on Flexible Automation*, pages 903–910. ISCIE, 1990.

[7] T.A. Lasky and T.C. Hsia. On force-tracking impedance control of robot manipulators. In *Int. Conf. on Rob. and Automation*, pages 274–280. IEEE, 1991.

[8] H. Seraji and R. Colbaugh. Adaptive force-based impedance control. In *Int. Conf. on Intelligent Robots and Systems*, pages 1537–1542. IEEE/RSJ, 1993.

[9] R.J. Anderson and M.W. Spong. Hybrid impedance control of robotics manipulators. *IEEE J. of Rob. and Automation*, 4(5):549–556, 1988.

[10] O. Khatib and J. Burdick. Motion and force control of robot manipulators. In *Int. Conf. on Rob. and Automation*, pages 1381–1386. IEEE, 1986.

[11] C.H. An and J. M. Hollerbach. The role of dynamic models in cartesian force control of manipulators. *The Int. Jal of Robotics Research*, 8:54–72, 1989.

[12] G. Alici and R.W. Daniel. Experimental comparison of model-based robot position control strategies. In *Int. Conf. on Intelligent Robots and Systems*, pages 76–83. IEEE/RSJ, 1993.

[13] O. Khatib. A unified approach for motion and force control of robot manipulators : The operational space formulation. *IEEE J. of Rob. and Automation*, 3(1):45–53, 1987.

[14] R. Volpe and P. Khosla. Experimental verification of a strategy for impact control. In *Int. Conf. on Rob. and Automation*, pages 1854–1860. IEEE, 1991.

[15] I.H. Suh, J.H. Hong, S.R. Oh, and K.B. Kim. Fuzzy rule based position/force control of industrial manipulator. In *Int. Workshop on Intelligent Robots and Systems*, pages 1617–1622. IEEE, 1991.

[16] H. Gaudin. *Contribution à l'identification in situ des constantes d'inertie et des lois de frottements articulaires en vue d'une application expérimentale au suivi de trajectoires optimales*. PhD thesis, Université de Poitiers, 1992.

Segmentation and Interpretation
of
Temporal Contact Signals

Brian S. Eberman and J. Kenneth Salisbury

MIT Artificial Intelligence Laboratory
Cambridge MA 02139

Abstract— The temporal structure of the force, or strain, signals from an internal force-torque sensor can be a rich source of information about robot/environment contact conditions. We present a procedure for representing and segmenting the individual signals with an auto-regressive model. This procedure produces segments of approximately constant spectrum, and is a powerful technique for detecting abrupt changes in noisy signals without prior calibration.

1. Introduction

It has long been recognized that manipulation is a problem involving discontinuous changes in contact conditions. The slot of a screw interacting with a screwdriver, the detent in a switch, the placing of an object on a table all have different contact conditions depending on which surfaces are mated. In addition, manipulating a grasped part will cause changing contact conditions at the grasp contact points.

Controlling the contact conditions requires the ability to recognize a given contact state, and recognition requires an adequate representation of the measured signals. Some interactions can be modeled by the constraints imposed by rigid bodies with friction. However, in more general environments this representation always becomes computationally complex and is often not even sufficient. Consider the difficulty of modeling the contact forces created by grasping a pile of paper clips, by sweeping a hand through mud looking for a rock, by shearing a sheep, or even by just snapping on a pen cap.

Humans manage these tasks by integrating the spatial-temporal output of four different tactile sensory channels with kinesthetic channels, vision, hearing, and prior knowledge. Using an unknown mechanism and representation, information is extracted on at least: 1) contact locations, 2) contact forces, 3) local tangential motions at the contact (slip), 4) surface texture, 5) contact geometry, and 6) tool motion constraints.

This is a difficult problem which will require a variety of different sensors and techniques. We have started looking at the temporal strain and position signals generated by idealized tool interactions. If manipulation is performed through a stick (or tool), the stick encodes the spatial-temporal contact forces into a single temporal signal. Humans are able to extract a great deal of information about the nature of the stick's contact from this signal [1].

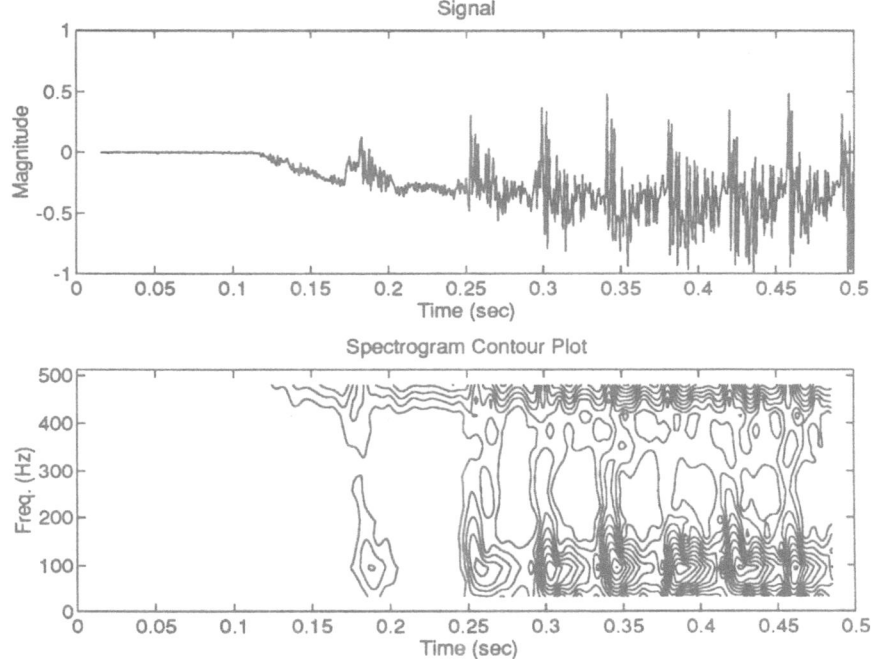

Figure 1. Spectrogram of a sequence of impact events. The figure shows a contour plot of the energy in frequencies from 32-480 Hz as a function of time. The signal was sampled at 1024 Hz. Thirty-two points windowed with a Hamming window were used for each Fast Fourier Transform (FFT). The FFT was computed for every new data point. Note the peaks that occur at each impact event and the short time scale of each event.

Our research is aimed at developing techniques for providing a robot with a force-torque sensor and position sensors the same interpretational capabilities. This should be possible, because, although the measurement channels are different the same information is being measured by both the robot and the human hand.

Our goal is to develop interpretation algorithms that:

1. Are based on the discontinuous nature of manipulation,

2. Are as free of calibration requirements as possible,

3. Are hierarchical where each level produces simplified features for the next highest level, and

4. Use geometry only at the highest level.

Because of the discontinuous way in which contacts are made and broken, the contact force signals consist of a sequence of regions separated by these contact

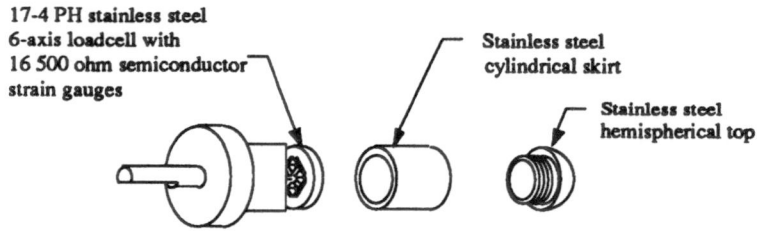

Figure 2. 6-axis fingertip force-torque sensor

events. Therefore, the algorithm should use these events to produce a sequence of force signal segments. Calibration is time consuming and can introduce systematic bias into procedures. By eliminating calibration, we hope to make the procedures easier to use and more robust. A multi-layered approach can be broken into many cooperating programs and processes for faster computational performance. Finally, the association of forces with geometry is computationally expensive in both cycles and memory. By using geometry only at the highest level, we hope to minimize the computational cost by providing higher level features than the raw force, and increasing the time between geometric calculations.

This paper reports on our experimental development of techniques for representing and segmenting scalar strain signal processes using an auto-regressive model. Vector processes are treated as independant scalar processes. Segmentation is performed with forms of the sequential likelihood ratio test, which is a statistical technique for testing for jump changes in stochastic processes. We concluded by discussing approaches to improve the techniques using estimated constraint surface models.

2. Intrinsic Contact Sensing

Intrinsic contact sensing is the measurement of information about the contact point from an internal, or intrinsic, force-torque sensor [2]. Our investigation used the sensor shown in figure 2. This sensor has a Maltese-cross connecting the outer shell to the base. The cross is instrumented with 8 strain-gauge half-bridges. The shell has a lightly damped natural frequency of approximately 700 Hz when the base is fixed and the shell free.

A tremendous amount of information, besides just the contact force, is available in the strain signals measured by such a force/torque sensor. Figure 1 shows a spectrogram from a single strain signal caused by a sequence of impacts. Each impact results in an energy at all frequencies locally around the event, and a persistent residual vibration at the sensor's natural frequency.

This information is carried not only in the values of the strains at a point in time, but also in the temporal structure or correlation of the strains through time. We are interested in capturing the temporal structure of the signals and detecting jump changes in that structure in order to detect events.

3. Scalar Signals

The correlation structure of a scalar (or vector) signal can be captured by modeling it as a autoregressive (AR) process. The scalar autoregressive model assumes the current strain signal $y(t)$ is generated by

$$y(t) = \sum_{i=0}^{n} a_i y(t-i) + v(t). \tag{1}$$

where $v(t)$ is a white noise process with unknown covariance Q, and the a_i are the lag autoregressive coefficients. When the contact conditions change, the parameters of this model are assumed to discontinuously jump.

An example of this type of model jump is shown in figure 3. In this figure, the sensor was dragged at constant velocity across a smooth plastic surface followed by a writing paper surface.

3.1. Automatic Segmentation

Segmentation of scalar signals based on the spectrum has a wide variety of applications and several techniques have been developed. Our technique is based on work on failure detection and speech segmentation [3, 4, 5]. We use a fast square-root algorithm for recursively computing the reflection coefficients of the process [6]. Then, our algorithm chooses the optimal model order using a minimum description length (MDL) length penalty [7]. Finally, sequential segmentation and model resetting is performed with the sequential likelihood ratio test [3].

The coefficients a_i fit an all pole spectrum to the data used in the estimation. Therefore, by using an appropriate segmentation procedure, a sequence of AR models can be generated which captures the spectrum over a window where the spectrum is approximately constant.

3.1.1. Fast Square-Root Algorithm and Model Order

Linear models, like equation 1 can be written in the form

$$y(t) = \Psi^T(t)\theta + v(t)$$

where $\Psi(t)$, a column vector, is formed from the lagged values of $y(t)$. In the case of equation 1 $\Psi^T(t) = [y(t-1), ..., y(t-n)]$ and $\theta^T = [a_1, ..., a_n]$. The least squares estimate at time t, $\hat{\theta}(t)$, is computed by sequentially solving

$$
\begin{aligned}
I(t) &= I(t-1) + \Psi(t)\Psi^T(t) \\
X(t) &= X(t-1) + \Psi(t)y(t) \\
I(t)\hat{\theta}(t) &= X(t).
\end{aligned}
$$

The equations can also be rewritten to avoid the linear accumulation, but this obscures the procedure.

124

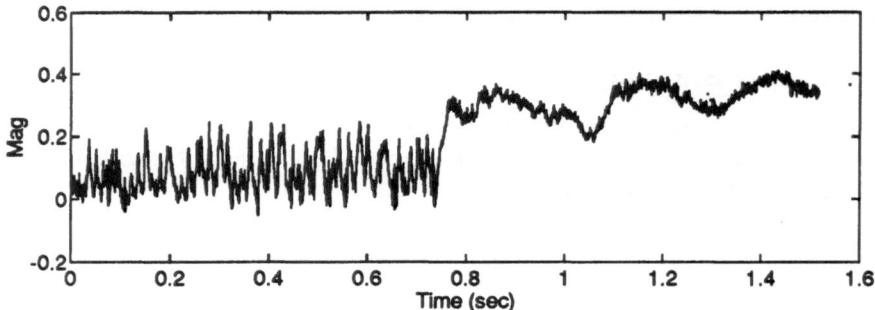

Figure 3. The figure shows one strain signal, sampled at 2700 Hz, that results from moving the probe across a relatively smooth plastic surface followed by a writing paper surface. The first texture was fit with sampled data poles at $[-.30 \pm 0.91i, 0.97, 0.79]$ at a variance of $2.2\,10^{-4}$ and the second was fit with poles at $[-0.30 \pm 0.88i, 1.0, 0.54]$ and a variance of $5.0\,10^{-5}$.

The calculations can be improved by generating an orthogonal set of regressors from $\Psi(t)$ and finding the parameters that correspond to this orthogonal set. These parameters are often called the reflection coefficients or linear predictor coefficients (LPC). The LPC's are computed by the following equations

$$R(t)R^T(t) = R(t-1)R^T(t-1) + \Psi(t)\Psi^T(t)$$
$$R(t)[\tilde{k}(t) \quad \bar{\Psi}(t)] = [X(t) \quad \Psi(t)]$$
$$k_j(t) = \tilde{k}_j(t)/R_{j,j}(t)$$

where $R(t)$ is the lower triangular square-root of $I(t)$ and $k(t)$ are the reflection coefficients. The first equation can be solved with order n^2 multiplies and n square-roots by applying a series of Householder transforms, which together form T, to solve

$$[R(t-1) \quad \Psi(t)]T = [R(t) \quad 0].$$

Ladder algorithms are slightly faster than this algorithm for AR models, but this method permits a general form for $\Psi(t)$.

Given the reflection coefficients $k(t)$, the optimal model order can be selected by minimizing the minimum description length over the number of parameters m

$$L_m(t) = -t\log(\hat{Q}_m) - b(m+1)\log(t) \tag{2}$$
where, $\tag{3}$
$$\hat{Q}_m(t) = \left[E(t) - \sum_{i=1}^{m} \tilde{k}_i^2(t)\right]/t \tag{4}$$

where $E(t) = E(t-1) + y(t)^2$ is the signal energy, and b is the parameter cost. The parameter cost is a design variable, we generally used a value of 2.4.

Asymptotically a value of 1 can be shown to give unbiased estimates of the optimal model size, however the estimated model order is relatively insensitive to b for small n. Finally, the residuals can be computed in the orthogonal frame via

$$\nu_j(t) = \nu_{j-1}(t) - \bar{\Psi}_j(t)\tilde{k}_j(t) \tag{5}$$

where $\nu_0(t) = y(t)$. The residuals are used for segmentation.

3.1.2. Segmentation

Sequential segmentation of both the AR model and the constraint surface models is performed using a sequential log-likelihood ratio test. This test can be shown to have the smallest delay to decision over all tests with a given error rate on a two hypothesis testing problem. If we assume that the error in the AR model is Gaussian, then the prediction residuals are white, zero-mean, Gaussian random variables with variance $\hat{Q}$. A statistical test which checks for changes in the variances of size $\alpha\hat{Q}$ was used.

Consider a hypothesis testing problem between two possible sources for an n dimensional stochastic process ν: $\mathcal{N}(0, Q)$ and $\mathcal{N}(0, \alpha Q)$ where $\mathcal{N}$ is the multivariate normal distribution. The log-likelihood ratio between these two hypotheses, for a sequence of measurements ν_r^t, is

$$
\begin{aligned}
l(\nu_r^t) &= \sum_{i=r}^{t} \log \mathrm{p}(\nu_i | \alpha Q) - \log \mathrm{p}(\nu_i | Q) \\
&= -\frac{1}{2}\sum_{i=1}^{t} n \log \alpha + \left(\frac{1}{\alpha} - 1\right)\nu(i)^T Q^{-1}\nu(i)
\end{aligned}
$$

The sequential log-likelihood ratio test maximizes this value over all values of r between 1 and t. This can be done by monitoring the test statistic $g(t)$ generated by the sequential rule

$$
\begin{aligned}
\gamma(t) &= -\frac{1}{2}\left(n \log \alpha + \left(\frac{1}{\alpha} - 1\right)\nu(t)^T Q^{-1}\nu(t)\right) \\
g(t) &= \max(0, g(t-1) + \gamma(t)).
\end{aligned}
$$

Under the no change hypothesis, $g(t)$ is a cumulative sum test with resetting at 0 since $\mathrm{E}(\gamma(t)) < 0$.

The last time g was 0 is the optimal change time. Changes are detected by thresholding g. Reasonable values of the threshold can be picked with a few experiments, or by noting that γ is a function of a chi-squared random variable. The fineness of the segmentation is controlled by the level of the decision parameter. This test is optimal if the two distributions for ν are known. Further details on this test and more powerful tests are given in [8].

In practice, we substitute the current value of $\hat{Q}$ for Q in the test and run two simultaneous tests. One test looks for changes of size αQ and the other test looks for changes of size Q/α. α is used as a design parameter. We used a value of 4 for α and 15 for the threshold in the AR segmentation.

3.2. Experimental Results

The experiments were carried out with the single degree-of-freedom device. An 8 inch glass-epoxy link connects a brushless DC motor, with an integral reflection torque sensor, to the sensor. The output torque from the motor is controlled to one part in 1000 using an integral controller. The motor's design and control is described in [9]. The glass-epoxy link provides a very stiff, well damped linkage. Test specimens can be placed under the sensor or on blocks at the end of its travel. Both the angle and force measurement have 12 bits of resolution.

To begin processing the signals, we applied a finite impulse response low-pass filter of width 32 and decimated the measurement at 2:1 to create a new measurement signal at 512 Hz. This eliminated most of the components of the signal that resulted from resonance of the sensor. We then experimented with different models for the measurement. We tried AR models which were extended with a mean, and with a slope and a mean. These additional terms caused difficulties. In some cases a first order AR process was more useful then a mean processes, and in other cases the mean was more useful. Since the model order selector cannot reorder the regressors, no one order for the regressors will work in both cases. There are two possible remedies.

First, additional models with different orderings can be run simultaneously and the optimal model can be chosen. This is the approach used in [8]. This has the advantage that a constant flat signal, which is generated in free space or while at rest, can be selected. Second, an eigenvalue decomposition of I can be performed to find the optimal regression directions. Unfortunately, both techniques are computationally expensive. Furthermore, during manipulation the mean only model will seldom be selected.

We also tried to use the difference between the lowpass signal and the raw measurement as an indicator of impacts. This high pass response does produce a distinct jump when impacts occur, but because of the sensor's poor damping the signal is blurred over a long time window. This makes time marking of the event difficult, especially when a sequence of impacts occurs.

The AR model does produce useful tests. In one experiment, the sensor was commanded to move at constant velocity over a polished aluminum plate, which had a single small scratch. A stylus was attached to the fingertip to decrease its effective size. Figure 4 shows the algorithm's response to the burr. All of the change information in this experiment is in the variance estimate. The burr produces signals which are improbable given the current estimate of the variance. This causes an accumulation in the decision statistic and after a few measurements a change is detected. The AR parameters for the low frequency signal are all at approximately 1.0 before and after the marked change caused by the burr.

The data shown in figures 1 and 5 was generated by commanding the sensor to move at constant velocity over a flattened hose-clamp. Each ridge in the hose-clamp produces a spike in the data. The AR model segmentation, detects a constant frequency spectrum consisting of poles at $[0.30 \pm 0.54i \quad -0.56]$. The

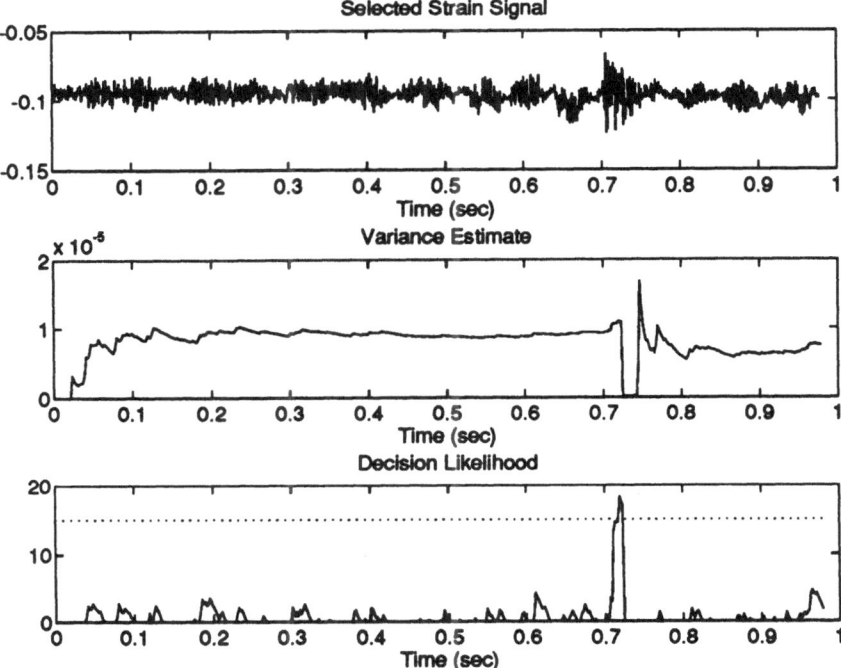

Figure 4. Burr Experiment. One of the eight strain signals is shown in the top figure. The burr occurs at approximately 0.75 seconds. The AR parameters for all the strains use a single AR parameters of just less than 1.0 (not shown). After regressing with this parameter, the variance estimate shown in the second figure is computed. The bottom figure shows the change likelihood and the threshold of 15. The beginning of the burr is marked correctly by the last time the likelihood was at zero. There is a lag of 2n steps after each detected change before a new variance estimate is produced.

additional reflection coefficients are introduced by the MDL model selection criteria which is computed for every new measurement. The change likelihood crosses the threshold at the beginning and end of the texture segment, because there is a marked change in the signal variance as the sensor enters and leaves the contact region.

Figures 6 and 7 shows a similar experiment with a roughness calibration plate. This experiment was performed by moving the sensor by hand and sampling with an effective (after decimation) rate of 1350 Hz. These two figures illustrate some of the problem sources for our experiments. First, even though the sensor was moved relatively slowly, the input signal has a frequency of approximately 40 Hz. This gives only 13 points per cycle at a 512 Hz sampling rate. Second, the signal is obscured by the senor's long vibration decay

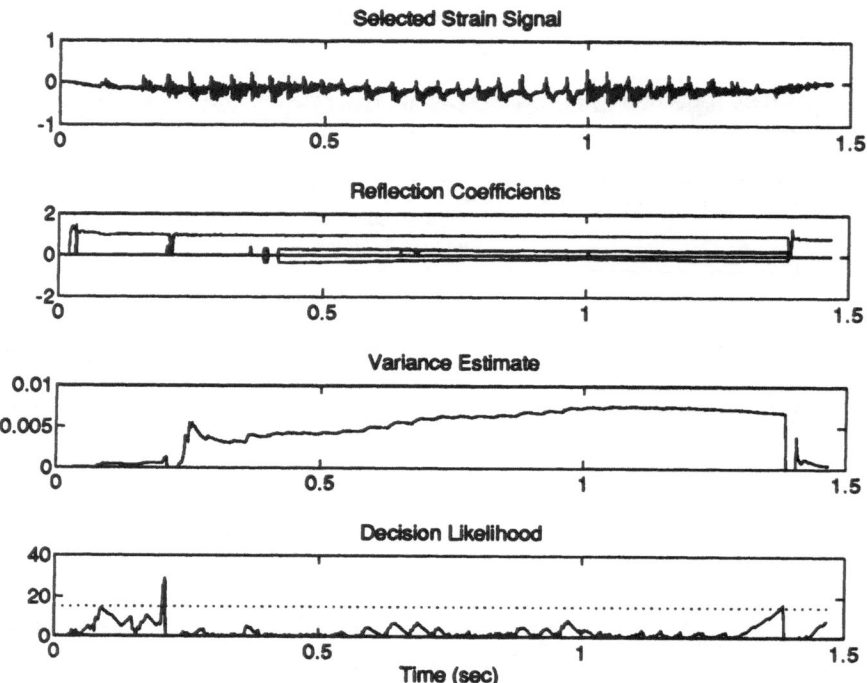

Figure 5. Texture Experiment. One of the eight strain signals is shown in the top figure for the same experiment illustrated in figure 1. The AR model is able to identify a constant frequency pattern starting at approximately 0.4 seconds represented by reflection coefficients [0.95, 0.21, -0.21, 0.23]. The likelihood test found three regions where the variance is significantly different (third figure). The bottom figure shows the change likelihood and the threshold of 15.

time. This is most evident in the second plot in figure 7. Therefore, a good system needs a high sampling rate (order 1-2 kHz) with well damped sensory characteristics across the entire frequency range.

4. Discussion

In general, the AR procedure was able to reliably pick out locations in the signal where the spectral shape or energy of the signal had appreciably changed, and is therefore a good representation for segmentation. This is a very useful feature in a layered interpretation scheme. Because it substantially reduces the work load on the higher layers by providing useful segments. The AR parameters are essential for this process, because they make the test significantly more sensitive to changes in energy by whitening the measurement process. For example, without a whitening scheme the burr signal would have been undetectable.

Whether the representation is rich enough to represent the myriad of differ-

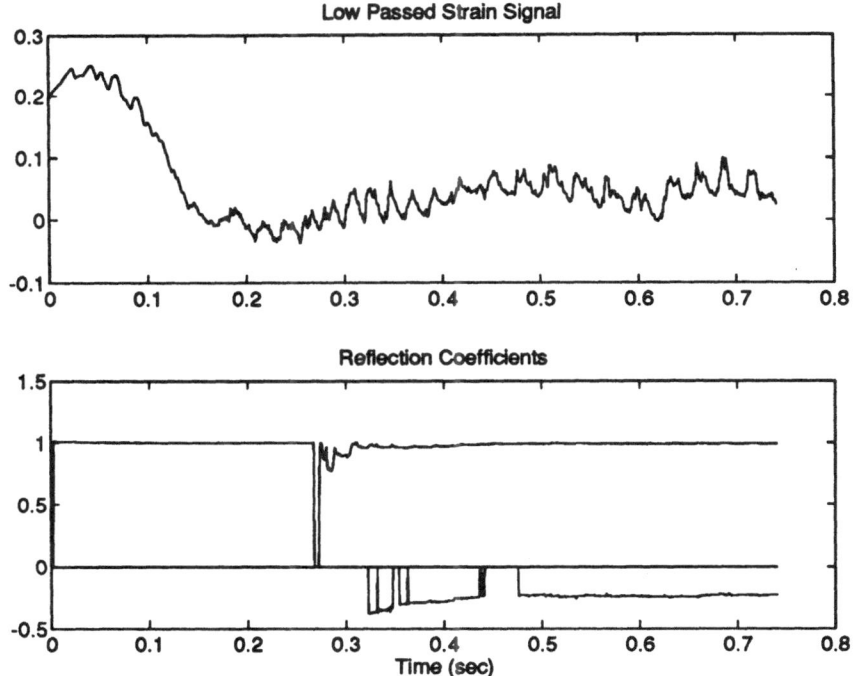

Figure 6. Calibration Plate Experiment. One of the eight strain signals is shown in the top figure. The sensor was moved, by hand, across a 12.5 μmRa turned surface roughness calibration plate. The plate contact beings at approximately 0.15 seconds. The AR model is able to identify a constant frequency pattern starting at around .32 seconds. The value of using an additional reflection coefficient changes depending upon the amount of noise in the signal.

ent signal forms that could occur during manipulation remains an open question. The procedure represents the power spectral density of a section of data. The length of the section is chosen based upon the data. However, the original signal is not recoverable from this representation because only its statistics are represented.

A second difficulty is that force interpretation, unlike vision or speech, has the unique attribute that the robot's actions directly effect the signal. The AR procedure models this by using a single reflection coefficient at or near 1.0. This essentially removes from the signal the very low frequency terms that correspond to the robot's motions. An alternative would be a quasi-static analysis of the strains. In theory, the low frequency term can be predicted given a model of the geometry, an estimate of the robot's location in configuration space, and measurements of the robot's applied torques. While this is a powerful method, it is very difficult computationally, and it requires a geometric model.

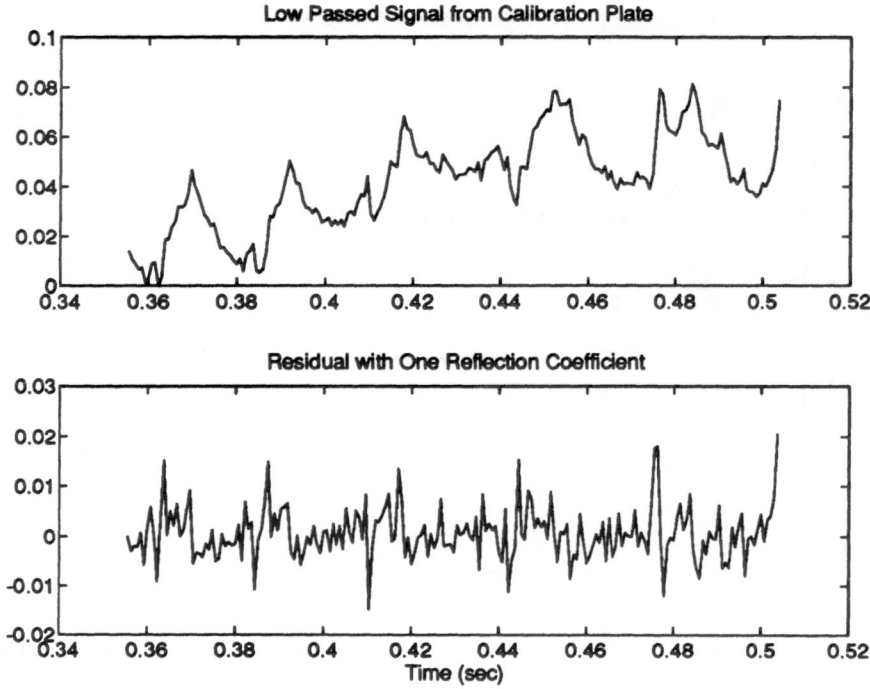

Figure 7. Calibration Plate Experiment. A section of the calibration plate strain signal and the residual after applying the first reflection coefficient is shown. The sinusoidal form of the signal is a result of the sinusoidal texture. The signal was sampled at 2700 Hz and decimated 2:1.

An alternative approach is to fit a regression relationship between the position (or desired position) of the robot and the strain measurements. This procedure fits a stiffness to a configuration space constraint surface patch. By performing the regression in a combined strain/position space the "stiffness" regression can be made non-singular and symmetric. Our preliminary experiments with this approach in 1 DOF are very promising.

5. Conclusion

The temporal structure of a contact signal provides a rich source of information about the contact conditions. An auto-regressive representation of the signals can be used to segment the signal into constant spectrum regions. This approach is useful for robustly detecting sudden changes in the signal without prior calibration, and for providing useful segments for higher level procedures.

Representing measured textures will require extending the representation to make it invariant with velocity. Adequate performance in measuring and recognizing texture patterns will require a sample rate of 1-2 kHz and a sensor with well damped performance characteristics.

This study is our first step toward building a layered algorithm for interpreting tactual data. Several other layers are required before an interpretation of the signals at the level of interacting parts is possible. In order for a layered approach to succeed the lowest layer must provide a large reduction in signal bandwidth. Our AR procedure can fulfill that function by segmenting the signal into interesting sections.

Acknowledgement

The authors would like to gratefully acknowledge the financial assistance of the Advanced Research Projects Agency of the Department of Defense under Office of Naval Research contracts N00014-92-J-1814 and N00014-91-J-4038; and the Office of Naval Research University Research Initiative Program under Office of Naval Research contract N00014-86-K-0685.

References

[1] R. H. LaMotte. Personal communication, 1993.

[2] Antonio Bicchi, J. Kenneth Salisbury, and David L. Brock. Contact sensing from force measurements. *International Journal of Robotics Research*, To be pulished, 1993.

[3] Michèle Basseville. Detecting changes in signals and systems - a survey. *Automatica*, 24(3):309–326, 1988.

[4] A. S. Willsky. A survey of design methods for failure detection in dynamic systems. *Automatica*, 1976.

[5] Regine Andre-Obrecht. A new statistical approach for the automatic segmentation of continuous speech signals. *IEEE Transactions on Acoustics, Speech, and Signal Processing*, 36(1):29–40, January 1988.

[6] Lennart Ljung. *Theory and Practice of Recursive Identification*. MIT Press, 1983.

[7] Jorma Rissanen. Stochastic complexity and modeling. *The Annals of Statistics*, 14(3):1080–1100, 1986.

[8] Brian Eberman and J. K. Salisbury. Application of change detection to dynamic contact sensing. Memo 1421, MIT Artificial Intelligence Laboratory, 1993.

[9] Michael Dean Levin. Design and control of a closed-loop brushless torque actuator. Technical report, MIT Artificial Intelligence Laboratory, AI-TR 1244, 1990.

Robot Force Control without Stability Problems

James Trevelyan

Technical Consultant, Automated Sheep Shearing Project
Senior Lecturer, Department of Mechanical Engineering,
University of Western Australia, Nedlands, 6009.

Abstract[1]—A type of open-loop force control scheme has been tested in a sheep-shearing robot. Stable, high speed surface following has been achieved on hard and soft surfaces without having to measure actual contact forces and without knowing the dynamics of the robot or the surface. A small hydraulic actuator holds the cutter against the surface of the sheep with a programmed force obtained by regulating the hydraulic pressure against a piston. Measurement of the cutter position relative to the robot provides guidance to modify the robot's position trajectory, and to compensate the hydraulic pressure for disturbance effects. Experience with this device has suggested a simpler approach to force control which might be helpful for future robot system designers.

1. Introduction

The tracking of an unknown surface with a robot has been the subject of many research projects – the sheep shearing robot represents a classical problem of this kind. Until 1989, we used sensors on the end-effector with a position control scheme. Then we realized that we needed to adopt a force control scheme to obtain best performance and shearing quality.

However, there are intrinsic stability problems associated with force control, particularly when operating with end-effectors on hard surfaces. A stable, high bandwidth force-feedback controller can be designed if one has an accurate model of the dynamics of the manipulator, the end-effector and the surface it is operating against. However, in practice it is not easy to model all these characteristics accurately enough. This has not stopped many researchers from trying. One way of avoiding these difficulties is to follow rules by which reliable control systems are usually designed. One good idea is to link an actuator as directly as possible to the device to be moved, and another is to avoid closed loop control if open loop control is good enough.

This paper describes how force control has been used for shearing sheep. This demonstrates how one can achieve fast, accurate surface following at high speed (up to 50 centimetres per second) on both hard and soft surfaces without stability problems and without having to know much about the dynamics of the robot or the surface. The

---

[1] The author acknowledges the support of his colleagues in preparing this paper, and in performing the work described, and financial support from the Wool Research and Development Fund which is administered by the Australian Wool Research and Development Corporation.

paper describes the principles behind the control system and the mechanical hardware.

The paper also shows why surface contact forces may be difficult or impossible to measure in practical situations, and why open-loop control neatly bypasses this problem. The paper suggests a different way of looking at force control problems which might be useful for designing simple and effective control systems which can be used with conventional robot manipulators.

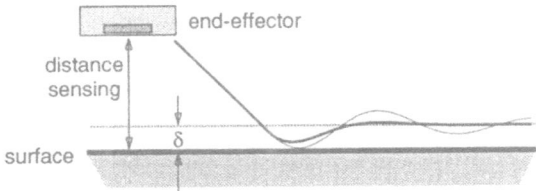

Figure 1 - refer to text

2. The Bandwidth Problem in Surface Tracking

To follow a surface, a robot has to guide its end-effector over the surface using error information from one or more sensors. Figure 1 illustrates this problem, and some simulation results which illustrate the results of attempting to follow the surface with a position control system. In the diagram, the robot end-effector has a distance sensor, and approaches the surface at a constant speed, traversing the surface towards the right so we can see its behaviour with time. The grey horizontal line shows the desired surface tracking distance δ. If we are not to hit the surface, this represents the maximum trajectory error allowed. Ideal results are shown for a control system with a closed loop performance simulated by a linear second order equation with damping ratios 0.18 (thin line) and 0.52 (thicker line).

It is possible to show that for reasonable performance, the minimum performance of the control system, represented by the bandwidth ω, must be approximately given by:

$$\omega \ \geq \ v / \delta \qquad\qquad (1)$$

where v is the maximum surface approach speed, ω is the controllable bandwidth (rad/sec) and δ is the maximum allowable tracking error.

A human arm has a bandwidth of 3 to 6 Hz, but this can be as low as 1 Hz when carrying a reasonable shearing tool at maximum reach. A well designed manipulator arm has a 6 to 10 Hz bandwidth. Larger (and heavier) actuators, and a more rigid structure are needed for a faster response. But the increased weight requires still more strength for equivalent stiffness, leading to compounding increases in weight, size and ultimately cost.

The approach speed limit above is related to the traverse speed limit across the surface by the relative slopes of the bumps. Assuming slopes up to 45°, the approach speed limit will be about the same as the traverse speed limit[2].

---

[2] This equation (1) applies equally well to a human arm. Try to move your finger across an irregular surface such as a rumpled rug or bedclothes, with your arm outstretched, keeping a constant height above the surface . Concentrating just on the gap between the finger tip and

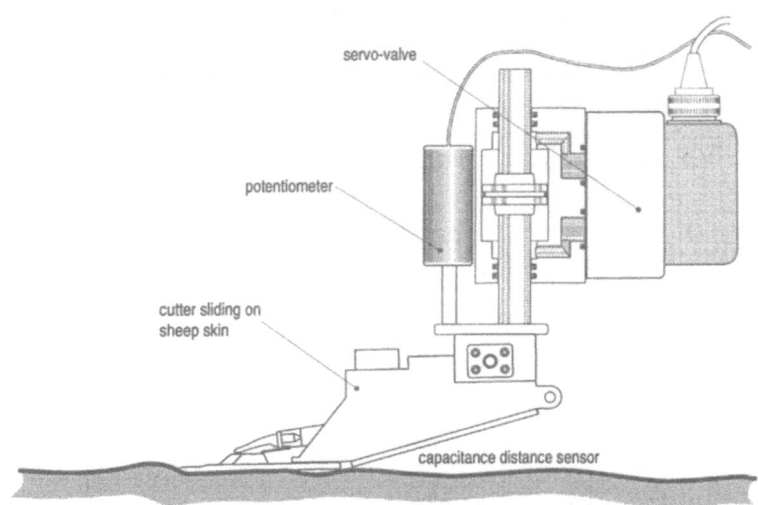

servo-valve

potentiometer

cutter sliding on
sheep skin

capacitance distance sensor

Figure 2 Position-controlled follower arrangement (till 1989)

A typical surface tracking distance for shearing sheep is about 2 millimetres and with
a manipulator bandwidth of 6 Hz a shearing speed of about 70 mm/sec can be achieved,
a long way short of a typical human shearer at up to 700 mm/sec. Eqn (1) suggests that
we would need a bandwidth of about 50 Hz to shear at a speed of 600 mm/sec. The validity of eqn (1) might be questioned because, on this basis, a human shearer ought to be
limited to a speed of between 10 and 40 mm/sec. Later sections explain this apparent
discrepancy.

Figure 2 shows a schematic diagram of the sheep-shearing robot end-effector (or
follower) before force control was adopted. The follower was carried at the end of the
manipulator arm (figure 7) and provided height adjustment for the cutter with the
required bandwidth of about 50 Hz. The piston position was controlled in response to
the capacitance distance sensors under the cutter, but the set height was calculated from
measurements of the electrical resistance between the comb and the skin of the sheep.
This latter sensor behaves more like a force sensor than a distance sensor, but can be
modelled as a distance sensor with a range of about 2 to 4 mm.

The entire end-effector can also be considered as an active sensor—the cutter position,
which follows the skin position through the control scheme just described, provides an
indication of the deviation of the cutter from the planned path being followed by the
manipulator. This signal is used to correct the planned path.

In effect, we have partitioned the robot into two parts. The first, the follower, has a
high bandwidth but limited range of adjustment in position. The arm has a lower

---

the surface, go as fast as you can without touching, keeping the gap at about 10 mm, but no
more than 20 mm. The maximum following error must be less than 10 mm. You should be
able to keep up a traverse speed of between 100 and 300 millimetres per second without
bumping into the surface. It may be less if you're tired at night or more if you're perky in the
morning. With an arm bandwidth of about 3 Hz, the equation predicts a speed of 180
mm/sec.

bandwidth, but very large range of position adjustment. The follower allows the robot to track surface irregularities for which the main arm alone would be too sluggish.

The maximum path error (for the arm) which can be accommodated is about half of the stroke of the piston. The stroke needs to be just long enough to accommodate dynamic arm positioning errors. Using equation (1) again, with an arm bandwidth of 6 Hz and a maximum shearing speed of 600 mm/sec, the stroke needs to be at least ±16 mm.

3. Position Control Experiments

The earliest machines used for shearing experiments provided a bandwidth of about 5 Hz for cutter height control [12: chap 2]. As predicted by eqn (1), the shearing speeds obtained by these machines were about 5 to 6 cm per second.

The surface follower mechanism was a central part of the design of the ORACLE and SM robots [12]. In principle, it is similar to other active end-effector devices described in the robotics literature such as precision positioners [10].

The main means of sensing the skin are electrical resistance and capacitance proximity sensors. The resistance sensing technique uses a small electric current which can pass between the metal teeth of the shearing comb and the skin. When the comb is in contact with the skin, this current is roughly proportional to the contact force, but is erratic and highly sensitive to changes in skin condition. Capacitance sensors measure the distance between the underside of the cutter and the skin. However, they need to be located some distance behind the cutter.

Much of our research effort was directed at improving methods of combination, or fusion, of two very different forms of data from capacitance proximity sensing and resistance contact sensing with fast computation rates.

The position control scheme was tested and refined between 1980 and 1987. By then we realized that there were fundamental difficulties with resistance sensing as our primary signal for keeping the cutter on the skin. Apart from wool contamination problems, we encountered stability problems similar to those encountered with force controlled robots. The resistance sensing technique behaves more like a force sensor than a displacement sensor in practice, and the compliance of the surface significantly affects the dynamic behaviour.

Figure 3 shows a typical model of a robot with a force sensor in the end effector. The major part of the manipulator is position controlled, and the end effector is attached, often with a compliant coupling as illustrated. Sometimes there is a force control element added, as shown.

A force controlled robot performs a similar task to surface following, except that the end-effector is in physical contact with the surface. In the models used by most researchers, the surface is considered to be compliant (figure 3) and the robot itself is compliant. Thus, a small position change, ∂x, can be related to a contact force change, ∂f, by the combined stiffness of the robot and the surface:

$$\partial f \approx k \, \partial x \qquad (2)$$
$$\text{where} \quad k = \left(\frac{1}{k_s} + \frac{1}{k_r}\right)^{-1} \qquad (3)$$

such that k_s is the effective surface stiffness and k_r is the effective robot stiffness.

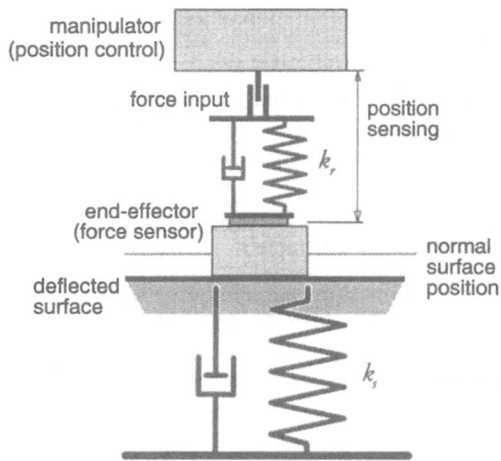

Figure 3 - refer to text

In the case of our sheep-shearing robots, the coupling to the manipulator (k_r) was effectively rigid. If the surface was soft, then the signal from the force sensor (resistance measurements) varied less with a given position change. Thus, when we used this signal as a feedback source, the effective gain was low. If the surface was hard, the gain was higher, leading to instability. A bouncing behaviour was often observed with a hard surface.

Because of this and other difficulties, shearing experiments with the SM robot rarely reached the maximum speed predicted by eqn (1).

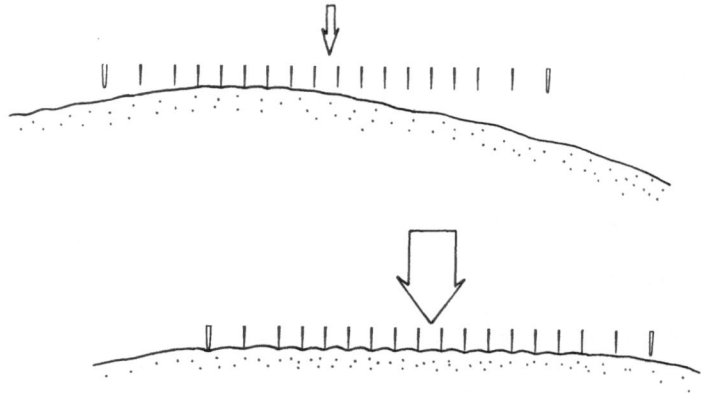

Figure 4 Effect of holding comb against skin with light pressure (above) or firm pressure (below)

4. Towards Force Control

From expert shearers we learned to press the base of the comb firmly down while shearing. This flattens the skin under the comb allowing all the teeth across the leading edge of the comb to be in contact with the skin. If some teeth are not in contact with the skin then wool fibres ahead of those comb teeth lean over away from the comb. The comb teeth can no longer penetrate these more densely packed fibres and shearing breaks down. Thus, it is important to press firmly onto the skin to be able to use the full width of the shearing comb (figure 4). Under typical shearing conditions the resistance sensing control method results in very light contact forces and on rounded parts of the sheep and only part of the comb width can be used. To shear well and efficiently, the comb contact force has to be controlled.

A major obstacle to effective force control is force measurement which can be difficult in a practical situation.

The comb experiences large and variable bending moments because the oscillating cutter is firmly pressed down onto its top surface. These bending moments are much larger than those arising from surface contact forces. They are also variable, depending on cutter sharpness and wool cutting conditions.

Even if we could measure the contact force on the comb teeth, we would find that the friction between wool fibres and the comb teeth would interfere with our measurements to the extent that accurate measurement would be impossible [12:chap 6]. Experiments conducted by Merino Wool Harvesting Pty Ltd[3] in Adelaide had shown that the combing forces can be large compared to typical contact forces applied to the cutter [12:chaps 7 and 14], [6]. As the comb moves forward the teeth penetrate between wool fibres; the tension force in the fibres pulls the comb teeth down onto the skin. In some circumstances, this effect can increase friction to the point where the comb snags the skin and the comb cannot be moved forwards without cutting the skin. A human shearer senses this and tips the comb back a little, lifting the comb tips.

Merino Wool Harvesting went on to experiment with force control but they experienced the same stability problems familiar to other robotics researchers. The stability of the control loop is critically dependent on the compliance of the surface and the tool. They could not obtain stable performance under all the different operating conditions found on the sheep. Further, the design of the cutter was constrained by the need for delicate strain gauge beams for force measurement.

5. Open-Loop Force Control

We chose an open-loop force generator to bypass these problems. In its simplest form, it is designed such that the cutter exerts a constant force on the sheep by maintaining a constant pressure difference between each side of the actuator piston. This is achieved by using the servo-valve to control oil flow through a restriction as shown in figure 5. The pressure across the restriction depends on the flow rate, so by maintaining a given flow rate, any desired pressure (up to some proportion of the hydraulic supply pressure) can be maintained. Sheep profile variations and arm positioning errors will cause the position of the piston to change with time, and this will cause oil to flow into and out of the cylinder ports. This affects the flow through the restriction, changing the pressure (and force) on the piston. To compensate for this, the position of the piston is measured, and the computer differentiates this signal to calculate velocity, and thus re-calculates the servo-valve setting so that the oil flow through the restriction remains constant. In this way, the pressure difference and force on the piston also remains constant.

To the conventional eye, this seems a strange way to use these components. However,

---

[3] A private company which developed a semi-automatic sheep shearing system using twin robots.

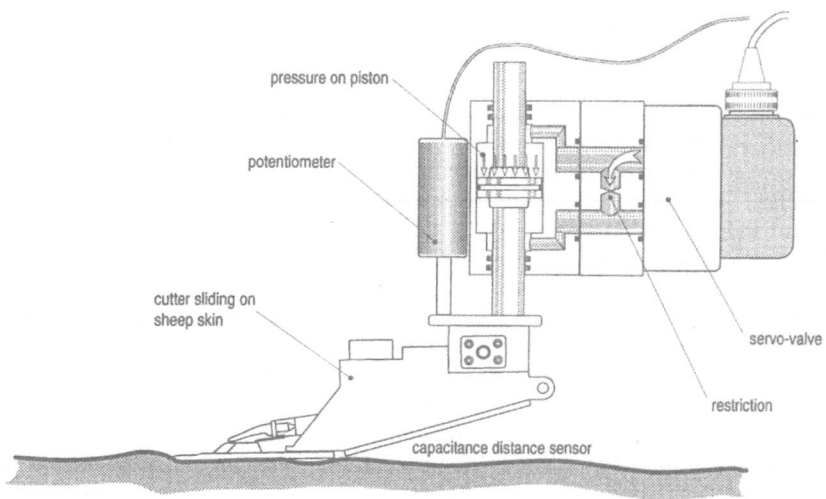

Figure 5. Follower arrangement with open-loop force control (post 1989)

it has important practical advantages.

First, this is an open-loop force controller which is inherently stable.

Second, we only need a single position transducer—we do not need to measure force or pressure. The position signal is used to achieve a further practical elaboration—the set force is varied with follower position under software control. This is useful for operations like dragging shorn wool away from the sheep.

Third, we retain the capability for very fast position changes. As soon as the sensor electronics begins to see signs of an imminent skin cut, the comb is flicked up by 3 mm in a few milliseconds. This seems to prevent the skin from becoming hooked on the leading tips of the comb, and may significantly reduce skin cuts.

More stable force control could also be achieved by inserting compliance between the end-effector and the manipulator. For example, Paul and Xu [4] demonstrated the use of a compliant wrist, but this cannot be used for high bandwidth active position control at the same time. In fact, we had also considered this approach in 1977, by holding the cutter against the sheep skin with a spring. We need a fast reaction to skin cuts when they are detected so a passive compliance cannot be used.

6. Experiments with Open-Loop Force Control

Shearing results have justified confidence in the open-loop control scheme. As well as providing smoother and faster robot performance and more efficient shearing, it has been possible to shear damp and urinated wool. This was quite impossible with the position control method used beforehand because the wool contamination makes the resistance sensing method inoperable.

We have demonstrated surface following at speeds of up to 600 mm/sec (about twice the intended maximum shearing speed), and the behaviour of the robot is stable on any type of surface. There is no bouncing or loss of stability on hard, non-compliant surfaces.

Friction between the piston rod and the cylinder seals was expected to be a significant problem. However, the device works surprisingly well with friction forces up to 60% of the programmed applied force. The current mechanical configuration needs to be improved to reduce the friction effects in order to complete our research on high speed shearing. We need this to find the friction level at which high speed shearing performance is significantly affected in order to design a cost-effective prototype robot system.

In retrospect, we can now understand the apparent failure of eqn (1) to predict the maximum speed attained by human shearers. The human shearer uses arm muscles to force the cutter against the sheep skin—a kind of open-loop force control. Thus the allowable tracking error is represented by a typical arm displacement (100–200 mm) rather than the 2 mm distance one might estimate from an initial understanding. This retrospective understanding, then, complies with eqn (1) and permits much higher maximum speeds. The typical maximum speed attained by a human shearer is about 700 mm/sec, though this occurs on only one or two of the 70 odd shearing movements needed to shear a sheep.

7. Discussion

Figure 3 illustrates how the compliant surface will be deflected by the force exerted by the end-effector, measured by a force sensor. The model illustrated is only intended to illustrate the effect of eqn (2). For surface tracking using force control, the maximum allowable position error is some fraction of the surface deflection under static conditions as illustrated.

By using eqn (2) to define the maximum tolerable force error in terms of a displacement error, equation (1) can be applied to force control problems. However, the stiffness of the surface and often the robot as well can vary considerably, so the stability margin of the controller also varies unless some form of adaptive control can be used. Many researchers have reported stability problems with force control: examples of just a few include Eppinger and Seering [3], Surdilovic and Vukobratovic [9], Elosegui et al [2], Stokic [8].

Most research projects on robot force control have adopted a closed-loop control approach. One or more sensors are used to measure force, and actuator feedback transducers measure position errors. The control schemes are typically a combination of force and position control. In practice, a robot will always require position control to move the end-effector from a parked position to begin the programmed task. During the execution of the task, a combination of force and position control is needed. The early work by Raibert and Craig [5] demonstrates such a combination, though there has been some dispute on the mathematical derivations [1].

Referring once again to figure 3, we see that many of the approaches adopted by different researchers can be modelled in the same, or similar ways. By doing this, we can draw some useful conclusions about effective force control methods.

Figure 6 illustrates a variety of force control scenarios, corresponding to different arrangements which have been used for force control. By classifying force control arrangements in this way, we can make observations about likely performance advantages. Note that the controller is not shown. By merely showing the arrangement of the end-effector, we can draw conclusions which will affect any controller. This may help to choose an appropriate con-troller for a given application.

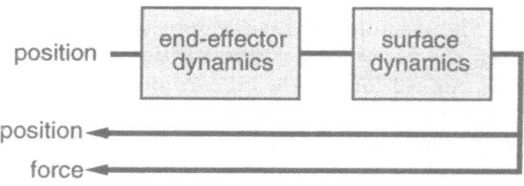

(a) Position input, position and force feedback

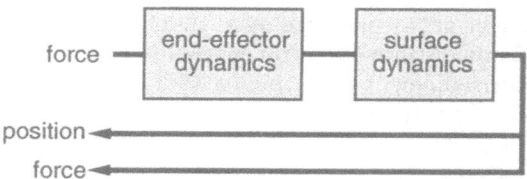

(b) Force input, force and position feedback

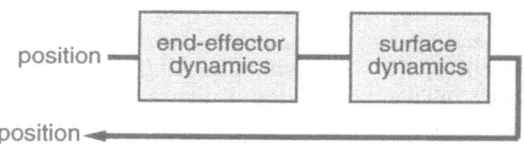

(c) Position input, position feedback, with
known compliance in end-effector

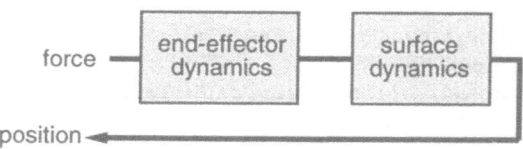

(d) Force input, position feedback

Figure 6 - refer to text

Figure 6a shows a typical conventional force feedback situation where the robot operates under position control, and the combined end-effector and surface compliance result in a force and position which are measured. Examination of the control-ability criterion reveals a fundamental weakness. Unless the characteristics of the end-effector *and* surface are known, then the behaviour cannot be predicted and the system may not be controllable. In terms of eqn (1), the allowable position error may be variable, so the speed limits may have to be constrained by the worst case. Adaptive control might be considered, but it might not be easy to implement.

Figure 6b illustrates a preferred form from a control-ability stand point, where the input to the system is a force. The characteristics of the surface are not so important, unless the dynamic characteristics are unusual. For example, if the damping in the end-effector and surface are both small, there may be large force oscillations between them. It may be impractical to provide a multiple axis force generator (for a good example, see Salisbury [7]) so this scheme may not be feasible in situations where the applied force cannot be provided from a single axis device.

Figure 6c shows how one can use position input and measurement provided the surface compliance is much less than the robot compliance. This approach was demonstrated by Paul *et al* [4] in which a passive compliance was added to the robot wrist. The deflection of the compliance against the operating surface is measured to achieve precise position control at the same time. By relating this to equations (1) and (2), we see that increasing robot compliance (reciprocal of k_r) increases the allowable displacement error, and hence increases the allowable working speed for a given control bandwidth.

Figure 6d represents the force control scheme adopted for the sheep-shearing robot. Since open-loop force control is used, there is no need to sense the applied force, and stability is guaranteed. This scheme is simple, and has proved to be robust and reliable in our experiments.

A good practice in control system design is to place the actuator as close to the intended point of application as possible. This means placing the actuator close, if possible, to the end-effector against the surface so the dynamics of the end-effector (see diagrams above) are simple, and possibly can be ignored. This rule also suggests that if we wish to control force, it is better to generate a force directly rather than using surface or robot compliance to generate contact force indirectly. Finally, if open-loop control is feasible, as it has been for sheep-shearing, it offers simplicity, robustness, and reliability—all of which are important practical advantages.

8. Conclusions

After reviewing the literature and our own experiences, one concludes that it is often impractical to control surface contact forces with force feedback control on a position controlled robot. This scheme is equivalent to figure 6a above. This has now been more widely recognized. The *Artisan* robot under construction at Stanford University is based on high quality force control at each joint[4] and is therefore a complex configuration based on 6b above.

Robot designers have much to gain from adopting simple approaches to design. We have to accept that position controlled manipulators are a well established technology—factory robots are now comparatively inexpensive and reliable. This suggests that robots which need to exert controlled forces need special end-effectors, if possible, rather than new arm or controller designs.

Careful consideration of the factors discussed in this paper could be helpful in designing such robots to have adequate performance without having to undertake extensive computer simulations or expensive research and development programmes.

---

[4]Khatib, O. Personal communication

142

References

[1] Duffy, J. (1990) The fallacy of modern hybrid control theory *inter alia. Journal of Robotic Systems* 7(2) 139-144.

[2] Elosegui, P., Daniel, R. W. and Sharkey, P. M. (1989) Joint servoing for robust manipulator force control, Report from Robotics Group, Department of Engineering Science, Oxford University.

[3] Eppinger, S, and Seering, W. (1987) Understanding bandwidth limitations in robot force control. *Proceedings of 1987 IEEE International Conference on Robotics and Automation.*

[4] Paul, R. P., Yangsheng Xu, and Xiaoping Yun (1990). The implementation of hybrid control in the presence of passive compliance. *Robotics Research: The 5th Symposium, MIT Press,* pp 193-200.

[5] Raibert, M. H. and Craig, J. J. (1981) Compliance and force control for computer controlled manipulators. *ASME Journal of Dynamic Systems, Measurement and Control* 102, 126-133.

[6] Rogers, K. J. (1990). Wool harvesting with robots. *Electrical and Electronic Engineering Transactions, Institution of Engineers, Australia,* EE10, (3), pp.207-15.

[7] Salisbury, K., Eberman, B., Levin, M., and Townsend, W. (1990). Design and control of an experimental whole-arm manipulator. *Robotics Research: The 5th Symposium, MIT Press,* pp 233-241.

[8] Stokic, D. M. (1991) Constrained motion control of manipulation robots—a contribution. *Robotica* 9 pp 157-163.

[9] Surdilovic, D. and Vukobratovic, M. (1993) Impact of target impedance on contact stability, Report from Fraunhofer Institute for Production Systems and Design technology IPK-Berlin, Germany.

[10] Taylor, R. H., Hollis, R. L., and Lavin, M. A. (1985). Precise manipulation with endpoint sensing, *Robotics Research: The 2nd International Symposium,* pp.59-69, MIT Press.

[11] Trevelyan, J. P. (1990). Replicating manual skills. *Robotics research: 5th International Symposium,* Tokyo, Japan, pp. 333-40, MIT Press.

[12] Trevelyan, J. P. (1992). *Robots for shearing sheep: shear magic.* Oxford University Press.

[13] Whitney, D. E. (1986). Real robots don't need jigs. *IEEE Conference on Robotics and Automation,* Vol. 2, pp. 746-52, San Francisco.

[14] Whitney, D. E. and Brown, M. L. (1987). Metal removal models and process planning for robot grinding. *17th International Symposium on Industrial Robots (ISIR).* pp. 19-29 - 44, Chicago.

A robotic cell for deburring of polygonal objects

J-P Merlet, N. Mouly, J-J. Borrelly, P. Itey

INRIA Sophia-Antipolis
BP 93, 06902 Sophia-Antipolis Cedex, France

Abstract— A robotic cell for deburring planar polygonal objects is described. The object comes from a conveyor and arrives in a random position on a parallel manipulator. Its center of mass is located through the measurements of a 6-componants force sensor, it is then grasped and a force-feedback scheme is used to fix the object on a special plate. Then a probing algorithm is used to discover the location of the vertices of the object with a minimal number of measurements. The coordinates of the vertices are then used to build an ideal reference model of the object which is fed to a force-feedback scheme which perform the deburring of the object.

1. Introduction

Surface following with force-feedback is an important robotics task useful for many applications: grinding, polishing, deburring. In this kind of tasks the tip of the grinding tool has to apply a constant force on an object and follow its contour with a velocity as close as possible from a given constant value. In most cases surface following is only a 2D problem as the tip of the grinding tool may be reduced to a point moving in a known plane.

Therefore many researchers have addressed this problem [1],[2]. Most of these works emphasize the problem of stability of the force-feedback scheme due to the high gain in the loop. Clearly stability is an important issue as soon as there is contact between the robot and the object. It has been shown that stability is deeply dependent on the sampling rate of the system (which must be the highest possible) and on the mechanical stiffness of the coupling of the robot and the surrounding.

The sampling rate is in general fixed for a given hardware. Interesting results have been obtained by modifying the stiffness of the robot, for example by using a micro-macro manipulators as described in [3],[4], [5],[6] in which case a parallel manipulator is used as a wrist.

But for the special case of the surface following problem stability may also be improved if some approximate model of the shape of the contour is known. This can be done by a learning algorithm [7]: during a first experiment no model of the object is known but a force-feedback scheme enables to follow the contour. Forces and position of the end-effector are recorded during the experiment. Then the position of the end-effector for which the force was in a given range are selected to construct a reference model of the contour (for example by computing splines whose control points are the selected positions). Using this learning algorithm experiments can be performed with a reference model and this model may even be refined by using the data of further experiments. The drawback of such an approach is that the learning algorithm is rather slow and the use of splines to build the reference model may be not appropriate in some case, for example for polygonal objects. In that case the reference model may be defined by the list of the coordinates of the vertices of the object.

2. Probing algorithm

2.1. Principle

The problem of determining the locations of the vertices of a polygon using a sensing device which gives local information on the shape of the contour is known under the name of *probing* [8]. Researchers involved in computational geometry have addressed this problem especially to design a sensing strategy enabling to find all the vertices of the object with a minimum number of measurements.

We have decided to use the probing algorithm described by Boissonnat [9]. In this algorithm the measurement starts from a point and is done in a given direction called the *ray* of the measurement. The sensing device is able to determine the coordinates of the intersection of the ray with the object together with the normal of the object at this intersection point and the full process is called a *probe*. The following assumptions are made: the object has no collinear edges, the coordinates of a point U belonging to the interior of the object are known and the operator is able to define a circle C_e centered in U which does not intersect any part of the object.

The sensing strategy is now described. For the first probe a random point M_1 on C_e is chosen and the ray is the line $M_1 U$, which insure that the probe will determine a point on the object. Together with the normal this point enables to determine the support line of an edge E_1 of the object. The second probe has as origin M_2, a point on C_e taken as the opposite point of M_1 with respect to U. This second probe enables to determine the support line of a second edge E_2 of the object. Let P_3 be the intersection point of the support lines of E_1, E_2. The third probe starts from a point M_3 intersection of the circle and the line $P_3 U$ and its ray is directed along $P_3 U$.

For this third probe two cases may occur (figure 1):
- E_1, E_2 are adjacent and therefore P_3 is a vertice of the object. The probe will detect that the contact point is the intersection of the support line and that the left and right normals are the normal to E_1 and E_2. From this fact the algorithm deduces that P_3 is a vertice of the object.
- E_1, E_2 are not adjacent: the contact point will be different from P_3 and the support line of a new edge E_3 of the object is discovered.

In the first case the full shape of the object between E_1, P_3, E_2 has been discovered and the fourth probe will start from M_4 the opposite point to M_3 and the ray will be directed toward P_3. This insure that a new edge E_3 will be discovered. In both cases a list of pair of edges is build which contains the edges for which the intersection point of the support line has not been tested

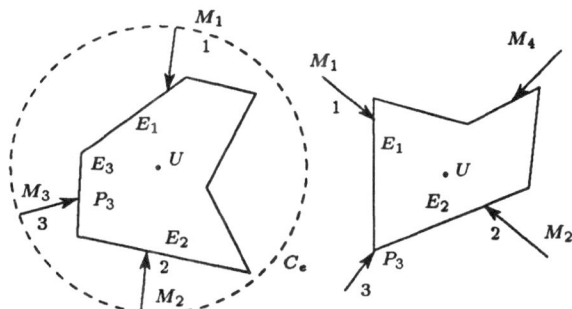

Figure 1. After completing the two first probes the third probe enables either to determine a vertice of the object (right) or a new edge (left).

as a potential vertice: for example after the third probe this list is { (E_1, E_3), (E_2, E_3) }. As soon as a pair is used to perform a probe it is removed from the list and the list is updated according to the result of the probe. When the list is empty all the vertices of the object have been determined. Clearly for a convex polygone only two probes are necessary to discover a vertice and therefore in that case the algorithm find all the vertices of a n-vertices polygon with $2n$ probes.

In some case for non-convex polygons a probe may not discover a vertice nor a new edge but it can be shown that in that case we can compute a probe which will discover either a new vertice or a new edge. Therefore for non-convex polygon it may happen that more than $2n$ probes are necessary but it may be shown that for a n-vertices polygon at most $3n - 3$ probes will be necessary.

2.2. Probing with a force sensor

The measurement device used in our cell is based on a force sensor and the tip of the robot moves along the ray of the probe.

The force sensor is able to detect the contact between the tip of the robot and the object: therefore the coordinates of the contact point in the robot frame is known. In order to determine the normal to the object at the contact point the following procedure is used: as soon as a contact has been detected the robot performs a small backward motion along the ray, then a small motion along the perpendicular to the ray in the left direction and then a motion in a direction parallel to the ray toward the object. A new contact point is found and the robot backtrack until it lie again on the ray and a similar procedure is used now on the right side of the ray to find a new point on the edge. After this procedure the coordinates of three points on the edge are known which enable to determine the support line of the edge with a good accuracy. Although this procedure will not work with very small edges in practice no problem have been encountered. Figure 2 shows two examples of experimental probing of complex polygons.

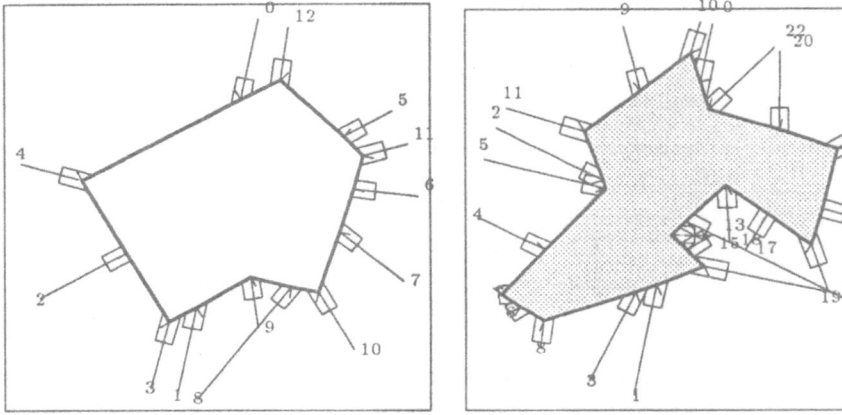

Figure 2. Two examples of application of the probing algorithm on complex polygons.

3. The cell

3.1. Constitution

Our robotic cell is constituted of three manipulators:

- a parallel manipulator called the "left hand" which receive the object on its end-effector plate in a random position but with an approximatively fixed orientation. This manipulator has a traction-compression force sensor in each of its 6 legs and through the 6 measurement of these sensors the external forces and torques acting on the end-effector may be computed.

- an IBM SCARA robot which perform the probing and then the surface following of the object. During these operations the object is fixed on a table lying in the vicinity of the robot and the force informations are given by a 6 componants force sensor mounted below the table.

- a 6 d.o.f. serial link manipulator AID which will pick the object on the end-effector of the left hand and transfer it on the table of the SCARA robot.

4. Picking of the object

4.1. Locating an object on the end-effector

Let $(O, \mathbf{x}, \mathbf{y}, \mathbf{z})$ be a frame $\mathcal{R}_l$ attached to the base of the left hand and $(C, \mathbf{x_r}, \mathbf{y_r}, \mathbf{z_r})$ be a frame $\mathcal{R}_l^r$ attached to the end-effector. By controlling the 6 links lengths the location of C and the orientation of the end-effector with respect to $\mathcal{R}_l$ can be modified at will. The posture of the end-effector of the left hand is assumed to be known and the forces and torques acting on the end-effector in the frame $\mathcal{R}_l$ can be computed.

For picking an object it is essential to determine the location of its center of mass (which may not be coincident with its geometrical center e.g. for non-homogeneous objects). Let M_x, M_y be the torques acting on the end-effector around the x, y axis, measured with respect to point C, F_z the force acting along the z axis and let x_m, y_m be the componants in $\mathcal{R}_l$ of the vector $\mathbf{CG}$ where G denotes the center of mass of the object. As the only force acting on the end-effector is the weight of the object we get:

$$M_x = -mgy_m \qquad M_y = mgx_m \qquad F_z = -mg \tag{1}$$

and therefore:

$$x_m = -\frac{M_y}{F_z} \qquad y_m = \frac{M_x}{F_z} \tag{2}$$

As the location of C in $\mathcal{R}_l$ is known it is now easy to determine the coordinate of the center of mass in $\mathcal{R}_l$. Therefore the force and torques measurements enable to compute the picking point of the object in $\mathcal{R}_l$.

4.2. Calibration procedure

Let $(O_A, \mathbf{x}, \mathbf{y}, \mathbf{z})$ be a frame $\mathcal{R}_a$ attached to the base of the AID robot. As the picking operation will be performed by this robot and as we have determined the picking point in a frame $\mathcal{R}_l$ linked to the left hand we have to determine the location of O with respect to $\mathcal{R}_a$ i.e. the componants of $\mathbf{O_A O}$.

Once again this operation is done by using the force measurements. Assume that the tip of the end-effector of the AID robot touch the end-effector of the left hand as some point P. Using the force measurements we are able to determine the componants of $\mathbf{OP}$. Using the direct kinematics of the AID we are also able to compute the componants of $\mathbf{O_A P}$. As $\mathbf{O_A O} = \mathbf{O_A P} - \mathbf{OP}$ we are theoretically able to compute $\mathbf{OO_A}$. As the force measurements may be noisy the procedure is repeated for a dozen of points and a least-square algorithm is used to determine $\mathbf{O_A O}$.

4.3. Picking

As soon as an object is put on the effector of the left hand the force along the z axis indicates that an object is present and the location of the center of mass is computed. If the picking point is in the reachable workspace of the AID it moves over the picking point, the end-effector being directed toward the object. In the opposite case the left hand moves toward the AID base until the picking point lie in the workspace of the AID.

The gripper is then opened and the AID moves toward the picking point along the z axis. This motion is stopped when a contact is detected by the force sensor of the left hand (therefore the height of the object has no importance). The gripper is then closed and the robot moves along the z axis with the object in the gripper. The correct gripping of the object is verified as the force along the z axis must go back to zero. In case of failure the procedure is repeated.

5. Transfer of the object and fixation on the table

As soon as the object has been picked the robot AID moves toward the table near the SCARA and put the object on the it. At this point the left hand and AID are free to receive and transfer a new object. Now the SCARA has to pick the object on the table. But during the transfer of the part toward the table and its setting its orientation may have changed. As the object will be fixed on the table through an insertion process its orientation shall be approximatively constant. Therefore the object has to be picked and its orientation corrected. This process begins as soon as the presence of an object is detected by the force sensor below the table. Once again the force sensor of the SCARA table is used to determine the location of the center of mass of the part in the SCARA frame (a calibration procedure has been previously performed to found the position of the center of the force sensor frame in the SCARA frame). Then a simple "pushing" algorithm [10] is used to correct the orientation of the object. A force-feedback scheme which has been described in [11] is then used to insert the pins of the object in some holes of the table.

6. Surface following

As soon as the object is fixed on the table a probing is completed and the force-feedback scheme can be used. A desired velocity V_d and a desired force F_d are given together with a direction for the deburring (e.g. clockwise). As soon as there is contact between the grinding tool and the object the use of the reference model enables to determine on which edge of the object the contact point is located together with the external normal unit vector $\mathbf{N}$ and tangent unit vector $\mathbf{V_T}$ of this edge . Note that the tangent vector is chosen according to the direction indicated for the deburring.

A proportional controller is used to adjust the contact force by generating a velocity $\mathbf{V_N}$ along the normal $\mathbf{N}$ of the object which is computed as:

$$\mathbf{V_N} = k_f(F_m - F_d)\mathbf{N} \tag{3}$$

where k_f is a constant positive gain and F_m the measured force. The amplitude of this velocity is thresholded to V. Then a velocity $\mathbf{V_T}$ along the tangent of the contour $\mathbf{T}$ is computed as:

$$\mathbf{V_T} = \sqrt{V^2 - ||\mathbf{V_T}||^2}\ \mathbf{T} \tag{4}$$

Near the vertices of the object the velocity is reduced to minimize the value of the change of the force: indeed contact may be lost at sharp corner (but in that case the reference model indicates in which direction the robot has to move in order to find the object) or big increase in force may occur at acute corner as the sampling rate of the force measurement is finite.

At the start of the experiment the tip of the end-effector is put at some fixed distance from an edge of the object and then the robot moves toward the object in the direction of the closest edge as determined by the reference model. As soon as contact is detected by the force sensor the force-feedback scheme is used. The task is completed and will be stopped as soon as the tip of the end-effector has moved along an edge different from the starting edge and is now at a small distance from the start point.

Experimental results of a deburring process are shown in figure 3 (followed contour) and in figure 4 (magnitude of the contact force). In this experiment

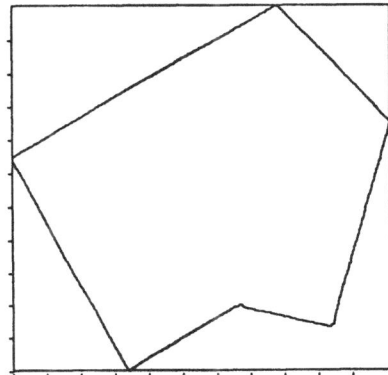

Figure 3. Result of the contour following task: followed contour.

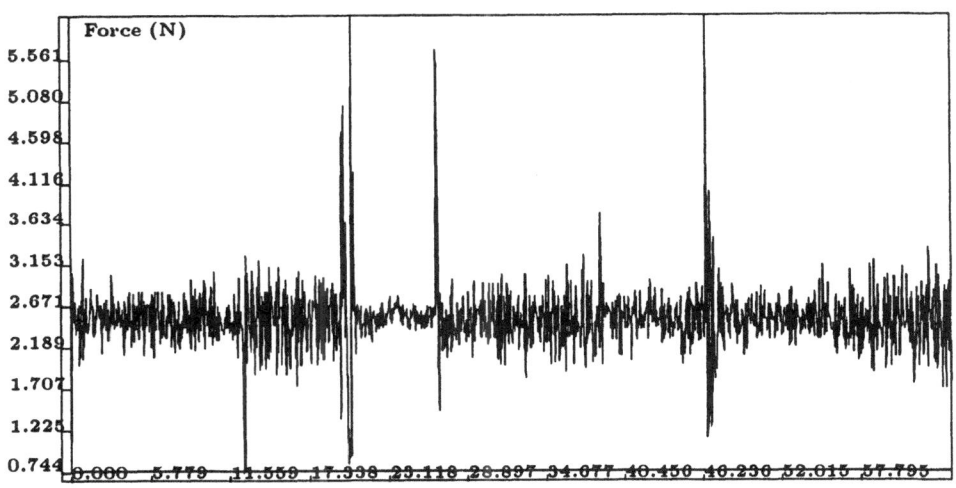

Figure 4. Result of the contour following task: measured force.

the velocity of the robot was 1 cm/s, the desired force was 2.5 N and the average contact force is about 2.506 N with a mean-square error of 0.355 (which is close from the mean-square of the sensor noise) i.e. very close from the desired value.

It may be seen that at some point there was important changes in the force signal which occur mainly when the robot moves in the vicinity of the vertices of the polygon.

7. Hardware

The AID and the left hand are controlled through a special robot controller based on a VME bus using two 68030 CPU boards which perform the servoing of the robots. Force conditioning is done by a special interface connected to the bus. The sampling rate of the force sensor is about 72 Hz with a sensibility of approximatively 50 gr (the force sensor was initially designed to enable the measurement of mass of about 100 kg). Motions orders are read from a memory board and the force measurement are written on this memory board. This memory is shared through a special PTVME board by a SUN workstation which compute the desired motion according to the force measurement which are read from the shared memory and write them on the memory.

Another robot controller is used to control the SCARA robot and acquire the force measurement from the table sensor. This controller use three 68020 CPU boards: one is used for the servoing of the robot, a second one manage the real-time aspect of the controller and the third one is used to execute a program written in a C-like robot language. Specific primitives for this language may be written in C on a SUN workstation and downloaded through a RS232 link in the memory of the controller.

The SUN workstation may modify at running time the behavior of the program through the use of specific primitives of the robot language which enable to read and write data on the memory of the controller through the serial link. Due to the low speed of this link the possibility is mainly used to get data from the controller in order to monitor the execution of the task. The two SUN workstations work independently. The force sensor is an AICO with a sampling rate of about 200 Hz and a force sensibility of about 5gr.

The probing algorithm is managed by the SUN workstation devoted to the SCARA robot and the program running on the SCARA controller just receive motion orders from the workstation, stop this motion as soon as a contact is detected and send the current position of the robot to the workstation which compute the motion necessary to perform the next probe. This is a rather slow procedure as all the information goes through the serial link and the average time for a probe is 7s. The surface following task is much more faster as this task is fully under the control of the SCARA controller, the serial link being used only to send position and force data to the workstation for debugging and recording purposes.

8. Conclusion

This experiment was designed for presentation during the IEEE Robotics Conference in Nice in 1992. Many researchers have seen it working (there was only one failure during the 25 presentations which have been made). We encounter mainly two kinds of problem: one was really a hardware problem and was due to the multiplicity of wires involved in this experiment. The other one was the problem of the management of the various tasks, their synchronization and dealing with the various events which may occur during the execution of the experiment (e.g. failure of a sub-task). One possibility to deal with this problem is to describe the task and the possible events in a synchronous language like ESTEREL which will generate a finite state automata which can be verified and deadlock discovered. Such an approach has been used successfully in simulation to manage an insertion task using a conveyor, two manipulators and the left hand [12],[13].

References

[1] Mason M.T. Compliant motion. In *in Robot , Motion-Planning and Control, Brady & al. Ed.*, Cambridge, 1982. The MIT Press.

[2] Planck G. and Hirzinger G. Controlling a robot's motion speed by a force-torque sensor for deburring problem. In *4th IFAC/IFIP Symp. on Information Control problems in manufacturing technology*, October 26-28, 1982.

[3] Reboulet C. and Robert A. Hybrid control of a manipulator with an active compliant wrist. In *3rd ISRR*, pages 76–80, Gouvieux, France, October 7-11, 1985.

[4] Khatib O. Inertial characteristics and dextrous dynamic coordination of macro/micro manipulator systems. In *7th CISM-IFToMM Symposium on Theory and Practice of Robots and Manipulators*, Udine, Italie, September , 1988.

[5] Merlet J-P. Force-feedback control of parallel manipulators. In *IEEE Int. Conf. on Robotics and Automation*, pages 1484–1489, Philadelphia, April 24-29, 1988.

[6] Reboulet C. and Pigeyre R. Hybrid control of a 6 d.o.f. in parallel actuated micro-macro manipulator mounted on a Scara robot. In *3rd ISRAM*, volume 3, pages 293–298, Burnaby, July 18-20, 1990. ASME Press Series.

[7] Merlet J-P. Manipulateurs parallèles, 3eme partie : applications. Research Report 1003, INRIA, March , 1989.

[8] Cole R. and Yap C. Shape from probing. *J. Algorithms*, 8:19–38, 1987.

[9] Alevizos P., Boissonnat J-D, and Yvinec M. On the order induced by a set of rays. application to the probing of non-convex polygons. Research Report 927, INRIA, November , 1988.

[10] Akella S. and Mason M. Posing polygonal object in the Plane by pushing. In *IEEE Int. Conf. on Robotics and Automation*, pages 2255–2262, May 12-14, 1992.

[11] Merlet J-P. Use of c-surface based force-feedback algorithm for complex assembly tasks. In *ISER*, Toulouse, June 25-27, 1991.

[12] Coste-Manière E. Utilisation d'Esterel dans un contexte asynchrone : une application robotique. Research Report 1139, INRIA, December 1989.

[13] Coste-Manière E. and Faverjon B. A programming and simulation tool for robotics workcells. In *Int. Conf. on Automation, Robotics and Computer Vision*, Singapore, September , 1990.

An Experimental Environment for Adaptive Robot Force Control

Louis L. Whitcomb,* Suguru Arimoto, Tomohide Naniwa

Department of Mathematical Engineering and Information Physics
University of Tokyo, Bunkyo-ku, Tokyo 113, Japan
Email: louis@arimotolab.t.u-tokyo.ac.jp

Fumio Ozaki

Energy and Mechanical Research Laboratory
Research and Development Center, Toshiba Corporation
4-1 Ukishima-cho, Kawasaki-ku, Kawasaki 210, Japan.
Email: 000013135207@tg-mail.toshiba.co.jp

Abstract—

Recent efforts in the development of an experimental robot system for the evaluation of the real-world performance of model-based force control algorithms are described. First, a model-based adaptive robot control algorithm for simultaneous position and force trajectory tracking of a robot arm whose gripper is in point contact with a smooth surface is reviewed. Second, a new experimental robot system for the testing of these new force control algorithms is described. Third, preliminary experiments with this arm show the performance of the new adaptive model-based force controller to be superior to its non-model-based counterpart.

1. Introduction

This paper reports recent efforts toward the development of an experimental robot system for the evaluation of the real-world performance of model-based force control algorithms. It is motivated by the recently reported class of model-based adaptive force control algorithms for robot arms [2].

For the problem of controlling robots whose motion is constrained by point contact with a smooth rigid environment, this new algorithm utilizes a sliding-mode control technique of a type first espoused in [10] for the case of free (non-contact) robot motion. Section 2 reviews the essential features of this force control algorithm. This new force control algorithm provides asymptotically exact tracking of both end-effector position *and* contact-force. Its performance can be mathematically proven with respect to the commonly accepted nonlinear

---

*We gratefully acknowledge the support of the Japanese Government agency Japan Society for the Promotion of Science (日本学術振興会) under a post-doctoral fellowship awarded to the first author.

rigid body dynamical equations of motion. Moreover, its adaptive extension can be shown to adaptively compensate for unknown plant parameters such as link and payload inertia, joint friction, and friction arising at the contact point between the tool tip and the surface.

In [7] the authors report satisfactory performance of this force control algorithm in numerical simulation studies. Of course, the practical advantages and disadvantages of control algorithms can most clearly be demonstrated by actual working implementations. Motivated by this, Section 3 reports on the mechanical and electrical components of the new system. Finally, Section 4 reports data from initial force control experiments with this new system which show the performance of the new adaptive model-based force controller to be superior to its non-model-based counterpart.

2. Geometry and Dynamics of Force Control

This section first reviews, in Sections 2.1, 2.2, and 2.3, the notation, geometry, and dynamical model for a point-contact force control model. Finally, Section 2.4 reviews the essential features of a previously reported nonlinear force controller for the point-contact force/position tracking problem. The reader is referred to [1, 7] for a detailed derivation.

2.1. Notation

The following notation is employed:

- n is a positive integer number of manipulator links.

- $\mathcal{J} = \mathbb{R}^n$ is n dimensional jointspace.

- $\mathcal{W} = \mathbb{R}^3$ is 3 dimensional euclidean workspace.

- $q(t) \in$ and $\dot{q}(t) \in \mathbb{R}^n$ are the robot joint position and velocity vectors respectively.

- $r(t)$, $\dot{r}(t)$, and $\ddot{r}(t) \in \mathbb{R}^n$ are the robot reference joint position, velocity, and acceleration vectors respectively.

- $f(t)$ and $f_r(t) \in \mathbb{R}^1$ are the magnitude of the component of the actual and reference tool tip forces normal to the rigid surface.

- $\nabla_y \, x(y) = \begin{bmatrix} \frac{\partial x(y)}{\partial y_1} \\ \vdots \\ \frac{\partial x(y)}{\partial y_n} \end{bmatrix}$ is the gradient of a scalar valued function $x(y)$ taken with respect to the independent $n \times 1$ vector y. $\nabla_y \, x(y)$ is an $n \times 1$ column vector.

- $D_y X$ is the generalized derivative of a scalar-valued (or vector-valued) function X taken with respect to the independent variable y. It is a row vector (or matrix). Example: If $x(y)$ is a scalar function of a vector y then $D_y x(y) = [\frac{\partial x(y)}{\partial y_1}, \ldots \frac{\partial x(y)}{\partial y_n}]$. Thus

$$
\begin{aligned}
\dot{x}(y) &= D_y x(y)\dot{y} \\
\ddot{x}(y) &= D_y x(y)\ddot{y} + \dot{y}^T D_y[(D_y x(y))^T]\dot{y}.
\end{aligned}
$$

2.2. Geometry of Point Contact

This section reviews perhaps the simplest robot force control problem - that of a robot manipulator consisting of n rigid links whose end-effector is in point-contact with a smooth rigid surface. We assume the availability of the following information and conditions:

1. The forward kinematic map to the end-effector contact point, $x : \mathcal{J} \mapsto \mathcal{W}$, a smooth function from jointspace to cartesian workspace, is exactly known.

2. A smooth function, $\phi : \mathcal{W} \mapsto \mathbb{R}^1$, which implicitly represents the rigid surface, is exactly known and possesses the following properties:

   - $\phi(x) = 0$ everywhere the surface.
   - $\phi(x) \neq 0$ everywhere off the surface.
   - $\|\nabla \phi(x)\| \neq 0$ on the surface.

   Define $\tilde{\phi}(q) = \phi \circ x(q)$ as the implicit representation of the surface in joint space.

3. The robot end-effector is (holonomically) constrained to be in contact with the surface, $\tilde{\phi}(q) = 0$. This assumption means that the endpoint of the robot end-effector is "captured" in the surface.

4. The reference position trajectory, $r(t)$ have been specified in advance such that $\tilde{\phi}(r(t)) = 0$.

5. The kinematic Jacobian, $J_x(q) = D_q x(q)$ is full rank everywhere on the surface.

The following sections, explicit dependence on the independent variable q and t will commonly be omitted. With these functions in hand, we see immediately that the robot is geometrically constrained in position, velocity and acceleration.

$$
\begin{aligned}
0 &= \tilde{\phi}(q) \\
0 &= D_q \tilde{\phi}(q)\dot{q} \\
0 &= D_q \tilde{\phi}(q)\ddot{q} + \dot{q}^T D_q[(D_q \tilde{\phi}(q))^T]\dot{q}
\end{aligned}
$$

Define the following useful functions:

1. $n(x) = \nabla \phi(x) \|\nabla \phi(x)\|^{-1}$ is the $n \times 1$ workspace surface normal vector at point x. The normal vector $n(x)$ should not be confused with the integer n, the number of links.

2. $\tilde{n}(q) = \nabla \tilde{\phi}(q) \|\nabla \tilde{\phi}(q)\|^{-1}$ is the $n \times 1$ jointspace surface normal vector at point q.

3. $P(q) = \tilde{n}(q)\tilde{n}(q)^{\mathrm{T}}$ is an $n \times n$ rank 1 matrix function whose image is the surface normal at point q and whose kernel is the surface tangent at q.

4. $Q(q) = I - P(q)$ is is an $n \times n$ rank $n - 1$ matrix function whose image is the surface tangent at point q , and whose kernel is the surface normal at q.

5. $v(q) = J^{\mathrm{T}}(q)n(x(q))$ is an $n \times 1$ vector which maps normal surface force magnitudes into corresponding joint forces through the transpose of the kinematic jacobian at point q.

2.3. Dynamics of Point Contact

For simplicity of exposition we assume the point contact between robot and surface is frictionless. For the case including various surface friction models, the reader is referred to [1, 7]. In the case of a frictionless surface, the contact force is constrained to act along the surface normal at the contact point, and can be represented as a scalar quantity $f(t)$ (in Newtons) acting along the the unit surface normal vector $n(x)$.

The equations of motion of a mechanical system in local coordinates presence of external forces arising from (i) the earth's gravitational potential, g, (ii) independently controlled torque actuators, τ, and (iii) a contact force normal to the contact surface $v(q)f(t) = J(q)^{\mathrm{T}}n(q)f(t)$ take the form

$$\ddot{q} = -M^{-1}(q)[C(q, \dot{q})\dot{q} + g(q) + v(q)f - \tau] \tag{1}$$

where $M(q)$ and $C(q, \dot{q})$ are the standard inertial and coriolis matrices respectively.

2.4. An Orthogonalized Sliding Mode Force Controller

The problem addressed in this review is the construction of a control law, $\tau(t)$, that causes the robot's position to track r asymptotically exactly, that is, $q \to r, f \to f_r$.

Building on experience with globally stable controllers for unconstrained trajectory tracking [10, 9, 14], we suspect that a controller derived from the following general form will do the job:

$$\tau_{force} = M(q)\ddot{r} + C(q, \dot{q})\dot{r} + g(q) - \tau_{pd} + v(q)f_r. \tag{2}$$

156

Where

$$\tau_{pd} = Ke; \quad K = [K_1, K_2]; \quad e = \left[\begin{array}{c} q - r \\ \dot{q} - \dot{r} \end{array} \right]$$

is ordinary PD state feedback.

The first reported force controller which is locally provably correct with respect to the full nonlinear robot model appears to have been reported by Arimoto et. al. in [2]. In [2] the authors employ the error coordinate system

$$
\begin{aligned}
\bar{e}(e) &= Q(q)(e_2 + \Lambda e_1) + \gamma \bar{v}(q) \Delta F \\
&= [Q(q)\Lambda, Q(q), \gamma \bar{v}(q)] \left[\begin{array}{c} e_1 \\ e_2 \\ e_3 \end{array} \right]
\end{aligned}
\tag{3}
$$

Where $\Lambda = \Lambda^{\mathbf{T}} > 0$, $\bar{v} = v/\|v\|^2$, γ is a positive scalar design parameter, and

$$
\begin{aligned}
e_1 &= q - r \\
e_2 &= \dot{q} - \dot{r} \\
e_3 &= \Delta F(t) = \int_0^t \Delta f(t) dt
\end{aligned}
$$

where $\Delta f(t) = f(t) - f_r(t)$.

This type of error coordinate system is often referred to as a "sliding mode" system, and was first reported in the robotics literature for the problem of unconstrained robot motion in [10]. The general idea is as follows: First choose an exponentially stable subspace of the state space, and define an error metric which "measures" the distance from this subspace. Second, construct a feedback controller which can be shown to drive the closed loop error dynamics (in an L^2 sense) onto the stable subspace. Finally, argue that the state of an exponentially stable system driven by an L^2 signal must also go to zero.

If the PD component of (2) is chosen in a time-varying "critically damped" fashion using (i) state feedback in the subspace tangent to the contact surface and (ii) integral force feedback in the normal subspace, then setting τ_{pd} in (2) to

$$\tau_{pd} = K_2 \bar{e}$$

causes the "distance" to the subspace satisfying the equality $\bar{e}(e) = 0$ that may be measured by

$$V = \frac{1}{2}[\bar{e}]^{\mathbf{T}} M \bar{e} + \frac{1}{2}\gamma \Delta F^2$$

to vary along motions of the closed loop system as

$$\dot{V} = -\bar{e}^{\mathbf{T}} K_2 \bar{e} + \bar{e}^{\mathbf{T}} \{ M(q)\dot{\eta} + C(q, \dot{q})\eta \}$$

where
$$\eta = P(q)\dot{r} + Q(q)\Lambda e_1 + \gamma\bar{v}(q)\Delta F$$
and
$$\bar{e} = e_2 + \eta.$$

Thus setting
$$\tau_{idef} = \tau_{force} - M(q)\dot{\eta} - C(q,\dot{q})\eta \qquad (4)$$
implies
$$\dot{V} = -\bar{e}^{\mathrm{T}} K_2 \bar{e}$$

It follows that all motion tends toward the set satisfying the equality $\bar{e} = 0$.

The problem with the error coordinate system (3) is that the subspace for which $\bar{e} = 0$ is *not* globally exponentially stable, in consequence of Q's rank deficiency. Thus the convergence of $\bar{e}(e)$ does not directly imply the convergence of e (state and force tracking error). This is in contrast to the case of unconstrained tracking. Indeed a simple counter example will demonstrate it is not possible obtain *global* asymptotic tracking with this type of controller. Thus a second regimen of arguments [1, 7], which fall beyond the scope of the present review, taking into account initial tracking error and surface geometry is required to conclude *local* asymptotic stability of the force and tracking errors.

Adaptive Extension

As with most model-based controllers, once the stability of the underlying non-adaptive control algorithm is assured, the extension to the adaptive case is straightforward. The reader is referred to [1, 7] for a detailed derivation.

3. Experimental Setup

In order to investigate the *practical* utility of this algorithm, we have recently completed an experimental robot system for comparing the performance of the new force controller with several well known existing force controllers, e.g. [15, 16, 5, 12, 4]. This section describes our new experimental system. It is organized as follows: Section 3.1 describes an experimental robot system for the testing of these new algorithms. Then, in Section 3.2, a new overload-protected force sensing wrist for robot force control applications is described.

3.1. The Toshiba Direct Drive Arm

The arm used for these experiments [3, 8] is a seven degree of freedom direct-drive arm developed at the Toshiba Corporation for advanced robot controls research. Each joint is equipped with a direct-drive DC motor, and high resolution 10^6 count laser optical encoder. For these force control experiments we require a minimum of three degrees of freedom. For mechanical safety and

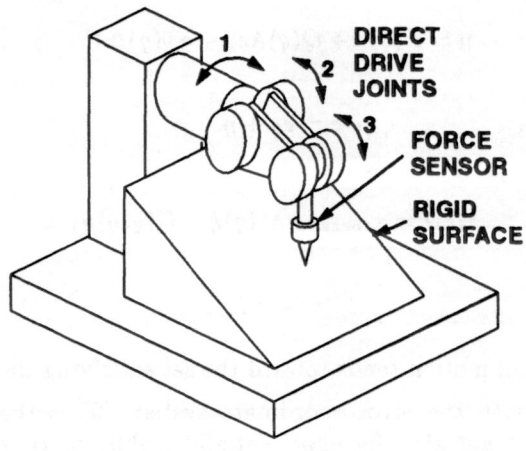

Figure 1. The Toshiba Direct Drive Arm.

simplicity, the original seven degree of freedom arm was been modified by removing a fragile 3-axis direct-drive wrist and locking one of the proximal joints. For these experiments, the arm had 3 active degrees of freedom.

The permanent-magnet DC brush motors are driven by conventional PWM power amplifiers, under the control of a dedicated real-time VME-based computer system and a PC-based development host. The real-time system employs 40Mhz M96002 floating-point digital signal processors, each capable of delivering 60 MFLOPS peak performance. The system is programmed in C, and uses commercially available compilers, symbolic debuggers, and development tools.

3.2. A New Force/Torque Limiting Wrist for Force Sensor Applications

The problem with commercial force-sensors for many robot applications is their combination of poor accuracy and low failure strength. Commercial force/torque sensors are sized by their maximum (full-scale) sensing capacity. Manufacturers often claim accuracy in the range of 0.1% to 0.5% of full scale. Practical experience suggests that somewhat poorer accuracies are often obtained. The failure strength (load at which the sensor is permanently damaged) for typical sensors are between 200% and 500% of full scale.

Roughly speaking, tool forces in contact with a stiff environment arise from three sources: (1) motor torque, (2) gravity, and (3) impact. The first two forces arise when a tool is in contact with an object, and the arm is being

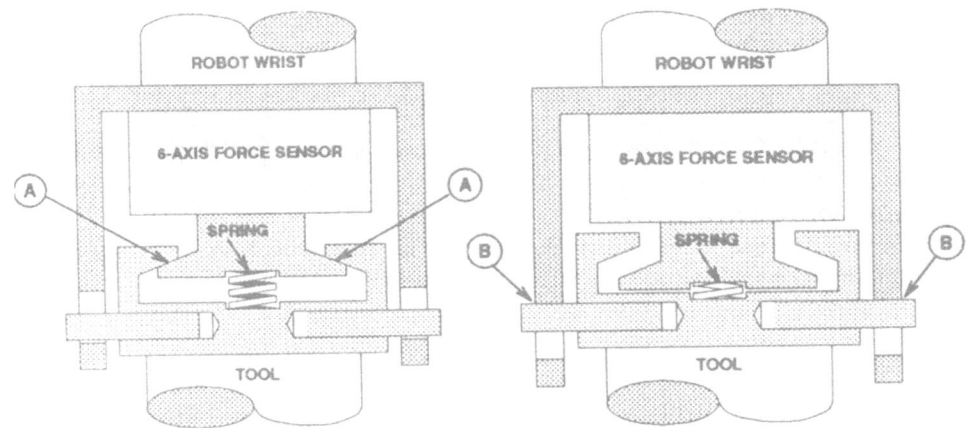

Figure 2. The Safety Wrist: A: (left) normal operating configuration. B: (right) in force overload configuration.

acted upon by forces from its motors and gravity. Maximum motor torque and gravity loads are, of course, easily computed. Impact forces are generated when a moving arm strikes a stiff object. They are nearly instantaneous and notoriously difficult to estimate. Experience shows that impact forces can easily exceed the other contact forces by several orders of magnitude.

Commercial force sensors large enough to survive peak impact loads will provide poor sensing accuracy in normal operation in consequence of their limited resolution. Thus it is often impossible to obtain required accuracy (small sensor required) while simultaneously being able to withstand peak loads (big sensor required).

Our solution to this problem is a new force sensor designed to safely withstand all three types of overloading. Figure 2 shows the "Safety Wrist" which incorporates a commercially available 6-axis force/torque sensor into a novel mounting designed to absorb forces and torques above a pre-determined threshold.

The wrist consists of a specially designed rigid "housing" which is attached to the robot arm, and a movable "tool mount" which is attached to the robot tool. It functions as follows:

1. **Normal Operation:** When applied forces and torques are below threshold, 100% of the tool force is transmitted to the force sensor via a preloaded spring (see Figure 2.A). The pre-loaded spring presses the tool mount into a bevel seat (labeled "A" in Figure 2.A) in the housing, which precisely locates the tool position. In normal operation the wrist is *completely rigid*.

2. **Overload Operation:** When applied forces and torques are above threshold, (see Figure 2.B), the pre-loaded spring is compressed as the tool

mount leaves the bevel seat, and is captured by mechanical stops in the housing (labeled "B" in Figure 2.B). All excess force is applied directly to the housing at point "B" and bypasses the sensor.

The safety wrist is designed to limit both applied forces and moments by passive mechanical "clipping" of forces and torques greater than a pre-set threshold. A built-in emergency-stop switch disables the robot servos when this force/torque threshold is exceeded. The threshold may be easily set by adjusting the spring pre-load without dis-assembling the unit.

4. Force Control Experiments

The practical advantages and disadvantages of various force control algorithms can most clearly demonstrated by actual working implementations. Proving the stability of a control algorithm and simulating its performance is essential, but does not demonstrate the utility of the controller on an actual practice. Of particular importance to the practicing engineer is the *comparative* performance of a new controller in contrast to various existing control algorithms. As an initial test of this new system, we have implemented and tested two force controllers. This section describes our preliminary experiments.

4.1. IDCFA: Adaptive Model Based Force Control

The first controller is the adaptive version of the controller described in Section 2.4, and will be called "IDCFA"[1]. Adopting the commonly used notation for robot dynamics in which a vector, θ, of unknown plant parameters is explicitly factored out [1, 7], this controller can be written

$$\tau_{idcfa} = W(\ddot{r} - \dot{\eta}, \dot{r} - \eta, \dot{q}, q)\theta - K_2\bar{e} + v(q)f_r \tag{5}$$

and the parameter update law is

$$\dot{\theta} = -\alpha W(\ddot{r} - \dot{\eta}, \dot{r} - \eta, \dot{q}, q)^{\mathbf{T}}\bar{e} \tag{6}$$

Where $\bar{e}$ and η are defined in Section 2.4 and α is a positive scalar valued adaptation gain.

4.2. PDF: Proportional-Derivative Position and Force Control

The second controller is a non-adaptive controller which does not use a plant model, and will be called "PDF" for proportional-derivative force controller. It takes the form

$$\tau_{idcfa} = -K_2\bar{e} + v(q)f_r. \tag{7}$$

---

[1] Acronym for "Inverse Dynamics Critically Damped Force Adaptive".

As with the IDCFA controller, the PDF controller has two distinct components. In the joint space subspace tangent to the rigid surface, the controller uses simple PD feedback. In the one-dimensional joint space subspace normal to the surface, the controller uses feedforward of the desired surface normal force, as well as feedback of the integral of the force error. The difference between the IDCFA and the PDF controller is that the PDF controller omits all model-based plant compensation. The IDCFA and PDF controller are identical when the IDCFA parameter vector θ has value zero.

4.3. Trajectories, Gains, and Plant Models

The feedback gain matrices used in all controllers were identical: they were selected empirically to give approximately critically damped response to the individual joints when in independent motion. In these experiments we set the adaptation gains to be as large a multiple of I as possible while preserving stability.

The experimenter also must choose initial values for the adaptive controller model parameter vector θ for each run. In all cases we initialized the IDCFA adaptive model parameters to zero.

The position feedback signals used in the experiments were obtained from 10^6 count optical encoders. The velocity signals were obtained by (i) numerically differentiating the position signals at 333 Hz and (ii) smoothing the resulting velocity signal with a recursive first-order low-pass digital filter with 10 ms time constant.

For the IDCFA controller's computational model we employed the *exact* Lagrangian dynamical equations, for fully general link inertia tensors (including the off-diagonal terms), without omission or approximation of a single term. The "C" source code equations were generated by a program written for the symbolic mathematics environment Mathematica. Included (except where noted) in the model-based controller implementations, though omitted for clarity from the equations of Section 2, is a coulomb and viscous friction [11] compensation term as well as a zero-torque DAC offset term [13] for each joint. Each joint therefore used 13 parameters. Two additional parameters are used for compensation of tool tip coulomb and viscous friction with the rigid surface, for a total of 41 parameters for the whole system. In the experiments presented herein all inertial and viscous friction model terms were enabled, and the coulomb and DAC-offset terms were disabled.

At present, we have not optimized the control algorithm in the least. For the model, we presently employ "C" code directly generated from Mathematica – with no optimization. We also employ a large, highly redundant parameter set – without using the commonly accepted technique of employing a smaller-dimensional set of base parameters, e.g. [6]. Even so, the M96000's fast floating point capability permits the control laws to be evaluated at 333Hz. In the future, a relatively modest amount of optimization could (without any loss of precision) improve this rate to at least 1000Hz.

Finally, force sensors are troublesome. While the robot is under servo control, we observed ambient noise in the force sensor readings on the order of 1-2 newtons. This was obtained by observing the computer processed force readings (in the world coordinate frame) while the robot tracked non-contact trajectories in unconstrained motion.

4.4. Performance

As an initial experiment, we selected a reference trajectory in which the robot tool tip describes a circle on the surface of a rigid, flat, aluminum plate. The face of the rigid surface was inclined about 70 degrees from the horizontal. The reference trajectory circle diameter was 0.2 meters. The reference trajectory speed was a constant 0.0628 meters/second. The robot tool tip thus completes a complete traversal of the circle in 10 seconds. The initial robot position error was about 0.05 meters. The initial robot velocity error was 0.0628 meters/second.

Figure 3 shows three graphs of actual IDCFA controller performance. The top graph shows the outline of the robot tool tip actual and reference position trajectories in the plane of the rigid surface. The middle graph shows the tracking errors in X and Y (in the plane) as well as Z (normal to the plane) as a function of time. The units are meters. These two graphs confirm that, under the IDCFA controller, the robot state quickly converges to provide steady state tracking errors of well under 0.005 meters. The bottom graph shows the actual and reference tool tip surface normal force. The reference force was a constant 40 newtons. The actual steady-state force tracking error under the IDCFA controller can be seen to remain within 10% of the reference.

Figure 4 shows the corresponding three graphs for the PDF controller performance. The top graph shows the outline of the robot tool tip actual and reference position trajectories in the plane of the rigid surface. The middle graph shows the tracking errors in X and Y (in the plane) as well as Z (normal to the plane) as a function of time. The top graph and middle graph show PDF position tracking performance which is dramatically different from that of the IDCFA controller. At steady state, the PDF controller provides tracking errors of 0.025 meters - approximately five times worse than that of the IDCFA controller. The bottom graph shows the actual and reference tool tip surface normal force. Again, the reference force was a constant 40 newtons. The actual PDF steady-state force can be seen to remain within about 20% of the reference — about 200% of that of the IDCFA controller.

These plots dramatically demonstrate the performance advantage of the model-based IDCFA controller over PDF control, and the ability of the adaptive controller to rapidly "learn" the plant parameters within a few seconds. We conclude that the poor performance obtained with the PDF controller is due to its inability to compensate for the robot's dynamics. While the PDF controller

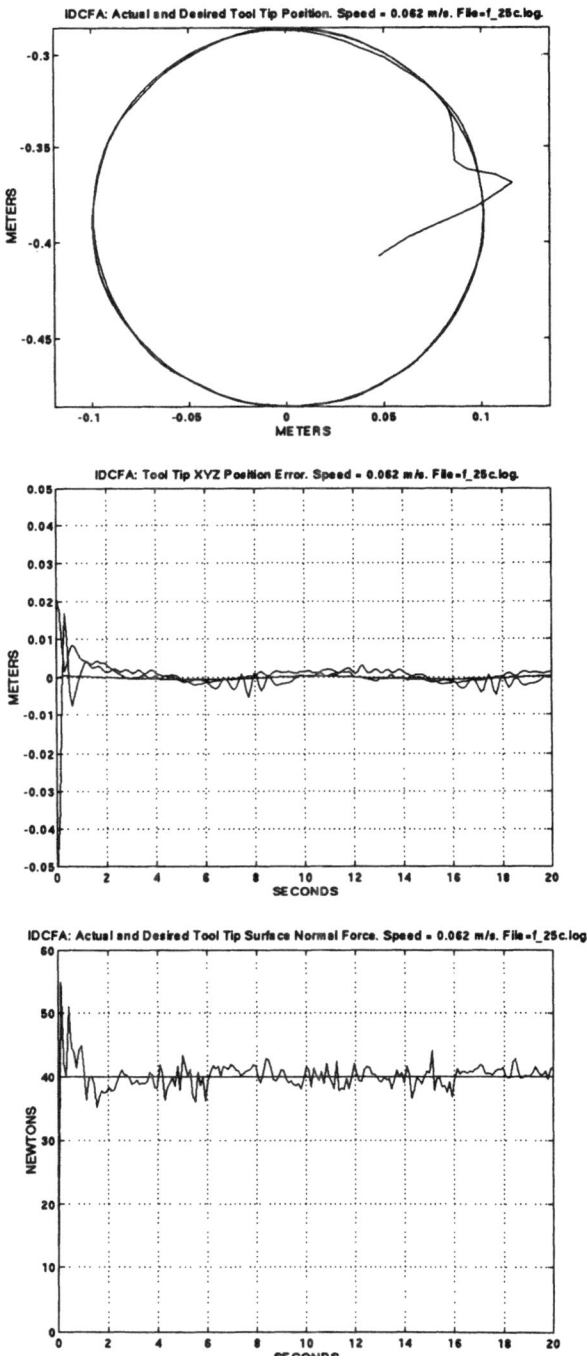

Figure 3. IDCFA: Adaptive model-based force control. Tool tip actual and reference surface trajectory (top). Tool tip X,Y, and Z tracking errors vs. time (middle). Actual and reference tool tip surface normal force vs. time (bottom).

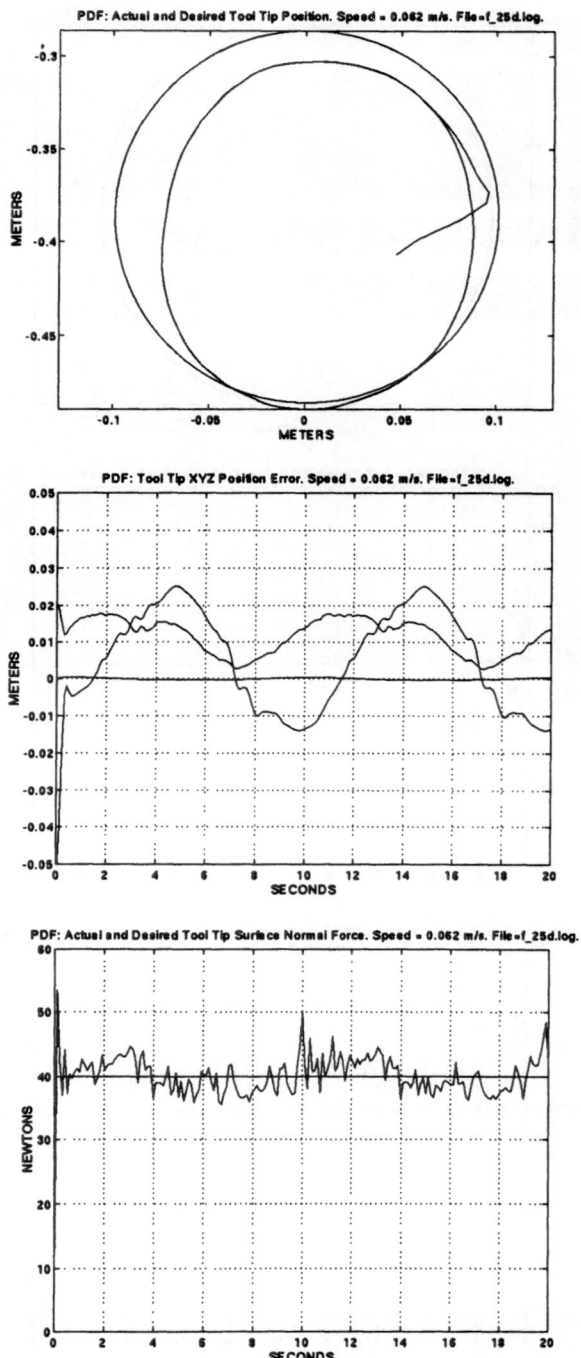

Figure 4. PDF: "PD" force control. Tool tip actual and reference surface trajectory (top). Tool tip X,Y, and Z tracking errors vs. time (middle). Actual and reference tool tip surface normal force vs. time (bottom).

is unable to compensate for these plant dynamics, the IDCFA controller is able largely to cancel these effects — thus providing superior tracking performance.

5. Conclusion

This paper has presented recent efforts toward model-based robot force control. A recently reported force control algorithm was first reviewed in Section 2. Section 3.1 then described the development of a high-performance direct-drive robot system for investigating the performance of advanced control algorithms. Section 3.2 described a new "safety wrist" force sensing wrist designed to limit both applied forces and moments by passive mechanical "clipping" of forces and torques greater than a pre-set threshold. This wrist design provides for precise rigid tool mounting, highly accurate force/torque sensing, and a rugged design which can withstand severe overloading without catastrophic sensor failure.

The preliminary experiments presented in Section 4 are, to the best of our knowledge, the first implementation of a provably correct nonlinear adaptive force controller on a high performance multi-axis direct-drive arm. The data illustrates several points. First, the position and force tracking performance of the IDCFA controller was observed to be superior to that of a conventional (non-adaptive, non-model-based) linear force control feedback law. Second, the IDCFA controller's parameter adaptation was observed to be rapid — the model parameters were seen to typically converge from a zero initial condition to steady-state within several seconds. Third, the IDCFA controller was observed to be stable under large initial position and velocity tracking error. Fourth, the mathematically complicated problem of transition from non-contact motion to contact motion (with resulting discontinuous change in the plant dynamical model structure) was observed not to be a problem in practice in this instance. The matter deserves careful attention. Finally, while the experiments presented herein demonstrated relatively slow tool-tip velocities, we have thus far observed the IDCFA controller to provide reliable, stable tracking at tool tip speeds of up to 1 meter/second. We believe that improvements in the implementation (e.g higher sample rate) will permit even higher tool tip speeds to be attained. A comprehensive report on our present experiments is in preparation.

Acknowledgments

Professor K. Asano, of the Tohoku Institute of Technology, has provided continued advice and encouragement. Mr. M. Obama of the Energy and Mechanical Research Laboratory at Toshiba Corporation provided invaluable assistance in instrumenting the new control system. Professor K. Osuka, presently at University of Osaka Prefecture, Japan, designed the original DD arm while with the Toshiba Corporation Energy and Mechanical Research Laboratory [3]. The authors are grateful for their generous contributions.

References

[1] S. Arimoto, Y. Liu, and T. Naniwa. Model-based adaptive hybrid control for geometrically constrained robots. In *Proc. IEEE Int. Conf. Robt. Aut.*, Atlanta, GA, USA, 1993.

[2] S. Arimoto, T. Naniwa, and T. Tsubouchi. Principle of orthogonalization for hybrid control of robot manipulators. In *Proceedings of IMACS'92 (Int. Symp. on Robotics, Mechatoronics, and Manufacturing Systems '92 Kobe)*, pages 16–20, Kobe, Japan, September 1992.

[3] H. Hashimoto, F. Ozaki, K. Asano, and K. Osuka. Development of a pingpong playing robot system using 7 degrees of freedom direct drive arm. In *Proceedings of IECON'87*, pages 608–615, Cambridge, MA, USA, 1987.

[4] O. Khatib. A unified approach for motion and force control of robot manipulators: The operational space formulation. *IEEE Journal of Robotics and Automation*, RA-3(1):43–53, 1987.

[5] S. Komada, K. Ohnishi, and T. Hori. Hybrid position/force control of robot manipulators based on acceleration controller. In *Proc. IEEE Int. Conf. Robt. Aut.*, pages 48–55, Sacramento, CA, USA, 1991.

[6] H. Mayeda, K. Yoshida, and K. Oshun. Base parameters of manipulator dynamic models. *IEEE Transactions on Robotics and Automation*, 6(3):312–321, June 1990.

[7] T. Naniwa, S. Arimoto, L. L. Whitcomb, Y. Liu, and J. Fujiki. Model-based adaptive hybrid control for geometrically constrained manipulators. In *Proc. of the Robotics Society of Japan Third Symposium*, Osaka, 1993. (in Japanese).

[8] F. Ozaki, H. Hashimoto, M. Muruyama, and H. Mayeda. Identification for a direct drive manipulator. In *Proceedings of IECON'90*, pages 421–426, Pacific Grove, CA, USA, 1990. IEEE.

[9] N. Sadegh and R. Horowitz. Stability and robustness analysis of a class of adaptive controllers for robotic manipulators. *The International Journal of Robotics Research*, 9(3):74–92, June 1990.

[10] J.-J. E. Slotine and W. Li. On the adaptive control of robot manipulators. *The International Journal of Robotics Research*, 6(3):49–59, Fall 1987.

[11] T. Tarn, A. Bejczy, X. Yun, and Z. Li. Effect of motor dynamics on nonlinear feedback robot arm control. *IEEE Transactions on Robotics and Automation*, 7(1):114–122, February 1991.

[12] R. Volpe and P. Khosla. An experimental evaluation and comparison of explicit force control strategies for robotic manipulators. In *Proc. IEEE Int. Conf. Robt. Aut.*, pages 1387–1393, Nice, France, 1992.

[13] L. L. Whitcomb. *Advances in Architectures and Algorithms for High Performance Robot Control.* PhD thesis, Yale University, 1992.

[14] L. L. Whitcomb, A. Rizzi, and D. E. Koditschek. Comparative experiments with a new adaptive controller for robot arms. *IEEE Transactions on Robotics and Automation*, 9(1):59–70, 1993.

[15] T. Yoshikawa and A. Sudou. Dynamic hybrid position/force control of robot manipulators — on-line estimation of unknown constraints. *IEEE Transactions on Robotics and Automation*, 9(2):220–226, April 1993.

[16] T. Yoshikawa, T. Sugie, and M. Tanaka. Dynamic hybrid position/force control of robot manipulators — controller design and experiment. *IEEE Transactions on Robotics and Automation*, 4(6):699–705, 1988.

Section 3
Visual Servoing

Visual servoing provides robots with ability to react to an unknown and/or dynamically changing environment without bodily contact. It is a key technology for increasing intelligence and autonomy of robotic systems.

Nelson and Khosla propose a visual tracking strategy that allows a camera held by a manipulator to track successfully features of moving objects while avoiding problems of focus, spatial resolution, field of view, and manipulator singularities. This strategy is obtained by specifying a sensor placement criterion and trying to keep its value as small as possible during the tracking operation. Experimental results are presented.

Espiau addresses the effect of camera calibration errors on the performance of a visual servoing algorithm which is based on an approximate value of interaction matrix relating the velocities of image points and the camera velocity. He gives two sufficient conditions for stability of the algorithm and several experimental results. From these results, it is claimed that accurate camera calibration is not necessary for the class of problems treated in the paper.

Noticing the limitations of the common visual servoing structure based on feedback and underlying axis position control, Corke proposes a new control strategy based on axis velocity control with estimated target velocity feedforward. The target velocity is estimated by using a Kalman filter under the assumption that the target has second-order dynamics with a non zero-mean Gaussian acceleration profile. The proposed strategy is verified experimentally.

Ravn, Andersen, and Sørensen investigate two automatic camera calibration methods that use multi-view 2-D calibration planes: one is a closed form solution method and the other is the two-stage calibration method of Tsai and Lenz. Their simulation and experimental results show that 2-D multi-view calibration gives results comparable to those obtained using 3-D calibration objects. They also present experimental comparison of five calibration methods from the view point of area determination.

Integrating Sensor Placement and Visual Tracking Strategies

Brad Nelson   Pradeep K. Khosla

The Robotics Institute
Carnegie Mellon University
Pittsburgh, PA 15213-3891, USA

Abstract—Real-time visual feedback is an important capability that many robotic systems must possess if these systems are to operate successfully in dynamically varying and/or uncalibrated environments. An eye-in-hand system is a common technique for providing camera motion to increase the working region of a visual sensor. Although eye-in-hand robotic systems have been well-studied, several deficiencies in proposed systems make them inadequate for actual use. Typically, the systems fail if manipulators pass through singularities or joint limits. Objects being tracked can be lost if the objects become defocused, occluded, or if features on the objects lie outside the field of view of the camera. In this paper, a technique is introduced for integrating a visual tracking strategy with dynamically determined sensor placement criteria. This allows the system to automatically determine, in real-time, proper camera motion for tracking objects successfully while accounting for the undesirable, but often unavoidable, characteristics of camera-lens and manipulator systems. The sensor placement criteria considered include focus, field-of-view, spatial resolution, manipulator configuration, and a newly introduced measure called *resolvability*. Experimental results are presented.

1. Introduction

Real-time visual feedback is an important capability that many robotic systems must possess if these systems are to operate successfully in dynamically varying and/or uncalibrated environments. In order to significantly increase the working region of a sensor providing real-time visual feedback, it is necessary to allow the visual sensor to move. An eye-in-hand system is a common technique for providing camera motion. Although eye-in-hand robotic systems have been well-studied, several deficiencies in proposed systems make them inadequate for actual use. Tracking regions must be severely constrained so that manipulators do not pass through kinematic singularities or joint limits. Objects whose depth from the camera might significantly vary can become defocused to such an extent that the object is lost, or the projections of all features on the object of interest cannot be constrained to fall on the image plane simultaneously. Camera spatial resolution may not allow a sufficiently high degree of tracking accuracy. Past eye-in-hand systems have only been able to deal with these situations by severely constraining object motion to ensure that these problems cannot arise.

We propose a tracking strategy that allows hand/eye systems to operate successfully without encountering many of the problems that previous eye-in-hand tracking systems have failed to address. Several factors are considered when controlling camera motion. Object motion induces appropriate camera motion to track the object of interest. If a camera with a fixed focal length lens is used, the eye-in-hand system can be used to move the camera closer to the task being performed in order to increase the spatial resolution of the sensor to a sufficient accuracy. Another benefit of moving the camera is to change the viewing direction with which the camera observes the object so that the spatial resolution in directions formerly along the optical axis is increased. However, the object being tracked must remain in focus and within the field-of-view of the camera. When servoing a camera mounted at a manipulator's end-effector, it is also important that the manipulator holding the camera maintains a "good" configuration far enough from kinematic singularities so that manipulator cartesian control algorithms are properly conditioned. All of these factors are allowed to potentially affect camera motion.

Camera motion can be induced by teleoperator input, as well. This allows a remote user looking only at the image produced by the camera to control camera motion without being concerned with manipulator singularities, joint limits, poorly focused objects, objects leaving the field of view of the camera, and poor spatial resolution. Systems of this type will prove useful for visually inspecting hazardous environments and for directing manipulation tasks being performed by other robots within these environments.

In the past, camera placement has been determined by considering such criteria as occlusions, field-of-view, depth-of-field, and/or camera spatial resolution off-line ([2], [9], and [11]). In none of these cases, however, is the camera actually servoed based on visual data. For dynamically changing manipulation tasks, the camera must move in real-time, so the placement of the camera must be determined quickly. Therefore, visual tracking algorithms can be effectively applied, and sensor placement criteria must be integrated into the tracking strategy. The configuration of the manipulator must also be taken into account in the control strategy.

A previously proposed visual servoing paradigm [6] is used as a framework for incorporating sensor placement criteria like those previously mentioned into an eye-in-hand robotic system. In this paper, the controlled active vision framework will first be used to derive a system model and controller for an eye-in-hand system. Dynamically determined sensor placement criteria will be presented, and it will then be shown how the control objective function can be augmented in order to introduce these various sensor placement criteria into the visual tracking control law. Our method of introducing different criteria into the control law results in hand/eye systems that can be programmed in an evolutionary way so that different behaviors can be easily introduced into the hand/eye controller. A brief description of the experimental system and presentation of experimental results complete the paper.

2. Modeling and Control of the Tracking System

We first present a system model and controller for visually tracking an object without considering any sensor placement criteria other than tracking the object. In Section 3, we will show how sensor placement criteria can be integrated into the system controller.

2.1. Modeling

To model the 3-D visual tracking problem, a pinhole model for the camera with a frame {C} placed at the focal point of the lens is used, as shown in Figure 1. A feature on an object with coordinates (X_o, Y_o, Z_o) in the camera frame projects onto the camera's image plane at (x,y). The manipulator holding the camera provides camera motion by moving the camera frame along its X,Y,Z axes. The eye-in-hand visual tracking system can be written in state-space form as

$$x_F(k+1) = A_F x_F(k) + B_F(k)u(k) + E_F d_F(k) \tag{1}$$

where $A_F = I_2$, $E_F = TI_2$, $x_F(k) \in R^2$, $d_F(k) \in R^2$, and $u(k) \in R^6$. The matrix $B_F(k) \in R^{2 \times 6}$ is

$$B_F(k) = T \begin{bmatrix} -\dfrac{f}{Z_o(k)s_x} & 0 & \dfrac{x(k)}{Z_o(k)} & \dfrac{x(k)y(k)s_y}{f} & -\left(\dfrac{f}{s_x} + \dfrac{x^2(k)s_x}{f}\right) & \dfrac{y(k)s_y}{s_x} \\ 0 & -\dfrac{f}{Z_o(k)s_y} & \dfrac{y(k)}{Z_o(k)} & \left(\dfrac{f}{s_y} + \dfrac{y^2(k)s_y}{f}\right) & -\dfrac{x(k)y(k)s_x}{f} & -\dfrac{x(k)s_x}{s_y} \end{bmatrix} \tag{2}$$

The vector $x_F(k) = [x(k) \; y(k)]^T$ is the state vector, $u(k) = [\dot{x}_c \; \dot{y}_c \; \dot{z}_c \; \omega_{xc} \omega_{yc} \omega_{zc}]^T$ is the vector representing possible control inputs, and $d_F(k)$ is the exogenous deterministic disturbances vector due to the feature's optical flow induced by object motion. The state vector $x_F(k)$ is computed using the SSD algorithm to be described in Section 2.3. In (2), f is the focal length of the lens, s_x and s_y are the horizontal and vertical dimensions of the pixels on the CCD array, and T is the sampling period between images. In order to simplify notation without any loss of generality, let $k = kT$. In addition, it is assumed that $Z_o \gg f$. This assumption holds because the focal length of our camera is 20mm, while Z_o is larger than 500mm.

Depending on the constraints placed on target motion and the objective of the visual tracking system, more than one feature may be required in order to achieve the system's goals. For example, for full 3D tracking in which it is desired to maintain a constant six degree of freedom transformation between the camera and the target, at least three non-collinear features are required [7]. To track an object constrained to move with motion in three dimensions, such as planar motion with rotations or 3D translational motion, at least two features are needed. A generalized state equation for a variable number of features can be written as

$$x(k+1) = Ax(k) + B(k)u(k) + Ed(k) \tag{3}$$

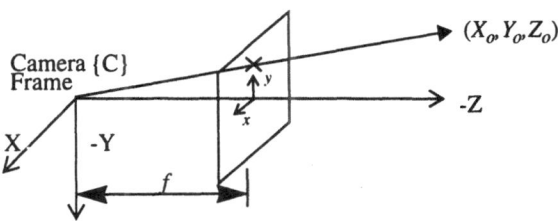

Figure 1. The pinhole camera model.

where M is the number of features required, $A=\mathbf{I}_{2M}$, $E=T\mathbf{I}_{2M}$, $x(k)\in R^{2M}$, $d(k)\in R^{2M}$, and $u(k)\in R^i$ ($i\in\{1,2,3,4,5,6\}$, the number of axes along which tracking occurs). The matrix $B(k)\in R^{2Mxi}$ is

$$B(k) = \begin{bmatrix} B_F^{(1)}(k) \\ \dots \\ B_F^{(M)}(k) \end{bmatrix} \qquad (4)$$

The superscript (j) denotes each of the feature points ($j\in\{1,2,...,M\}$). Thus, the size of B is dependent on the number of non-zero cartesian control inputs and the number of features required, which the system designer determines based on task requirements. The vector $x(k)=[x^{(1)}(k)\ y^{(1)}(k)...\ x^{(M)}(k)\ y^{(M)}(k)]^T$ is the new state vector, and $d(k)$ is the new exogenous deterministic disturbances vector.

2.2. Control

The control objective of an eye-in-hand visual tracking system is to control camera motion in order to place the image plane coordinates of features on the target at some desired position, despite object motion. The desired image plane coordinates could be changing with time, or they could simply be the original coordinates at which the features appear when tracking begins. The control strategy used to achieve the control objective is based on the minimization of an objective function at each time instant. The objective function places a cost on differences in feature positions from desired positions, as well as a cost on providing control input, and is of the form

$$F(k+1) = [x(k+1)-x_D(k+1)]^T Q [x(k+1)-x_D(k+1)] + u^T(k)Ru(k) \qquad (5)$$

This expression is minimized with respect to $u(k)$ to obtain the following control law

$$u(k) = -(B^T(k)QB(k)+R)^{-1}B^T(k)Q [x(k)-x_D(k+1)] \qquad (6)$$

The weighting matrices Q and R allow the user to place more or less emphasis on the feature error and the control input. Their selection effects the response and stability of the tracking system. The Q matrix must be positive definite, and R must be positive semi-definite for a bounded response. Although no standard procedure exists for choosing the elements of Q and R, general guidelines can be found in [6].

In Section 3, this basic controller formulation will be extended to include additional objectives rather than just feature tracking. These additional objectives create behaviors which cause the system to improve the spatial resolution of the visual data provided and to avoid undesirable camera-lens and manipulator characteristics. By introducing the objectives into the system through the control objective function, different behaviors can be easily tested to ensure the desired collective system response is achieved. Another advantage of introducing sensor placement objectives into the controller in this manor, is that the hand/eye system parameters directly affect the magnitude of the control response to the different criteria. Thus, each behavior's effect on the overall system response can be determined independent of the particular hand/eye system.

2.3. Measurement of feature positions

The measurement of the motion of the features on the image plane must be done continuously and quickly. The method used to measure this motion is based on optical flow techniques and is a modification of the method proposed in [1]. This technique is known as Sum-of-Squares-Differences (SSD) optical flow, and is based on the assumption that the intensities around a feature point remain constant as that point moves across the image plane. A more complete description of the algorithm and its implementation can be found in [6].

3. Dynamic Sensor Placement Criteria

Several different criteria can be used to influence camera motion. This section presents several dynamically determined sensor placement criteria, as well as a technique for effectively integrating all of the criteria into the visual tracking control law.

3.1. Measure of Focus

Keeping features in focus is important to the success of the SSD optical flow algorithm. Several techniques for measuring the sharpness of focus are investigated in [3]. One problem with traditional focus measures is that they are dependent on the scale of the feature. When adjusting the focal ring, feature size changes only slightly so scaling effects can be ignored. However, in dynamic sensor placement strategies, changing the depth by moving the camera is the only way to bring a fixed focal length camera-lens system into focus. This means that greater changes in scale must be tolerated, and thus, traditional focus measures prove inadequate. Because of this, a Fourier transform based focal measure is used which analyzes the high frequency content of the feature in the most recent image to determine whether the object being tracked is within the depth-of-field of the camera.

A well accepted model of the point spread function of a camera-lens system is represented by a Gaussian distribution. Since defocusing corresponds to greater attenuation of high frequencies, the Gaussian distribution representing the point spread function of a camera-lens model becomes wider as the sharpness of focus decreases, and the Fourier transform of the camera lens system, which is also Gaussian, becomes narrower. One would therefore assume that the high frequency magnitudes of the Fourier transform of the feature window would become smaller. While this is true, it is also true that thresholds at which these high frequency magnitudes become small enough to indicate defocused features is dependent on the feature as well. This makes it difficult to determine whether the feature is significantly defocused, or whether the feature actually contains relatively few high frequency components.

In computing the fast Fourier transform of the 16x16 feature, windowing effects must be considered. The window introduces certain effects in the Fourier transform which, for a step edge, causes the Fourier transform to exhibit sinc-like behavior, as shown in Figure 2. As the sharpness of focus decreases, a Gaussian point spread function indicates that the Fourier transform becomes narrower as high frequencies become attenuated. Windowing effects become less pronounced and the Fourier transform looks less like a sinc function, as illustrated by Figure 3.

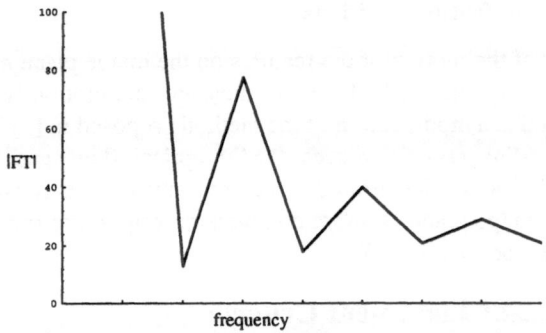

Figure 2. FT magnitude of a windowed step edge in sharp focus.

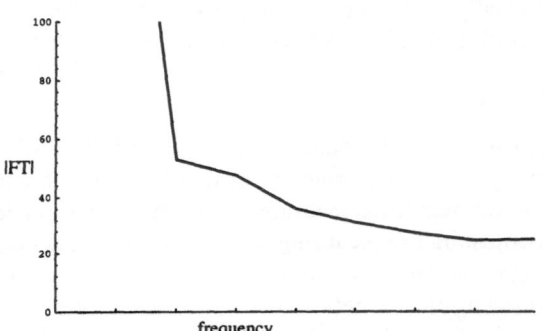

Figure 3. FT magnitude of a windowed defocused step edge.

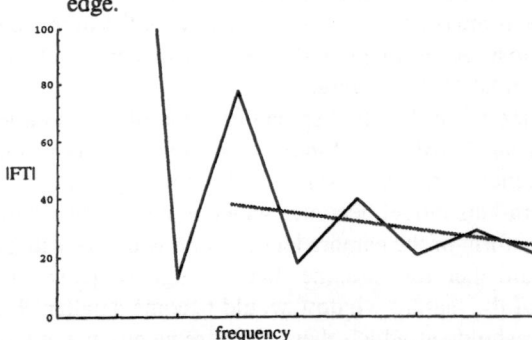

Figure 4. Least-squares line fit of FT high frequency magnitudes.

We propose to measure the degree of focus by observing how closely the high frequency components of the feature's Fourier transform approximate a sinc function. When features are sharply focused, a least-squares line fit of the high frequency components results in a large residual error shown by Figure 4. As features become less focused, the residual error decreases as the high frequency components fit a line well. Figure 5 shows the measure of focus as the object depth varies from 20cm to 80cm. At 46cm, the feature is in sharpest focus and the measure is very high. At 20cm, the linear scale of the feature is approximately 1.7 times the original scale, and at 80cm

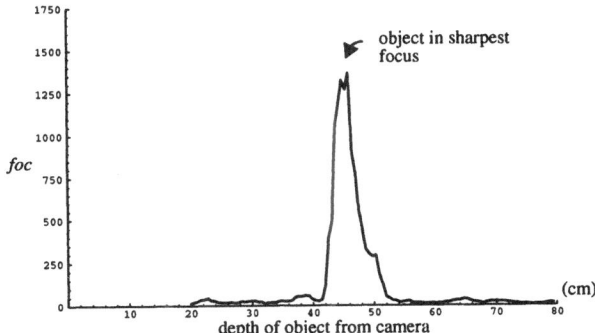

Figure 5. Proposed measure of focus of a feature on an object whose depth varies 60cm.

the scale is approximately 0.43 the original scale, but the measure is clearly independent of the overall scale change factor of four. Experimental results indicate that this measure is quite robust.

This measure can be introduced into the control objective function as

$$F(k+1) = [x(k+1) - x_D(k+1)]^T Q [x(k+1) - x_D(k+1)] + u^T(k)Ru(k) + \frac{U}{(foc)} \quad (7)$$

resulting in a control law of the form

$$u(k) = -(B^T(k)QB(k) + R)^{-1} \left[B^T(k)Q[x(k) - x_D(k+1)] - \frac{U}{2(foc)^2} \nabla^T_{u(k)}(foc) \right]$$
$$(8)$$

3.2. Spatial Resolution

A spatial resolution constraint is necessary for ensuring that objects are being observed with the maximum possible spatial resolution. For autonomous visually guided manipulation, this also allows parts being visually servoed to be brought near enough their goal so that final mating can be successfully accomplished by force control. To incorporate spatial resolution constraints into the eye-in-hand system, it is assumed that maximum spatial resolution is always desired. Thus, the depth of the object from the camera, $Z_o(k)$, is to be minimized. Introducing the spatial resolution constraint into the objective function results in an objective function and control law of

$$F(k+1) = [x(k+1) - x_D(k+1)]^T Q [x(k+1) - x_D(k+1)] + u^T(k)Ru(k) + VZ_o^2(k+1) \quad (9)$$

$$u(k) = -(B^T(k)QB(k) + R)^{-1}x \quad (10)$$
$$\left[B^T(k)Q[x(k) - x_D(k+1)] + VZ_o(k)T\begin{bmatrix} 0 & 0 & 1 & 0 & 0 & 0 \end{bmatrix} \right]$$

3.3. Field of View

In order to ensure that the projections of features being observed do not exceed the boundaries of the image plane causing the features to be lost, it is necessary to introduce a field of view constraint. A potential function can be created on the image

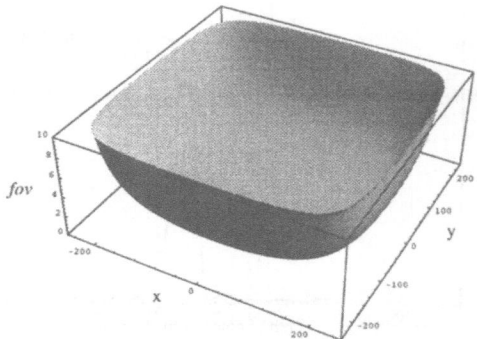

Figure 6. Field-of-view measure on the image plane.

plane, which maintains a constant value away from the image plane boundaries and approaches infinity at the boundaries. The function can be represented analytically as

$$fov = \prod_{i=1}^{n} \left(1 - \frac{x_i^2(k+1)}{x_{Bi}^2} \right) \tag{11}$$

where n is the number of system states, and x_i are the states of the system, which is equivalent to the feature coordinates. The term x_{Bi} represents the bound on the absolute value of the particular state, which is either the maximum x or y coordinate on the image plane.

Figure 6 illustrates the potential function graphically. As camera or object motion causes feature projections to approach the edges of the CCD array, the gradient of this function in cartesian space can be analytically calculated, and appropriate camera motion directions determined to cause the feature projections to move away from the boundaries.

The field of view measure can be introduced into the control objective function as

$$F(k+1) = [x(k+1) - x_D(k+1)]^T Q [x(k+1) - x_D(k+1)] + u^T(k)Ru(k) + \frac{W}{(fov)} \tag{12}$$

resulting in a control law of the form

$$u(k) = -(B^T(k)QB(k) + R)^{-1}x$$
$$\left[B^T(k)Q[x(k) - x_D(k+1)] - \frac{W}{2(fov)^2} \nabla^T_{u(k)}(fov) \right] \tag{13}$$

3.4. Manipulator Configuration

A significant problem with eye-in-hand systems is the avoidance of kinematic singularities and joint limits. In [4], an efficient technique for avoiding singularities and joint limits while visually tracking is presented. A manipulability measure of the form

$$w(q) = \left(1 - e^{-k \prod_{i=1}^{n} \frac{(q_i - q_{imin})(q_{imax} - q_i)}{(q_{imax} - q_{imin})^2}} \right) (det(J(q)))^2 \tag{14}$$

is used to avoid singularities along redundant or unconstrained tracking axes. In (14), k is a user defined constant, n is the number of joints, q_i is the ith joint angle, q_{imin} and q_{imax} are the minimum and maximum allowable joint values, respectively, for the ith joint, and $J(q)$ is the Jacobian matrix of a non-redundant manipulator. This measure is a modification of one proposed by Tsai [10], which multiplies Yoshikawa's [12] measure of nearness to singularities by a penalty function which indicates the distance to the nearest joint limit. In [4], experimental results are presented which show that tracking regions of hand/eye systems can be significantly extended by incorporating singularity and joint limit avoidance into the tracking strategy.

The objective function and control law for including the manipulability measure is represented by

$$F(k+1) = [x(k+1) - x_D(k+1)]^T Q [x(k+1) - x_D(k+1)] + u^T(k)Ru(k) + \frac{S}{w'(q(k))} \quad (15)$$

$$u(k) = -(B^T(k)QB(k) + R)^{-1}\left[B^T(k)Q[x(k) - x_D(k+1)] - \frac{S}{2w'(q)^2}\nabla^T_{u(k)}w'(q)\right] \quad (16)$$

3.5. Augmenting the Controller Function for Achieving Multiple Objectives

Some of the previously proposed placement measures can result in systems that do not function properly when implemented individually. For example, the spatial resolution constraint would tend to drive the camera toward the object to reduce the depth to zero. It becomes necessary to include a focus constraint or a field-of-view constraint to ensure that the depth does not become too small. An advantage of the controller formulation we propose is that all of these measures can be easily combined into a system that attempts to satisfy all system constraints collectively. To include all of the positive semi-definite measures previously discussed into a single control law, the objective function given by (5) becomes

$$F(k+1) = [x(k+1) - x_D(k+1)]^T Q [x(k+1) - x_D(k+1)] +$$

$$+ u^T(k)Ru(k) + \frac{S}{w'(q(k))} + \frac{U}{(foc)} + VZ_0^2(k+1) + \frac{W}{(fov)} \quad (17)$$

This results in a control law of the form

$$u(k) = -(B^T(k)QB(k) + R)^{-1}\left[B^T(k)Q[x(k) - x_D(k+1)] - \frac{S}{2w'(q)^2}\nabla^T_{u(k)}w'(q) - \right.$$

$$\left. - \frac{U}{2(foc)^2}\nabla^T_{u(k)}(foc) + VZ_0(k)T[0\ 0\ 1\ 0\ 0\ 0] - \frac{W}{2(fov)^2}\nabla^T_{u(k)}(fov)\right] \quad (18)$$

The terms representing the gradients of the focus measure foc and manipulability w can be approximated numerically in order to determine the current camera velocity which maximally increases their values. The field-of-view measure fov can be calculated analytically. Relative weights are placed on the different criteria functions by S, U, V, and W. If the cartesian axes along which visual tracking takes place are different from axes along which any sensor placement criteria may influence, it is necessary to slightly alter B_F given in (2) in order to properly use this control law. The columns of B_F which correspond to cartesian axes along which visual tracking should not occur are simply set to zero. This inhibits visual tracking along these axes, but allows the cost term $u^T R u$ to influence sensor motion along non-tracking axes due to dynamic sensor placement criteria. This also ensures that a six dimensional control input results from (18).

3.6. Resolvability

When an eye-in-hand system is used to provide visual input for a second manipulator performing a manipulation task, the sensor placement measure *resolvability* can be used in place of the spatial resolution constraint presented in Section 3.2. This measure provides a technique for estimating the relative ability of various visual sensor systems, including single camera systems, stereo pairs, multi-baseline stereo systems, and 3D rangefinders, to accurately control visually manipulated objects.

The term *resolvability* refers to the ability of a visual sensor to resolve object positions and orientations. The *resolvability ellipsoid* [5] illustrates the directional nature of *resolvability*, and can be used to direct camera motion and adjust camera intrinsic parameters in real-time so that the servoing accuracy of the visual servoing system improves with camera-lens motion. The object centered Jacobian mapping from task space to sensor space is an essential component of the sensor placement measure. A singular value decomposition of this mapping provides the six-dimensional *resolvability* measure, which can be interpreted as the system's ability to resolve task space positions and orientations on the sensor's image plane.

Figure 7 shows the *resolvability* in depth of an object 10cm in length lying in a plane parallel to the image plane. As one might expect, the plot shows that when considering the limited size of the camera CCD it is preferable to decrease depth as much as possible in order to increase *resolvability* in depth, rather than to increase the focal length of the lens. This directs the system to move the camera closer to the object rather than to increase the focal length of the lens. However, this measure does not account for the depth-of-field of the lens. When the focus measure presented in Section 3.1 is considered, the depth of the object cannot decrease beyond a particular value, depending on lens properties, before the object becomes severely defocused and untrackable. Thus, a tradeoff between focal length, depth, and depth-of-field must be achieved when considering *resolvability* in depth. Our proposed control strategy which allows multiple camera-lens-manipulator behaviors to influence camera-lens motion allows this tradeoff to be achieved. Visual servoing experiments demonstrate that *resolvability* can be successfully used to direct camera-lens motion in order to increase the ability of a visually servoed manipulator to precisely servo objects [5].

4. Experimental Results

The visual tracking algorithm described previously has been implemented on a robotic assembly system consisting of three Puma 560's called the Rapid Assembly System. One of the Pumas has a Sony XC-77RR camera mounted at its end-effector. The camera is connected to a Datacube Maxtower Vision System. The Pumas are controlled from a VME bus with two Ironics IV-3230 (68030 CPU) processors, an IV-3220 (68020 CPU) which also communicates with a trackball, a Mercury floating point processor, and a Xycom parallel I/O board communicating with three Lord force sensors mounted on the Pumas' wrists. All processors on the controller VME run the Chimera real-time operating system [8].

The experimental results were obtained with a strategy using motion about the three camera axes for visual tracking, motion along the three axes for singularity/joint limit avoidance, and motion along the camera's optical axis (Z) for increasing spatial resolution and maintaining focus. Two features on the target object were tracked so that the position and orientation of the object on the image plane could be maintained as the target moved, subject to the other sensor placement criteria. The object being tracked moved 25cm in a direction parallel to the camera's X axis. Without sensor placement criteria, the depth of the object should increase as illustrated by the dashed line in Figure 8. However, the spatial resolution constraint causes the depth to actually decrease, until the decreased focal measure causes the camera to increase the depth in order to improve focus, as shown in Figure 9. Without sensor placement criteria affecting depth and focus, the object would have become defocused to such an extent

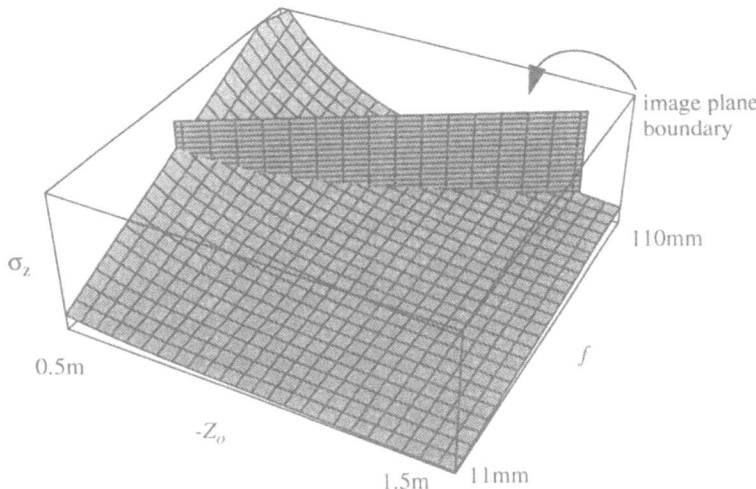

Figure 7. *Resolvability* of depth σ_z versus depth of object and focal length for two features 10cm apart lying on an object in a plane parallel to the image plane.

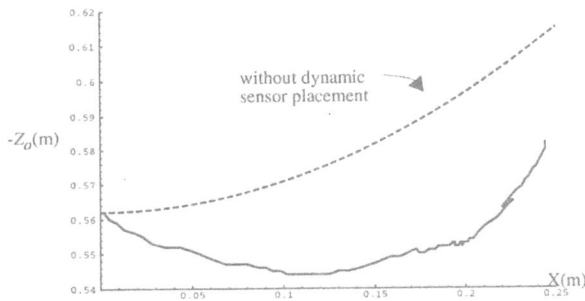

Figure 8. Depth of object from camera frame origin versus distance object translates along X.

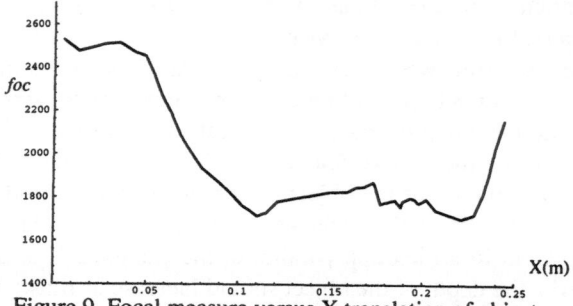

Figure 9. Focal measure versus X translation of object.

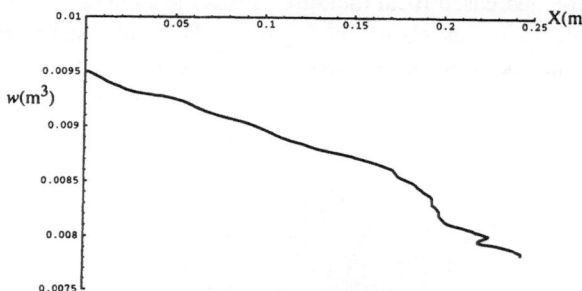

Figure 10. Manipulability versus X translation of object.

that the SSD tracker would have failed after the object moved less than 10cm. Instead, object motion of 25cm is successfully tracked.

Another criterion which, in this instance, causes the depth to increase is the singularity/joint limit avoidance criterion. Figure 10 shows that the manipulability measure decreases due to camera translation caused by the spatial resolution constraint. The decrease in manipulability occurs because the second and third joints of the Puma 560 become nearly aligned, and the first joint nears a joint limit.

5. Conclusion

The working region of a camera providing visual feedback for robotic manipulation tasks can be significantly increased by combining visual tracking capabilities with dynamic sensor placement criteria. The controlled active vision paradigm provides a useful tracking framework for integrating different sensor placement criteria into an eye-in-hand tracking system's control law. A system which accounts for focus, spatial resolution, and manipulator configuration has been experimentally verified.

This tracking strategy is also useful when teleoperator input based on image observation alone is provided to a hand/eye system remotely. The operator is relieved of the need to concern him or herself with the possibility of causing the manipulator to pass through a kinematic singularity or joint limit. Recent experiments have successfully demonstrated the capability to teleoperate a hand/eye system at Carnegie Mellon's Advanced Manipulators Lab in Pittsburgh, Pennsylvania from Sandia National Laboratories in Albuquerque, New Mexico. The teleoperator in Albuquerque inspected an object in Pittsburgh by moving the camera mounted on a Puma 560 around the object while the system simultaneously visually tracked the object. The operator's feedback (in Albuquerque) came solely from images relayed from the hand/eye camera (in Pittsburgh). Although delays of up to one second sometimes occurred during the image transfer, the hand/eye system at Carnegie Mellon continued to operate successfully and provided useful visual input to the teleoperator without entering singular configurations or violating joint limits.

Future work will include a greater number of dynamic sensor placement criteria in the control law, such as occlusion avoidance, and will allow the observed object to be visually servoed by a second manipulator for automatic assembly while using the newly introduced *resolvability* sensor placement measure. A motorized zoom/focus/aperture lens has recently been incorporated into the system in order to further enhance system capabilities, and control issues related to dynamically reconfigurable sensors are currently being explored.

Acknowledgments

This research was supported by the U.S. Army Research Office through Grant Number DAAL03-91-G-0272 and by Sandia National Laboratories through Contract Number AC-3752D.

References

[1] P. Anandan, "Measuring Visual Motion from Image Sequences," Technical Report COINS-TR-87-21, COINS Department, University of Massachusetts, 1987.

[2] C.K. Cowan and A. Bergman, "Determining the Camera and Light Source Location for a Visual Task," *Proc. of the 1989 IEEE Int. Conf. on Robotics and Automation*, 509-514, 1989.

[3] E. Krotkov, "Focusing," *International Journal of Computer Vision*, (1), 223-237, 1987.

[4] B. Nelson and P.K. Khosla, "Increasing the Tracking Region of an Eye-in-Hand System by Singularity and Joint Limit Avoidance," *Proc. of the 1993 IEEE Int. Conf. on Robotics and Automation*, 3:418-3:423, 1993.

[5] B. Nelson and P.K. Khosla, "The Resolvability Ellipsoid for Visual Servoing," Technical Report CMU-RI-TR-93-28, The Robotics Institute, Carnegie Mellon University., 1993.

[6] N.P. Papanikolopoulos, B. Nelson, and P.K. Khosla, "Monocular 3-D Visual Tracking of a Moving Target by an Eye-in-Hand Robotic System," *Proc. of the 31st IEEE Conf. on Decision and Control (31stCDC)*, 3805-3810, 1992.

[7] N.P. Papanikolopoulos and P.K. Khosla, "Robotic Visual Servoing Around a Static Target: An Example of Controlled Active Vision," *Proc. of the 1992 American Control Conference*,1489-1494, 1992.

[8] D.B. Stewart, D.E. Schmitz, and P.K. Khosla, "The Chimera II Real-Time Operating System for Advanced Sensor-Based Control System," *IEEE Trans. Sys., Man and Cyb.*, 22, 1282-1295, 1992.

[9] K. Tarabanis, R.Y. Tsai, and P.K. Allen, "Automated Sensor Planning for Robotic Vision Tasks," *Proc. of the 1991 IEEE Int. Conf. on Robotics and Automation*, 76-82, 1991.

[10] M.J. Tsai, *Workspace Geometric Characterization of Industrial Robot*, Ph.D. Thesis, Ohio State University, Department of Mechanical Engineering, 1986.

[11] S. Yi, R.M. Haralick, and L.G. Shapiro, "Automatic Sensor and Light Positioning for Machine Vision," *Proc. of the 10th Int. Conf. on Pattern Recognition*, 55-59, 1990.

[12] T. Yoshikawa, "Manipulability of Robotic Mechanisms," *Robotics Research 2*, eds. H. Hanafusa and H. Inoue, 439-446, MIT Press, Cambridge, MA, 1985.

Effect of Camera Calibration Errors on Visual Servoing in Robotics

Bernard Espiau

INRIA Rhône-Alpes
46, avenue Félix Viallet
38031 Grenoble Cedex
France

Abstract— This paper addresses the problem of the influence of camera calibration errors on the performances of a control algorithm. More precisely, we examine the effects of errors in the values of intrinsic and extrinsic camera parameters on the stability and the transient behavior of a visual servoing scheme. After having recalled the basic principles of the visual servoing approach, we derive a general form for the interaction matrix associated with a set of image points, which makes all the contributions of camera parameters explicit. We then give briefly some examples of analytical stability results in very simple cases. Since any further theoretical analysis appears extremely difficult, we then present a purely experimental study of the problem. After having given a few conclusions taken from a simulation step, we select and comment some results of the experimental approach based on a robot/camera testbed. We observe an apparently good robustness of the method. General comments on the approach and guidelines for future work are given in the conclusion.

1. Introduction

The calibration of cameras in robotics is generally presented as an unavoidable operation. This is true when accurate metric information is required, for example in 3D reconstruction or in image-based metrology. In that case, the needed calibration stage requires clever experimental procedures, which takes time, and are often tricky to implement.

One may however wonder if such procedures are indeed really necessary in all vision-based robotics applications. A partial answer is given by researchers from computer vision which claim that a lot of interesting things may be done without any explicit calibration, by only using the matching of several image points and working in projective spaces (see for example [3], [8], [5]). This approach is to be used in a stereoscopic or multi-imaging framework.

Another possible point of view, topic of this paper, is a consequence of the visual servoing philosophy. Let us recall that, in that last case, an array of measurements taken from a set of images is used as an error function in order to compute at each time a control vector. This last is applied to the system (robot and cameras) and makes it moving in order to finally reach a desired situation directly depicted in the images. The knowledge of 3D issues is often unnecessary. In turn, it is important that the controlled system exhibit an acceptable behavior in terms of reliability and efficiency. Besides, since the system is in closed-loop, the effects of some kinds of errors and disturbances may be interestingly compensated for in a implicit way. More precisely, the problem becomes now the one of *robustness* with respect to modelling uncertainties, for example calibration errors or bad knowledge of 3D geometry.

The aim of this paper is to study, mainly in an experimental way, the influence of such uncertainties on the behavior of a vision-based robot servo-loop. The paper is structured as follows: in a first part, we recall the principles (cf [1], [4], [6]) of the visual servoing approach: modelling of the interaction and basics of control. Then, we derive a general form for the jacobian matrix of the problem associated with image points, called interaction matrix, in which the contribution of every uncertainty explicitly appears: intrinsic and extrinsic parameters, 3D coordinates. Since their exists a sufficient matrix positivity condition for stability, we try to use it in order to evaluate the robustness of the system. Unfortunately, it appears that, due to its highly non-linear and coupled character, this condition becomes quickly untractable except in simple cases. Furthermore, since it is only sufficient, it does not provide the user with enough practical information. This is why we then conduct an experimental study of the influence of selected parameter uncertainties on the efficiency of the control scheme. Using a four points target, we mainly examine the behavior of the tracking error norm for various initial conditions and in different cases of intrinsic and extrinsic parameter errors. The paper ends with some further considerations about robustness of visual servoing.

2. Principles of Visual Servoing

We do not describe this now classical approach in depth. All details, theoretical analysis, and the study of more general cases may be found in [1], [7] and [2]. Let us recall here only the issues which are necessary and sufficient for understanding the following.

2.1. Modelling

Let us consider a camera handled by a robot, the configuration of which is parametrized by ψ, $dim(\psi) = 6$. It is assumed that we lie out of kinematics singularities, i.e. $\frac{\partial \bar{r}}{\partial \psi}$ is nonsingular, which $\bar{r} \in SE3$, the group of displacements.

We now observe in a image a set of "signals" $s(\bar{r}) = \{s_i , i = 1..n\}$, and, given desired values s^*, we would like to have

$$s - s^* = 0, \tag{1}$$

These signals are usually geometrical features like point coordinates, line parameters, etc... When constraints (1) are compatible, they define a *virtual linkage* (cf [1]). Differentiating (1) with respect to $\bar{r}$ (and not to ψ owing to the nonsingularity) gives:

$$\dot{s} = \frac{\partial s}{\partial \bar{r}} \dot{\bar{r}} = 0 \tag{2}$$

It may be shown ([7]) that, in (2), $\frac{\partial s_i}{\partial \bar{r}}$ may be considered as a screw, H_i, called *interaction screw*. $\dot{\bar{r}}$ is the velocity screw of the camera, $\tau = \{V, \Omega\}$. Like in a real linkage, free motions lie in the subspace spanned by the set $\{H_1, \ldots, H_n\}$. The associated matrix form is $\frac{\partial s}{\partial \bar{r}} = L^T$, called *interaction matrix*. It gathers all the information related to the interaction between robot and environment, i.e., here, to the variation of the image signals with respect to camera motion.

2.2. Control

2.2.1. A Simple Approach

We have now to design a control scheme allowing to realize the constraint (1). Let us consider here that the control variable is simply the velocity screw τ. Let us also assume that dim $(s) = 6$ and that L^T is nonsingular. By setting $e = s - s^*$, the exponential convergence of every component of e is ensured by choosing the control:

$$\tau = -g L^{-T} e, \tag{3}$$

where g is a positive scalar, since then: $\dot{e} = \frac{\partial s}{\partial \bar{r}} \tau = -ge$. However, and this is our concern in this paper, since L^T depends on camera parameters and on some 3D environment features which may be badly known, we may only use in the control an approximated expression denoted $\hat{L}^T$. Therefore:

$$\tau = -g \hat{L}^{-T} e \tag{4}$$

and

$$\dot{e} = -g L^T \hat{L}^{-T} e \tag{5}$$

By writing

$$\frac{d}{dt}(\frac{1}{2}\|e\|^2) = e^T \dot{e} = -g e^T L^T \hat{L}^{-T} e \tag{6}$$

we see that a sufficient condition for having $\|e\|$ decreasing is that $L^T \hat{L}^{-T}$ be a positive matrix (condition C1). See [7] for a more general study.

2.2.2. Case of an Hybrid Task

A common case is rank $(L^T) = p < 6$. The solution to (1) is then non unique. As shown in [7], a nice way of setting the problem consists in defining an *hybrid task* satisfying (1) while tending to minimize a cost function h_s representing a trajectory tracking. We then set:

$$e = W^\dagger e_1 + \alpha(I_6 - W^\dagger W)\frac{\partial h_s}{\partial \bar{r}} \tag{7}$$

where α is a positive scalar, W a matrix such that $R(W^T) = R((\frac{\partial e_1}{\partial \bar{r}})^T)$ and

$$e_1 = D(s - s^*) \tag{8}$$

where D is a $p \times n$ conbination matrix to be found, such that DL^T be of full rank. Let us now give again a sufficient convergence condition. It may be shown ([7]) that, if

$$DL^T W^T > 0 \qquad (9)$$

then $\frac{\partial e}{\partial \bar{r}}$ is positive. By choosing a control scheme of type (4), i.e.

$$\tau = -g\left(\frac{\hat{\partial} e}{\partial \bar{r}}\right)^{-1} e, \qquad (10)$$

then convergence is ensured by choosing $\left(\frac{\hat{\partial} e}{\partial \bar{r}}\right)^{-1} = I_6$. Then:

$$\frac{d}{dt}\left(\frac{1}{2}\|e\|^2\right) = -ge^T \frac{\partial e}{\partial \bar{r}} e < 0 \qquad (11)$$

A combination matrix D satisfying (9) is then $D = WL$ and W is a matrix the columns of which are a basis of the subspace $\{H_1, \ldots, H_n\}$".

Since, as previously, L may be badly known, the condition (9) becomes:

$$W\hat{L}L^T W^T > 0 \qquad (12)$$

that is to say (condition C2)

$$\hat{L}L^T \geq 0 \qquad (13)$$

Finally, by noting that $\hat{L}L^T = \hat{L}L^T(\hat{L}^{-T}\hat{L}^T)$, we see that conditions C1 and C2 take analogous forms. We will use that later.

3. The Interaction Matrix for Image Points

We restrict in the following our study to the case of image points. L^T is then the optical flow matrix. We derive in this section a general form for L^T in which the camera calibration parameters will appear explicitly.

Let R_0 be a reference frame, R_c the camera frame with optical centre O_c. Let $s = (u\ v)^T$ be the retina coordinates vector of the projection of a point M, with coordinates $\bar{X}^T = (X\ Y\ Z\ )^T$ in R_0, where Z is the depth. By using projective coordinates, we have $q = PM$, where $q = \delta \begin{pmatrix} s \\ 1 \end{pmatrix}$ and $M = \begin{pmatrix} \bar{X} \\ 1 \end{pmatrix}$, P being the perspective projection matrix:

$$P = AP_c D \qquad (14)$$

where:

- A is a matrix of retina coordinates change:

$$A = \begin{pmatrix} -fk_u & fk_u\ cot\theta & u_0 \\ 0 & -\frac{fk_v}{sin\theta} & v_0 \\ 0 & 0 & 1 \end{pmatrix} = \begin{pmatrix} A_{11} & \alpha \\ 0\ 0 & 1 \end{pmatrix} \qquad (15)$$

the parameters of which, called *intrinsic* are the focal length (f), the axes scale factors (k_u and k_v), the coordinates of the intersection between the optical axis and the retina plane (u_0 and v_0), and the angle between image frame axes (θ). By setting $\alpha_u = f k_u$ et $\alpha_v = f k_v$, we see that A is defined by 5 independent parameters (distortion is not considered here).

- $P_c = (I_3 \mid 0)$ is the normalized perspective projection matrix
- D is a 4×4 matrix defining the frame change $R_0 \rightarrow R_c$:

$$D = \left(\begin{array}{ccc} R_{c0} & & [O_c O_0]_c \\ 0 \quad 0 \quad 0 & & 1 \end{array} \right) \tag{16}$$

The 6 related parameters are called *extrinsic*. Knowing that, it is rather easy (see [9]) to derive the following general form for the 2×6 interaction matrix corresponding to an image point:

$$L^T = A_{11} L_0{}^T (\bar{X}') \Theta \tag{17}$$

where $L_0{}^T$ is the simplest possible form of the interaction matrix for a point:

$$L_0{}^T(\bar{X}') = \left(\begin{array}{cccc} -1/Z' & 0 & Y'/Z'^2 & X'Y'/Z'^2 \\ 0 & -1/Z' & Y'/Z'^2 & 1 + Y'^2/Z'^2 \end{array} \right.$$

$$\left. \begin{array}{cc} -(1 + X'^2/Z'2) & Y'/Z \\ -X'Y'/Z'^2 & -X'/Z' \end{array} \right) \tag{18}$$

with

$$\left(\begin{array}{c} \bar{X}' \\ 1 \end{array} \right) = D \left(\begin{array}{c} \bar{X} \\ 1 \end{array} \right) = DM \tag{19}$$

being the coordinates of $\bar{M}$ in R_c.

Since the lines of $L_0{}^T$ are screws, Θ is the associated change-of-frame operator:

$$\Theta = \left(\begin{array}{cc} R_{c0} & -R_{c0} \, As[O_c, O_0]_0 \\ 0 & R_{c0} \end{array} \right) \tag{20}$$

where $As[.]$ denotes the skew-symmetric matrix associated with a vector.

We may remark that an error on Θ means that we are using a wrong control frame. More generally, we can now see in (17) the possible origins of uncertainties:

- 6 extrinsic parameters in Θ and $L_0{}^T$,

- 3 intrinsic parameters in A_{11},

- 3 3D variables for any point M in L_0.

4. A Few Results from a Tentative Analytic Study

In ref [9], very simple examples are investigated. Let us briefly present some of the obtained results, as an illustration of the type of study which can be conducted from equation (17) and conditions C1 and C2.

Example 1 Let us consider 3 points, the 3D coordinates of which are used in the control computation. Extrinsic parameters are known, and approximate values $\hat{f}$, $\hat{k_u}$, $\hat{k_v}$ and $\hat{\theta}$ are used in the control. We have also $u_0 = v_0 = \hat{u_0} = \hat{v_0} = 0$ and $\theta = \pi/2$ for simplification. Then, a sufficient stability condition is that all the signs of $\hat{f}$, $\hat{k_u}$, $\hat{k_v}$ are right and that $\sin\hat{\theta} > 0$.

Example 2 Let us consider 3 points having the same third coordinate at the equilibrium, Z^*, which is supposed to be the only unknown. We also assume that L^T is constant around the equilibrium state (small motions). Using this approximation and the control (4), it is easy to see that a NSC for the stability is $Z^*/Z > 0$.

These two examples might allow us to expect rather interesting properties of robustness of the control with respect to calibration errors. However they remain quite simple. Moreover we did not succeed in performing a more relevant theoretical investigation, since studying positivity conditions C1 or C2 becomes quickly difficult in the general case. Furthermore, these conditions are only sufficient and, in practice, it often appears that stability margins are wider than indicated by the conditions. This is why an experimental study was conducted, the results of which are given in the following.

5. Experimental Analysis

5.1. Summary of Simulation Results

The results of an extensive simulation study are reported in [9]. Since we focus on experimental results, we only recall here the main conclusions, obtained in the same conditions as the truely experimental study (see later).

Intrinsic Parameters The influence of offset errors is small. A scaling error of ratio 2 or 3 on the focal length changes the convergence time in the same ratio, but does not modify the shape of the quasi-exponential transient step. The effect of an axis orientation error becomes significant only when it reaches unrealistic values.

Extrinsic Parameters Errors on the camera frame may have a perceptible effect when they are large enough. An attitude error (from 30 deg) induces by coupling effects a modification of the whole transient step. Concerning the position, sensitivity to an error in the Z component is greater than to errors in X or Y.

5.2. Experimental Results

As a testbed we used a camera mounted on the end effector of a 6 d.o.f high precision robot. The actual intrinsic parameters were obtained using an accurate calibration algorithm (cf [1]) : $u_0 = 383$, $v_0 = 234$, $fk_u = fk_v = 1149$. The working frame is the estimated camera frame (i.e. $D = Id$). The target is a set of 4 points which form a 30 cm $\times$ 30 cm square. Initial and final position are shown figure 1. The control, of type (4), used either a value of $\hat{L}^T$ updated at each time or its value at the equilibrium. We only present here the last case.

In each case, we plot in the left window the components of $e = s - s^*$ and in the right the evolution of $\|e\|$. We have successively:

- Figure 2: ideal case (no calibration errors)

- Figure 3: $\hat{f} = f/2$

- Figure 4: $\hat{f} = 2f$

- Figure 5: $(O_c\hat{O}_o)_Z = (O_cO_o)_Z + 5$ cm.

- Figure 6: rotation error of 5 deg around the Z axis

- Figure 7: combined errors: 20% on f, 5 deg around every rotation axis and 5 cm on every translation axis

6. Conclusion

The obtained results show that, for the class of visual servoing problems we consider, there is in general no need for an accurate calibration step which, moreover would be valid for a given working domain only. It appears that approximated values of intrinsic parameters are sufficient and that, for example, we may use directly the parameters provided by a camera supplier. This is all the more interesting in practice that working conditions may change. For example, it is known that changing aperture because of lighthing variations, or using lenses with variable focal length modifies the values of several intrinsic parameters.

It is also to be noticed that the influence of errors on extrinsic parameters may be significant when they are large. However the overall stability margin remains wide and is largely compatible with a rough calibration.

A last thing is that we have to find the target specification, s^*. An idea coherent with the proposed approach is to place the camera in the desired situation and to *measure* the values of s^* before starting the regulation step. This is quite better than computing it for example from a 3D desired position, because, in that last case, calibration errors would lead to errors on the resulting position which might be unacceptable. On the contrary, even large tracking errors during the transient step are quite acceptable provided that the reached steady state value is the right one. This is why, if it is wished to avoid any accurate calibration step, we recommend finally to use both reasonnably approximated parameters in the control and experimental measurement of the desired values of image features. This last is in fact a calibration of the application instead of a camera calibration, which is often easier and more efficient.

Now, a guideline for further studies in this framework is to consider the case of stereovision, as evoked in the introduction, where the use of "auto" or "weak" calibration is a recent trend. Robustness studies remain to be done in that case.

References

[1] F. Chaumette. La relation vision-commande: théorie et application à des tâches robotiques. *Thèse, Université de Rennes*, France, juillet 1990

[2] B. Espiau, F. Chaumette, P. Rives. A New Approach to Visual Servoing in Robotics. *IEEE Trans. on Robotics and Automation*, vol 8 no 3, june 1992

[3] O. D. Faugeras. What can be seen in three dimensions with an uncalibrated stereo rig? *European Conference on Computer Vision*, Genova, Italy, 1992

[4] J.T. Feddema, O.R. Mitchell. Vision-guided servoing with feature-based trajectory generation. *IEEE Trans. on Robotics and Automation*, vol 5 no 5, october 1989

[5] R. Mohr, L. Quan, F. Veillon, B. Boufama. Relative 3D Reconstruction Using Multiple Uncalibrated Images. *Research Report LIFIA/IMAG RT84*, Grenoble, France, 1992

[6] P. Rives, F. Chaumette, B. Espiau. Visual Servoing Based on a Task-function Approach. *1st International Symposium on Experimental Robotics*, Montreal, Canada, june 1989

[7] C. Samson, M. Le Borgne, B. Espiau. Robot Control: the Task Function Approach. *Clarendon Press, Oxford University Press*, Oxford, UK, 1990

[8] A. Shashua. Projective Structure from two Uncalibrated Images: Structure from Motion and Recognition. *AI Memo no 1363*, MIT, Cambridge, USA, sept 1992

[9] B. Espiau. Asservissement visuel en calibration incertaine. *Research Report, INRIA Rhône-Alpes*, to appear, fall 1993

Acknowledgements

The author strongly thanks François Chaumette for having performed the experimental tests and for his helpful suggestions, and Aristide Santos for his help in the simulations.

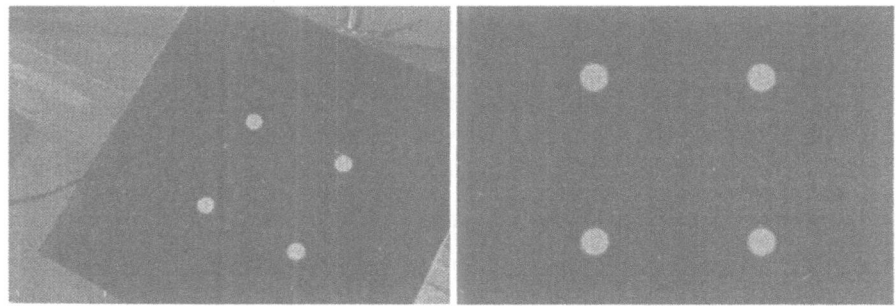

Figure 1. Initial and final (desired) images(512 × 730 pixels)

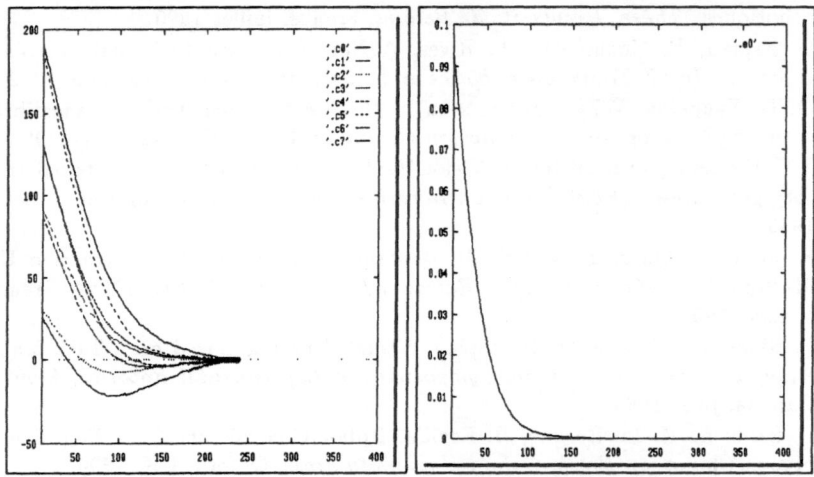

Figure 2. No calibration error

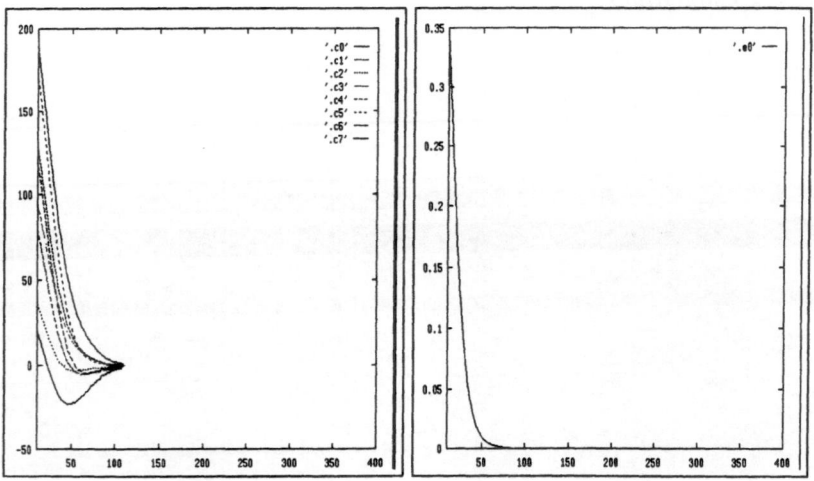

Figure 3. The used focal length is half the true one

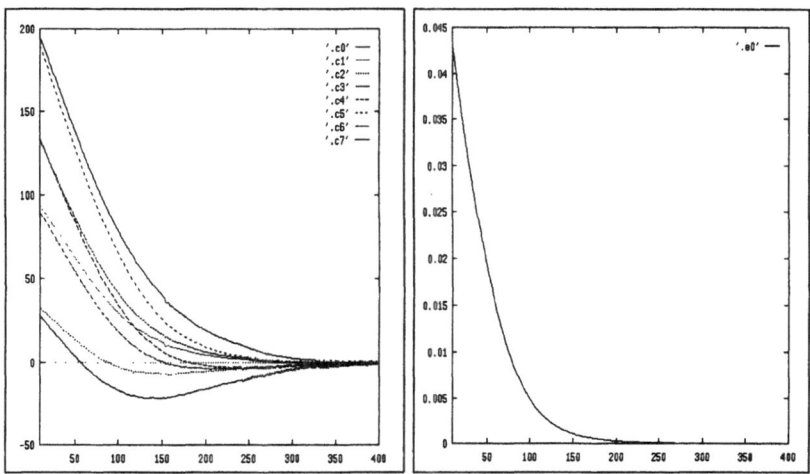

Figure 4. The used focal length is twice the true one

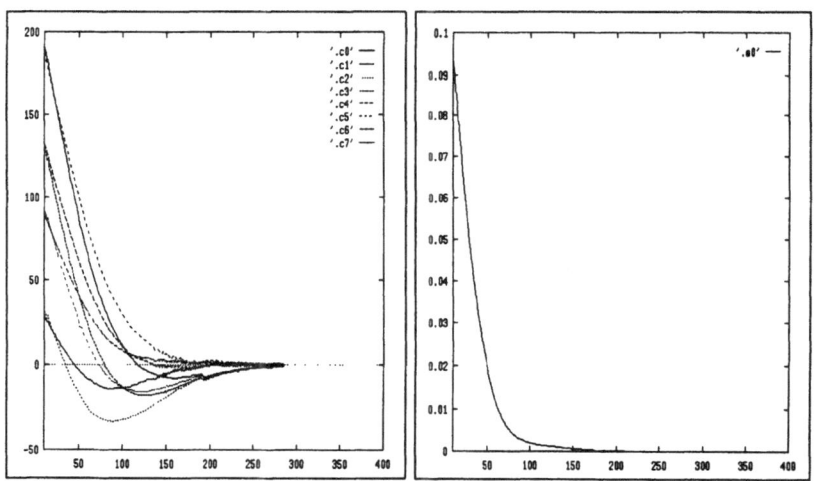

Figure 5. Error on the z component

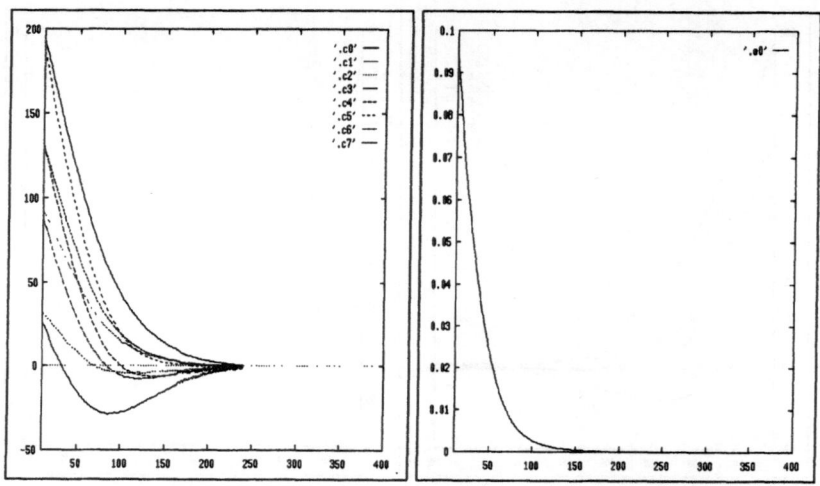

Figure 6. Rotation error around the z axis

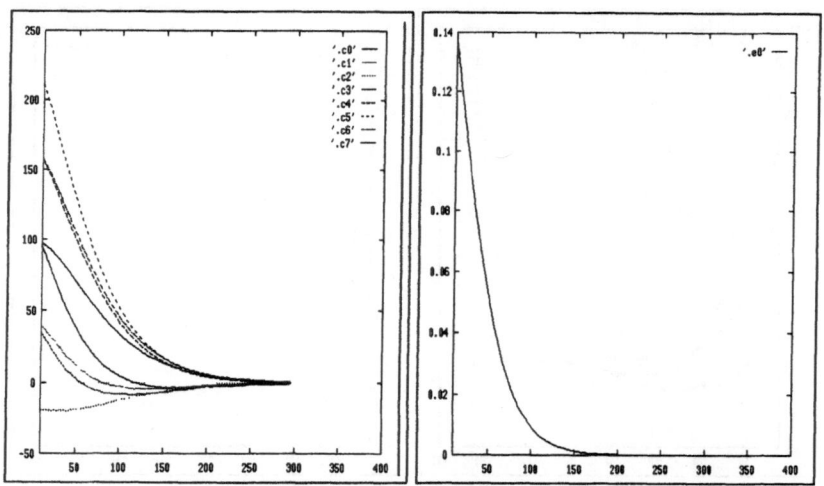

Figure 7. Combined Errors

Experiments in high-performance robotic visual servoing

Peter I. Corke

CSIRO Division of Manufacturing Technology
Locked Bag 9
Preston, 3072. Australia
pic@mlb.dmt.csiro.au

Abstract— This paper discusses some important control design issues for robot visual servoing, in particular high-performance tracking. The limitations of pure visual feedback control are discussed, and an approach based on axis velocity control and estimated target velocity feedforward is proposed. Experimental results are obtained with a 50 Hz low-latency vision system and an end-effector mounted camera to close the robots position loop.

1. Introduction

In conventional robots the pose of the tool tip is inferred from the measured joint angles, and knowledge of the kinematic structure of the robot. The *accuracy* of a robot is the error between the measured and commanded poses. Discrepancies between the kinematic model assumed by the controller, and the actual robot serve to reduce accuracy which fundamentally limits the usefulness of location data derived from a CAD model. Sources of kinematic discrepancy may include manufacturing tolerances in link length, axis alignment as well as link deformation due to load. For high accuracy applications these parameters must be precisely determined for the individual robot. *Repeatability* is the error with which a robot returns to a previously taught point and in general is superior to accuracy – being related directly to axis servo performance.

The speed of a robot is also significant since the economic justification is frequently based upon cycle time. Although machines now exist capable of high tool-tip accelerations, the overall cycle time is dependent upon many other factors such as settling time. Accurate control is needed to rapidly achieve the desired location, with minimum overshoot. At high accelerations the electromechanical dynamic characteristics of the manipulator become significant, and stiff (massive) links are required. An alternative approach may be to use less stiff manipulators and endpoint position feedback.

Vision has been used as a sensor for robotic systems since the late 1970's [16], and systems are now available from major robot vendors that are highly integrated with the robot's programming system. However these systems are static, and operate on a 'look then move' basis, with an image processing time of the order of 1 second. With such 'open loop' operation the achievable accuracy depends directly on the accuracy of the visual sensor and the manipulator and its controller. An alternative to increasing the accuracy of these subsystems is to use a visual feedback loop, which increases the overall accuracy of the system – the principle concern to the application.

Visual servoing is the use of vision as a high speed sensor, giving many measurements or samples per second to close the robot's position loop. The principle of negative feedback ensures that errors or non-linearities within the system are greatly reduced in magnitude. The camera may be stationary or held in the robot's 'hand', the latter providing endpoint relative positioning information directly in the Cartesian or task space. Endpoint feedback presents opportunities for greatly increasing the versatility, accuracy and speed of robotic automation tasks.

Section 2 briefly reviews relevant prior work, and some of the significant issues in visual servo system implementation. Section 3 examines the limitations of the common visual servoing structure based on feedback and underlying axis position control. Section 4 introduces a new control structure, based on axis velocity control with estimated target velocity feedforward. Section 5 briefly describes the experimental system and presents some experimental results in robot fixation and gaze control for complex target motion.

2. Prior work

The literature regarding dynamic effects in visual servo systems is scant compared to that addressing the kinematic problem, that is, how the robot should move in response to observed visual features. Effects such as oscillation and lag tend to be mentioned only in passing [7, 11] or ignored. For instance recent results by Wilson *et al.* [15] shows clear evidence of lag, of the order of 100ms, in the tracking behaviour. The earliest relevant report is by Hill and Park [10] in 1979 who describe visual servoing of a Unimate robot and some of the dynamics of visual closed-loop control.

An important prerequisite for high-performance visual servoing is a vision system capable of a high sample rate, low latency and with a high-bandwidth communications path to the robot controller. However many reports are based on the use of slow vision systems where the sample interval or latency is significantly greater than the video frame time. If the target motion is constant then prediction can be used to compensate for the latency, but the low sample rate results in poor disturbance rejection and long reaction time to target 'maneuvers'. Unfortunately most off-the-shelf robot controllers have low-bandwidth communications facilities. The communications link between vision system and robot is typically a serial link [7, 15]. For example with the Unimate VAL-II controller's *ALTER* facility, Tate [13] identified the transfer function between input and manipulator motion, and found the dominant dynamic characteristic below 10Hz was a time delay of 0.1s. Bukowski *et al.* [2] describes a novel analog interconnect between a PC and the Unimate controller, so as to overcome that latency.

In earlier work with a custom robot controller and vision system [6, 4] a detailed model of the visual servo system was established. The system is multi-

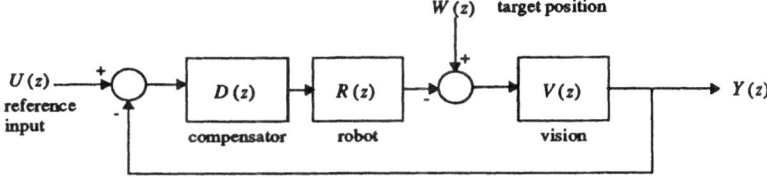

Figure 1. Block diagram depicting operation as a regulator with target position as a disturbance input.

rate but may be approximated by a single-rate model

$$\frac{K_{lens}}{z^2(z-1)} \tag{1}$$

operating at the vision sampling interval of 20ms. The pole at $z = 1$ is due to the software velocity control, and the the two poles at the origin represent 40ms of lumped delay due to pixel transport, communications and servo response. The lens introduces a target distance dependent gain, K_{lens}, due to perspective.

3. Limitations of feedback control

A visual servo system may be used to keep the target centered in the camera's field of view. Such motion is referred to as *fixation*, and exists as a 'low-level' unconscious mechanism in animal vision. In control systems terminology this system is a regulator, maintaining a constant centroid location on the image plane. Motion of the target is actually a non-measurable disturbance input to the regulator system as shown in Figure 1. $W(z)$ is the unknown target position, $U(z)$ the reference input or desired image plane centroid location, and $Y(z)$ is the actual image plane centroid location.

With purely proportional control, the system (1) can achieve acceptable rejection of step disturbances, but the response to a sinusoid shows considerable lag (around 30°). That system has one open-loop integrator and is thus of type 1, and will have zero steady-state error to a step input. A ramp input, or higher order input, results in finite error as observed in simulation and experiment. Some common approaches to achieving improved tracking performance are

1. Increase the loop gain so as to minimize the magnitude of any proportional error. However this is not feasible for this system since the poles leave the unit circle as gain is increased.

2. Increase the 'type' of the system, by adding open-loop integrators. Without additional compensation, additional integrators result in a closed-loop system that is unstable for any finite loop gain.

3. Franklin [8] shows that the steady-state ramp error may be reduced by closed-loop zeros close to $z = 1$, though this causes increased overshoot and poor dynamic response.

4. Introduce feedforward of the signal to be tracked [12]. For the visual servo system, the input is not measurable and must be estimated, as discussed in the next section.

PID and poleplacement controllers can be designed which increase the type of the system to type 2, and give satisfactory (settling time ≈ 300ms) dynamic characteristics. PID control is ultimately limited by noise in the derivative feedback. Pole placement controllers were designed using the procedure of Åström and Wittenmark [12]. The acceptable closed-loop pole locations are constrained if a stable series compensator is to be realized, and the design procedure involves considerable experimentation.

Increasing the system type affects only the steady state error response, the short term error response is still dictated by the closed-loop pole locations. Given the constraints on closed-loop pole location as mentioned, a new approach is required to achieve high-performance tracking control.

4. A new control structure

In earlier work [4,6] the camera's Cartesian velocity demand was a function of sensed position error

$$\underline{\dot{X}}_d = \mathbf{V} \left[\begin{array}{cc} {}^iX - {}^iX_d & {}^iY - {}^iY_d \end{array} \right] \tag{2}$$

where $\mathbf{V}$ selects how the robot's pose, $\underline{X}$, is adjusted based on observed and desired image plane centroid location $({}^iX, {}^iY)$. In that earlier work the camera was translated to keep the target centered in the field of view. In this paper, to achieve higher performance the last two axes of the robot are treated as an autonomous camera 'pan/tilt' mechanism, since the wrist axes have the requisite high performance.

4.1. Velocity control

In the earlier work [6,4], and much other reported work, the visual servo loops are built 'on top of' underlying position control loops. This is probably for the pragmatic reason that most robot controllers present the abstraction of a robot as a position controlled device. An alternative, axis velocity control, has a number of advantages:

1. Axis position control uses redundant position feedback; from the axis position sensor and from the vision system. In reality these two position sensors measure different quantities due to dynamics in the structure and drive-train of the manipulator, but it is end-effector position that is the relevant quantity in an application.

2. In a non-structured environment precise positions have less significance than they would in say a manufacturing cell, and may be difficult to determine with precision. Observe that we control cars and ships in an unstructured environment, not by a sequence of precise spatial locations, but rather a velocity which is continually corrected on the basis of sensed position error, similar to (2).

3. The Unimate's position loop causes particular difficulty for visual servoing since its sample interval of 14 or 28ms is different to the 20ms visual sample interval. This complicates analysis and control [4].

There are two approaches to implementing Cartesian velocity control. In earlier work the desired Cartesian velocity (2) was integrated, then joint position, $\underline{q}$, solved for

$$\underline{X}_d = \int \underline{\dot{X}}_d \, dt \tag{3}$$

$$\underline{q}_d = \mathcal{K}^{-1}(\underline{X}_d) \tag{4}$$

To exploit axis velocity control the Cartesian velocity can be *resolved* to joint velocity [14]

$$\underline{\dot{q}}_d = \mathbf{J}(\underline{q}) \, \underline{\dot{X}}_d \tag{5}$$

Initial experiments using axis velocity control made use of the Unimate's underlying velocity control loops. Modifying the digital servo firmware allowed the velocity demand to be set at any time by the host. Experiments showed that the Unimate velocity loops had relatively low gain, which limited the maximum velocity to around 50% of the fundamental limit due to amplifier voltage saturation. The loops also had considerable DC offset which led to position drift. To counter this, software velocity control loops have been implemented which command motor current directly. These loops run at video field rate, 50Hz, and the axis velocity is estimated using a 3-point derivative of measured joint angles.

$$\hat{\dot{\theta}}_i = \frac{3\theta_i - 4\theta_{i-1} + \theta_{i-2}}{2T} \tag{6}$$

Motor current is proportional to velocity error

$$I_m = K_v(\dot{\theta}_d - \hat{\dot{\theta}}) \tag{7}$$

4.2. Target state estimation

To counter the target disturbance in this velocity controlled system it is necessary to feedforward the target velocity. Since this is not measurable an estimate must be used. The target is assumed to have second-order dynamics and a zero-mean Gaussian acceleration profile. The problem then is to estimate 5 states (2 target, 3 robot and vision system from (1)) from a single output, but controllability analysis shows that the complete state is not observable. To achieve observability it is also necessary to measure the joint angle, θ_m. In simple terms, from the camera's point of view, it is impossible to determine if the observed motion of the target is due to target or robot motion.

The image plane coordinate of the target is given by

$$^i X_t = \alpha f(\theta_c - \frac{^C x_t}{^C z_t}) \tag{8}$$

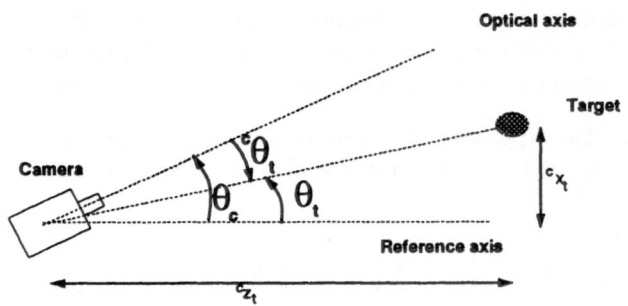

Figure 2. Angle notation used.

where α is the CCD pixel pitch, f the lens focal length, $^C z_t$ and $^C x_t$ are the target distance and displacement, and θ_c the camera pointing angle derived from axis encoder and gear ratio. $^c\theta_t = {}^C x_t / {}^C z_t$ is the target's angle with respect to an arbitrary axis fixed in space. The angle notation is shown in Figure 2. An estimate of the target's angle is given by

$$\overline{\theta_t} = {}^c\theta_t - \frac{{}^i X_t}{\alpha f} \tag{9}$$

The target angle estimates obtained from the binary vision system, $\overline{\theta_t}$ are contaminated by spatial quantization noise, and simple differencing cannot be used to estimate velocity. A Kalman filter is therefore used to reconstruct the target's angle and velocity states, based on noisy observations of target position, and to predict the velocity one step ahead, thus countering the inherent delay of the vision system.

Equation (9) and the target states (position and velocity) are computed for each of the pan and tilt axes. Target state for each axis comprises angle and rotational velocity

$$\underline{X_k} = \begin{bmatrix} \theta_k & \dot{\theta}_k \end{bmatrix}^T \tag{10}$$

In discrete-time state-space form the target dynamics are

$$\underline{X}_{k+1} = \Phi \underline{X}_k + \omega_k \tag{11}$$
$$y_k = \mathbf{C}\underline{X}_k \tag{12}$$

where $\underline{\omega}_k$ represents state uncertainty and y_k is the observable output of the system, the target's pan or tilt angle. For constant-velocity motion the state-transition matrix, Φ, is

$$\Phi = \begin{bmatrix} 1 & T \\ 0 & 1 \end{bmatrix} \tag{13}$$

The observation matrix $\mathbf{C}$ is

$$\mathbf{C} = [1 \ 0] \tag{14}$$

where T is the sampling interval, in this case 20ms, the video field interval.

For each axis the predictive Kalman filter [12] is given by

$$\mathbf{K}_{k+1} = \mathbf{\Phi P}_k \mathbf{C}^T (\mathbf{CP}_k \mathbf{C}^T + \mathbf{R}_2)^{-1} \tag{15}$$

$$\hat{\underline{X}}_{k+1} = \mathbf{\Phi}\underline{X}_k + \mathbf{K}_{k+1}(y_k - \mathbf{C}\underline{X}_k) \tag{16}$$

$$\mathbf{P}_{k+1} = \mathbf{\Phi P}_k \mathbf{\Phi}^T + \mathbf{R}_1 \mathbf{I}_2 - \mathbf{K}_{k+1} \mathbf{CP}_k \mathbf{\Phi}^T \tag{17}$$

where $\mathbf{K}$ is a gain, $\mathbf{P}$ is the error covariance matrix and $\mathbf{I}_2$ is 2×2 identity matrix. $\mathbf{R}_1$ and $\mathbf{R}_2$ are input and output covariance estimates, and are used to tune the dynamics of the filter. This filter is predictive; that is $\hat{\underline{X}}_{k+1}$ is the predicted value for the next sample interval.

The Kalman filter equations are relatively complex and time consuming to execute in matrix form. Using the computer algebra package MAPLE the equations were reduced to an efficient scalar form, and the corresponding 'C' code automatically generated for inclusion into the real-time system. This reduces the computational load by more than a factor of 4. Suitable values for $\mathbf{R}_1$ and $\mathbf{R}_2$ are determined empirically.

4.3. Control structure

A schematic of the variable structure controller is given in Figure 3. It operates in two modes, gazing or fixating. When there is no target in view, the system is in gaze mode and maintains the current gaze direction by closing a joint position control loop.

In fixation mode the system attempts to keep the target centered in the field of view. The velocity demand comprises the predicted target velocity feedforward, and the target image plane error feedback so as to center the target on the image plane. Wrist axis cross-coupling is explicitly corrected. When the target is within a designated region in the center of the image plane, integral action is enabled, so the feedback law is

$$\dot{\theta}_{fb} = \left\{ \begin{array}{l} P(^iX - {}^iX_d) \\ P(^iX - {}^iX_d) + I \int (^iX - {}^iX_d)dt \end{array} \right. \tag{18}$$

The transition from gaze to fixation involves initializing the state estimates and covariance of the Kalman filter. The initial velocity estimate is based on a first order position difference. No vision based motion is allowed until the target has been in view for six consecutive fields in order for the state estimates to converge satisfactorily.

Figure 4 shows details of the timing, particularly the temporal relationship between sampling of the image and joint angles. The robot's joint angles are sampled by a task during the vertical blanking interval. Immediately prior to the blanking interval the camera's electronic shutter is opened for 2ms, and the image captured in that interval is processed during the next field time. The actual instant of camera sampling is not signaled by the camera, but has been determined experimentally. The short exposure time is needed to approximate an ideal sampler, as assumed by the control structure. Unlike for fixed camera systems, motion blur is not an issue since the fixation motion keeps the target

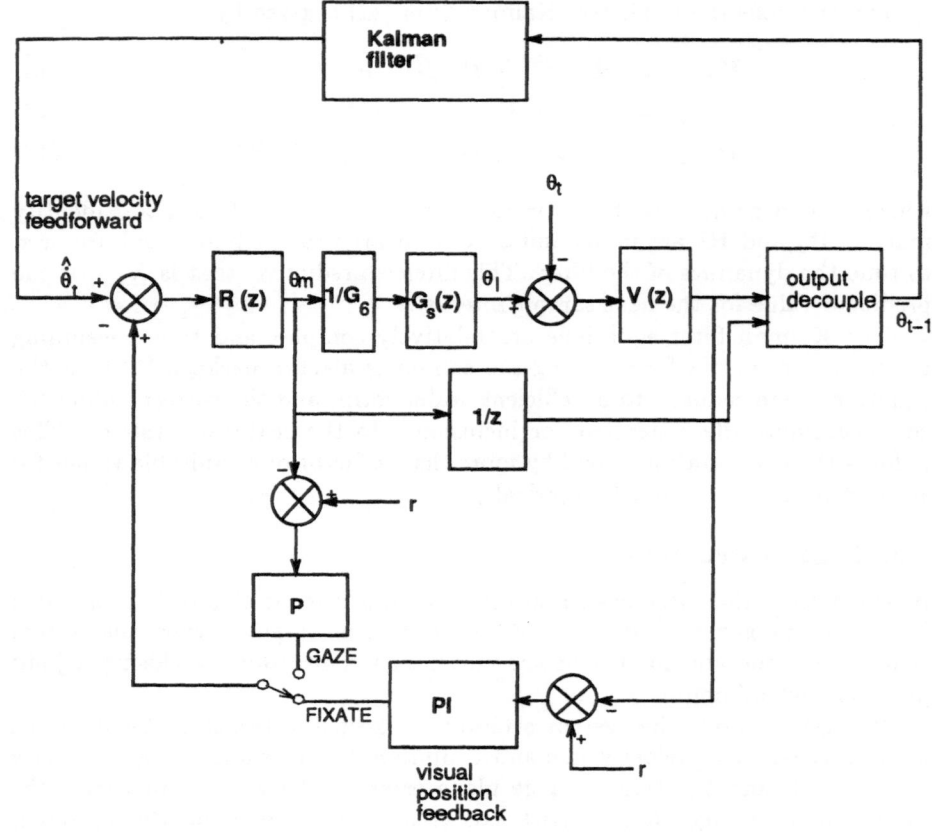

Figure 3. Pan/tilt control structure.

fixed with respect to the camera. There is some finite time difference between sampling the joint angles and the pixel exposure. This does not appear to be a significant issue, but could be overcome by the use of a simple predictor, though this has not been investigated.

5. Experiment

5.1. System hardware

The system is built on the VME-bus, with a Motorola MV147 68030 processor running at 25MHz. The CPU is responsible for feature post-processing, robot control, user interface and data logging. To date, a single processor has been found adequate for all these tasks. The robot is a Unimate Puma 560 with the joint servos interfaced to the VME-bus via a custom board. The Unimation Mark 1 servo system introduces some difficulties, in particular communications overhead and a sample rate that is non-integrally related to the video sample rate.

The image processing subsystem is based on Datacube pipeline processing

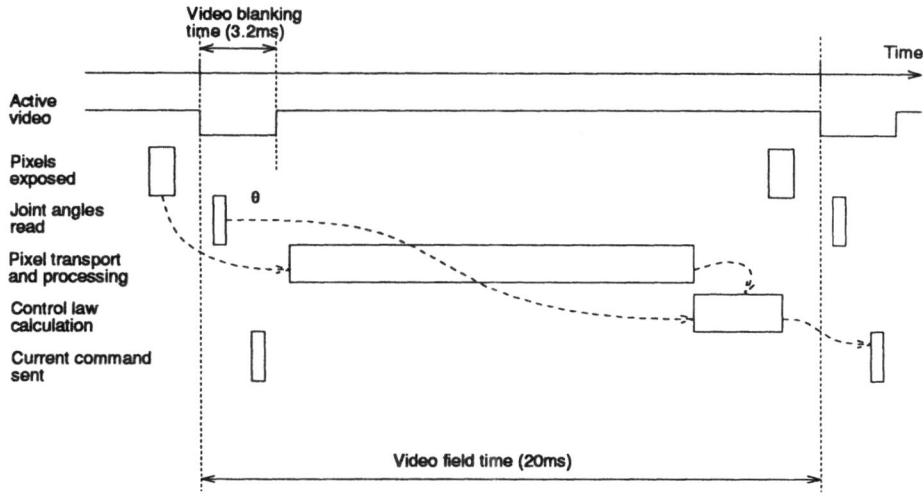

Figure 4. Details of system timing.

modules, VME-bus boards that perform various operations on digital video data. The inter-module video data paths are patch cables installed by the user. The boards are controlled by the host computer via the VMEbus, whilst the video data paths, known as MAXBUS, run at 10Mpixels/s. The incoming video stream is digitized then thresholded by a lookup table which maps pixels to one of two grey levels, corresponding to the binary values *black* or *white*. Binary median filtering on a 3x3 neighbourhood is used to eliminate one or two pixel noise regions which may overwhelm the host processor. Another framestore is used by the run-time software to display real-time overlay graphics and performance data.

The APA-512+ [1] (for Area Parameter Accelerator) is a hardware unit designed to accelerate the computation of area parameters of objects in a scene. The APA binarizes incoming video data and performs a single pass connectivity (simple-linkage region growing [9]) analysis, and subsequent computation of moments upto second order, perimeter and bounding box. The APA performs very effective data reduction, reducing a 10Mpixel/s stream of grey-scale video data input via MAXBUS, to a stream of feature vectors representing objects in the scene, available via onboard shared memory.

Field-rate processing is achieved by treating the interlaced data as two consecutive frames of half vertical resolution. This eliminates de-interlacing latency required for full frame processing, as well as providing a higher sample rate. The CCD camera employs field shuttering so a video frame comprises two fields exposed a field time apart, providing another argument in favor of field-rate processing. CCIR format video is used with a field rate of 50Hz.

5.2. System software

The visual servoing kernel software runs under the VxWorks real-time multi-tasking operating system. The kernel is loaded into memory first, and provides all the infrastructure for vision and robot control. Internally the kernel comprises around 14 tasks for various housekeeping activities. The applications programs are loaded subsequently and access the kernel via a well-defined function call interface. Robot control is performed using the ARCL robot software package [5].

Early experimental work in visual servoing showed that quite simple applications rapidly became bloated with detailed code dealing with the requirements of vision and robot control, graphical display, diagnostics and so on. Considerable work has gone into the design of the current system, which provides a powerful and clean interface between the kernel software encompassing visual-feature extraction and robot control, and the user's application program. [3] Convenient control of the kernel and application is obtained with an interactive control tool running under OpenWindows on an attached workstation computer via remote procedure calls. The inbuilt data logging facility was used to record the experimental results shown in the next section.

5.3. Experimental results

A number of experiments have been conducted to investigate the performance of the fixation controller. To achieve a particularly challenging motion for the tracking controller a turntable has been built whose rotational velocity can be constant, or a reversing trapezoidal profile. The resultant target motion has a complex Cartesian acceleration profile.

In the experiment the turntable is rotating at 4.2rad/s, and Figure 5 shows the image plane coordinate error in pixels. It can be seen that the the target is kept within ±12 pixels of the reference. The joint velocity of up to 2 rad/s is close to the sustained maximum of 2.3 rad/s due to voltage saturation. The previous, feedback only, strategy results in large following errors and a lag of over 40°.

Figure 5 also shows the measured and feedforward joint velocity for the pan axis. The feedforward signal provides around half the velocity demand, the rest being provided by the PI feedback law. This shortfall may be due to unmodeled servo dynamics or the second-order Kalman filter which is based on the assumption of constant velocity target motion.

Motion of the first three robot joints introduces a disturbance input to the fixation controller. This effect could be countered by feeding forward the camera pan/tilt rates due to the motion of those joints. Such a structure would be similar to the human eye in that the eye muscles accept feedback from the retina, as well as feedforward information from the vestibular system, giving head lateral and rotational acceleration, and position information from the neck muscles.

Figure 6 shows experimental results with the same controller attempting to track a ping-pong ball thrown across the robot's field of view. The measured

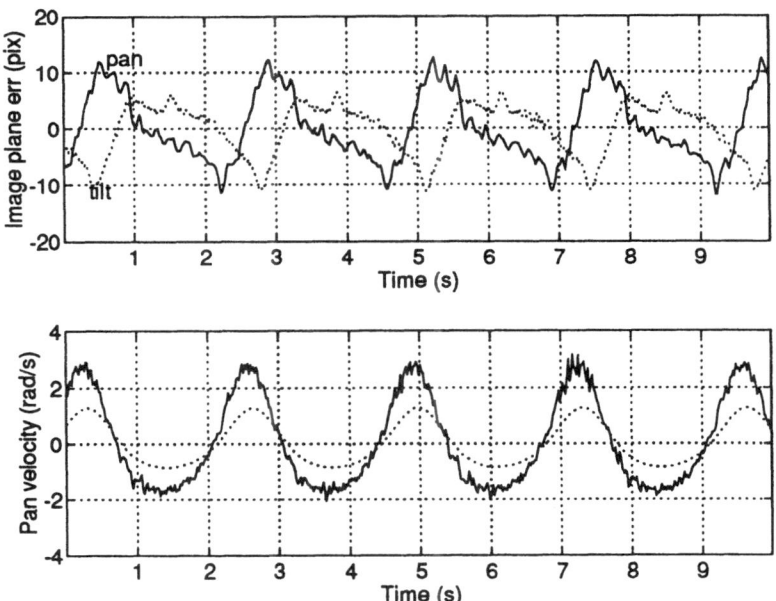

Figure 5. Tracking an object on turntable revolving at 4.2rad/s – experimental results. Top graph is the image plane error in pan (solid) and tilt (dotted) directions. Bottom graph is the measured link velocity (solid) and the estimated target velocity feedforward (dotted).

joint velocities of the robot are shown, along with the centroid position error. The robot does not move until the ball has been in the field of view for a few frametimes in order for the Kalman filter to converge. The apparent size of the ball (area) can be seen to vary with its distance from the camera.

6. Conclusion

A number of factors that limit the usefulness of robots in conventional applications have been reviewed. Visual servoing is proposed as a solution to the problems of part location and cycle time reduction by utilizing end-effector relative position feedback. The limitations of pure feedback control and control based on underlying axis position control have been discussed. A new control strategy, based on axis velocity control, and estimated target velocity feedforward is introduced and verified experimentally.

High performance fixation control is an important component of many active vision strategies. Given the advances in image processing and feature extraction hardware, the limiting factor in visual servoing is now the sensor frame rate.

204

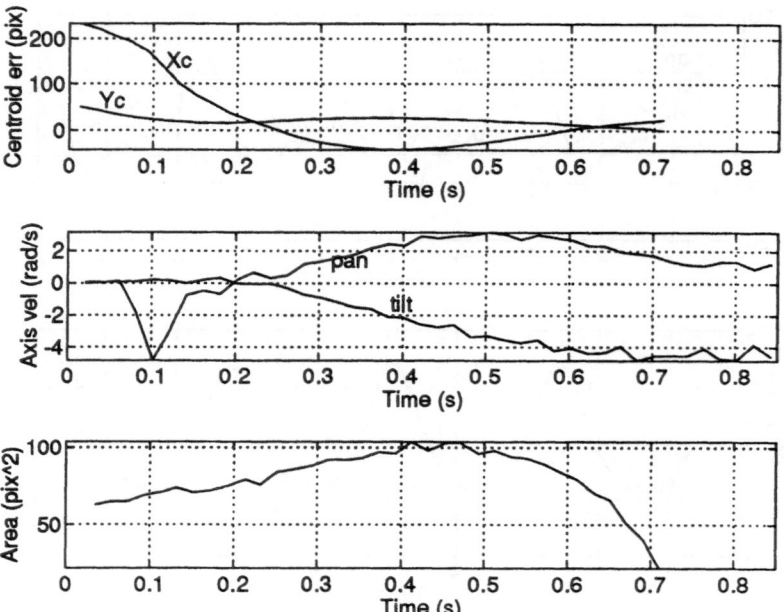

Figure 6. Tracking a flying ping-pong ball — experimental data. Showing centroid error, actual link velocity and observed area of target.

References

[1] Atlantek Microsystems, Technology Park, Adelaide. *APA-512+ Area Parameter Accelerator Hardware Manual*, April 1993.

[2] R. Bukowski, L.S. Haynes, Z. Geng, N. Coleman, A. Santucci, K. Lam, A. Paz, R. May, and M. DeVito. Robot hand-eye coordination rapid prototyping environment. In *Proc. ISIR*, page 16.15 to 16.28, October 1991.

[3] P.I. Corke. An experimental facility for robotic visual servoing. In *Proc. IEEE Region 10 Int. Conf.*, pages 252–256. IEEE, 1992.

[4] P.I. Corke and M.C. Good. Dynamic effects in high-performance visual servoing. In *Proc. IEEE Int.Conf. Robotics and Automation*, pages 1838–1843, Nice, 1992.

[5] P.I. Corke and R.J. Kirkham. The ARCL robot programming system. In *Robots for Competitive Industries*, pages 484–493, Brisbane, July 1993. Mechanical Engineering Publications.

[6] P.I. Corke and R.P. Paul. Video-rate visual servoing for robots. In V. Hayward and O. Khatib, editors, *Experimental Robotics 1*, volume 139 of *Lecture Notes in Control and Information Sciences*, pages 429–451. Springer-Verlag, 1989.

[7] J.T. Feddema. *Real Time Visual Feedback Control for Hand-Eye Coordinated Robotic Systems*. PhD thesis, Purdue University, 1989.

[8] G.F. Franklin and J.D. Powell. *Digital Control of dynamic systems*. Addison-Wesley, 1980.

[9] R. M. Haralick and L. G. Shapiro. Survey. image segmentation techniques.

Computer Vision, Graphics, and Image Processing, 29:100–132, 1985.

[10] J. Hill and W. T. Park. Real time control of a robot with a mobile camera. *9th International Symposium on Industrial Robots*, pages 233–246, March 1979.

[11] W. Jang, K. Kim, M. Chung, and Z. Bien. Concepts of augmented image space and transformed feature space for efficient visual servoing of an "eye-in-hand robot". *Robotica*, 9:203–212, 1991.

[12] K. J. Åström and B. Wittenmark. *Computer Controlled Systems: Theory and Design.* Prentice Hall, 1984.

[13] A. R. Tate. Closed loop force control for a robotic grinding system. Master's thesis, Massachusetts Institute of Technology, Cambridge, Massachsetts, 1986.

[14] D.E. Whitney and D. M. Gorinevskii. The mathematics of coordinated control of prosthetic arms and manipulators. *ASME Journal of Dynamic Systems, Measurement and Control*, 20(4):303–309, 1972.

[15] W.J. Wilson. Visual servo control of robots using Kalman filter estimates of relative pose. In *Proc. IFAC 12th World Congress*, pages 9–399 to 9–404, Sydney, 1993.

[16] N. Zuech and R.K. Miller. *Machine Vision.* Fairmont Press, 1987.

Auto-calibration in Automation Systems using Vision

Ole Ravn, Nils A. Andersen and Allan Theill Sørensen

Institute of Automatic Control Systems,
Technical University of Denmark, Build. 326,
DK-2800 Lyngby, Denmark, e-mail: or@sl.dth.dk

Abstract— This paper discusses aspects of automatic camera calibration in automation systems. Three different scenarios are analysed: an AGV-system, a warehouse automation system, and a robot system. The properties of two methods for automatic camera calibration using multi-view 2-D calibration planes are investigated. It is found that the achievable calibration accuracy is comparable to methods using 3-D calibration objects.

1. Introduction

An important feature of automation systems is the achieved flexibility, which can be used when the automated process changes. The use of vision as a sensor in automation systems add considerable flexibility to the overall performance of the systems. Especially the start-up, shutdown and exception handling procedures can be made more robust and flexible. (Andersen et al., 1991). Vision is a multi-purpose sensor reconfigureable by software for different measurements. This versatility can be used to minimize the number of physical sensors needed. In automation systems this flexibility may be crucial for the robust performance of the system.

The determination of the transformations between world coordinates and camera coordinates is a problem related to the use of vision as a sensor. These transformation functions can be determined by measuring the position and orientation of the camera directly and using supplied information of the intrinsic camera parameters. However accurate measurements of the camera position and orientation are difficult and skilled staff is normally required.

The alternative solution which is described in this paper is auto-calibration. Based on measurements using the instrumentation of the automation system itself (camera, encoders, odometers etc.) and a known test pattern the auto-calibration procedure can derive the transformation functions. An effective auto-calibration procedure can greatly enhance the applicability of the automation system.

The problem of auto-calibration is very important and has previously been covered by numerous authors, e.g. (Ganapathy, 1984), (Tsai, 1987), (Lenz and Tsai, 1988), (Haralick, 1989), (Weng et al., 1992), (Faugeras et al., 1992), (Crowley et al., 1993), (Harris and Teeder, 1993).

This paper is divided into two main parts. Section 2 gives insight into the problems related to calibration and demonstrates the considerations of auto-calibration using three scenarios. Furthermore the basic transformation equations are stated and two auto-calibration algorithms are described. One being the algorithm of Tsai & Lenz extended to handle multi-views, the second a closed form solution which solves the calibration problem by multiple views of a single calibration plane. In section 3 a number of experiments are described showing the performance of the algorithms. An experiment is performed on a robot laboratory-setup measuring the true area of an object using the different auto-calibration procedures.

2. Auto-calibration

A good auto-calibration procedure should make it easy for an operator to calibrate the system without special knowledge of vision systems and with as little extra instrumentation as possible.

To demonstrate the considerations necessary in relation to choice of calibration tile, calibration procedure, accuracy etc., 3 scenarios will be considered in this part of the paper:

1. *The AGV case.* In this case the camera is mounted on an Autonomous Guided Vehicle and the task is to calibrate the correspondence between the vehicle coordinate system and the camera coordinate system.

2. *The warehouse case.* In this example the cameras are mounted near the ceiling in a large hall and the task is to calibrate the cameras individually and to make the transition from one field of view to another as smooth as possible.

3. *The robot case.* In this case the camera is mounted over the robots working area and the task is to calibrate the transformations between the robot coordinate system and the camera coordinate system.

Aspects of auto-calibration that should be considered include:

- Absolute versus relative calibration.
- Choice of calibration object.
- The degree of automation in the calibration procedure in relation to the use of unskilled labour.
- The achievable accuracy related to the additional work involved.

Generally the calibration object should be easy to produce and manageable in size for one person. As accurate 3-D objects are difficult to produce and maintain, only 2-D objects are considered in this paper. Using a 2-D calibration object instead of a 3-D object is possible when multiple views are taken of the 2-D object. This could be done moving (intelligently) the tile between two measurements or by having two tiles in the field of view. The tiles should be positioned appropriately with respect to each other but the relative position and orientation need not be known. Accurate 2-D objects can be produced using photographic processes or using a laser printer. We use a laser printed pattern mounted on an aluminium plate. It should be noted that the accuracy of laser prints are not equivalent to 300 dpi resolution, but generally somewhat less, especially in the direction of the feed of the laser printer. Measurements on a tile generated by our Brother HL8-PS laser printer show that the deviation in the feeding direction is below 3/100 of an inch. We have concluded that the accuracy of the laser print was sufficient in relation to our needs and the demand on the easy production of the tile was more important. We are considering a tile with 9x6 black circles having a radius of 1 cm. The position of these circles can be found with subpixel accuracy using a centre of gravity calculation. The inaccuracy of the determination of the centre due to the perspective transformation is considered unimportant.

To achieve a good calibration by multiple views of a 2-D calibration object two aspects are important. The calibration points should cover a considerable part of the field of view and the views should include changes in the tilt around both the x- and y-direction, (Harris and Teeder, 1993).

In the case of scenario 1, the camera is mounted on a mobile platform and the objective is to calibrate the extrinsic parameters (orientation and position in relation to the coordinate system of the platform) and the intrinsic parameters of the camera. The calibration object should be stationary during the calibration operation but the vehicle can be repositioned and the relative position of the camera is measurable using information from the odometers of the AGV. An A4 size tile may not be sufficient for filling the field of view. This scenario is well suited for having a pre-calibrated camera and just calibrating on-line the extrinsic parameters.

In the warehouse example the tile would be unmanageable large if it was to fill the whole field of view, but the tile could be repositioned in the field of view between measurements, the camera being stationary. If possible it could be an advantage to calibrate the intrinsic parameters of the camera prior to mounting the camera in the ceiling and then only calibrate the extrinsic parameters from the calibration tile. If the tile is to small to be viewed accurately calibration points on the warehouse floor could be marked and their position measured once. Refitting (and re-calibration) of the cameras is much easier in this case.

In the robot scenario a direct calibration of all parameters could be used. Normally it is possible to position the calibration tile such that most of the field of view is filled thus giving a good measurement.

2.1. Mathematical background

The goal of the camera calibration is to find the relationship of the 3-D points in a world coordinate system and 2-D picture points in the camera, see figure 1.

The parameters of this relationship are often grouped into intrinsic parameters describing the camera (focal length, lens distortion, etc.) and extrinsic parameters describing the position and orientation of the camera. The advantage of this distinction is the possibility of determining the intrinsic parameters in a calibration setup leaving only the extrinsic parameters to the calibration procedure in the automation system.

Calibration consists of determination of a set of extrinsic parameters, $(\phi, \theta, \Psi, t_x, t_y, t_z)$, and a set of intrinsic camera parameters, $(f_x, f_y, C_x, C_y, \kappa_1, \kappa_2, \kappa_3, ...)$. f_x and f_y are the focal lengths of the camera, (C_x, C_y) is the optical centre and κ_i are distortion coefficients.

The transformation between world coordinates (x_w, y_w, z_w) and undistorted camera coordinates (X_u, Y_u) is described by:

$$X_u = C_x - f_x \frac{t_x + r_1 x_w + r_2 y_w + r_3 z_w}{t_z + r_7 x_w + r_8 y_w + r_9 z_w} \tag{1}$$

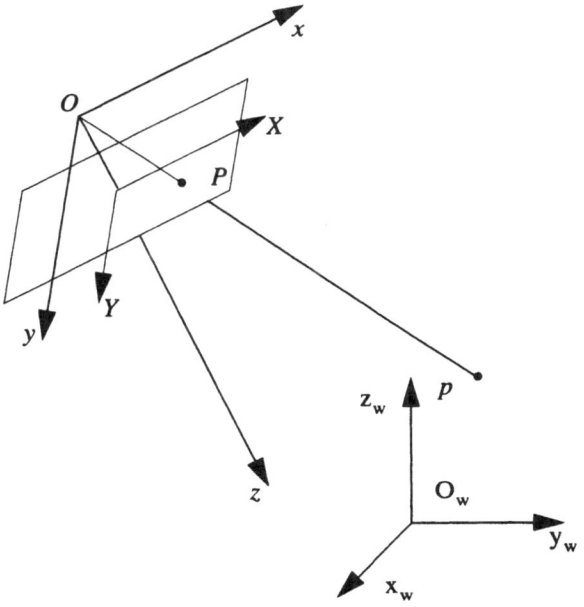

Figure 1. Connection between world coordinates camera coordinates and image coordinates.

$$Y_u = C_y - f_y \frac{t_y + r_4 x_w + r_5 y_w + r_6 z_w}{t_z + r_7 x_w + r_8 y_w + r_9 z_w} \qquad (2)$$

Where

$$R = \begin{bmatrix} r_1 & r_2 & r_3 \\ r_4 & r_5 & r_6 \\ r_7 & r_8 & r_9 \end{bmatrix} = \begin{bmatrix} c_\theta c_\psi & c_\theta s_\psi & -s_\theta \\ -c_\phi s_\psi + s_\psi s_\theta c_\psi & c_\phi c_\psi + s_\phi s_\theta s_\psi & s_\phi c_\theta \\ s_\phi s_\psi + c_\phi s_\theta c_\psi & -s_\phi c_\psi + c_\phi s_\theta s_\psi & c_\phi c_\theta \end{bmatrix} \qquad (3)$$

$$T = \begin{bmatrix} t_x \\ t_y \\ t_z \end{bmatrix} \qquad (4)$$

are the rotation matrix and translation vector describing the transformation from 3-D world coordinates to 3-D camera coordinates. (ϕ, θ, Ψ) are the rotation angles around the x- y- and z-axis of the camera-coordinate system. The rotations are performed first around the z-axis, then around the current y-axis and at last around the current x-axis. After the rotation the coordinate-system is translated along T.

The distorted screen coordinates (X_f, Y_f) are found from the undistorted as:

$$X_f = X_u(1 + \kappa_1 r^2 + \kappa_2 r^4 + \kappa_3 r^6 + \ldots) \qquad (5)$$

$$Y_f = Y_u(1 + \kappa_1 r^2 + \kappa_2 r^4 + \kappa_3 r^6 + \ldots) \qquad (6)$$

where

$$r = \sqrt{X_u^2 + Y_u^2} \qquad (7)$$

(1) - (7) describes the connection between world coordinates (x_w, y_w, z_w) and screen coordinates (X_f, Y_f).

2.2. Auto-calibration methods

In this section two auto-calibration procedures are considered. The first method, which is described in detail, is a closed form solution solving the calibration problem by multiple views of a single 2-D calibration tile. The other method is the two-stage calibration method of Tsai & Lenz. The advantage of the first method is the closed form property enabling the solution by just solving two sets of linear equations, but the drawback is that lens distortion is not considered. The second method considers radial lens distortion but non-linear optimization techniques are necessary to solve the calibration.

2.2.1 Closed form algorithm for multi-view 2-D calibration plane

One calibration point give two independent equations using (1) and (2). It can be shown that it is possible to obtain 8 independent equations using coplanar points, i.e. points from a 2-D calibration plane. Choosing the world coordinate system with the x_w- and y_w-axis in the calibration plane, i.e. $z_w=0$ and ensuring $t_z \neq 0$ one point gives the following two linear equations:

$$\begin{bmatrix} 1 & 0 & x_{wi} & y_{wi} & 0 & 0 & X_{fi}x_{wi} & X_{fi}y_{wi} \\ 0 & 1 & 0 & 0 & x_{wi} & y_{wi} & Y_{fi}x_{wi} & Y_{fi}y_{wi} \end{bmatrix} \cdot \begin{bmatrix} k_1 \\ k_2 \\ : \\ k_8 \end{bmatrix} = \begin{bmatrix} -X_{fi} \\ -Y_{f1} \end{bmatrix} \qquad (8)$$

with $(k_1...k_8)$ defined as

$$k_1 = \frac{f_x t_x - t_z C_x}{t_z} , \; k_2 = \frac{f_y t_y - t_z C_y}{t_z} , \; k_3 = \frac{f_x r_1 - r_7 C_x}{t_z} , \; k_4 = \frac{f_x r_2 - r_8 C_x}{t_z} \qquad (9)$$

$$k_5 = \frac{f_y r_4 - r_7 C_y}{t_z} , \; k_6 = \frac{f_y r_5 - r_8 C_y}{t_z} , \; k_7 = \frac{r_7}{t_z} , \; k_8 = \frac{r_8}{t_z} \qquad (10)$$

With at least 4 points from a 2-D calibration plane equation (8) gives a set of simultaneous equations from which $(k_1...k_8)$ can be determined.

Combining the expressions for $(k_3...k_8)$ and the orthonormality condition on the rotation matrix the following equations can be derived

$$\begin{bmatrix} k_3 k_8 + k_4 k_7 & k_5 k_6 & k_5 k_8 + k_6 k_7 & k_7 k_8 \\ 2(k_3 k_7 - k_4 k_8) & k_5^2 - k_6^2 & 2(k_5 k_7 - k_6 k_8) & k_7^2 - k_8^2 \end{bmatrix} \cdot \begin{bmatrix} v_1 \\ v_2 \\ v_3 \\ v_4 \end{bmatrix} = \begin{bmatrix} -k_3 k_4 \\ k_4^2 - k_2^2 \end{bmatrix} \qquad (11)$$

The variables v_1-v_4 are defined as

$$v_1 = C_x \ , \ v_2 = \frac{f_x^2}{f_y^2} \ , \ v_3 = \frac{f_x^2}{f_y^2}C_y \ , \ v_4 = \frac{f_x^2}{f_y^2}C_y^2 + C_x^2 + f_x^2 \qquad (12)$$

Combining two or more views of the 2-D calibration tile equation (13) can be solved and using the solution the determination of the intrinsic parameters is straight forward. These views must have different values of ϕ and θ.

If more than four point and/or more than 2 views are used the uncertainty on the estimates can be reduced solving the two overdetermined linear equation systems using a standard SVD method.

2.2.2 Method of Tsai & Lenz

The calibration algorithm of Tsai & Lenz (Tsai, 1987), (Lenz and Tsai, 1988) is based on the Radial Alignment Constraint, utilizing the fact that radial distortion occurs only along radii from the optical centre.

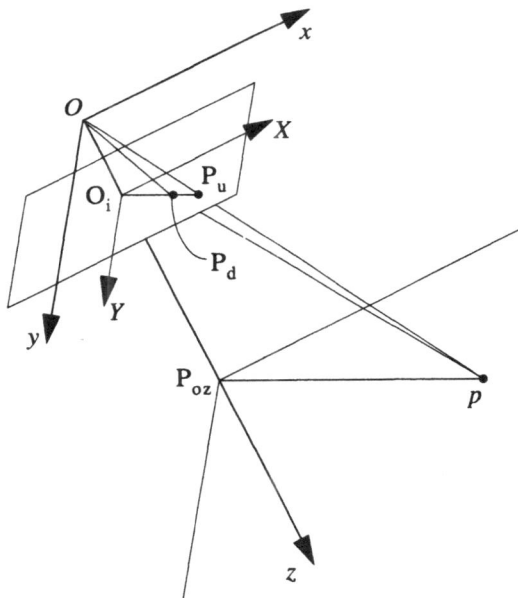

Figure 2. RAC-constraint on the imaging geometry.

This means that (see figure 2)

$$\overline{O_i P_d} \parallel \overline{O_i P_u} \parallel \overline{P_{oz} p} \qquad (13)$$

O_i being the intersection of the optical axis and the image plane, P_d the distorted image point, P_u the undistorted image point, p the 3-D calibration point and P_{oz} the projection of p on the optical axis. Based on this constraint the calibration problem is decoupled, such that 5 extrinsic parameters (ϕ, θ, Ψ, t_x, t_y) are determined separately by a closed-form solution. In the second stage of the algorithm t_z, f_x and κ_1 are determined. For a given κ_1 it is possible to determine t_z and f_x by a closed form and the resulting κ_1 is chosen to minimize the distance between estimated undistorted image coordinates inferred from the calibration points and the estimated undistorted image coordinates inferred from the measured distorted image points.

This basic algorithm assumes knowledge of (C_x, C_y), but to determine also the optical centre a global optimization is necessary to find the centre coordinates that optimizes the goodness of fit between camera model and observed image coordinates.

It is possible to calibrate a camera using this method with a single view of a single 2-D calibration plane if the camera has lens distortion.

The algorithm is dependant on knowledge of the aspect ratio of the camera. This can be determined from hardware data of camera and vision system.

3. Experiments

Several experiments have been carried out in the laboratory verifying that the auto-calibration procedures described above work well in the lab set-up's. Results about the accuracy of the calibrations will be reported. The experiments are:

- Determination of the intrinsic parameters for two cameras using a precise photogrammetric setup at the Institute of Surveying and Photogrammetry. These parameters are considered as a reference.
- Determination of the same parameters using the method reported in (Tsai, 1987), (Lenz and Tsai, 1988). Comparison of results using single and multiple views.
- Determination of the same parameters using the closed form multi-view method.

Simulation results using 2-D multiview calibration objects are compared to simulation results using 3-D calibration objects reported by (Weng et al., 1992).

On our robot system an experiment relating to vision based area determination using the intrinsic parameters derived above has been carried out.

3.1. Simulated calibrations

Simulations have been carried out to determine the achievable accuracy of the method of Tsai & Lenz extended to multi-views and the closed-form calibration procedure described previously. The simulations have been performed adding zero-mean uncorrelated gaussian noise to the calibration images. The algorithms have been tested for 2 values of the gaussian noise and 3 values of the camera lens distortion. For each combination of noise and distortion 100 calibrations have been carried out and the reported results are mean values and standard deviations. Each calibration is based on two views with the following extrinsic parameters:

$$(\phi, \theta, \Psi, t_x, t_y, t_z) = (30°, 0°, 0°, 0, 0, 0.7m) \qquad (14)$$

and

$$(\phi, \theta, \Psi, t_x, t_y, t_z) = (0°, 30°, 0°, 0, 0, 0.7m). \qquad (15)$$

The simulation results are shown in tables 1 and 2.

For the distortion free lens, the results of the two calibration methods are comparable. The standard deviation of the focal length determined by the closed form method seems smaller than those obtained by the Tsai & Lenz method. With lens distortion the closed form method keeps a small standard deviation but with a wrong estimate

meas. noise std.dev. (pixels)	dist. 10^{-8} pix^{-2}	focal length, f_x pixels	aspect ratio, $\frac{f_y}{f_x}$ (%)	C_x, pixels	C_y, pixels	radial dist. 10^{-8} pix^{-2}
0.2	0	1511.0(9.8)	-	260.0(3.9)	299.8(3.7)	0.0(0.7)
0.5	0	1509.5(20.2)	-	260.4(8.6)	300.8(8.2)	0.2(1.5)
0.2	6.1	1510.5(7.9)	-	260.4(5.3)	299.8(4.5)	6.1(0.7)
0.5	6.1	1506.7(22.0)	-	260.1(11.8)	300.0(11.0)	6.1(1.8)
0.2	60.5	1510.9(6.1)	-	259.8(1.3)	300.2(2.3)	60.4(0.8)
0.5	60.5	1507.3(14.9)	-	260.6(3.7)	299.6(5.9)	60.4(2.2)
True		1511.9	96.2	260.0	300.0	-

Table 1: Simulation, Tsai & Lenz multi-view. Standard deviations in parentheses.

meas. noise std.dev. (pixels)	dist. 10^{-8} pix^{-2}	focal length, f_x pixels	aspect ratio, $\frac{f_y}{f_x}$ (%)	C_x, pixels	C_y, pixels	radial dist. 10^{-8} pix^{-2}
0.2	0	1512.2(4.3)	96.22(0.10)	260.0(3.0)	300.1(2.2)	-
0.5	0	1512.2(13.8)	96.19(0.21)	259.5(8.5)	300.3(5.8)	-
0.2	6.1	1518.1(5.0)	96.14(0.10)	259.4(3.7)	299.7(2.4)	-
0.5	6.1	1519.5(14.3)	96.10(0.27)	261.0(8.6)	300.9(6.0)	-
0.2	60.5	1569.0(5.6)	95.57(0.11)	259.8(3.5)	300.1(2.5)	-
0.5	60.5	1567.7(13.9)	95.54(0.29)	260.5(9.2)	300.2(6.3)	-
True		1511.9	96.2	260.0	300.0	-

Table 2: Simulation, Closed form multi-view. Standard deviations in parentheses.

of the focal length. With low distortion cameras with a maximum distortion of 1-2 pixels the error is around 0.4% but even for high distortion cameras with distortion around 10-20 pixels the error is less than 4%. The simulation results shows that the accuracy of the image centre determination is independent of the distortion. For lenses with high distortion better results are achieved by the Tsai & Lenz method than by the closed form method. This is due to the fact that the camera model of Tsai & Lenz incorporates lens distortion.

Tsai & Lenz gives a standard deviation of about 0.5% of the true focal length. The closed form method gives about 0.3% for distortion free cameras. The standard deviation for both methods in determination of the image centre is 3-4 pixels or about 0.25% of the focal length (which seems to be an appropriate measure of the relative error of the centre).

(Weng et al., 1992) reports similar simulations with two methods. One without distortion modelling and one which incorporates both radial and tangential distortion. Both methods use accurate 3-D knowledge of the calibration points. They use a gaussian noise of standard deviation 0.07 pixels in the determination of the distorted image points. The achieved standard deviation on determination of the focal length is 0.5% and the standard deviation on the image centre is 0.2%-0.4% compared to the focal length. For a camera with a maximum distortion of 3-4 pixels the achieved accuracy of the focal length is about 3% for the linear camera model and about 0.5% for the model incorporating lens distortion.

Our results indicate that it is possible to obtain the same accuracy using 2-D multi-view calibration as can be obtained with a 3-D calibration method.

3.2. Experimental calibration results

Real calibrations have been carried out with two cameras: Grundig FA85-I with a 16mm lens and Philips CCD806 with a 6mm lens. Both cameras have been calibrated at the Institute of Surveying and Photogrammetry using a 3x2x3m 3-D calibration object with 100 calibration points positioned with an accuracy of less than 0.1mm.

In our laboratory we have recorded 40 views of our calibration tile with each camera. Combining these views the cameras have been calibrated using the method of Tsai & Lenz and the closed form method. The Tsai & Lenz method is used on combinations of 1, 2, 4 and 40 views, while the closed form method, which needs at least to views have been used on combinations of 2, 4 and 40 views. The calibrated parameters are shown in table 3.

Two series of calibration by the Tsai & Lenz method are shown for the Philips camera. The first one uses the aspect ratio found by the calibration of the Institute of Surveying and Photogrammetry while the second series (marked 'bad asp.rat.') uses the value that is deduced from the hardware data of camera and vision system.

Comparing the calibrated focal length of the Grundig camera shows comparable results for the two methods. The two methods give results that are within 1.5 standard deviation of the measurements. The focal length found by the Institute of Surveying and Photogrammetry is smaller than the focal lengths found by the two methods. This may be explained by a small distortion of the calibration plane caused by inaccurate printing by the laser printer. This occurs only in one direction and the closed form method compensates by reducing the aspect ratio compared to the true value, while the method of Tsai & Lenz which has a fixed aspect ratio must compensate by increasing the focal length.

The calibrated optical centres for the Grundig camera are all within 1 standard deviation of the results from the Institute of Surveying and Photogrammetry.

For the Philips camera the calibrated focal length found by the closed form method is now lower than the true value found by the Institute of Surveying and Photogrammetry. This may be due to the large distortion of this camera which is unmodelled by this method. The Tsai & Lenz method again shows results that are within 1 standard deviation from the true value.

		focal length, f_x pixels	aspect ratio, $\dfrac{f_y}{f_x}$ (%)	C_x, pixels	C_y, pixels	radial dist. 10^{-8} pix^{-2}
		Grundig camera				
Inst. Surv.		1520.7(3.9)	96.19	295.6(8.3)	261.4(9.4)	5.58(1.46)
CF 2 v's		1514.2(12.5)	96.00(0.14)	287.1(7.1)	247.9(18.3)	-
CF 4 v's		1528.7(9.7)	95.70(0.16)	287.2(6.9)	250.3(6.0)	-
CF 40 v's		1528.9(-)	95.70(-)	286.8(-)	250.1(-)	-
TL 1 v's		1605.1(99.1)	96.20(-)	318.3(51.2)	259.8(66.9)	5.43(2.85)
TL 2 v's		1529.0(10.9)	96.20(-)	301.6(5.7)	244.1(16.0)	4.63(2.40)
TL 4 v's		1544.0(10.4)	96.20(-)	297.0(2.4)	244.1(5.3)	5.45(1.78)
TL 40 v's		1544.2(-)	96.20(-)	296.6(-)	244.1(-)	5.80(-)
		Philips camera				
Inst. Surv.		803.0(1.5)	97.91	260.2(2.6)	253.5(3.5)	24.4(1.7)
CF 2 v's		792.4(13.1)	97.27(0.57)	268.7(7.5)	250.1(2.5)	-
CF 4 v's		790.9(5.0)	97.26(0.32)	268.7(5.6)	255.2(4.1)	-
CF 40 v's		790.4(-)	97.25(-)	268.7(-)	255.3(-)	-
TL 1 v's		822.8(32.3)	97.91(-)	273.1(9.2)	251.2(19.0)	25.9(0.9)
TL 2 v's		811.6(9.4)	97.91(-)	271.3(6.5)	252.7(4.0)	25.5(2.3)
TL 4 v's		812.3(0.9)	97.91(-)	268.5(0.9)	255.3(0.6)	25.5(2.1)
TL 40 v's		812.2(-)	97.91(-)	268.4(-)	255.3(-)	25.5(-)
TL 1 v's	[a]	804.8(134.6)	101.69(-)	267.2(15.8)	242.4(21.2)	23.9(1.2)
TL 2 v's	[a]	850.7(37.9)	101.69(-)	276.5(47.3)	250.5(13.3)	14.0(8.7)
TL 4 v's	[a]	843.6(7.2)	101.69(-)	269.0(5.4)	275.0(20.2)	23.9(17.6)
TL 40 v's	[a]	843.7(-)	101.69(-)	268.7(-)	264.0(-)	21.5(-)

Table 3: Experiments. Standard deviations in parentheses.
CF: Closed form method, TL: Tsai & Lenz method. a: Bad aspect ratio.

The two methods show very good agreement between the calibrated values of the optical centre for the Philips camera. Compared to the centre found by Institute of Surveying and Photogrammetry the values are still within 1 standard deviation.

The results for single view calibration using the method of Tsai & Lenz shows large standard deviations of the calibrated parameters. The worst results are seen with the Grundig camera which has a lens of relative low distortion. This is because the distortion of the calibration points must be utilized to determine the image centre (and thus the rest of the parameters) from a single view of a single plane. Even in the case of the Philips camera which has a higher distortion the standard deviation of the focal length determined by single view is 3.9%.

The results for the Tsai & Lenz method using a wrong aspect ratio of the Philips camera show large standard deviations compared to the same results obtained by the right aspect ratio. The aspect ratio is considered known in the Tsai & Lenz algorithm, but these results show that the reliability of the achieved results depend highly on a precise knowledge of the aspect ratio. In this case the aspect ratio deduced from the hardware data of the camera and vision system was not sufficiently accurate for the algorithm to produce reliable results.

3.3. Object area determination

This section describes the determination of the area of an object moved in the workplane of a robot. The object is viewed by a camera and based on the calibration data for the camera the true area of the object can be determined. The camera is placed 110 cm from the centre of the plane and with a tilt of 35° from vertical. Area determination based on five calibration methods are compared. The first is a calibration using the direct linear transformation (DLT), (Schalkoff, 1989), where the calibration points are defined by moving the tool-centre of the robot in the workspace. The second is a calibration using the single view Tsai & Lenz method with a calibration tile in the workplane of the experiment. The three last calibration methods uses pre-calibrated intrinsic parameters from the previous experiment, respectively Tsai & Lenz 40 views, closed form method 40 views and the values from the Institute of Surveying and Photogrammetry. Based on these intrinsic parameters the extrinsic parameters are calibrated by viewing a calibration tile in the workplane of the experiment. The object is a white triangle on a black background. It is detected in the image by a thresholding operation and based on the calibration parameters the real area of each pixel is determined. Figure 3 shows 14 determinations of the area using the 5 methods.

The position of the triangle has been uniformly distributed in the working plane with the low numbers corresponding to the far end of the plane. The standard deviation is 4% for the DLT method and less than 1% for the 4 other methods. The large error of the DLT method is probably due misalignment of the working plane defined by the robot and the real working plane. The area determination is highly dependant on a correct classification of the object pixels at the periphery which means that the threshold value has high influence on the resulting area. This is clearly illustrated by the fact that changing the threshold value from 100 to 140 results in 7% decrease of the calculated area. This shows that except for the DLT the errors introduced by the calibration is much smaller than the error caused by the inherent uncertainty of the

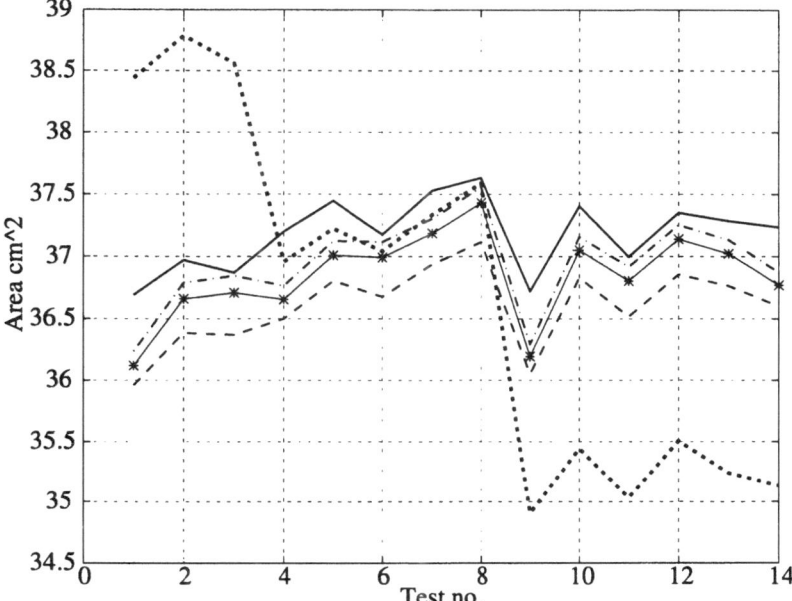

Figure 3. Calculated areas using 5 calibration methods: DLT (dotted), Tsai & Lenz single view (solid), Tsai & Lenz 40 views (dash dotted), closed form method 40 views (dashed) and Inst. of Surv. (solid with asterisk)

pixel classification. This is probably also the case for the DLT if the robot is allowed to define the working plane of the experiment.

4. Conclusion

The aim of this work has been to investigate auto-calibration procedures for an experimental robot used for hand-eye coordination research and an AGV equipped with a vision system. We have investigated two methods for single plane multi-view calibration and compared them with methods using accurate 3-D calibration objects. The comparison has been made by simulation and experiments. Our results show that 2-D multi-view calibration gives results comparable to those obtained using 3-D calibration objects. Our experiments with the method of Tsai & Lenz show that the intrinsic parameters found by singleplane/single-view are very inaccurate even with high distortion lenses. We have also shown that the method of Tsai & Lenz is very sensitive to errors in the information on the aspect ratio. Our experiment with determination of areas have shown that the investigated calibration methods have an accuracy which is often better than the inherent uncertainty of the image processing algorithms.

5. Acknowledgements

The authors wish to acknowledge the work done by Keld Dueholm for performing the calibration calculations of the 3-D calibration at the Institute of Surveying and Photogrammetry. We would also like to acknowledge the work done by Lone Steen Kristensen and Jan Damsgaard performing the area determination experiment.

6. References

Andersen, N. A., Ravn, O., and Sørensen, A. T. (1991). Real-time vision based control of servomechanical systems. In *Proceedings of the Second International Symposium on Experimental Robotics*.

Crowley, J. L., Bobet, P., and Schmidt, C. (1993). Dynamic calibration of an active stereo head. In *4th ICCV*, pages 734–741.

Faugeras, O. D., Luong, Q.-T., and Maybank, S. J. (1992). Camera self-calibration: Theory and experiments. In Sandini, G., editor, *Proceedings of the Second European Conference on Computer Vision*, pages 321–334. Springer-Verlag.

Ganapathy, S. (1984). Decomposition of transformation matrices for robot vision. *Pattern Recognition Letters*, 2(6):401–412.

Haralick, R. M. (1989). Determining camera parameters from the perspective projection of a rectangle. *Pattern Recognition*, 22(3):225–230.

Harris, C. G. and Teeder, A. (1993). Geometric camera calibration for vision-based navigation. In Charnley, D., editor, *1st IFAC International Workshop on Intelligent Autonomous Vehicles*, pages 77–82.

Lenz, R. K. and Tsai, R. Y. (1988). Techniques for calibration of the scale factor and image center for high accuracy 3-D machine vision metrology. *IEEE Transactions on Pattern Analysis and Machine Intelligence*, 10(5).

Schalkoff, R. J. (1989). *Digital Image Processing and Computer Vision*. John Wiley & Sons, Inc.

Tsai, R. Y. (1987). A versatile camera calibration technique for high-accuracy 3D machine vision metrology using off-the-shelf TV cameras and lenses. *IEEE Journal of Robotics and Automation*, RA-3(4).

Weng, J., Cohen, P., and Herniou, M. (1992). Camera calibration with distortion models and accuracy evaluation. *IEEE Transactions on Pattern Analysis and Machine Intelligence*, 14(10).

Section 4
Sensing and Learning

The papers in this section deal with sensing and learning, which are the prerequisite of intelligent behavior of robots.

Sikka, Zhang, and Sutphen propose an approach to tactile-based object manipulation, which is called tactile-servo in analogy to visual servo. Several features (the zeroth-, first-, and second-order moments, the centroid, and the axis of minimum inertia) of the two-dimensional stress distribution image from the tactile sensor on a robot finger are used to develop a tactile servo algorithm for the task of rolling a cylindrical pin on a planar surface by the finger. Experimental results show the validity of the approach.

Dario and Rucci describe an approach to robotic tactile perception which is anthropomorphic in terms of both sensor configuration and tactile signal processing. A robot finger having three kinds of tactile receptors (a static array sensor, dynamic piezoelectric film sensors, and thermal sensors) is developed and a neural network architecture for adjusting the contact forces and the finger position so as to enhance the sensed tactile feature is presented.

Yang and Asada present an approach to learning a compliance control parameters for robotic assembly tasks. Learning is achieved by the adaptive reinforcement algorithm based on a task performance index specified for the given assembly task. Experimental results for a simple box palletizing task indicate that a robot can efficiently learn a damping control law through repeated attempts.

Nishikawa, Maru, and Miyazaki propose a method for detecting occluding contours and occlusion by moving a stereo camera actively. In this method a stereo camera is moved laterally a very short distance and from the resulting change of image seen by one camera, an occluding contour model is formed. This model is then used to restrict the region of possible matched points seen by the other camera. Experimental results show the effectiveness of the method.

Tactile Servo: Control of Touch-Driven Robot Motion *

Pavan Sikka, Hong Zhang and Steve Sutphen

Department of Computing Science
University of Alberta
Edmonton, Alberta T6G 2H1 Canada

Abstract— In this paper, a new approach to tactile sensor–based object manipulation is proposed. The approach makes use of the simple observation that the progress of a manipulation task can be characterized by the tactile images produced by tactile sensors mounted on the fingertips of a robot hand. In analogy to image–based visual servo, a control scheme based upon features derived from tactile images is used to control the movement of a robot arm. The approach is applied to derive a control scheme for the task of rolling a cylindrical pin on a planar surface using a planar robot finger equipped with a tactile array sensor. The tactile features used in the control scheme are derived from a theoretical and experimental study of the variation of the normal stress distribution as a result of applied force. The experiment demonstrates that information from array tactile sensors can be used in a simple, direct and effective manner to control manipulation tasks. This approach is currently being extended to other tasks involving a dextrous multi–fingered robot hand.

1. Introduction

Tactile sensing refers to the sense of touch and provides rich information about contact when a robot interacts with its environment to perform manipulation or exploration. Tactile sensing has been studied extensively in recent years [1]. Much research has been concerned with the design of tactile sensors using different transduction technologies [2, 3] and with object recognition using touch [4], in which the objective is to determine the identity and/or physical description of an object that is being explored.

The possibility of controlling the motion of a robot manipulator using tactile feedback has been explored recently by several researchers. Houshangi and Koivo [5] conducted a simple empirical study in which the robot determined the angle between a cylindrical peg and a planar surface using the tactile sensor output from the robot hand. No attempt was made however to model the tactile sensing process or generalize the experiment to other contact situations. Berger and Khosla [6] formulated an algorithm to perform edge tracking using tactile feedback. The approach was novel in that tactile information was used to actively control the motion of the robot. A hierarchical control scheme was used for the task with an inner control loop consisting of a hybrid controller [7]

---

*This research was partially supported by NSERC under Grant No. OGP0042194, and by the Foundation for Promotion of Advanced Automation Technology.

driving the robot manipulator and an outer control loop using the information from the tactile sensor to obtain the next command for the hybrid controller.

In contrast, minimum research has been done in using tactile sensing for robot manipulation tasks. In a manipulation operation such as rolling a cylindrical object between the fingers of a dextrous hand, the identity as well as the geometric description of the object being manipulated is known, and the objective of the task is to cause purposeful motion of the object. This is different from robot motion control using tactile feedback such as the edge following task described above [6] in that the object being manipulated is in motion whereas the edge that is being tracked remains stationary. The understanding of active object manipulation using tactile sensing is important since it often provides the most direct, intuitive, and sometimes the only observation of the progress of a task.

2. Tactile Servo

In this paper, a new approach to tactile-based object manipulation is proposed. It is referred to as tactile servo in analogy to image-based visual servo [8]. The approach makes use of the simple observation that the progress of a manipulation task can be characterized by the tactile images produced by the tactile sensors mounted in the robot hand that actively manipulates the object. Tactile servo has been applied successfully to the manipulation task of pushing a rolling pin on a planar surface using a planar robot finger equipped with a tactile sensor array. The work is being extended to other tasks involving a dextrous multi–fingered robot hand.

In general when performing a manipulation task, the task can be divided into a sequence of stages or states. The successful completion of the task can be represented as a set of transitions among the states. For a robot contact task such as manipulation, if the contact can be monitored by tactile sensors, each state in the task sequence can be characterized by the tactile image acquired by the tactile sensors. To perform the necessary transitions between states, therefore, it is only necessary for the robot control system to control the motion of the robot so that the desired goal image is obtained.

Given any task, it is important to be able to compute the sequence of tactile signatures or features that define the states. This process can be expressed by a function of the form:

$$\mathbf{F}_i = f(\mathbf{X}_i) \quad \text{for } i = 1, 2, \ldots, n \tag{1}$$

where $\mathbf{X}_i$ define the various states of the object in terms of its position and orientation with respect to the robot hand, and $\mathbf{F}_i$ are the corresponding feature vectors that characterize the tactile images associated with the n states. The dimension and representation of $\mathbf{F}_i$ and the number of states n vary from task to task. In the operation of pushing a cylindrical object with a planar end-effector equipped with tactile sensor as in the case of rolling a pencil over the table top (see Figure 1), for example, $\mathbf{F}_i$ can be represented by a rectangle whose boundaries represent the area of contact.

Similar to [8], a Jacobian matrix, $\mathbf{J}_s$, which defines how the change in tactile image is related to the motion of the robot end-effector, must be derived. That is

$$\mathbf{J}_s = \frac{\partial \mathbf{F}}{\partial \mathbf{X}} \tag{2}$$

is required. With $\mathbf{J}_s$, the desired feature values $\mathbf{F}^d$ and the actual feature values $\mathbf{F}^a$ at the j-th step are compared and the differential motion in the joint space of the robot manipulator can be calculated as

$$d\Theta = \mathbf{J}_\theta^{-1} d\mathbf{X} = \mathbf{J}_\theta^{-1} \mathbf{J}_s^{-1} d\mathbf{F} \tag{3}$$

where $d\mathbf{F}$ represents conceptually $\mathbf{F}^d - \mathbf{F}^a$, the difference in image features, and $\mathbf{J}_\theta$ the Jacobian matrix relating the Cartesian differential change vector to differential change in joint space. $d\Theta$ is integrated to derive the joint trajectory that needs to be followed.

3. Example Task : Rolling a pin

To illustrate the approach outlined above, consider the task of rolling a pin along a given direction. This is shown in Figure 1. In this task, the tactile sensors can be used to characterize the progress as long as contact is maintained between the finger and the pin. In order to use tactile servo to perform this task, we need to identify (a) the various configurations X_i for this task, (b) the corresponding features F_i in the tactile image, and (c) the change in each image feature corresponding to a change in the configuration ($\mathbf{J}_s$).

The various states of the manipulator/task correspond to the position and orientation of the finger, as defined by a homogeneous transformation from the base of the arm. As the task progresses, the change in state can be characterized by the change in the position and the orientation of the arm and this change is represented by the six-dimensional vector $d\mathbf{X} = [dx \ dy \ dz \ \delta x \ \delta y \ \delta z]^T$ where the first three components correspond to differential translations and the last three components correspond to differential rotations.

The tactile features to be used in the tactile servo depend upon the forces acting on the rolling pin and the tactile sensor. These forces result in relative displacement between the sensor surface and the rolling pin. This displacement leads to a distribution of forces within the area of contact and these forces in turn result in the normal stress distribution measured by the sensor. To determine the features to be used in controlling the task of rolling a pin, we need to characterize the relationship between the forces applied to the pin, the displacement of the pin and the sensor, and the stresses measured by the tactile sensor.

3.1. Normal Force along Z

Figure 2 shows the output of the tactile sensor for different forces applied on the pin along Z in Figure 1. The darker areas in the image correspond to larger forces as detected by the sensor. The image on the right corresponds

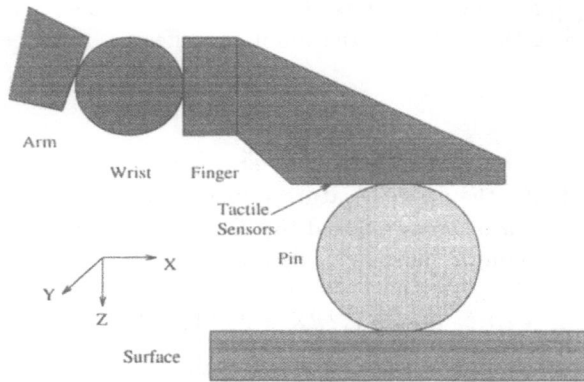

Figure 1. *Planar view of a robot arm equipped with a sensorized finger rolling a pin.*

Figure 2. *The two-dimensional stress distributions obtained experimentally from the tactile sensor in which the darkness indicates the pressure intensity.*

to a larger applied normal force. The force applied along Z presses the pin into the elastic layer and the elastic layer deforms within the area of contact to conform to the shape of the rolling pin. Since no force is applied along Y, the stress distributions are symmetrical about the line of action of the applied force. Increasing the magnitude of the applied force presses the pin further into the elastic layer, thus increasing both the maximum stress and the area of contact, as can be seen from the tactile images in Figure 2.

3.2. Tangential Force along X

As mentioned above, the force along Z presses the pin and the sensor together while the force along X causes a shearing effect in the area of contact. It also causes a turning moment in the area of contact and this has the effect of rotating the pin about Y into the surface of the sensor. All of this causes the resulting stress distribution to be skewed about the center of contact along X.

Figure 3. *The stress distributions obtained experimentally from the tactile sensor.*

3.3. Tangential Force along Y

Figure 3 shows the stress distributions obtained from the tactile sensor when a normal force acts along Z and a tangential force is applied along Y. The image on the left corresponds to a very small applied tangential force while the image on the right corresponds to a larger applied tangential force. As before, the darker areas in the images correspond to larger values of force as detected by the tactile sensor. In this case, the force along Z presses the pin against the sensor while the force along Y causes the pin to rotate about X into the sensor surface. This skews the stress distribution about the center of contact along Y. However, due to the geometry of the pin and the sensor, the effect in this case is a lot more dramatic than for the case of a force applied along X.

4. Contact Model

In order to identify features that can be used in tactile servo control of the task of rolling a pin, we consider a theoretical model of the contact between the tactile sensor and the rolling pin. The results obtained from this model are then compared with experimental data obtained using a rolling pin and a tactile sensor. This information is finally used to derive the control law for this task.

The tactile sensor is covered by a thin, elastic layer of rubber. This layer provides protection for the sensor and also aids in forming and maintaining contact between the sensor and the rolling pin. However, this layer also acts as a mechanical filter so that the stress distribution in the area of contact is transformed in a complicated manner to the stress distribution that is actually measured by the sensor [9].

Figure 4 shows the model used in analyzing the relationship between the forces acting on the rolling pin and the resulting stress distributions measured by the sensor. A major problem associated with this model is that analytical solutions are difficult or impossible to obtain for many situations. Numerical

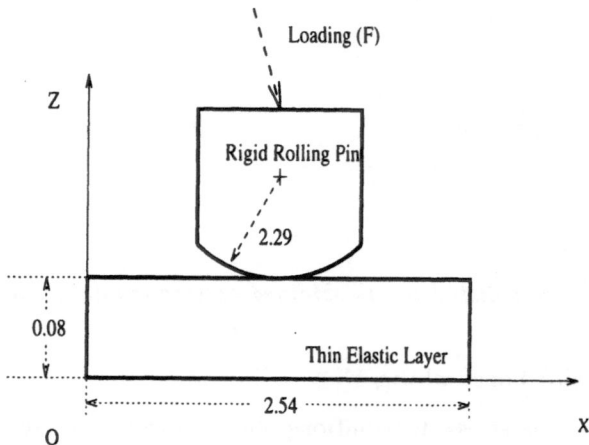

Figure 4. *The simplified two-dimensional model used in analyzing the contact between a rolling pin and the tactile sensor.*

methods are often used to obtain the solutions and the finite-element method is one of the most general methods for numerically solving contact problems in elasticity. We use the finite-element method to model the contact and to study the effect of applying different forces on the pin. The problem is simplified to two-dimensions by assuming plane-strain, i.e., the model shown is assumed to extend to infinity along Y and the forces applied to the pin do not vary with Y. The details of this approach are presented in [9].

This model can be used to analyze forces applied along Z and X. However, when a force is applied along Z and Y, we can no longer assume plane-strain which requires, in this case, that the loading be uniform along Y and should be applied in the $X - Z$ plane. In order to analyze this situation, a three-dimensional model needs to be solved. However, when the radius of the pin is large, the pin can be approximated by a flat rectangular bar. We make use of this approximation to analyze the stresses resulting from contact between a flat rectangular bar and the tactile sensor. Similarly, when the radius of the pin is small, the contact between the pin and the sensor can be approximated by line contact.

5. Identifying Features and Determining J_s

The results presented in the previous section characterize the stress distributions that are obtained for different applied forces. In order to perform tactile servo, we need to identify features in these stress distributions that relate in a unique and unambiguous manner to the position and orientation of the finger. In [9], it was found that the various moments associated with the stress distribution vary in a monotonic manner with the applied forces.

The features used in this study are the zeroth-, first- and second-order moments, the centroids, and the axis of minimum inertia of the stress distribution. These are defined below:

$$M_0 = \int \int \sigma_{zz}(x,y)dxdy$$

$$M_x = \int \int y\sigma_{zz}(x,y)dy, \quad M_y = \int \int x\sigma_{zz}(x,y)dx$$

$$x_c = \frac{M_y}{M_0}, \quad y_c = \frac{M_x}{M_0}$$

$$M_{xx} = \int \int y^2 \sigma_{zz}(x,y)dx\,dy$$

$$M_{yy} = \int \int x^2 \sigma_{zz}(x,y)dx\,dy$$

$$M_{xy} = \int \int xy\sigma_{zz}(x,y)dx\,dy$$

$$\tan(2\theta_p) = \frac{-2M_{xy}}{M_{xx} - M_{yy}}.$$

Here, $\sigma_{zz}(x,y)$ represents the distribution of contact force acting along Z and distributed in the $X - Y$ plane, M stands for the moment, (x_c, y_c) is the centroid, and θ_p is the angle between the axis of minimum inertia and X.

In order to obtain $\mathbf{J}_s$, a relation between the above features and the six components of the manipulator tip position needs to be obtained. However, due to the geometry of the task, information about all the components of the manipulator tip position (also the object position, since the two are in contact) cannot be obtained from the tactile sensors. For example, since the contact covers the entire sensor area along Y, it is not possible to obtain any information about displacement along Y. The relation between the remaining components of the manipulator tip position and the features mentioned above is considered next.

x_c is the X-component of the location of the contact centroid and is related to the X-component of the object location. M_0 is an estimate of the total force acting along Z and can thus be used as an estimate of the object position along Z. M_x, the first-order moment of the stress distribution about the X-axis, is related to the relative orientation of the object about the X-axis and can thus be used as an estimate of the rotation about the X-axis. Similarly, M_y, the first-order moment of the stress distribution about the Y-axis, is related to the relative orientation of the object about the Y-axis and can thus be used as an estimate of the rotation about the Y-axis. Finally, since the shape of the contact is rectangular, θ_p is related to the orientation of the pin in the $X - Y$ plane and can be used as an estimate of the rotation about the Z-axis.

It was found experimentally that M_y is not a very reliable indicator of the rotation about Y due to noise and the contact conditions. Hence, for the purpose of this experiment, it was not taken into consideration. Taking into account the above discussion, the following relationship between manipulator displacements and the corresponding changes in the selected features is considered:

$$
\begin{bmatrix} dx_c \\ dM_0 \\ dM_y \\ d\theta_p \end{bmatrix} = \mathbf{J}_s \begin{bmatrix} dx \\ dz \\ \delta x \\ \delta z \end{bmatrix}.
\tag{4}
$$

A constraint imposed by the plane strain assumption is that the effect of forces on the stress distribution can be studied only in the $X - Z$ plane. Due to this, for example, it is not possible to study the simultaneous effect of displacements along X, Y, and Z. 3–D models are required to study such effects. This constraint makes it difficult to obtain any theoretical estimates of some of the off–diagonal elements of $\mathbf{J}_s$. However, it can be argued that most of the off–diagonal elements should be small when compared to the diagonal elements. For example, dz, δx, and δz should contribute relatively small values to dx_c as compared to the contribution of dx. Similarly, dx, δx, and δz contribute relatively small values to dM_0 when compared to dz. Hence, as an initial approximation, $\mathbf{J}_s$ is assumed to be diagonal.

$$
\mathbf{J}_s = \begin{bmatrix} 1 & 0 & 0 & 0 \\ 0 & k2 & 0 & 0 \\ 0 & 0 & k3 & 0 \\ 0 & 0 & 0 & 1 \end{bmatrix}.
\tag{5}
$$

We now present some theoretical and experimental data to demonstrate the variation of these chosen features with the applied forces (and hence displacements). Figure 5(a) shows the variation of M_0 with the applied normal force, as obtained from the finite-element model. As can be seen from the figure, this relationship is monotonic and linear. Figure 5(b) shows the variation of M_0 with increasing normal force as obtained experimentally from the tactile sensor. This relationship is monotonic as well and is similar to the one obtained from the model.

Similarly, Figure 6(a) shows the variation of M_x with the applied tangential force. As can be seen from the figure, this relationship is also monotonic and linear. Figure 6(b) shows the variation of M_x with applied tangential force as obtained experimentally from the tactile sensor. In this case, the relationship is monotonic but non-linear. It also exhibits a saturation effect due to the fact that the sensor has a maximum force that it can measure. However, if we take this saturation effect into account, then the relationship is very nearly linear until the applied force is large enough to cause the sensor to saturate.

6. Experiments

The algorithm outlined in the previous section was implemented for the simple operation of pushing a cylindrical peg along a planar surface, using a planar end-effector mounted on a PUMA 260 industrial robot. An Interlink piezoresistive tactile array sensor was used on the planar end-effector. This sensor

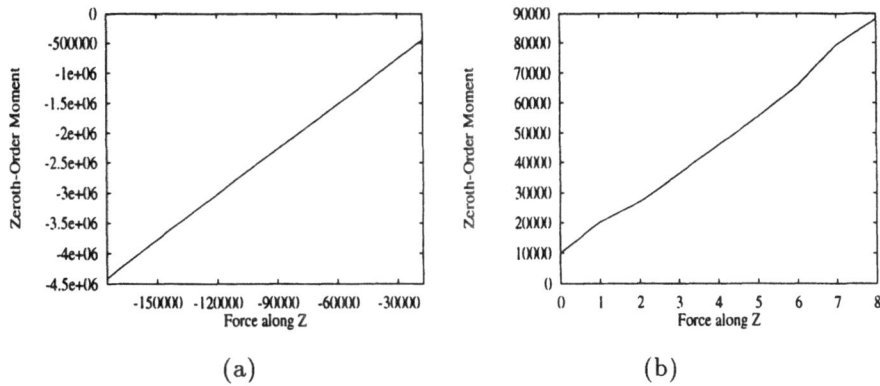

(a)   (b)

Figure 5. *The variation of M_0 with increasing normal force, (a) obtained from the finite-element model and, (b) normal force, obtained experimentally.*

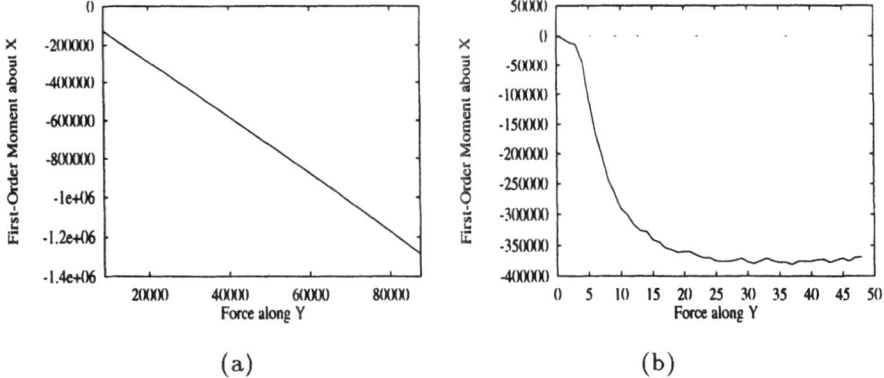

(a)   (b)

Figure 6. *The variation of M_x with increasing tangential force, (a) obtained from the finite element model and, (b) obtained experimentally.*

contains 16x16 elements over an area of 1 square inch. Both the sensor data processing and real–time robot control were performed in a Sun-3/VME based robot controller. The experimental setup is shown in Figure 7. A data collection rate of 200 Hz from the tactile array was achieved, while the position servo loop was running at 400 Hz. The planar table surface along which the peg rolls was also covered with a layer of soft rubber, in order for the robot to be able to perform the task without considering the contact stability problem. The rubber layers on the table top and tactile sensor also facilitate firm contact.

In the experiment, the orientation of the table top was only known approximately and the robot must adjust the orientation of its end-effector to accommodate the changes in the orientation of the table top. In this case, due to the compliance of the rubber layer, the desired feature values $\mathbf{F}_i$ are de-

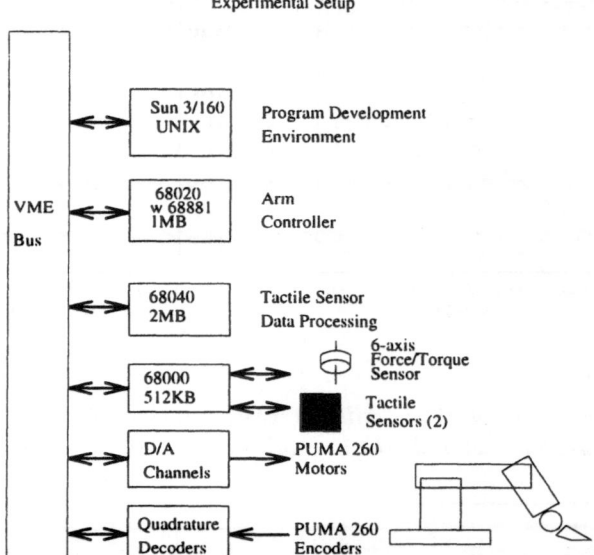

Figure 7. *The Experimental Setup.*

fined by the contact area between the peg and the tactile sensor, which should
be over an rectangular area determined by the dimension of the tactile sensor
and the radius of curvature of the peg. Discrepancy between the actual tactile
feature values and the desired feature values is caused by the uncertainty in
the orientation and by other external disturbances, and motion of the robot
end-effector is adjusted according to Equation (3).

The robot was controlled to move back and forth along X. This was
achieved by alternately setting the desired value of x_c to be the two ends of
the tactile sensor. The tactile sensor has 16 columns and the desired value of
x_c is set to be 2. Once this value is achieved to within a given tolerance, the
desired value is changed to 15. The maximum speed along X was 2 cm/s. The
desired value for M_0 was set to 15000 while the desired value for M_y was 0.
The desired value for θ_p was also set to 0. Figures 8 to 11 show the variation
of various parameters during the experiment.

Figure 8(a) shows the variation of M_0 as a function of time. The desired
position for this feature was set at 15000 in the controller. Figure 8(b) shows the
corresponding position of the fingertip in the robot base frame. The disturbance
in this figure is the result of lifting the pin off the table and then putting it
back on the table. Figure 9(a) shows the variation of x_c with time. The section
upto about 900 shows the movement of the finger from one end of the sensor
to the other along X. During the remaining part of the figure, the pin was
dragged along X as it was being rolled. This becomes clear from figure 9(b)
which shows the movement of the fingertip in robot base frame. The movement
in the first part is about 3cm while in the second part it is about 9cm, the extra
movement being a result of the dragging. Figure 10(a) shows the variation of

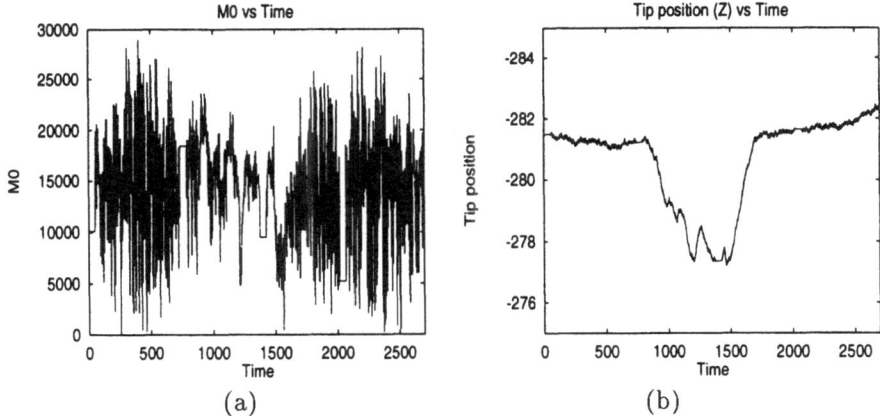

Figure 8. *(a)* M_0 *as a function of time during the experiment. (b) The Z position of the finger tip as a function of time (as obtained from the encoder counts) during the experiment.*

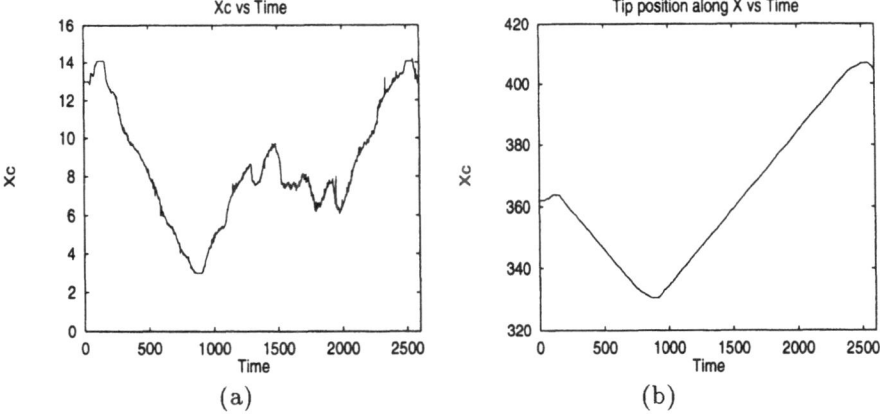

Figure 9. *(a)* x_c *as a function of time during the experiment. (b) The X position of the finger tip as a function of time (as obtained from the encoder counts) during the experiment.*

M_y as a function of time. The desired position for this feature was set at 0 in the controller. Figure 10(b) shows the variation in the orientation of the tip with time. The orientation of the tip is taken to be the angle between the Z-axis of the frame attached to the tactile sensor and the Z-axis of the robot's base frame. The figure indicates that the orientation was initially 155 degrees, then it slowly changed to about 165 degrees and then changed back to 155 degrees. The change is a result tilting the pin along X and then placing it back on the table. Figure 11(a) shows the variation of θ_p as a function of time. Finally, Figure 11(b) shows the variation of the fingertip orientation with respect to X in the robot base frame. The change from about 176 degrees to

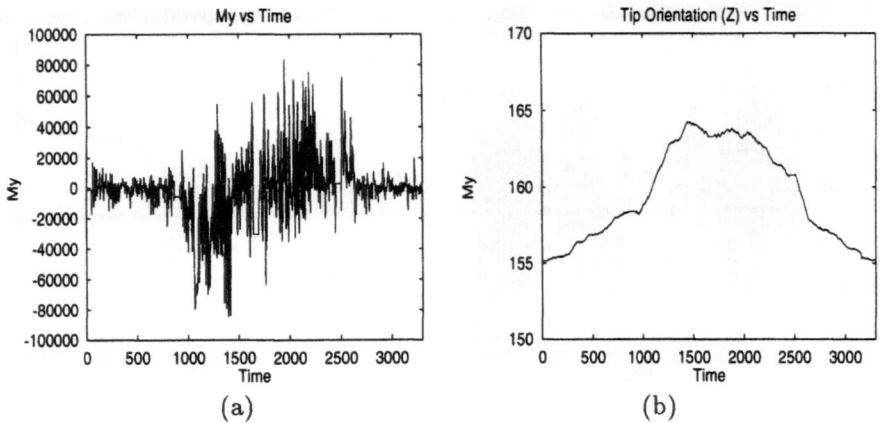

Figure 10. *(a) M_y as a function of time during the experiment. (b) The orientation of the fingertip with respect to Z as a function of time (as obtained from the encoder counts) during the experiment.*

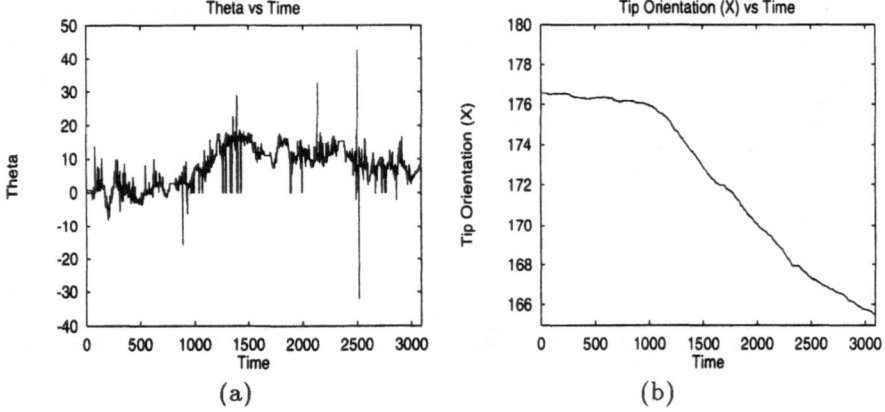

Figure 11. *(a) θ_p as a function of time during the experiment. (b) The orientation of the fingertip with respect to X as a function of time (as obtained from the encoder counts) during the experiment.*

about 166 degrees represents a rotation of the pin about Z of the robot base frame. In all the above figures, the time is shown in terms of the number of sampling periods. Since the sampling period in this case is 2.5ms, a value of 1000 on the time axis corresponds to 2.5s.

7. Conclusions

Tactile servo was introduced as a method to control tasks involving contact, in a manner similar to visual servo where vision plays an important role in controlling the motion of the robot manipulator. The task of rolling a cylindrical pin was used to illustrate tactile servo in detail. The finite-element method was used to model the contact between a cylindrical indenter and a tactile sensor. Using the results obtained from the finite-element analysis as well as experiments with a tactile sensor, a set of features to be used in the tactile servo was identified. These features were then used to develop the tactile servo algorithm. Experimental results for control using tactile servo were also presented.

Tactile servo is a simple and effective method for control in the performance of tasks involving contact. Tactile sensors provide rich information about the contact and tactile servo uses this information directly in controlling the task. The experiments showed that the performance was satisfactory even though the contact model was very approximate and the output of the tactile sensor was very noisy. An alternative method to recover the contact centroid and the equivalent force and moment at the centroid is based on "Intrinsic" Contact Sensing and is presented in [10].

This paper represents a first step in developing effective methods to control contact tasks using touch. The control algorithm is very simple and contact model is very simplistic and approximate. It is important to develop a more complicated model of contact that is able to bring out the interdependence amongst the different features. Such a model can then be used to develop more sophisticated control algorithms.

References

[1] P. Dario. Tactile Sensing for Robots : Present and Future. Oussama Khatib, John J. Craig, and Tomas Lozano-Perez, editors, *The Robotics Review 1*. The MIT Press, 1989.

[2] L. Harmon. Automated Tactile Sensing. *International Journal of Robotics Research*, 1(2):3–31, 1982.

[3] H. R. Nicholls and M. H. Lee. A Survey of Robot Tactile Sensing Technology. *International Journal of Robotics Research*, 8(1):3–30, 1989.

[4] Peter K. Allen. *Robot Object Recognition Using Vision and Touch*. Kluwer, Boston, 1987.

[5] N. Houshangi and A. J. Koivo. Tactile Sensor Control for Robotic Manipulations. *IEEE 1989 International Conference on Systems, Man, and Cybernetics*, pp. 1258–1259, 1989.

[6] Alan D. Berger and Pradeep K. Khosla. Using Tactile Data for Real-Time Feedback. *International Journal of Robotics Research*, 10(2):88–102, 1991.

[7] M. H. Raibert and J. J. Craig. Hybrid Position/Force Control of Manipulators. *Journal of Energy Resources Technology*, 103(1):126–133, 1981.

[8] L. Weiss, A. Sanderson, and C. Neuman. Dynamic Sensor-based Control of Robots with Visual Feedback. *IEEE Journal of Robotics and Automation*, RA-3(5):404–417, October 1987.

[9] Pavan Sikka, Hong Zhang, and Roger W. Toogood. On Modeling Tactile Sensors. Technical Report TR 93–08, Department of Computing Science, University of Alberta, 1993.

[10] A. Bicchi, J. K. Salisbury, and D. L. Brock. Contact Sensing from Force Measurements. *International Journal of Robotics Research*, 12(3):249–262, June 1993.

A Neural Network-Based Robotic System Implementing Recent Biological Theories on Tactile Perception

Paolo Dario and Michele Rucci

ARTS Lab
Scuola Superiore S. Anna
Via Carducci 40, 56127 Pisa
Italy

Abstract - The combination of recent improvements in robotic tactile sensors and the development of new neural network paradigms and tools currently allows to investigate of tactile perception problems in robotics by mimicking last findings in neurophysiology. In this paper we describe an antropomorhic approach to robotic tactile perception, which is characterized by a strong biological inspiration both in the hardware tools and in the processing methodologies. As a first step toward this goal, an antropomorphic robotic finger including three different kinds of tactile receptors (static, dynamic and thermal) has been designed and fabricated, and the problem of sensory-motor control for feature enhancement micro-movements has been investigated. A neural network architecture is proposed for what we believe to be the basic problem of tactile perception, that is the autonomous learning of sensory-motor coordination. Experiments are described, system performances are analyzed and possible applications are outlined.

1. Introduction

Tactile perception is extremely important in robotics, in particular when the robot must operate in unknown environments. An antropomorphic robot should be able to properly perceive if a contact with the surrounding world has been achieved, and to grasp, manipulate and analyze objects of interest, all operations that require tactile capabilities. Nevertheless, the importance of tactile perception has been often disregarded by researchers in the field of robotic and neurosciences, who devoted most of their efforts to vision. As a consequence, whereas journals and conferences specifically devoted to visual perception are now available, papers focusing on tactile perception are fewer still spread among various journals. The main reasons for such a limited interest are technological problems (dexterous hands have turned out to be difficult to control, and tactile sensors have serious problems in terms of performance and reliability) and the difficulties of processing tactile data and of integrating sensorial and motor processes [1] [2] [3] [4]. However, recent improvements concerning robotic tactile sensors technology and the development of new neural network paradigms and methods may allow to investigate tactile perception problems as a complement, or even an alternative to vision for the study of perception. Furthermore, recent neurophysiological findings have further elucidated how in primates different tactile stimuli are analyzed by means of separate systems and how these systems cooperate toward the complete tactile perception of the scene, thus disclosing new perspectives to robotics research.

Towards the goal of replicating human capabilities, we are pursuing an approach which is strongly inspired by the biological world both in terms of sensor configuration and of tactile signal processing. In particular, we focus on the development of neural network architectures for sensory-motor control procedures. Both the use of neural networks paradigms and the active interaction with the scene have recently received attention in the field of robotic tactile perception after the observation [5] that some tactile low-level problems are ill-posed in the sense of Hadamard [6] [7]. In fact, as research in the field of visual perception has shown, ill-posed problems can be solved both by means of regularization techniques (which can be well implemented by artificial neural network paradigms based on optimization criteria), and by active interaction with the environment. By gathering additional information from the scene ill-posed problems can often be transformed in well-posed and stable ones [8].

It should be observed that the human approach to the solution of ill-posed problems is based on both methods. Humans usually have a high *a priori* knowledge of the surrounding world in terms of its physical structure (surfaces are usually characterized by a spatial continuity and even discontinuities in depth or surface orientations are smooth almost everywhere [9]). Furthermore, the active interaction with the environment seems to be crucial for them as well as for most other biological beings.

In the field of tactile perception active approaches play a major role, due to the fact that touch is an intrinsically active sensorial modality, and that the tactile stimulation of sensorial receptors requires motor actions to achieve the contact with the external world. When humans perceive tactile sensations, fingers not only come into contact with external surfaces, but they *touch* these surfaces. Even without making use of an explicit conscious control, we are able to orient our fingers and to adjust contact forces so as to optimize the stimulation of tactile receptors and the extraction of tactile features. Such *feature enhancement micro-movements*, which involve the fine adjustment of sensor position and pressure after that the contact with the explored surface has been established, are extremely important when the perception system is provided with tactile sensors whose reliability is not high, as is the case for most current robotic systems. In fact, a proper adaptation of the position and pressure exerted by of the tactile sensors allows the sampling of signals that better discriminate among the explored tactile features.

In this paper we present our overall approach to robotic tactile perception. As a first experimental tool, we have developed a multifunctional tactile probe (a robotic finger) which includes different kinds of tactile receptors. Furthermore, we have addressed the problem of correlating the variations of the sensed tactile pattern to motor actions as the basis for executing purposive feature enhancement micro-movements.

The following section includes a brief review of most recent neurophysiological findings on the human tactile perception system. In section 3 the robot finger is illustrated by pointing out its similarity with biological tactile sensors, and the neural network architecture for sensory-motor coordination is described. Finally, in section 4, system performances are analyzed.

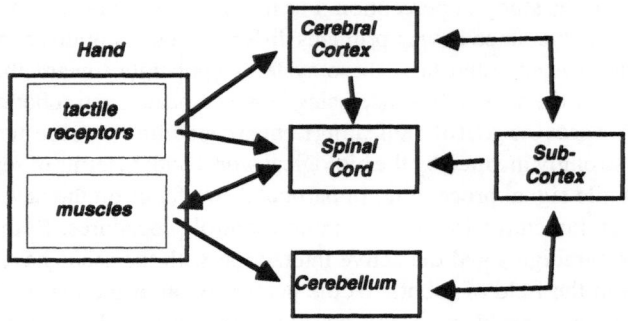

Figure 1. Information flow in primate's tactile perception system The receptor signals are analyzed by both the cerebral cortex and the spinal cord. The output arrows from the muscles indicate proprioceptive pathways, the input ones are motor pathways.

2. Recent Neurophysiological Findings on Human Tactile Sensing

Grasping and manipulating objects, as well as exploring and recognizing their physical properties by means of touch, is a task seemingly effortless for humans. Nevertheless, as shown in Fig.1, sensory control involves a large number of processing stages, and different kinds of signal are processed at the same time in different parts of the system. Tactile data provided by the tactile receptors of the hand (which is one of the part of the body with the highest concentration of tactile receptors) are given to the cerebral cortex where they are analyzed and integrated, and to the spinal cord where they close the sensory-motor loop with sensory reflexes. Furthermore, proprioceptive data are analyzed by the spinal cord and the cerebellum. It is immediately clear from such a scheme that touch is an intrinsically active sensory modality and that the study of tactile perception cannot be separated from the analysis of motor control.

Several receptors exist in the human skin, which differ for the kind of response and for the tactile parameter to which they are most sensitive. Such receptors are usually classified according to the size of the receptive fields and to the speed of adaptation of the response produced when a static stimulation is applied. According to the former classification, tactile receptors are denominated as Type I if they have a small receptive

Table I: Characteristics of human tactile receptors

Receptor	field	Type	Frequency
Meissner	3-4 mm	RAI	8-64 Hz
Merkel	3-4 mm	SAI	2-32 Hz
Pacinian	>20 mm	RAII	64-400Hz
Ruffini	>10 mm	SAII	1-16 Hz

field and as Type II if the receptive field is sensibly larger (as is intuitive, the width of the receptive field increases as the receptor is located deeper in the skin). According to the speed of the response to a static stimuli classification, receptors are split in other two categories: Slowly Adapting (SA) and Rapidly Adapting (RA). The estimated characteristics of the most common human tactile receptors are shown in Table 1 [10].

In the last decade the sensitivity of various receptors to different parameters has been investigated by several researchers. Recently, by means of a careful analysis of psychophysical tests and of neurophysiological recordings, Johnson and Hsiao [11] have proposed a global framework of the neural mechanisms involved in the perception of shape and texture. They have proposed that the SAI, RAI and Pacinian (PC) afferent systems could have complementary roles in tactile perception: in particular, they suggest that the SAI system is the primary spatial system and is responsible for tactual form and roughness perception when the fingers contact a surface directly, and also for the perception of external events through the distribution of forces across the skin surface. The PC system is responsible for the perception of external events that are manifested through transmitted high frequency vibrations. Finally, the RAI system, which has a lower spatial acuity than the SAI system but a higher sensitivity to local vibration, is responsible for the detection and representation of localized movements between skin and surface as well as for surface form and texture detection when surface variations are too small to engage the SAI system.

3. Mimicking Biological Tactile Sensing in Robots

Research on robotic tactile perception has been hampered hardly by the limited performance of available sensors, and by their low reproducibility and reliability. However, it is our opinion that the tools developed so far (tactile sensors, force control, multifingered hands), although not yet sufficient to solve completely the theoretical problem of artificial tactile sensing, can provide useful results and implement practical tools if a different approach were followed. In this sense, we believe that most robotic sensors should not be regarded as absolutely precise measurement instrumentation (this is strictly necessary only for some industrial robots and for high precision manufacturing), but rather as sources of fairly qualitative information to be exploited by the control system in a "clever" way. After all, as neurophysiological research indicates, also biological receptors seem to possess very low degree of accuracy and quantitative reliability.

The neural network paradigm allows to approach perception and sensory processing tasks by learning how to optimize the use of the available information. This is exactly the way in which most biological systems work, rather than trying to solve systematically perceptual problems by a priori modelling or using absolute sensory information. Our approach is thus focused on mimicking the strategies followed by the evolutive process as far as possible, as regards both the architecture and sensor displacement of the hardware tool, and the signal processing methodologies [12] [13].

3.1. An antropomorphic robotic finger

The system we propose is based on the functional organization of the tactile perceptual system in primates proposed by Johnson and Hsiao and described in section 2. Due to

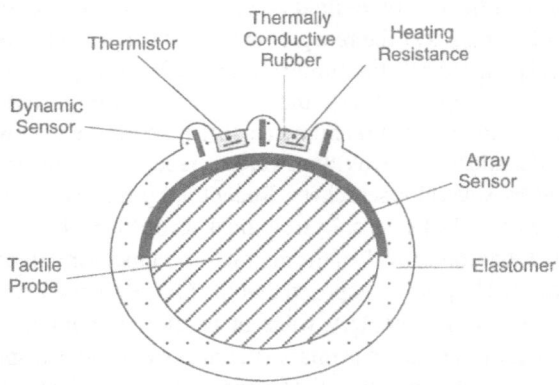

Figure 2. A section of the multifunctional tactile probe showing the location of the different tactile receptors.

the fact that their fiber number accounts for about 70% of the whole number of afferent tactile fibers [10] we focused mainly on the SAI and RA systems. We have thus developed a multifunctional tactile probe which includes several types of tactile receptors and which acts as a sort of robotic finger. A section of the finger illustrating the receptors layout is shown in Fig.2.

The functionality of Merkel receptors is simulated by an Artificial Slowly Adapting I (ASAI) system including an array sensor which is located under a compliant layer made of silicon rubber. The array sensor, based on piezoresistive semiconductor polymer technology, has been designed in our laboratory and manufactured by Interlink Europe [14]. The distribution of the sensing elements has been studied so as to mimic the distribution of Merkel corpuscles in the human fingerpad. As shown in Fig. 3, 16x16 sensing elements - which correspond to the intersection points of the conductive bars - are located in an area of 2.92 x 3.07 cm^2. In the central portion of the sensor the receptors are uniformly spaced, so as to produce a sort of "tactile fovea" where spatial acuity is maximized. The spacing between adjacent bars tends to increase in the periphery in order to achieve a wider sensing field with a limited number of sensing elements. The sensor is scanned by an analog multiplexing unit with an image rate equal to 50 Hz, and it provides a force range approximately equal to 0.1-20 N.

The RA system is emulated by an Artificial Rapidly Adapting (ARA) System which includes as receptors three piezoelectric film strips embedded into the compliant layer. Due to their technology, these strips, which are only few microns wide, are sensitive to high frequency (10-1,000 Hz) stimuli related to the strains induced in each strip by vibrations and contact. Such dynamic sensors are associated with an appropriate charge amplifying unit which acquires data with a sampling frequency of 1 Khz.

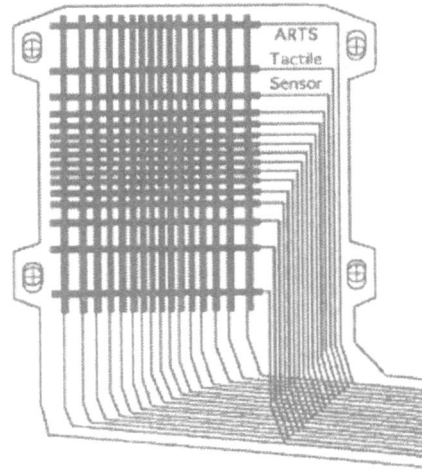

Figure 3. The ARTS Tactile Sensor.

In addition to the ARA and the ASAI receptors the tactile probe include several heat receptors located between the dynamic sensors, each including a miniaturized thermistor and a heating resistence. By means of these sensors also thermal properties of the examined surface can be detected.

In order to allow an active exploration of the scene, the tactile sensing pad has been mounted on a PUMA manipulator. Resultant contact forces are monitored in real-time by means of Force-Torque sensor located at the robot wrist [15] [2].

3.2. A neural network-based architecture for learning motor coordination

Feature enhancement micro-movements, aimed at adjusting the contact forces and the finger position so as to enhance the sensed tactile feature, can be described as shown in Figure 4. Let S_S (x_S,y_S,z_S) be the reference frame fixed with the explored surface and S_p (x_p,y_p,z_p) the reference frame fixed with the tactile probe. Since some compliance is provided by the rubber layer incorporated at the fingertip surface, the finger reference frame can be moved along the three degree of freedom γ, θ and z. All movements along these degrees of freedom have an immediate counterpart in the sensed tactile pattern, due to consistent variations of the sensed forces.

It should be considered that in order to execute skilled movements, such as feature enhancement micro-movements, a robotic system should be able to predict the consequences that each action could have on sensorial data, that is it should possess a link among sensorial and proprioceptive representation frames. Sensory-motor coordination is the first step toward the perception of the environment [16].

The difficulties of the mathematical modelling of robotic systems, the possible variations of the system due to ageing and damages, and also the unpredictability of

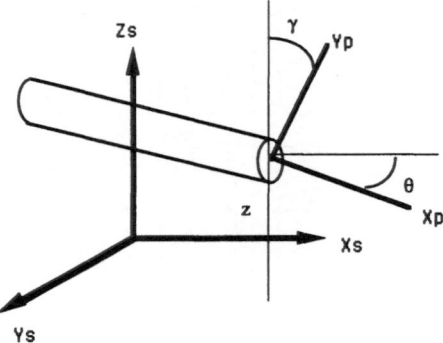

Figure 4. Feature enhancement movements.

the environment contribute to promote the autonomous development of such a correlation by means of learning. As proposed by Piaget [17], autonomous learning can be achieved with the circular reaction scheme, by producing randomly generated movements and associating the outputs of separate perceptual systems. Unfortunately, apart from some notable exception, mainly focusing in the visuo-motor case, [18] [19] [20] not many studies have been carried out on this subject.

By extending the concept of circular reaction, the system can learn its tactile-proprioceptive coordination in a totally autonomous fashion, that is without any external intervention. Let the initial position of the finger (given by the system proprioceptive apparatus) be P_1 and let T_1 be the sensed tactile pattern: a random movement generator moves

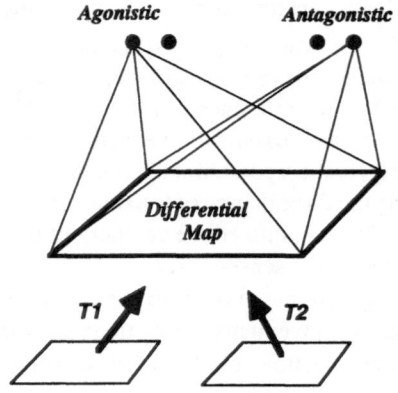

Figure 5. Scheme of the neural architecture (see text for details).

the finger into a new position P_2, thus producing a differential movement $P = P_2 - P_1$ and a differential tactile pattern $dT = T_2 - T_1$. By means of the mapping function F, an estimate $P'=F(dT)$ of the actual differential movement P is calculated and the function F is recursively refined based on the error between P' and P.

In this paper we focus on the analysis of feature enhancement movements involving rotations of the robotic finger, that is along the γ and θ degrees of freedom. The system architecture that we developed, shown in Fig. 5, includes a tactile map, where the differential tactile pattern is produced, and a layer of weights which activates two couples of agonistic-antagonistic actuators, one for each degree of freedom.

Let T_1 be the initial tactile pattern and T_2 the final pattern, acquired with finger positions $P_1=(\theta_1, \alpha_1)$ and $P_2=(\theta_2, \alpha_2)$, respectively. The values of each tactile pattern have 256 levels and are included in the range (0-1). The differential pattern is evaluated as

$$dT = \frac{T_1 - T_2}{2} + 0.5 \tag{1}$$

By applying the differential pattern to the motor map, the activation of the agonistic - antagonistic pairs ($j\theta+$, $j\theta-$ and $j\alpha+$, $j\alpha-$), are estimated as

$$\begin{cases} j\theta^+ = \dfrac{1}{A}\sum_{i,j} dT(i,j)\, w_\theta^+(i,j) \\ j\alpha^+ = \dfrac{1}{A}\sum_{i,j} dT(i,j)\, w_\alpha^+(i,j) \end{cases} \tag{2}$$

for the agonistic actuators, and as

$$\begin{cases} j\theta^- = \dfrac{1}{A}\sum_{i,j} dT(i,j)\, w_\theta^+(i,j) = 1 - j\theta^+ \\ j\alpha^- = \dfrac{1}{A}\sum_{i,j} dT(i,j)\, w_\alpha^+(i,j) = 1 - j\alpha^+ \end{cases} \tag{3}$$

for the antagonistic ones, where

$$A = \sum_{i,j} dT(i,j) \tag{4}$$

The training session is not separated by the operative phase, but on-line hebbian learning is performed at any time even when the motor actions are not randomly generated. As proposed by the circular reactions scheme the real position of the finger is evaluated by means of the proprioceptive system:

$$\begin{cases} \overline{j\theta^+} = \dfrac{d\theta + \theta_{sup} - \theta_{inf}}{2(\theta_{sup} - \theta_{inf})} \\ \overline{j\alpha^+} = \dfrac{d\alpha + \alpha_{sup} - \alpha_{inf}}{2(\alpha_{sup} - \alpha_{inf})} \end{cases} \tag{5}$$

where $\theta_{sup} = \alpha_{sup} = 1.57$ and $\theta_{inf} = \alpha_{inf} = -1.57$. Weight changes are evaluated so as to reduce the error among the estimated position and the real proprioceptive one:

$$\begin{cases} \delta w_\theta^+(i,j) = \varepsilon(\overline{j\theta^+} - j\theta^+)\,(dT(i,j)-0.5)\left\|\dfrac{d\theta}{2(\theta_{sup}-\theta_{inf})}\right\| \\ \delta w_\alpha^+(i,j) = \varepsilon(\overline{j\alpha^+} - j\alpha^+)\,(dT(i,j)-0.5)\left\|\dfrac{d\alpha}{2(\alpha_{sup}-\alpha_{inf})}\right\| \end{cases} \tag{6}$$

and, finally, weights are updated by means of an additive rule:

$$\begin{cases} w_\theta^+(i,j,t+1) = w_\theta^+(i,j,t) + \delta w_\theta^+(i,j) \\ w_\alpha^+(i,j,t+1) = w_\alpha^+(i,j,t) + \delta w_\alpha^+(i,j) \end{cases} \qquad (7)$$

4. Results

In order ot test our system with different tactile conditions we fabricated a set of aluminum cones which were used for providing the required tactile patterns. A low angle (3°) was used for the cones generatrix so as to assume a constant curvature along all the contact area. Motor actions are performed by the PUMA manipulator while the forces and torques values are carefully controlled so as to avoid critical conditions. Data at ten different curvatures values, with a curvature radius starting from 30 mm to 10 mm, were sampled with increment of 3 mm while keeping the contact force along the surface normal approximately equal to 13 N, and the finger perpendicular to the cone height and tangent to the cone surface. Tactile patterns were preprocessed by filtering them with a gaussian filter (s=2) in order to reduce sensor noise, and they were scaled in the range (0-1). After more than 2×10^4 iterations, each one involving a randomly generated motor action and the estimate of the final position, as well as the updating of the synaptic weights, we obtained a mean relative error equal to 4.7%. In particular, the mean absolute errors were 0.01 along the θ degree of freedom, and 0.03 along α.

The effect of learning can be appreciated also by analyzing pictorially the layer of weights. In Fig.6 the weights of the agonistic actuators before and after a few thousand iterations are illustrated. The grey-level of each image pixel is proportional to the weight strength. It is clear from the image structure that the weights of different actuators tend to be arranged in a complementary manner, becoming increasingly more selective to different parts of the differential tactile pattern.

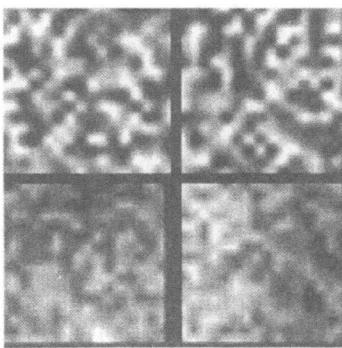

Figure 6. Weight maps for agonistic muscles before training (above) and after several thousand iterations (below).

5. Conclusions

In the last two decades, experiments with kittens have shown that a proper correlation between sensorial perception and motor action seems to be vital for the development of correct perceptual capabilities. We have analyzed the development of such a correlation in the context of force and position fine adjustment for a robotic system. We believe that sensory-motor correlation is the basic step toward the execution of intelligent and purposive motor actions. In fact, complex operations such grasping and manipulation cannot be executed without formulating a priori hypotheses on the possible effect of movements.

The experiments that we have shown in this paper is a part of a general approach to robotic tactile perception that we are currently following, which relies on the integration of tactile data acquired with different receptors, and on the active interaction with the scene based on the integration of different sensory systems, such as the tactile and proprioceptive systems. We believe that such approach can become a powerful tool to revisit the field of tactile perception.

Acknowledgements

This work has been supported by the Special Project on Robotics of the National Research Council of Italy and by MURST. One of the authors (M. Rucci) has been supported by a fellowship from "Istituto per la Ricostruzione Industriale" (IRI).

References

[1] R. S. Fearing and J.M. Hollerbach "Basic solid mechanics for tactile sensing", Int. J. Robotics Research , 4, (3):40-54, 1985.

[2] P. Dario and G. Buttazzo, "An anthropomorphic robot finger for investigating artificial aactile perception", Int. J. Robotics Research, 6, (3):25-48, 1987.

[3] R. Howe, N. Popp, P. Akella, I. Kao, M. Cutkosky, "Grasping, manipulation and control with tactile sensing", Proc. IEEE Conf. on Robotics and Automation, Cincinnati, 1258-1263, 1990.

[4] I. Amato "In search of the human touch", Science, 258: 1436-1437, 1992.

[5] Y. C. Pati, P. S.Krishnaprasad, M. C. Peckerar "An analog neural network solution to the problem of early taction", IEEE Trans. on Robotics and Automation, 8, (2):196-212, 1992.

[6] M. Bertero, T. Poggio and V. Torre, "Ill-posed problems in early vision", Proc. of the IEEE 76, 869-889, 1988.

[7] A. N. Tikhonov and V. Y. Arsenin, Solution *of Ill-Posed Problems*, Winston and Wiley, Washington DC, 1977.

[8] J. Aloinomos, I. Weiss and A. Bandopadhay, "Active vision", Int. J. Comput. Vision 2, 333-345, 1988.

[9] D. Marr, *Vision*, Freeman & C., New York, 1982.

[10] J. R. Phillips, K. O. Johnson and S. S. Hsiao "Spatial pattern representation and

transformation in monkey somatosensory cortex", Proc. Natl. Acad. Sci. USA, 85, 1317-1321, 1988.

[11] K. O. Johnson and S. S. Hsiao "Neural mechanisms of tactual form and texture perception", Annu. Rev. Neurosci. 15, 227-50, 1992.

[12] M. Rucci and P. Dario "Active exploration procedures in robotic tactile perception", Workshop on Intelligent Robotic Systems, Zakopane, July 1993.

[13] M. Rucci and P. Dario "Active exploration of objects by sensorimotor control procedures and tactile feature enhancement based on neural networks", Proc. Int. Conf. on Advanced Mechatronics, Tokyo, 445-450, 1993.

[14] Force Sensing Resistor, Interlink Inc., Santa Barbara, California.

[15] R. Bajcsy, "What could be learned from one-finger experiments", Int. Sym. Robotics Research, 509-527, 1983.

[16] A. Hein, "Prerequisite for development of visually guided reaching in the kitten", Brain Resarch 71, 259-263, 1974.

[17] J. Piaget *The grasp of consciousness: action and concept in the young child*, Cambridge, MA, Harvard University Press, 1976.

[18] M. Kuperstein, "Adaptive visual-motor coordination in multi-joint robots using parallel architecture", Proc. IEEE Conf. on Robotics and Automation, 1595-1602, Raleigh, 1987.

[19] M. Kuperstein, "INFANT neural controller for adaptive sensory-motor coordination", Neural Networks 4, 131-145, 1991.

[20] P. Gaudiano and S. Grossberg, "Vector associative maps: unsupervised real-time error-based learning and control of movement trajectories", Neural Networks 4, 147-183, 1991.

Reinforcement Learning of Assembly Robots

Boo-Ho Yang and Haruhiko Asada

Department of Mechanical Engineering
Massachusetts Institute of Technology
Cambridge, Massachusetts, U.S.A.

Abstract— This paper presents a new approach to learning a compliance control law for robotic assembly tasks. In this approach, a task performance index of assembly operations is defined and the adaptive reinforcement learning algorithm [1] is applied for real-time learning. A simple box palletizing task is used as an example, where a robot is required to move a rectangular part to the corner of a box. In the experiment, the robot is initially provided with only predetermined velocity command to follow the nominal trajectory. However, at each attempt, the box is randomly located and the part is randomly oriented within the grasp of the end-effector. Therefore, compliant motion control is required to guide the part to the corner of the box while avoiding excessive reaction forces caused by the collision with a wall. After repeating failures in performing the task, the robot can successfully learn force feedback gains to modify its nominal motion. Our results show that the new learning method can be used to learn a compliance control law effectively.

1. Introduction

Compliance control is an essential technique to cope with inherent uncertainty of task environments in robotic assembly. In many assembly tasks such as peg-in-hole insertion or box palletizing, compliance control allows a robot to correct its planned trajectory by sensing the interaction with the environment. One main stream of methodology to construct compliance control is the force feedback gain method, where contact force is measured and fed back to modify the planned trajectory. In damping control [2], for example, the robot measures contact force and modifies its nominal velocity by multiplying the force vector by an admittance matrix. Many other approaches such as stiffness control [3], hybrid control [4] [5] and impedance control [6] have been developed in the past several decades.

For efficiently programming the force feedback compliance controllers, many methods have also been developed. Hirai and Iwata used a geometric model of the environment to derive damping control parameters [7]. Peshkin also proposed a method for deriving an admittance matrix by optimizing the bounded forces condition and the error reduction condition of the manipulation [8]. In general, as described above, deriving the compliance control parameters requires intensive efforts to thoroughly understand the complex assembly process. Furthermore, as robot systems become more sophisticated with many sensors and complex control programs, the on-site debugging and tuning of the implemented controller become more time-consuming and error-prone, being a bottle neck of robotic application. It is therefore desired that robot systems can automatically tune and acquire the complaince control parameters from on-site experiences.

In this paper, we will present a new approach to learning and tuning the compliance control parameters through repeated experience. Our approach is based on the adaptive reinforcement learning control method [1]. In this learning method, the robot is initially provided with no control knowledge except the predetermined velocity command. The robot then repeatedly attempts to perform the task and gradually betters the task performance through the attempts. Contact force signals are interpreted as learning signals and used to update the feedback control parameters. To demonstrate the utility and the validity, this learning method is applied to a problem of learning damping control parameters in a simple box palletizing task.

2. Box Palletizing Task

2.1. Task Description

Box palletizing has been used as a typical example in the compliance control community. The main feature of this task is that, unlike peg-in-hole insertion, a single linear force feedback law suffices for the task [9]. Although our method can learn a nonlinear control law by simply using a nonlinear neural network to represent the controller [1], we use a linear controller for the purpose of illustrating the learning behavior more clearly.

Figure 1 shows the environment of a simple box palletizing task. A robot is required to locate a rectangular part at one corner inside the large box. The robot is initially provided with a nominal velocity command $(\hat{v}_x, \hat{v}_y, \hat{\omega})$ to track the straight planned trajectory from the initial position $(\hat{x}_I, \hat{y}_I, \hat{\theta}_I)$ to the nominal goal position $(\hat{x}_G, \hat{y}_G, \hat{\theta}_G)$. If there were no uncertainty in the task environment, the problem would be quite simple. The robot would have only to follow the nominal velocity to locate the rectangular part at the corner of the box. The real-world assembly task environments, however, inherently include the uncertainty, and the real position of the corner (x_G, y_G, θ_G) is not exactly known to the robot. Furthermore, slips of the part within the grasp of the endeffector may cause an error in sensing the orientation of the part. The robot is, therefore, required to change the nominal motion as soon as the part collides with a wall of the box so that the robot can align the part with the wall and then guide it along the wall to reach the corner of the box (x_G, y_G, θ_G) as shown in Figure 1.

2.2. Damping Control

Damping control [2] is a technique to correct a predetermined velocity command by feeding back the force and moment signals. Assuming that the process of the box-palletizing is static, a simple linear damping control law suffices to perform the task. Therefore, in this paper, we use a 3-dimensional damping control law as our target controller to be learned.

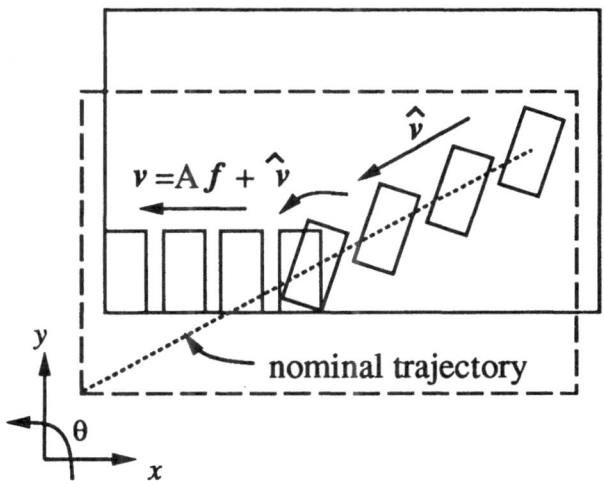

Figure 1 Box palletizing task

Let f be the force and moment signals $(f_x, f_y, N)^T$ from the environment and v be the velocity and the angular velocity commands $(v_x, v_y, \omega)^T$. The 3-dimensional damping control law is then expressed as:

$$v = Af + \hat{v} \tag{1}$$

where $\hat{v} = (\hat{v}_x, \hat{v}_y, \hat{\omega})^T$ is a given, constant nominal velocity and A is an admittance matrix(3×3) shown by

$$A = \begin{bmatrix} a_{11} & a_{12} & a_{13} \\ a_{21} & a_{22} & a_{23} \\ a_{31} & a_{32} & a_{33} \end{bmatrix} \tag{2}$$

Thus the objective of learning is to find the optimal values for a_{ij} from repeated experience.

3. Adaptive Reinforcement Learning

3.1. Reinforcement of the Assembly Task

Learning is a process of updating the controller's parameters for the purpose of improving the controller's performance. To achieve the purpose, the controller's performance has to be carefully defined in terms of measurable outputs from the process based on the task objective and constraints, and the measured performance has to be utilized as a learning signal. Reinforcement learning is one of the learning methods that explicitly use the performance evaluation, termed *reinforcement*, as the learning signal to modify the controller's parameters.

In robotic assembly, the objective of compliance control is to locate a part at an unknown, desired position with minimum interactions with the environment. In the box palletizing task, for example, the robot is required to perform the task while keeping the received reaction forces from the walls as small as possible. At the same time, in order to align the part to the corner of the box, the robot is also required to push the part against the walls in the direction of the planned velocity $\hat{v}$. Based on these consideration, we define the controller's performance r(called *reinforcement* from now on) for each time step as follows. At a time step t, the controller in eq.(1) receives outputs from the process $\boldsymbol{f}(t) = (f_x(t), f_y(t), N(t))^T$ and generates the velocity command $\boldsymbol{v}(t) = (v_x(t), v_y(t), \omega(t))^T$, resulting in the reaction forces $\boldsymbol{f}(t+1)$. The reinforcement of the control action for the time step is then defined as:

$$r(t) = -[k_f \|\boldsymbol{f}(t+1)\|^2 + k_v \|\hat{v} - \boldsymbol{v}(t)\|^2] \qquad (3)$$

where k_f and k_v are given coefficients. Obviously, the optimal controller can be achieved by maximizing this reinforcement. As previously stated, the objective of the learning is to use this reinforcement as a learning signal to update the controller's parameters so that the updated controller would receive a higher reinforcement. Note that, since $\boldsymbol{f}(t+1)$ is a function of $\boldsymbol{f}(t)$ and $\boldsymbol{v}(t)$, the reinforcement $r(t)$ can be considered as a function of $\boldsymbol{f}(t)$ and $\boldsymbol{v}(t)$, such as $r(t) = r(\boldsymbol{f}(t), \boldsymbol{v}(t))$. If the function were known, the learning problem would be rather simple. One has only to calculate the gradient of r in terms of $\boldsymbol{v}$ and modify the control parameters in the direction to increase r. However, since, the relationship between the control action $\boldsymbol{v}(t)$ and its result $\boldsymbol{f}(t+1)$ is generally unknown, the reinforcement function cannot be formulated explicitly and, as a result, the gradient of the reinforcement cannot be calculated for the gradient following approach.

3.2. Related Works

For the purpose of following the unknown gradient, many reinforcement learning approaches have been presented. In general, reinforcement learning needs some mechanism to vary or perturb its control action in order to explore or discover a better control. Barto, Sutton and Anderson, for example, proposed a stochastic approach of reinforcement learning [10], where the mechanism of the perturbation is implemented by random generation of the control actions from a probability density function. Although most of the research such as above has been devoted to control problems of discrete action space, where the controller is represented by a mapping table and the task of learning is to find the best action among a finite number of candidates, several prominent algorithms have been proposed for problems of learning continuous control actions. Gullapalli, for example, developed an algorithm of learning real-valued functions using a multilayered neural network with stochastic real-valued(SRV) units [11]. The SRV unit generates its output stochastically from a probability function and adjusts a set of parameters involved in the unit based on the received reinforcement value for a higher reinforcement. A similar algorithm has been independently proposed by Williams [12].

These algorithms for stochastic reinforcement learning have strong advantages in the sense that the controller can be improved without explicitly computing the gradient of the reinforcement or even storing the past information of the reinforcement. However, due to the trial-and-error behavior of the learning, the improvement of the controller is extremely slow. Furthermore, since no analytical tool of proving the convergence of these algorithms is yet available [12], the validity can be discussed only by simulation experiments.

In order to overcome those problems of the stochastic reinforcement learning algorithms, an indirect, model-based approach has been suggested [13]. This indirect approach, similarly to the generalized supervised learning approach [14], uses an additional differentiable neural network to model the unknown reinforcement function and identifies the model from input-output observations. The gradient information of the reinforcement function for training the controller is then acquired by differentiating the identified model in terms of control actions. This approach accelerates the controller's improvement of performance significantly. However, this indirect approach requires that the plant model be accurate enough not only in terms of its outputs, but also in terms of the derivative of the output with respect to control actions. This requirement is not always feasible since none of the existing training algorithms assures the accuracy of the model's derivative. Namely, regardless of the accuracy of the plant model, generally measured by the sum of squared errors of the model's outputs, the model's derivative is not guaranteed to be accurate enough. Furthermore, in the real world, the presence of uncertainties and measurement noise has a significant effect on the accuracy of the derivative of the model, especially in the early stage of learning and adaptation. Consequently, this type of indirect reinforcement learning algorithms have the risk of amplifying the inaccuracy of the model's derivative and supplying incorrect gradient information to the controller.

In order to solve these problems, we have proposed the adaptive reinforcement learning algorithm [1]. In the learning algorithms, we use a neural network to model the unknown reinforcement function similarly to the indirect approach discussed above. However, instead of using the model's derivative, we use the model's output itself to estimate the gradient of the reinforcement by employing a correlation technique. As a result, the proposed learning algorithm can always increase the controller's performance even without relying on the the model's derivative. In the following subsection, we briefly review this learning algorithm and describe how to apply this algorithm to the box-palletizing problem.

3.3. Adaptive Reinforcement Learning Algorithm

Figure 2 shows the overall learning system. First, the unknown reinforcement function $r = r(\boldsymbol{f}, \boldsymbol{v})$ is modeled as $\hat{r} = \hat{r}(\boldsymbol{f}, \boldsymbol{v})$ using an adaptive Gaussian

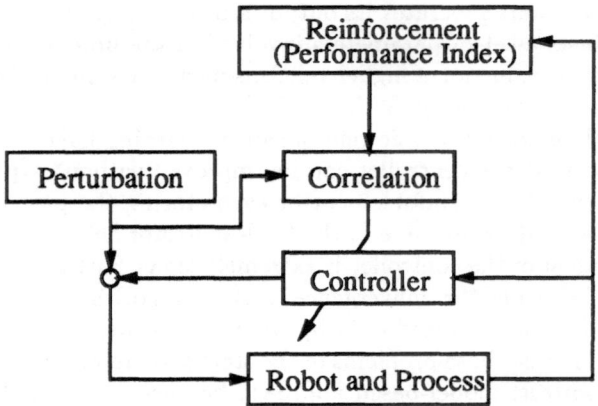

Figure 2 Schematic of the adaptive reinforcement learning system

network. The Gaussian network consists of a linear combination of Gaussian functions with a set of predetermined centers, and the linear coefficients are adaptively updated at every time step by using the recursive least squares algorithm. The main idea of the learning algorithm is that the controller's output is intentionally deviated and the deviation is correlated with the resultant reinforcement using the reinforcement model for the purpose of estimating the gradient of the reinforcement.

At every time step, learning occurs in the following manner. At given process outputs $\boldsymbol{f}$, a control action $\boldsymbol{v}$ is commanded from the controller defined by an admittance matrix A in eq.(1). An action deviation $\Delta \boldsymbol{v}$ is generated from a probability density function and then added to the control command. The real control command $\boldsymbol{v} + \Delta \boldsymbol{v}$ is executed by the robot and a reinforcement $r(= r(\boldsymbol{f}, \boldsymbol{v} + \Delta \boldsymbol{v}))$ is provided from the environment as a result of the control action. At the same time, the reinforcement the controller would receive with $\boldsymbol{v}$ is estimated by using the model $\hat{r} = \hat{r}(\boldsymbol{f}, \boldsymbol{v})$. The proposed learning algorithm to update the admittance matrix A of the controller is:

$$\Delta \boldsymbol{a} = \alpha \frac{\partial \boldsymbol{v}}{\partial \boldsymbol{a}}^T (r(\boldsymbol{f}, \boldsymbol{v} + \Delta \boldsymbol{v}) - \hat{r}(\boldsymbol{f}, \boldsymbol{v})) \Delta \boldsymbol{v} \tag{4}$$

where $\boldsymbol{a}$ is a 9 × 1 vector consisting of all the elements of the admittance matrix and α is the learning rate. In this algorithm, the second term $(\partial \boldsymbol{v} / \partial \boldsymbol{a})^T$ can be calculated easily if the control law is linear as in our example. Even in the case of a nonlinear controller, if we use a multilayer neural network to represent the controller, the term can be calculated by backpropagating derivatives from the control action output units to the weights of the controller network. The main feature of this algorithm is to use the model's output itself to estimate the gradient of the reinforcement by correlating the perturbation to the difference between the real reinforcement and the estimated reinforcement. With mild constrains about the selection of the deviation Δv, no matter how inaccurate the derivative of $\hat{r}$ is, the learning algorithm is proven to improve the performance of the controller deterministically at every time step(a proof is given in [1]), resulting in a fast, reliable convergence to the optimal control law.

4. Experiment

In this section, we demonstrate the benefits and performance of the proposed algorithm to learn the optimal damping control law on a simple robotic assembly task such as the box-palletizing task described in the previous sections. Before learning, the robot is provided with a initial admittance matrix. In our experiment, we used zeros for all the elements of the matrix. Namely, the robot is not able to feed force signals back to modify the planned velocity in the beginning of the learning and, as a result, repeats failures in performing the task. After accumulating the experience, however, the robot gradually learns the optimal values of the admittance matrix. The detail and the results of the experiment are following.

4.1. Setup

Figure 3 shows the experimental setup of the robotic box-palletizing system. The arm is a six degree of freedom direct-drive arm. Since a minimum of three degrees of freedom are required, we locked the unnecessary three joints. A six degree of freedom force sensor is attached to the tip of the arm and a rectangular part is fixed to the force sensor. A box is located on the worktable where the arm is fixed.

The robot system is controlled by a real-time computer system, which consists of a 68030 single board computer with a floating-point subprocessor and a variety of interface boards on a VME backplane network running the VxWorks operating system. All the programs, written in C, were developed on a Sun workstation. The host computer is connected to the VME backplane through an Ethernet.

A local feedback loop was constructed for velocity control so that velocity commands from the damping controller can be achieved smoothly and quickly. In other words, since the arm is controlled in a low speed operation, we do not need to cope with the dynamics of the arm and the environment.

4.2. Procedure

We first initialized the Gaussian network of the model $\hat{r}(f, v)$ before learning. Although the network is adaptively updated from input-output observation at every learning step, the initialization is important to avoid the divergence in the early stage of learning caused by the lack of the prediction power. In order to initialize the Gaussian network, we acquired 1000 sample data by placing the part uniformly on the xy plane, imposing control actions randomly and measuring the reinforcements. All the variables and parameters such as

Figure 3 Experimental setup of the box palletizing system

the coefficients and covariance matrix of the Gaussian network are initialized from these data.

At each learning trial, the position of the large box is first randomly deviated from its nominal position $(\hat{x}_G, \hat{y}_G, \hat{\theta}_G) = (0m, 0m, 0rad)$ along x and y axes. The rectangular part is also located at the initial position $(\hat{x}_I, \hat{y}_I, \hat{\theta}_I) = (0.5m, 0.5m, 0rad)$ and randomly oriented. These deviations and the orientation are unknown to the robot. The robot then starts moving the part from the initial position at $t = 0$ based on the damping controller given in eq.(1). The trial stops either when the operation time reaches 10 seconds or when the reflection force from the walls in x or y direction exceeds a threshold $F_{max} = 5N$. Then, the part is reset to the initial position and a new learning trial starts with new random box position and part orientation.

At every time step t in a learning trial, the controller outputs the velocity command $v(t)$ based on the force measurement $f(t)$. A small perturbation $\Delta v(t)$ is also generated based on a uniform probability density function. The robot then executes the real control action $v(t) + \Delta v(t)$ and receives reaction forces $f(t + 1)$, resulting in the reinforcement r given in eq.(3). The robot also estimates the reinforcement at $v(t)$ using the reinforcement model $\hat{r}(f(t), v(t))$. By plugging these values into the learning algorithm shown in eq.(4), the elements of the admittance matrix are updated. We continue this procedure until the robot can consistently succeed in the box palletizing task.

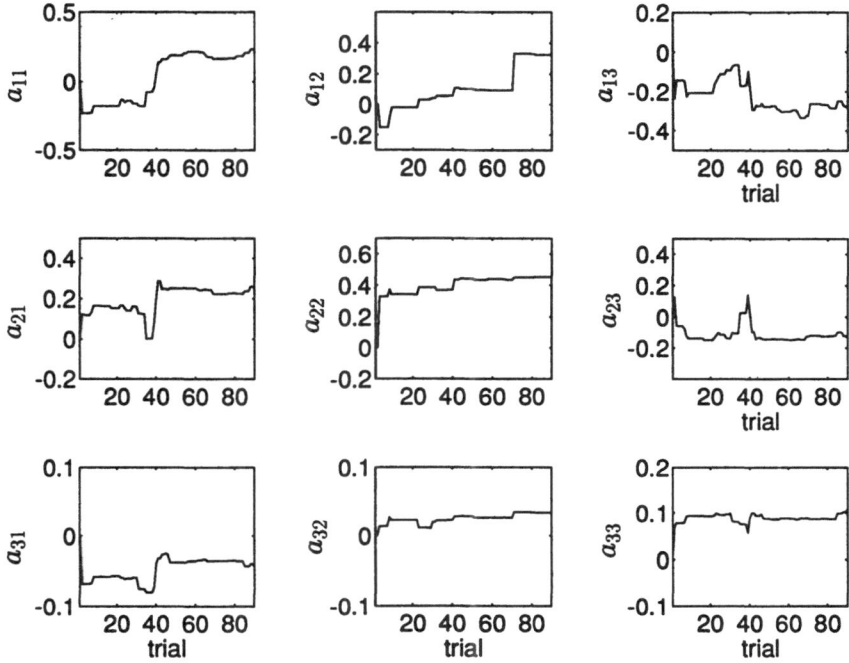

Figure 4 Learning curves of the gains over 90 trials. The unit of each parameter is $m/(sN)$.

4.3. Results

We repeated the learning process 90 times. The learning curves of the parameters in the damping control law over the 90 trials are shown in Figure 4. In the beginning of the learning process, the robot could not change the motion of the part even after colliding with a wall due to the lack of the damping control knowledge and received an excessive reaction force from the wall, resulting in a failure. At the 40th trial, the robot acquired some meaningful values for the compliance gains a_{ij} and started succeeding the box palletizing task. However, the learning curves did not reach a plateau and the reaction forces during the trial were still high, almost reaching the threshold. After the 40th trail, the learning curves began converging and the controller could consistently succeed in performing the task with small reaction forces.

5. Conclusion

A new approach of learning a compliance control law in a robotic assembly task was presented. The new learning algorithm, the adaptive reinforcement learning method, was briefly reviewed and applied to a box palletizing task. In the learning experiment, the robot was initially provided with no control

254

knowledge except the predetermined velocity command. However, the robot gradually betters the task performance through the repeated attempts. The experimental results indicate that a robot can efficiently learn a damping control law through repeated attempts.

References

[1] B.-H. Yang and H. Asada, "Adaptive Reinforcement Learning and Its Application to Robot Control", DSC-Vol. 49, Advances in Robotics, Mechatronics, and Haptic Interfaces, ASME Winter Annual Meeting, 1993

[2] D. E. Whitney, "Force Feedback Control of Manipulator Fine Motions", ASME J. of DSMC, vol. 99, no. 2, pp 91-97, 1977

[3] J. K. Salisbury, "Active Stiffness Control of a Manipulator in Cartesian Coordinates", Proc. of the 19th IEEE Conf. on Decision and Control, pp. 95-100, 1980

[4] M. T. Mason, "Compliance and Force Control for Computer Controlled Manipulators", IEEE Trans. on Systems, Man, and Cybernetics, vol. SMC-11, no. 6, pp. 418-432, 1981

[5] M. Raibert and J. Craig, "Hybrid Position/Force Control of Manipulators", ASME J. of DSMC, vol. 102, no. 2, pp. 126-133, 1981

[6] N. Hogan, "Impedance Control: An Approach to Manipulation: Part I-III", ASME J. of DSMC, vol. 107-1, pp. 1-23, 1985

[7] S. Hirai and K. Iwata, "Derivation of Damping Control Parameters Based on Geometric Model", Proc. of IEEE Int. Conf. on Robotic and Automation, vol. 2, pp. 87-92, 1993

[8] M. A. Peshkin, "Programmed Compliance for Force Corrective Assembly", IEEE Trans. on Robotics and Automation, Vol. 6, No. 4, August, 1990

[9] H. Asada, "Teaching and Learning of Compliance Using Neural Nets: Representation and Generation of Nonlinear Compliance", Proc. of IEEE Int. Conf. on Robotics and Automation, 1990

[10] A. G. Barto, R. S. Sutton and C. W. Anderson, "Neuronlike Elements That Can Solve Difficult Learning Problems", IEEE Trans. of Systems, Man, and Cybernetics, vol. 13(5), p.p. 835-846, 1983

[11] V. Gullapalli, "A Stochastic Reinforcement Learning Algorithm for Learning Real-Valued Functions", Neural Networks, Vol. 3, pp. 671-692, 1990

[12] R. J. Williams, "Simple Statistical Gradient-Following Algorithm for Connectionist Reinforcement Learning", Machine Learning, 8, 1992

[13] P. J. Werbos, "Generalization of Back Propagation with Application to a Recurrent Gas Market Model", Neural Networks, 1, 1988

[14] M. I. Jordan and D. E. Rumelhart, "Forward Models: Supervised Learning with a Distal Teacher", Cognitive Science, 16, 1992

Detection of Occluding Contours and Occlusion by Active Binocular Stereo

Atsushi NISHIKAWA, Noriaki MARU and Fumio MIYAZAKI

Faculty of Engineering Science
Osaka University
Osaka 560, Japan

Abstract—We propose a reliable method to detect occluding contours and occlusion by active binocular stereo with an occluding contour model that describes the correspondence between the contour of curved objects and its projected image. Applying the image flow generated by moving one camera to the model, we can restrict possible matched points seen by the other camera. We detect occluding contours by fitting these matched points to the model. This method can find occluding contours and occlusion more reliably than conventional ones because stereo matching is performed by using the geometric constraint based on the occluding contour model instead of heuristic constraints. Experimental results of the proposed method applied to the actual scene are presented.

1. Introduction

In triangulation such as binocular stereo, the measurement error is often generated on the contour of a curved object because a point on the contour seen by one camera cannot be sometimes seen by the other camera. These view-dependent contours are called as *contour generators*, and their projected image as *occluding contours*(see Figure 1).

The occluding contour is an important source of geometric information about its 3-D shape as follows:

① Surface orientation along the occluding contour can be uniquely computed from image data[1].

② Qualitative surface shape(such as elliptic, hyperbolic, parabolic) can be estimated from the sign of the curvature of the occluding contour[2].

③ Surface curvature can be estimated from the deformation of the occluding contour under known viewer motion[3].

Therefore, it is very useful for a variety of robotic applications — such as navigation, motion planning, object recognition and manipulation — to develop a reliable method of detecting the occluding contour. Few researchers, however, try to solve the problem of *detecting* occluding contours in the image.

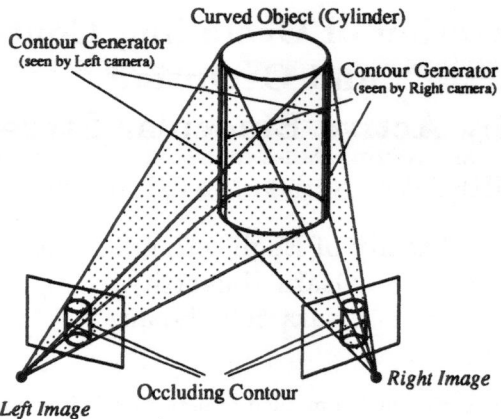

Figure 1. Relation between occluding contour and contour generator

Recently, Vaillant et al.[4] and Cipolla et al.[5] have independently proposed an effective method to detect occluding contours in the image. They use more than three views to identify occluding contours by fitting a series of projected images to the equation which describes the relationship between the occluding contours and the surface of the object. The key point of these approaches is how to obtain a series of the projected images corresponding to the occluding contours. Vaillant et al. solved this problem by using heuristic constraints such as figural and neighboring continuity to propagate the results of trinocular stereo. Cipolla et al. obtained a series of the projected images by tracking edges on the EPI(Epipolar Plane Image) taken from a monocular moving camera in a short interval. The former method, however, may often produce correspondence errors because the heuristic constraints are not always true for every scene, and the latter method needs a lot of computation because it deals with a large amount of image data.

We propose a new method which overcomes their drawbacks by active binocular stereo[7][8] with an occluding contour model that describes the correspondence between the contour of curved objects and its projected image. Applying the image flow generated by moving one camera to the occluding contour model, we can restrict possible matched points seen by the other camera. We detect occluding contours by fitting these matched points to the model. This method can find occluding contours and occlusion more reliably than conventional ones because stereo matching is performed by using the geometric constraint based on the occluding contour model. Experimental results on the actual scene are presented to demonstrate the effectiveness of this method.

2. Occluding Contour Model

We will derive an *occluding contour model* that describes the correspondence between the contour of curved objects and its projected image.

2.1. Geometry of Stereo Camera

First of all, we explain the stereo camera geometry and define some symbols. Let us consider the standard stereo rig composed of more than three cameras(focal length f) with their optical axes mutually parallel and perpendicular to the baseline.

For convenience, the cameras are numbered from the left as $0, 1, 2, \cdots$ (see Figure 2). We denote the focal point of camera i by o_i, and the distance between o_0 and o_i by b_i. We now consider the right-handed coordinate system of which the origin is at the focal point of camera $0(o_0)$, and the x axis is along the baseline and the z axis is along the optical axis of camera 0, and the y axis is perpendicular to both x and z. Also consider the 2-D coordinate system on the image plane of each camera as shown in Figure 2.

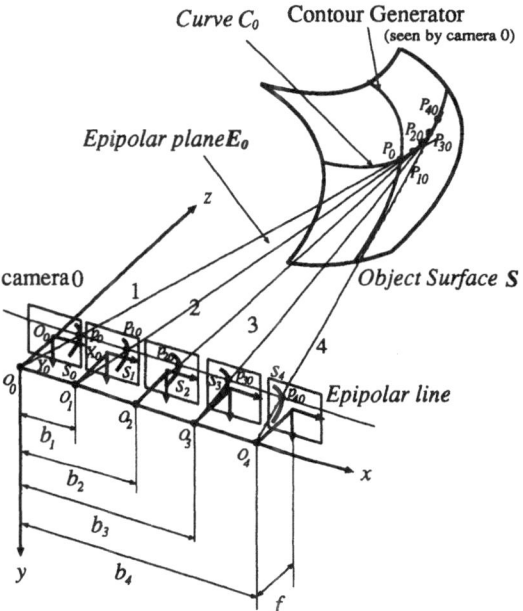

Figure 2. Geometry of stereo camera and object

2.2. Geometry of Surface and Its Projection

Now we suppose that the stereo rig is looking at a smooth object surface S. The camera i looks at the contour of the surface S, which produces the occluding contour s_i(see Figure 2).

Let p_0 be a point on the occluding contour s_0 seen by camera 0, and P_0 be a point on S corresponding to p_0(i.e., p_0 is the image of P_0). Consider the epipolar plane of p_0 with respect to another camera $i(\neq 0)$, denoted by E_0 (E_0 is identical for arbitrary i). Suppose that E_0 intersects the occluding contour

s_i at a point p_{i0} which is the image of a point P_{i0} on S. Now we consider a curve C_0, which is the intersection of the plane E_0 with the object surface S. It is easily seen that the tangent at P_0 to C_0 is the optical ray of camera 0, and the tangent at P_{i0} to C_0 is the optical ray of camera i (see Figure 2). Now consider the orthogonal projection of these optical rays and the curve C_0 onto the xz plane H as shown in Figure 3. If the object surface is locally smooth in the vicinity of the point P_0, we can locally approximate the curve C_0 as a circle, that is, we can consider the *osculating circle* to curve C_0 at P_0 in the plane H. Denoting the osculating circle by C, the radius of C by r_C, the center of C by $O_C(x_C, z_C)$, and X coordinate of the projection of P_0, P_{i0} by X_0 and X_i respectively, we have

$$x_C = \frac{X_i}{f} z_C + b_i \pm \frac{\sqrt{f^2 + X_i^2}}{f} r_C \qquad (1)$$

The double sign of Eq.(1) depends on the relative position between O_C and P_0. Denoting x coordinate of P_0 by $x_0(\neq x_C)$, the double sign of Eq.(1) is plus for $x_0 < x_C$, and minus for $x_0 > x_C$.

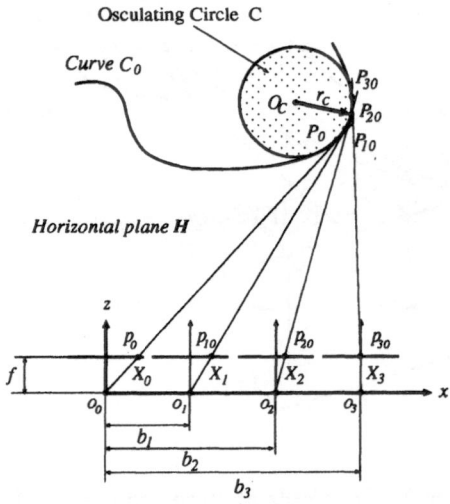

Figure 3. Occluding contour model

In this article, the geometric relation represented by Eq.(1) is called "*Occluding Contour Model*". In this model, the radius of curvature r_C is non-negative. If $r_C = 0$, Eq.(1) is consistent with the well-known perspective projection model because the third term vanishes. Namely, the *occluding contour model* represents occluding contour if r_C is positive, and regular edge (e.g., surface marking, discontinuity in depth or orientation) if r_C zero. Therefore, the *occluding contour model* is considered as an extension of the perspective projection model.

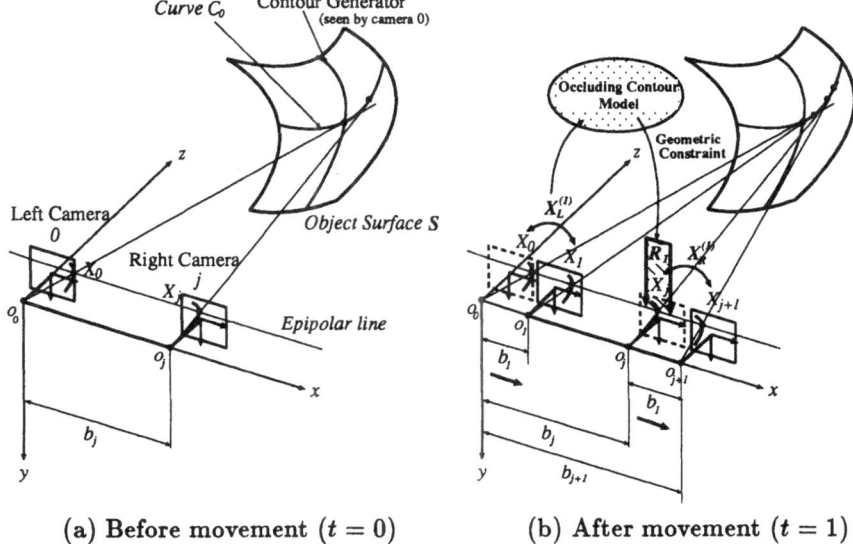

(a) Before movement $(t = 0)$ (b) After movement $(t = 1)$

Figure 4. Stereo correspondence algorithm

3. Stereo Correspondence Algorithm

In this section, we will explain the active stereo correspondence algorithm with the occluding contour model.

3.1. Summary of the Algorithm

We assume that 1) the environment is static, 2) the calibration of the stereo camera is done accurately, 3) the camera motion is known exactly, 4) the range of scene depth is known in advance.

Under these assumptions, we describe a summary of the algorithm as follows :

① Consider the stereo rig of which the left camera is camera 0 and the right camera is camera j as shown in Figure 4(a). Set $t \leftarrow 0$.

② Set $t \leftarrow t + 1$. Move the stereo camera laterally along the baseline to the right so that the distance of movement of the cameras from the initial position may be b_t(see Figure 4(b)).

③ Calculate the paired feature points $X_L^{(t)} = \{X_0, X_t\}$(left camera), $X_R^{(t)} = \{X_j, X_{j+t}\}$(right camera) respectively.

④ Limit the search range of a point to be matched seen by the right(left) camera by applying the pair $X_L^{(t)}$ $(X_R^{(t)})$ to the occluding contour model (see Section 3.2). —For example, we can use $X_L^{(1)}$ generated by the left camera to restrict possible matched points seen by right camera j within the range $\boldsymbol{R}_1$ as shown in Figure 4(b).

⑤ Find the matched point and occlusion, by counting the number of candidates of the matched point in the limited search range, defined as n_t. If $n_t = 0$, the point is in *occlusion* because it means that there is no possible matched point in the search range. If $n_t = 1$, the candidate will be considered as the matched point because it means the correspondence is unique. If $n_t \geq 2$, it means the correspondence is still ambiguous; this resulting ambiguity of the correspondence will be resolved by the successive camera movement(see Section 3.2). —For example, if there is only one point X_j in $\boldsymbol{R}_1$, we will consider that $\boldsymbol{X}_L^{(1)}$ matches X_j, that is, $\boldsymbol{X}_L^{(1)}$ matches $\boldsymbol{X}_R^{(1)}$ as shown in Figure 4(b).

⑥ Back to ②. (Repeat until no more matched points and occlusion are found)

The important points of this method are as follows:

- The principle to limit the search range of possible matched points based on the occluding contour model.

- The strategy of movement of the stereo camera (Decide b_t, lest the correspondence should be ambiguous in calculating $\boldsymbol{X}_L^{(t)}$ and $\boldsymbol{X}_R^{(t)}$.)

We will explain the former point in the next section. The latter point will not be discussed in this article (see [7][9]).

3.2. Geometric Constraint Based on Occluding Contour Model

We will derive the geometric constraint based on the occluding contour model, which is used to limit the search range of possible matched points (see algorithm ④). Now we derive the constraint for restricting possible matched points seen by the *right* camera based on the *left* pair $\boldsymbol{X}_L^{(t)}$.

3.2.1. In Ideal Case – the case that there is no positioning error of feature points –

Let us first assume that there is no positioning error of feature points. In this case, the stereo correspondence problem in our algorithm will be rewritten as follows:

Limit the search range of the matched point X_j seen by camera j based on the given pair $\{X_0, X_t\}$, where $j > t > 0$.

We have three equations by substituting X_0, X_t, X_j into Eq.(1) respectively. Considering these equations as a system of equations with three unknowns

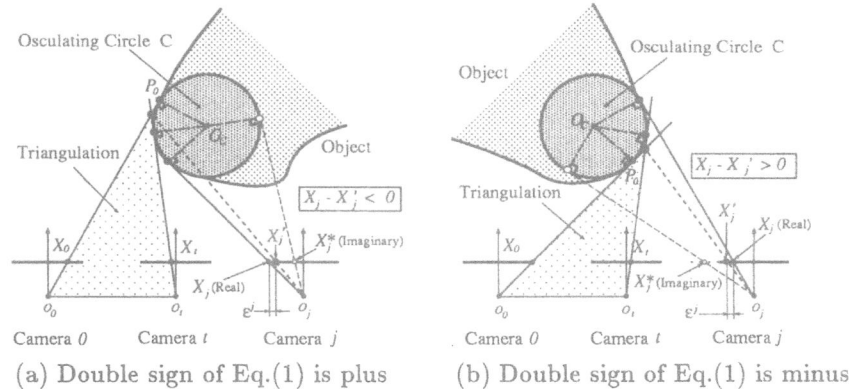

(a) Double sign of Eq.(1) is plus   (b) Double sign of Eq.(1) is minus

Figure 5. Geometric constraint based on occluding contour model

(x_C, z_C, X_j), and assuming that $f^2 \gg X_i^2\ (^\forall i)$, $z_C \gg r_C$, we obtain the following form with respect to X_j:

$$X_j = X_j' \pm \varepsilon^j \tag{2}$$

$$X_j' = X_0 - \frac{b_j}{b_t} D_t \tag{3}$$

$$\varepsilon^j = r_C \frac{\frac{b_j}{b_t}\left(\frac{b_j}{b_t} - 1\right) D_t^2}{2\{(f + k_t)z_C \mp X_j' r_C\}} \quad (\geq 0) \tag{4}$$

where

$$k_t = \frac{1}{2f}\{X_0^2 - \frac{b_j}{b_t}(X_0^2 - X_t^2)\}$$

$$z_C = \frac{1}{D_t}\{fb_t - (l_0 - l_t)r_C\}$$

$$D_t = X_0 - X_t \ (> 0)$$

$$l_t = \mp\sqrt{f^2 + X_t^2}$$

This relationship is derived in [9].

In Eq.(2), X_j' expresses the position X_j when $r_C = 0$ (in the case of regular edge), and ε^j means the magnitude of the gap between X_j and X_j' caused by bounded curvature of the object surface (see Figure 5).

As can be seen from Eq.(4), the magnitude of the gap ε^j increases when the distance from the camera to the object z_C decreases and the radius of curvature r_C increases, that is, when the ratio r_C/z_C increases.

In fact, we cannot estimate X_j directly from Eq.(2) because neither the double sign of Eq.(2) (i.e.,the direction of the gap) nor r_C are known. We now notice the relationship $z_C \gg r_C$ (described above). In this case, the positive constant $\gamma(> 0)$ which satisfies the following inequality exists:

$$\frac{r_C}{z_C} \leq \gamma \ll 1 \tag{5}$$

Then the maximum of ε^j is given by

$$\varepsilon^j_{max} = \frac{\frac{b_j}{b_t}\left(\frac{b_j}{b_t} - 1\right)D_t^2\gamma}{2\{(f + k_t) - |X'_j|\gamma\}} \quad (> 0) \tag{6}$$

Hence, we can limit the search range of X_j as follows:

$$X'_j - \varepsilon^j_{max} \leq X_j \leq X'_j + \varepsilon^j_{max} \tag{7}$$

3.2.2. In Real Case - the case that there is some positioning error of feature points -

Now we assume that there is some positioning error of feature points. In this case, denoting the measurement of $X_k(^\forall k)$ by $\hat{X}_k$, the stereo correspondence problem in our algorithm will be rewritten as follows:

> *Limit the search range of the matched point $\hat{X}_j$ seen by camera j based on the given pair $\{\hat{X}_0, \hat{X}_t\}$, where $j > t > 0$.*

Let $\tilde{X}_j$ be an estimate of X_j obtained by introducing $\hat{X}_0$ and $\hat{X}_t$ into Eq.(2) (superscript "~" means a calculation value from noisy data). The positive constant $\Delta^j_{max}(> 0)$ which satisfies the following inequality exists theoretically:

$$|\hat{X}_j - \tilde{X}_j| \leq \Delta^j_{max} \tag{8}$$

We now denote the maximum of positioning error of feature points by δ. By means of a simple error analysis, we obtain

$$\Delta^j_{max} = 2\left(\frac{b_j}{b_t} + 1 + 3\frac{\varepsilon^j_{max}}{D_t}\right)\delta \tag{9}$$

Hence, from Eq.(7),(8), we can limit the search range of $\hat{X}_j$ as follows:

$$X'_j - \varepsilon^j_{max} - \Delta^j_{max} \leq \hat{X}_j \leq X'_j + \varepsilon^j_{max} + \Delta^j_{max} \tag{10}$$

Notice that the magnitude of the search range of $\hat{X}_j$: $2(\varepsilon^j_{max} + \Delta^j_{max}) + 1$ decreases when the distance of movement b_t increases. It means that the ambiguity of the stereo correspondence may be resolved by the successive camera movement (see algorithm ⑤,⑥).

4. Detection of Occluding Contours

We can obtain a series of feature points $\hat{X}$ by applying the algorithm described in Section 3. When the number of the component of $\hat{X}$ is more than four, we can estimate the three parameters of osculating circle $(\tilde{x}_C, \tilde{z}_C, \tilde{r}_C)$ and the double sign of Eq.(1) by fitting $\hat{X}$ to Eq.(1) with the least square approximation.

As described in Section 2.2, the condition for $\hat{X}$ to be on the occluding contour is that r_C is nonzero(positive). Now we consider the probability for r_C to be positive, denoted by $P\{r_C > 0\}(\in [0,1])$. Assuming that the positioning error of feature points follows a uniformed distribution between $-\delta$ and δ, the variance of estimation error of $\tilde{r}_C$, denoted by $\sigma_{r_C}^2$, is approximately given [9] by

$$\sigma_{r_C}^2 = \frac{\delta^2}{3} \sum_k \left(\frac{\partial F}{\partial X_k}(\hat{X}) \right)^2 \tag{11}$$

where $r_C = F(X)$; $F(X)$ is a function of X.

Intuitively, the larger $K = \tilde{r}_C/\sigma_{r_C}(> 0)$, the higher $P\{r_C > 0\}$. Mathematically, the Chebyshev inequality [6] announces that

$$P\{r_C > 0\} > P\{|\Delta r_C| < \tilde{r}_C\} = 1 - \frac{1}{K^2} = P_{inf}$$

where Δr_C is the estimation error of r_C.

In this article, if $\hat{X}$ satisfies the following inequality (12), each component of $\hat{X}$ is considered as the point on the occluding contour. Otherwise, each of them is considered as the regular edge point.

$$P_{inf} > T \tag{12}$$

where T is a threshold that indicates the probability of the assertion that we have chosen. In practice, $T = 0.8(80\%)$.

5. Experiment

The proposed method has been applied to the actual scene.

5.1. Experimental Procedure

5.1.1. Experimental Setup

The two CCD cameras(baseline length: 120(mm), focal length: 16(mm)) are mounted on the slider and can move along the baseline in the precision of 0.001(mm). Stereo image sequences(370 × 370 (pixel), 8(bit) respectively) are taken by the stereo camera and digitized into the memory of the host computer(Sun SPARC station 1). Figure 6 shows the stereo image taken at an initial position. In this experiment, five stereo images(i.e.,ten images) are taken by moving the cameras four times and the amount of movement b_t is selected as Table 1.

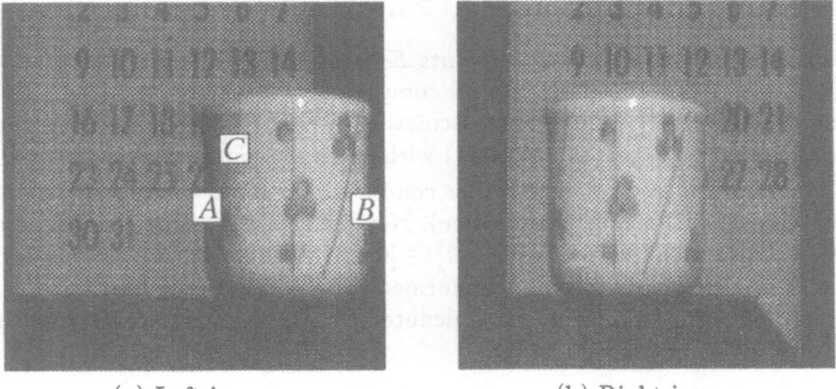

(a) Left image   (b) Right image

Figure 6. Stereo image taken at time $t = 0$

Table 1. Camera position

(Left Camera)		(Right Camera)	
t	b_t(mm)	t	b_t(mm)
0	0.000	5	120.008
1	2.015	6	122.023
2	4.021	7	124.029
3	8.003	8	128.011
4	16.000	9	136.008

5.1.2. Feature Extraction

In this experiment, zero crossing points acquired by $\nabla^2 G$ filter with the standard deviation $\sigma = 1.5$(pixel) are used as the feature points. We detect the position of zero crossing points in sub-pixel precision: $\delta = 0.1$(pixel).

5.1.3. The Scene

The scene used in this experiment is composed of a Japanese tea cup and a calendar as shown in Figure 6. The distance from the stereo rig to the cup is about 500(mm), and that to the calendar is about 1200(mm). As the tea cup(radius:about 40(mm), height:about 80(mm)) is approximately cylindrical, the projection of its side contours is considered as the occluding contour. Notice that there are a lot of occlusion caused by the cup (a part of calendar is occluded by the cup). We set $\gamma = 0.2$(see Eq.(5)).

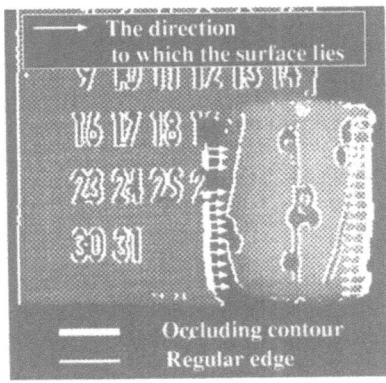
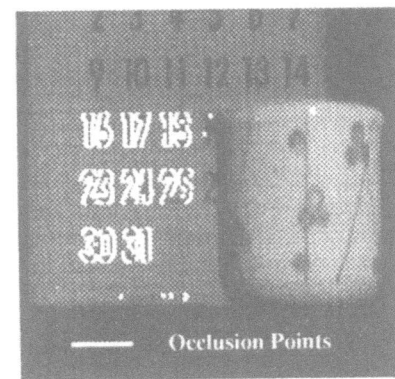

(a) Occluding contours   (b) Occlusion points

Figure 7. Experimental results

5.2. Results and Discussion

The experimental results are showed by presenting the original image (Figure 6(a)) overlaid with detected and labeled feature points as shown in Figure 7.

In Figure 7(a), thick lines indicate occluding contours and fine lines indicate regular edges and arrows attached thick lines indicate the direction to which the surface lies; the direction is determined by the double sign of Eq.(1) as shown in Figure 5. We can easily see that the occluding contour of the cup is almost detected correctly. On the other hand, the top parts of the left side contour are not classified as occluding contour. The reason is that the contour of characters on the calendar was extracted as feature points instead of the contour of the cup; therefore, this classification is correct.

In Figure 7(b), the brightest points indicate occlusion points which have no corresponding point in Figure 6(b). It is seen that almost all of the occlusion points caused by the cup are also detected correctly.

Table 2 shows the value of the following parameters for a set of three points A,B,C: the actual radius of curvature r_C, the estimated radius $\tilde{r}_C$, the value of the deviation σ_{r_C}, and the infimum of the probability P_{inf}. We can see that the estimated radii of curvature agree with the actual ones in the precision

Table 2. Estimated radius of curvature

Point	$r_C(\mathrm{mm})$	$\tilde{r}_C(\mathrm{mm})$	σ_{r_C}	$P_{inf}(\%)$
A	39.04	38.53	15.69	83.4
B	39.04	33.86	12.73	85.9
C	0.00	5.01	13.76	0.0

of about 5(mm). Note that the actual value of radius r_C is not always true because r_C were measured using calipers in the precision of 0.05(mm).

In this experiment, the number of feature points, denoted by N_a, is 4624 and the total computational time (except the feature extraction process) is about 23.7(sec). *Detection Rate*$(= N_c/N_a)$ and *Correct Answer Rate*$(= N_c/N_d)$ are 89.2(%), 98.7(%) respectively; where N_c means the number of occluding contour points detected *correctly* and N_d means the total number of detected occluding contour points (including mistakes).

6. Conclusion

In this article, we propose a reliable method to detect occluding contours and occlusion by moving the stereo camera actively. This method is more effective for every scene than conventional ones because the geometric constraint based on the occluding contour model instead of heuristic constraints are used to obtain a series of the projected images corresponding to occluding contours. The occluding contour model is an extension of the perspective projection model in the sense that not only regular edges but also occluding contours are explicitly formulated as the projection of the scene. Our method opens potential applications to robotics especially in the non-polyhedral environment.

References

[1] H. G. Barrow and J. M. Tenenbaum. "Interpreting line drawings as three-dimensional surfaces". *Artif. Intell.*, 17:75–116, 1981.

[2] J. J. Koenderink. "What does the occluding contour tell us about solid shape ?". *Perception*, 13:321–330, 1984.

[3] A. Blake and R. Cipolla. "Robust estimation of surface curvature from deformation of apparent contours". In *Proc. of the European Conf. on Computer Vision*, pp. 465–474, 1990.

[4] R. Vaillant and O. D. Faugeras. "Using extremal boundaries for 3-D object modeling". *IEEE Trans. on Pattern Anal. Mach. Intell.*, 14(2):157–173, 1992.

[5] R. Cipolla and A. Blake. "Surface shape from the deformation of apparent contours". *Int. J. Computer Vision*, 9(2):83–112, 1992.

[6] C. Derman, L. J. Gleser and I. Olkin. *A guide to probability theory and application*. Holt, Renehalt and Winston,Inc., pp. 227, 1973.

[7] N. Maru, A. Nishikawa, F. Miyazaki, and S. Arimoto. "Active detection of binocular disparities". In *Proc. IEEE/RSJ Int. Workshop on Intell. Robots and Systems*, pp. 263–268, 1991.

[8] N. Maru, A. Nishikawa, F. Miyazaki, and S. Arimoto. "Active binocular stereo". In *Proc. IEEE Int. Conf. on Computer Vision and Pattern Recognition*, pp. 724–725, 1993.

[9] A. Nishikawa, N. Maru, and F. Miyazaki. "Detection of occluding contours by using active stereo vision". *The Trans. of the Institute of Electronics, Information, and Communication Engineers*, J76-D-II(8):1654–1666, 1993(in Japanese).

Section 5
Dynamic Skills

Among varied tasks performed by human beings or other animals, not a few of them require dynamic skills. The papers in this section deal with dynamic behaviors in which impact dynamics between the robot and the environment plays an important role.

Gregorio, Ahmadi, and Buehler report ongoing efforts toward realizing practical legged robots with improved stability properties and autonomous operation. They have built a planar one-legged robot powered by electric motors, instead of hydraulic actuators used in Raibert's work. Raibert's controller is modified to reduce the power requirements and is shown experimentally to achieve stable running.

Rizzi and Koditschek deal with a task called juggling, to bat two freely falling balls into stable periodic vertical trajectories with a single robot arm using a real-time stereo camera system. The paper concentrates on a nonlinear state estimator to recover the position and velocity of falling balls directly from the image plane measurements of a stereo camera pair.

Experiments with an Electrically Actuated Planar Hopping Robot

P. Gregorio, M. Ahmadi and M. Buehler

Centre for Intelligent Machines
McGill University, Department of Mechanical Engineering
Montréal, Québec, Canada, H3A 2A7

Abstract—We have built a planar one-legged robot to study the design, actuation, control and analysis of autonomous dynamically stable legged robots. Our $15kg$ robot is powered by two low power $80W$ DC electric motors, yet it operates in a stable and robust fashion and currently achieves a top running speed of $1.2m/s$. Both design and control borrow heavily from Raibert's work whose robots are actuated by powerful hydraulic actuators. Surprisingly, even with our drastically reduced power available for actuation, only the hopping height controller had to be modified to achieve stable running. As a basis for further improvements we introduce a scalar "locomotion time" variable, which maps one locomotion cycle onto a fixed interval, independent of operating conditions. A dynamical model and a computer simulation have been developed which accurately predict the robot's behavior.

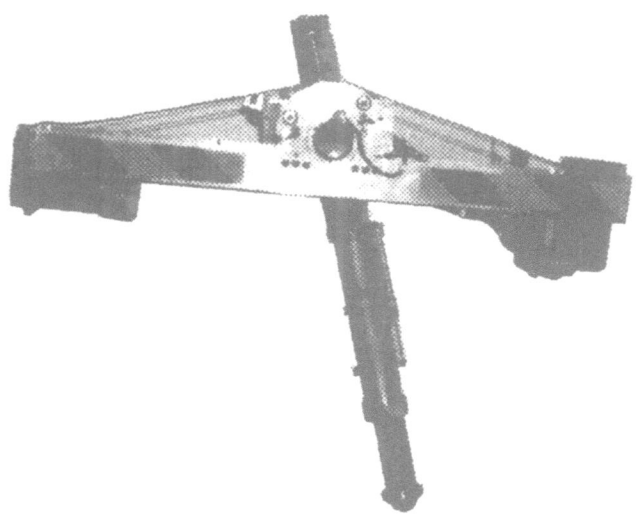

Figure 1. Photograph of the Prismatic Planar Hopping Robot.

1. Introduction

1.1. Motivation

Legged robots which are capable of dynamic operation and balance promise to exhibit similar mobility, efficiency and dexterity as their biological counterparts.

Such robots would be able to operate in a large range of environments and surface conditions and promise to be the mobile platform of choice compared to wheeled systems. However, before legged robots become practical, improved stability properties and autonomous operation are essential. To this end, we focus on electrical actuation, instead of hydraulics, since it is more suitable for indoor autonomous robots, and is a cleaner, safer and a less expensive technology.

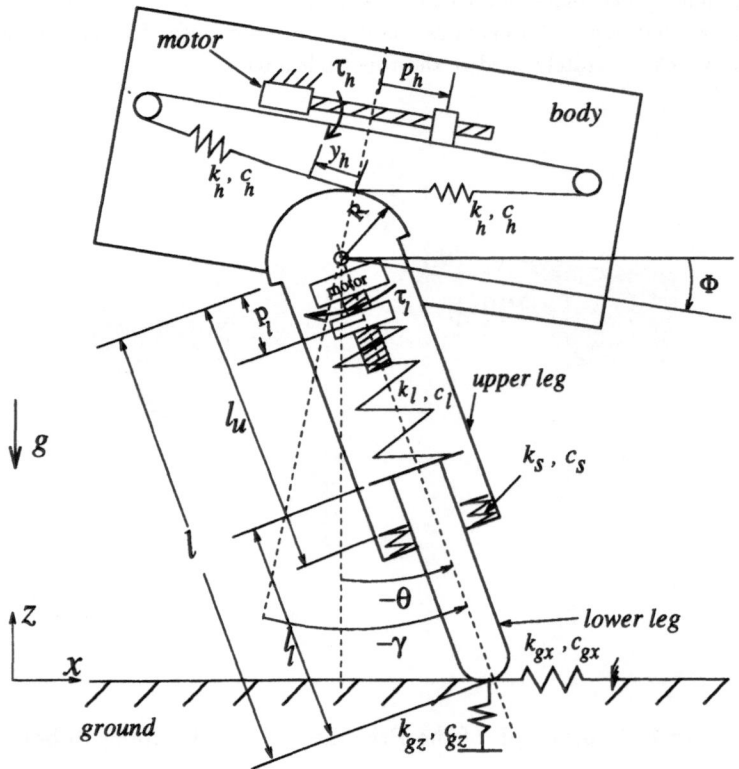

Figure 2. Simplified model of the Planar Hopper

The pioneering work in the field of dynamically stable legged robots has been performed by Raibert [12] whose robots employ powerful hydraulic actuators. His research describes a set of control laws which he was able to apply, with minor modifications, to all of his machines with great success. How much of this control could be applied to machines with much less powerful electric actuators? As our first legged prototype, we designed a planar one legged robot similar to Raibert's planar one-legged hopper, and optimized electric actuation systems for leg and hip.

We found that the the thrust controller for hopping height needed to be modified [10], in order to distribute the power requirements throughout the stance phase. With just this modification, our planar hopper was able to run up to $1.2m/s$. This fact underlines the strength and generality of Raibert's controllers.

In order to apply the same principle of distributing the power requirements throughout the flight phase, we propose to replace the step input command for the leg swing with a continuous path, and implement ground speed matching at touchdown and liftoff, while keeping Raibert's basic foot placement algorithm unchanged. This modification should result in further reduction of power requirements and increase in running speed.

As a next step, we are going to investigate the control of a passive dynamic hip to further decrease the power requirements. This would entail adding a spring in series with the hip actuator such that the unforced response would be close to the running motion. Passive dynamic hip motion has been investigated in the past [14, 6]. Success in harnessing its power for running machines via active control remains elusive. Nature, however, has ample demonstrations of its utility [3, 2, 1].

1.2. Related Work

Experimental research on actively balanced running machines began with Matsuoka [5] who built a planar one-legged hopping machine to study repetitive hopping in humans. His machine was sliding on a 10 degree inclined plane and thrusted by an electric solenoid. The first walking machine with active balance was built by Miura and Shimoyama [7]. It operated with stiff legs, similar to humans on stilts. Our work has been inspired by past research of Marc Raibert [12] who has led the field of dynamically stable legged locomotion during the past decade. He built a variety of running robots, starting with a planar one-legged machine [11], followed by a 3D one-legged, a two-legged planar robot, and a quadruped. His latest robots include a 3D two-legged robot, where each leg has four actuated degrees of freedom. Except for the first one-legged planar hopper, which was pneumatically actuated, his designs are actuated by powerful hydraulic actuators and rely on pneumatics for the leg spring only. This permitted Raibert to focus on robot design and control without being limited significantly by actuator constraints. The strategy was eminently successful and he built many different running robots based on an almost standardized set of custom parts. It is important to realize that all of his robots are controlled by some derivative of the tri-partitioned decoupled control developed for the original one-legged planar hopper [12]. Papantoniou [9, 8] has also built a planar hopper actuated by two electric motors and is using a derivative of Raibert's controller. He employs a clutch which engages the spinning motor at maximum leg compression to maximize energy transfer to the leg. To make this approach work, the stance time needed to be increased considerably to $470ms$, resulting in a mode of operation he terms "compliant walk". There are many more research labs working on dynamic locomotion. An extensive review of research in legged locomotion can be found in [13].

The paper is organized as follows. Section 2 discusses the mechanical design issues relevant to our robot design. Our simulation model is described and verified by comparison to experimental runs in Section 3. Section 4 describes our modifications to Raibert's controller for electric actuation. In Section 5 we present and discuss experimental running data, and discuss further improvements in Section 6.

2. Design

2.1. General Considerations

The planar, one-legged robot currently being studied is an elaboration of our electrically actuated, vertical hopping robot. The robot is comprised of two systems. The prismatic leg serves to excite and maintain a harmonic, vertical oscillation. Mechanical design considerations for the leg are discussed in [10]. In order to allow forward running, the leg attaches to the robot body through an actuated hip. This hip actuation system must allow for the control of the leg angle essential to stable forward running. Since only the relative angle between the body and leg can be actuated, it is desired to have a high ratio of body inertia to leg inertia so that the leg can swing with minimum body pitching. Conversely, low body mass is desirable to reduce overall energy requirements. These requirements dictated a horizontally long body with concentrated masses located at the ends. This morphology is evident in Figure 1.

The hip torque requirements for leg swing during forward running (up to $55Nm$) necessitate a gear ratio of about 30:1 between our 80W motor with $1.78Nm$ stall torque and the hip. Moreover, the design should allow for the insertion of a spring in series with the hip motor. Our design uses a high efficiency ballscrew (similar to the one for leg actuation) coupled to the motor through a miniature HTD belt and unity ratio pulleys. This arrangement allows the motor to be placed underneath the ballscrew and results in a body length below our desired maximum of $1m$. The ball nut is attached to cables which run over idlers and connect to a pulley attached to the leg and centered at the hip as illustrated in Figure 2. This basic configuration was optimized as discussed in the following section. As implemented, this system provides a gear ratio of 30:1 between motor and hip and allows for simple inclusion of compliance in the cables. Even for direct actuation, the Spectra cables provide enough compliance to protect the ballscrews from excessive impulsive forces at touchdown or during failures.

x	body horizontal position (m)
z	body vertical position (m)
l_c	leg C.G. to hip displacement
m_l	leg total mass (kg)
k_s	stopper model stiffness (N/m)
c_s	stopper model damping $(N/m/s)$
θ_m	hip motor angle angle (rad)
m_{hs}	hip ball nut mass (kg)
J_{hs}	hip ball screw inertia (kgm^2)
F_h	hip spring force (N)
r_h	leg ball screw lead (m/rad)

Table 1. Nomenclature

2.2. Detail Design

The geometric model depicted in Figure 2 presents the parameters relevant to the design of the actuated hip and Table 1 contains the basic nomenclature. Although the current robot uses direct actuation for leg angle servoing, compliance may be added in the future and our design optimization takes both direct and compliant actuation cases into account. An identical model is used to analyze both cases, but for direct actuation we assign a large stiffness to the hip springs k_h. Validation of this assumption was confirmed through simulation.

The design problem is to select the design parameters under our control, namely the pulley radius, the ballscrew lead and the spring stiffness. We specify a periodic desired hip angle and require that the driving motor remains within its specified torque-speed limits. To simplify the analysis we assume that the body is stationary. A sinusoidal path is considered for the hip angle θ which is close to the leg motion in the passive running case. Based on the equations of motion of a ball screw,

$$\tau_h = r_h F_h - \alpha_h \ddot{p}_h \qquad (1)$$

with

$$\alpha_h = J_{hs}/r_h + r_h m_{hs},$$

the expression for the hip string force

$$F_h = 2k_h(y_h - p_h) + 2c_h(\dot{y}_h - \dot{p}_h) \qquad (2)$$

and a small angle approximation $sin(\theta) \approx \theta$ we obtain for the hip dynamics

$$J_l \ddot{\theta} = RF_h - C_t\dot{\theta} - m_l g l_c \theta. \qquad (3)$$

Solving (3) for a prescribed sinusoidal path $\theta = \theta_0 \sin(\omega t)$ with $\theta_0 = 0.35$ and $\omega = 4\pi rad/sec$ requires a hip torque

$$\tau_h = A\sin(\omega t) + B\cos(\omega t) \qquad (4)$$

where A and B are functions of the system design parameters, primarily the lever arm R, the stiffness k_h and ball screw lead r_h (equivalently L in mm/rev).

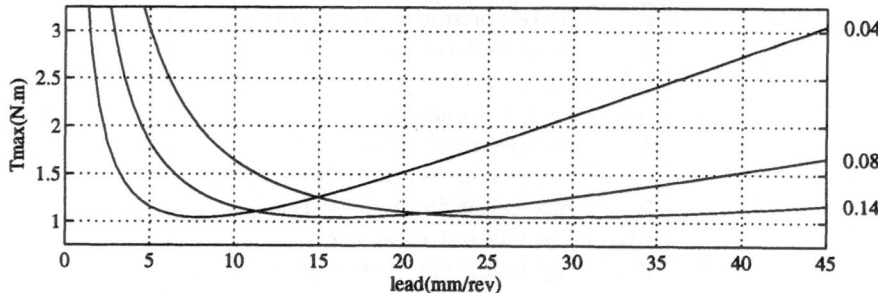

Figure 3. Maximum required torque vs L and different R for direct actuation case

We can now evaluate the motor's torque–speed requirements for this motion $\tau_h(\omega)$, where the maximum torque depends on the design parameters in a simple fashion,

$$\tau_{max} = \sqrt{A^2 + B^2}.$$

This problem cannot be cast as a straightforward standard numerical minimization problem, since there are many non-standard constraints. For example, the motor characteristics require that the maximum torque be a function of the velocity. Furthermore ball screws are available only in coarsely discrete leads and lengths and their inertial parameters are only available via lookup tables.

The generated hip force F_h should be kept small since it affects the overall size and weight of our structures. Since we plan to use elastomeric springs, the spring stiffness was limited to $5kN/m$ due to energy storage limitations. The ball screw travel should be less than $10cm$ to limit the inertia, mass and body length. The design is further complicated by the necessity to meet the constraints for both direct and compliant actuations. The compliant case is facilitated by large pulley radius R, resulting in smaller torque requirements, but this increasing R, when combined with the optimum lead (Figure 3), increases motor velocity, acceleration, ball screw travel and lead in the direct actuation case.

The final design parameters of lead $L = 16mm$, pulley radius $R = 0.08$ and spring constant $k_h = 3600$ are an attempt to satisfy all the constraints concerning pulley radius, ballscrew lead, maximum motor torque and velocity, as well as the resulting ballscrew travel. This set of parameters minimizes the motor torque-speed requirements for the compliant hopper by matching the natural frequency of the system with the frequency of the vertical oscillations of about $2Hz$ [10].

3. Simulation

In parallel with our experimental work, we developed a planar six degree of freedom (seven d.o.f if the hip is compliant) simulation based on the model

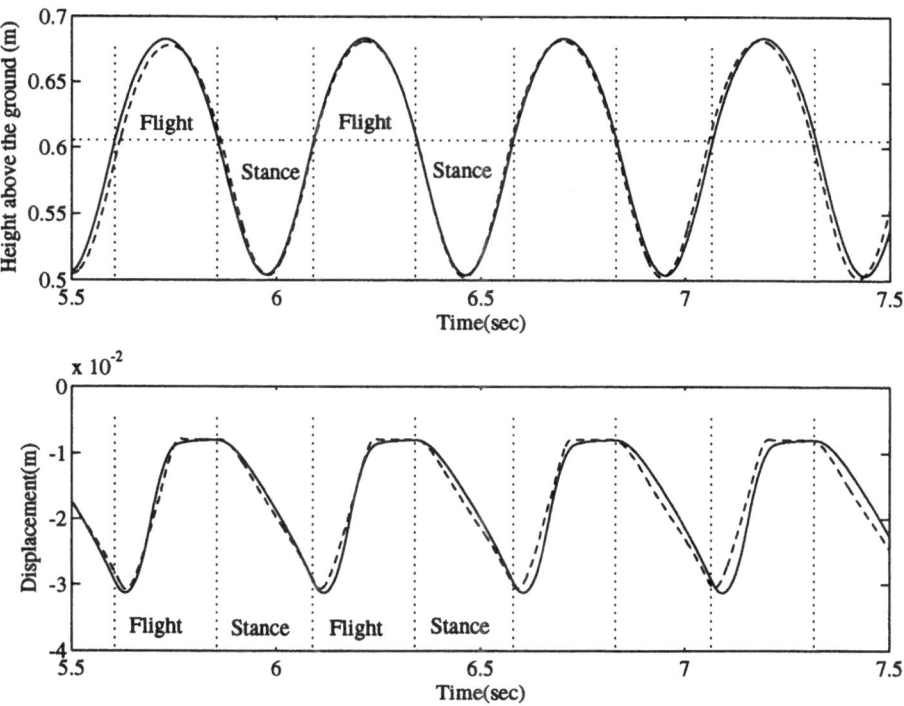

Figure 4. Hopping height and actuator displacement comparisons. Solid Line: Simulation, Dashed line: Experiment

depicted in Figure 2. The vertical dynamic model from [10] has been expanded to include the full planar dynamics. In contrast to our analytical model which contains distinct stance and flight models, the simulation employs only the flight model, and implements the ground interaction during stance via an external force applied at the toe generated by the ground spring-damper model. In Figure 4, a comparison of the vertical body height and the leg actuator length obtained from simulation and an experimental run validates our model.

The simulation is set up such that we can develop and verify control code, and subsequently transfer it to the runtime environment with minimal modification. Moreover, the validated model can now be used for analytical performance studies and controller development.

4. Control

We took Raibert's three part control algorithm [12] and applied minimal modifications as described below.

Vertical Motion (Hopping height):
"Apply a constant position step to the leg actuator at time of maximum com-

pression."

At maximum compression, the ground force acting against the leg actuator is approximately $500N$, which makes this an impractical strategy with low power electric actuators. Instead we use a different open loop controller described in [10],

$$\tau = \bar{\tau}(1 - \frac{\omega}{\tilde{\omega}}),$$

which specifies a scaled version of the maximum torque-speed curve of our motor given by

$$\tau \leq \hat{\tau}(1 - \frac{\omega}{\tilde{\omega}})$$

with $0 \leq \bar{\tau} < \hat{\tau}$ in the first quadrant (stance), and a high gain PD controller to return the actuator during flight. This strategy is exactly implementable and applies thrust continuously during the stance phase. For our experiments we chose $\bar{\tau} = 1Nm$ with $\hat{\tau} = 1.78Nm$.

Forward speed (Foot placement algorithm, FPA):
Raibert's FPA

$$x_f = \frac{\dot{x}T_s}{2} + \kappa_{\dot{x}}(\dot{x} - \dot{x}_d)$$

was adjusted for treadmill running, with $\dot{x}$ replaced by $\dot{x}_{TM} + \dot{x}$. In addition, an "integral" term to servo the robot around the center of the treadmill was added. The FPA implementation is thus given by

$$x_f = \frac{\dot{x}_{TM}T_s}{2} + \kappa_x(x - x_d) + \kappa_{\dot{x}}(\dot{x} - \dot{x}_d) \tag{5}$$

where x is the robot's position.

Body Attitude (Pitch control):

$$\tau = \kappa_p(\phi - \phi_d) - \kappa\dot{\phi}$$

This controller was left untouched.

5. Experiments

The hopping robot illustrated in Figure 1 was constrained to move in a vertical plane through a virtual motion system (VMS). The VMS is a planarizer consisting of horizontal and vertical ball slides and a revolute joint. All three degrees were instrumented with optical encoders to measure the horizontal and vertical position and the pitch (body angle). An electric $5hp$ treadmill was installed beneath the VMS allowing unlimited horizontal travel for robot running.

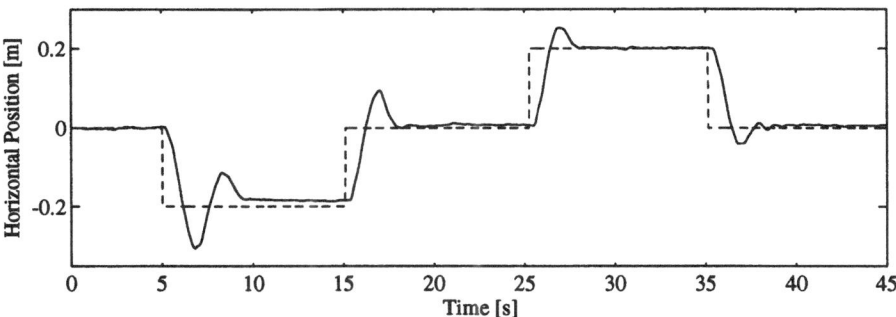

Figure 5. Horizontal position regulation. *The dashed line is the horizontal setpoint while the solid line indicates the robot position.*

5.1. Position Tracking

The simplest form of planar motion is vertical hopping with a horizontal disturbance. In Figure 5 we see the response of the hopping robot to step inputs in desired horizontal position. Control of horizontal position is achieved through the modified FPA discussed above. Stable position control is evident for horizontal perturbations on the order of one third the leg length.

5.2. Velocity Tracking

Beyond position regulation, velocity tracking is the essence of planar running. Since our robot runs on a treadmill, any steady state speed error would result in an increasing position error and the robot would quickly exceed the limits of treadmill size. Our modified version of the FPA (5) introduces an integral term to eliminate the velocity steady state error and thus limit the position error with respect to the treadmill center. Therefore the treadmill speed in Figure 6 is roughly equivalent to the robot running speed.

Figure 6 shows small scale variations in the treadmill speed due to toe impact at touchdown (upper solid curve) and the limited bandwidth of the treadmill velocity controller. Deviations from the desired position at treadmill centre are limited to 0.2m (lower dashed curve) while the robot speed varied between 0 and $1.05m/s$. The total distance traversed was $26m$.

5.3. Pitch Control

As the robot leg swings throughout locomotion, the body also pitches freely. Due to its large mass and inertia, any instability in the body pitching motion can quickly destabilize the entire running cycle. Thus control of body pitching during stance is critical to stable running.

Figure 7 shows the cyclic oscillations of the robot body (lower solid curve) as the forward speed increases to $0.97m/s$. Despite the large variation in speed the robot's pitching oscillation is kept within only $2deg$ of the horizontal.

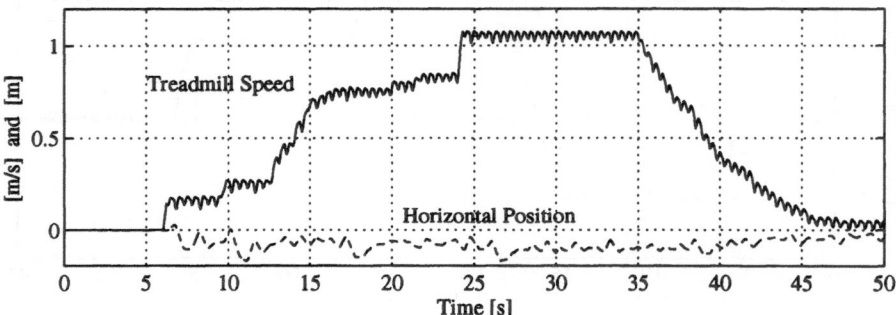

Figure 6. Velocity tracking for forward running. *The treadmill speed is manually regulated from zero, vertical hopping, to 1.05m/s and back to zero (upper, solid curve). The lower dashed curve shows the robot deviation from the centre of the treadmill.*

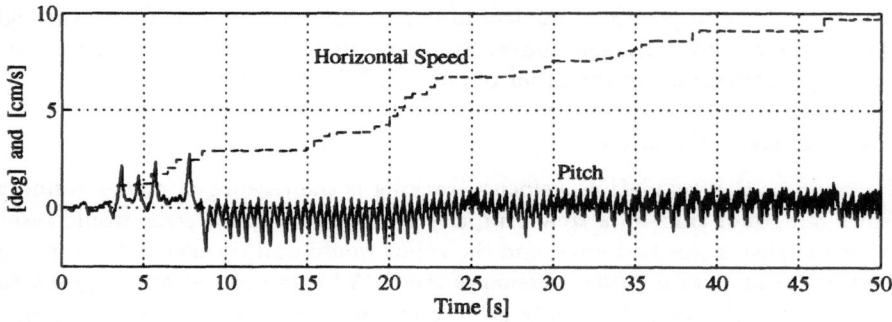

Figure 7. Body pitching during forward running. *The cyclic pitching motion of the robot body about the hip is kept to an amplitude of 2deg (solid, lower curve) while the robot accelerates on the treadmill to 0.97m/s (upper dashed curve).*

6. Improvements

From the previous sections, it is apparent that Raibert's control, with minor modifications, results in very good performance even on our electrically actuated robot. This attests to the fundamental nature of the algorithm, and, to some extent, to the sound design principles employed in our robot. In this section, we are describing our ongoing efforts toward further improving the robot's efficiency necessary for autonomous operation.

6.1. Locomotion Time

A scalar variable indicating the fraction of flight or stance time that has been elapsed, would be of general utility for control. It will be used, for example,

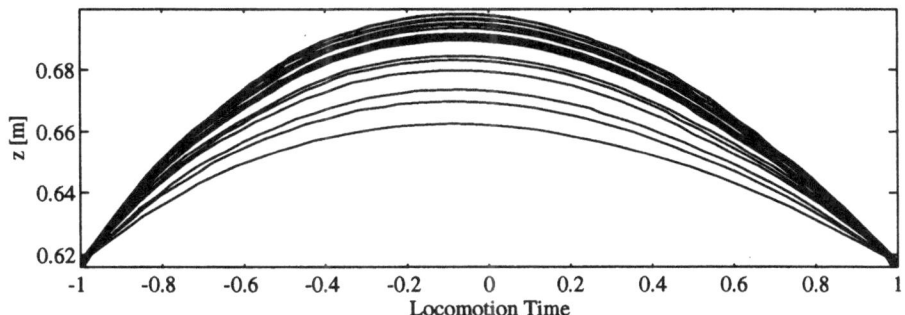

Figure 8. Locomotion Time vs. z for vertical hopping.

in our online trajectory calculations for ground speed matching and power minimization. Such a measure is

$$\epsilon \stackrel{\triangle}{=} \frac{\dot{z}}{\sqrt{2g(z - z^*) + m\dot{z}^2}}. \tag{6}$$

This is the same measure as the "phase angle" introduced in [4] which was critical for the stable coordinated control of the "two-juggle," a robotic juggling task with two pucks. For vertical hopping, with touchdown and liftoff position at z^* the locomotion time (6) maps a flight (or stance) phase onto the interval $(-1, 1)$ between liftoff and touchdown, with $\epsilon = 0$ at apex (max. compression). Moreover, if we assume zero dissipation during flight phase, its derivative is constant. Figure 8 shows the locomotion time computed online for an experimental run.

6.2. Speed Matching

At present, after liftoff, a high gain PD controller servos the leg toward its desired touchdown angle, as calculated from the foot placement algorithm (5). Since convergence must be assured for even short flight times, high gains are necessary, resulting in unnecessarily jerky motion for longer flight times and large transient hip torques. Furthermore, the leg is servoed to zero touchdown angular velocity, and upon ground contact is forcefully accelerated backwards. Both of these drawbacks can be alleviated using a reference leg trajectory during flight based on locomotion time. At liftoff and at touchdown, a parabola is matched to actual (liftoff) and desired (touchdown) leg trajectory. The parabolas are joined by a straight line tangent to both. Figure 9 compares the proposed reference trajectory to the actual flight phase path for forward running at $0.7 m/s$.

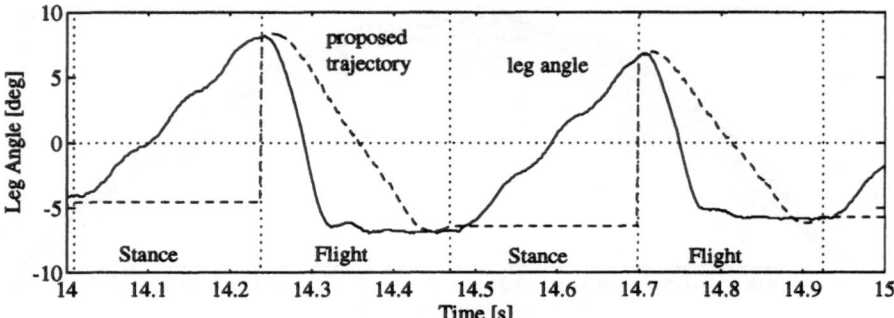

Figure 9. Actual leg swing and proposed reference trajectory. *The solid curve shows the cyclic swinging motion of the robot leg about the vertical position, zero leg angle, for running at 0.7 m/s. During the flight phase the leg angle is servoed to a desired touchdown value that will assure stable running. The broken line indicates the proposed reference trajectory. This trajectory terminates at the same touchdown setpoint but additionally matches the leg swing rates at liftoff and touchdown.*

6.3. Passive Dynamics

During flight, the hip motor has to decelerate sharply, accelerate and then decelerate the leg, which requires peak power of our motors at $1m/s$. Most of this return swing motion of the leg can be provided by a properly dimensioned spring in series with the hip actuator. This has been proposed and experimentally tested in passive (unstable) mode by Raibert [14]. Our robot has been designed to accommodate such springs. We are currently implementing them and will report on this research in the near future.

7. Conclusions

We have presented the design of an electrically actuated planar one-legged hopping robot, as a research vehicle for autonomous legged systems. Raibert's controller was modified to reduce the power requirements for the leg actuation and was shown experimentally to achieve stable running up to 1.2 m/s. To reduce power requirements further, we proposed a locomotion time based flight trajectory which also implements ground speed matching. Finally, we are currently pursuing the application of controlled passive dynamic hip motion.

We are also in the process of replacing the prismatic leg joint with an articulated leg. This will, in addition to a reduction in friction, also improve the robot's dexterity, and greatly improve its reliability. Finally, the friction properties of revolute joints are much more constant over time, which will simplify the application of analytical results based on theoretical models. In the future we will also emphasize the derivation of analytical results to eliminate or at least minimize the need for empirical gain settings, which requires excessive amounts of time, even for our current relatively simple machine.

References

[1] R. McN. Alexander. Three Uses for Springs in Legged Locomotion. *Int. J. Robotics Research*, 9(2), 1990.

[2] R. McN. Alexander, N. J. Dimery, and R. F. Ker. Elastic Structures in the back and their role in galloping in some mammals. *J. Zoology (London)*, 207, 1985.

[3] R. McN. Alexander and A. Vernon. The mechanics of hopping by kangaroos. *J. Zoology (London)*, 177, 1975.

[4] M. Buehler, D. E. Koditschek, and P. J. Kindlmann. Planning and Control of Robotic Juggling and Catching Tasks. *Int. J. Robotics Research*, (accepted: 17. Dec. 91).

[5] K. Matsuoka. A mechanical model of repetitive hopping movements. *Biomechanisms*, 5:251–258, 1980.

[6] T. McGeer. Passive dynamic walking. *Int. J. Robotics Research*, 9(2), 1990.

[7] H. Miura and I. Shimoyama. Dynamic walk of a biped. *Int. J. Robotics Research*, 3:60–74, 1984.

[8] K. V. Papantoniou. Control architecture for an electrical, actively balanced multi-leg robot, based on experiments with a planar one leg machine. In *Proc. Int. Fed. Automatic Control*, pages 283–290, Vienna, Austria, 1991.

[9] K. V. Papantoniou. Electromechanical design of an actively balanced one leg electrically powered robot. In *Proc. IEEE Conf. Intelligent Systems and Robots*, Osaka, Japan, 1991.

[10] H. Rad, P. Gregorio, and M. Buehler. Design, modeling and control of a hopping robot. In *Proc. IEEE/RSJ Conf. Intelligent Systems and Robots*, Yokohama, Japan, Jul 1993.

[11] M. H. Raibert. Dynamic stability and resonance in a one-legged hopping machine. *4th Symp. Theory and Practice of Robots and Manipulators*, 1981.

[12] M. H. Raibert. *Legged Robots That Balance*. MIT Press, Cambridge, MA, 1986.

[13] M. H. Raibert. Legged robots. In P. H. Winston and S. A. Shellard, editors, *Artificial Intelligence at MIT*, pages 149–179. MIT Press, Cambridge, MA, 1990.

[14] M. H. Raibert and C. M. Thompson. Passive dynamic running. In V. Hayward and O. Khatib, editors, *Experimental Robotics I*, pages 74–83. Springer-Verlag, NY, 1989.

Dynamic Stereo Triangulation for Robot Juggling

A. A. Rizzi *and D. E. Koditschek [†]

University of Michigan, Artificial Intelligence Laboratory
Department of Electrical Engineering and Computer Science

Abstract— We have devised a nonlinear state estimator to recover the position and velocity of falling balls directly from the image plane measurements of a stereo camera pair. Avoiding an explicit triangulation step in the estimation procedure allows the continued use of one camera's data even when the other camera may be occluded. This paper presents a rudimentary analysis, some simulation results, and data from a working implementation that suggest the potential utility of the idea.

1. Introduction

Figure 1 depicts the system architecture underlying a juggling robot that we have discussed at length in previous publications [8, 6]. This paper concerns a new algorithm we have introduced into our signal processing module, the block labeled "Linear Observer" in the figure, which filters the data issued by the stereo camera pair we use as "eyes." Previously, this module incorporated a linear state estimator for Newtonian free flight in order to obtain reliable information concerning the position and velocity of the bodies to be juggled. For its input, this linear observer requires measurements in cartesian coordinates of the bodies' positions. Thus, the "Vision Coordinator" block in our system architecture has included an algebraic triangulation step. In the present paper we discuss recent experiments with a new signal processing method that eliminates the algebraic triangulation procedure. Camera output is directly filtered by a nonlinear state estimator. Our empirical experience with this new scheme is sufficiently promising to motivate our further efforts — presently in progress — to prove its stability.

The immediate motivation in our lab for this revision in the signal processing architecture has a very practical origin. It is almost guaranteed that a camera will "lose a ball" during each flight phase of a juggle: balls temporarily travel out of one or the other the field of view; one ball temporarily occludes the view of the second. Such losses are generally recoverable, and we have discussed in previous publications certain "control of attention" mechanisms that significantly aid in the recovery [7]. However, as matters stood until recently, once one camera had lost a ball, the valid data from the other was ignored, as

---

*Supported in part by IBM through a Manufacturing Graduate Fellowship and in part by the National Science Foundation under grant IRI-9123266.

[†]Supported by the National Science Foundation in part through a Presidential Young Investigator Award and in part under grant IRI-9123266.

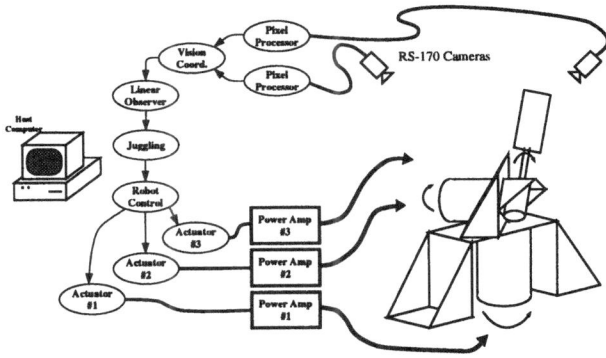

Figure 1. The Spatial Juggler

there was no new stereo image pair input to triangulate. Thus there was no resulting cartesian estimation error input to the linear observer (which consequently ran as a pure predictor until the camera system had recovered again). Despite its frequent disregard of valid data, this system managed to work reasonably well. It juggled one and two balls for hours at a time [6]. But it was clear to us that a more rational data management scheme could be found. Why throw out both cameras' outputs when only one is blinded?

As an alternative approach, consider a state estimation scheme that can work directly from the cameras' image plane measurements. Now, the signal processing algorithm can continue to work with the information from one camera while substituting predicted data in place of the other camera's information[1]. But the transformation from cartesian space to image plane pairs is nonlinear and this entails a nonlinear state estimator. Happily, because of the special algebraic structure of the projective transformation, we have been able to devise a working nonlinear estimator to do the job. Moreover, there seems good reason to hope that a formal proof of its efficacy can be provided.

These ideas extend naturally to situations wherein sporadic observations of a dynamical object from n-cameras must be integrated to form an evolving view of its cartesian motion[2]. More generally, we presume that many instances of the sensor fusion problem — the problem of deriving from multiple sources of overlapping but independently distorted data more accurate information than may obtain from any one alone — may be recast as nonlinear observer problems. Thus, one might refer to this approach as on-line dynamical data-fusion. Here,

---

[1]Even in the previous scenario, one might imagine running the triangulator on partially valid and partially predicted data in a similar manner. Indeed, we plan to run comparative experimental studies of this kind in the near future. However, as will be argued below, the present approach seems to generalize more readily to broader problems of sensor fusion. Moreover, there seems to be a much better chance of making theoretical sense of it as will be observed in the third section of this paper.

[2]This includes the particularly interesting case of a monocular camera, $n = 1$ being used to estimate the spatial motion of an object with known dynamics — see Ghosh [5] for a more careful setting of this problem.

we explore a particularly simple instance.

2. Juggling Apparatus

Our juggling system, pictured in Figure 1, consists of three major components: an environment (the ball); the robot; and an environmental sensor (the vision system). We now sketch the operation of this system sufficiently to place the current problem in a meaningful context.

2.1. The Need for Continuous Flight Information

Following Bühler *et al.* [3], we command a robot to "juggle" by forcing it to track a reference trajectory generated by a distorted reflection of the ball's continuous trajectory. This policy amounts to the choice of a map m from the phase space of the body to the joint space of the robot. A robot reference trajectory,

$$r(t) = m(b(t), \dot{b}(t)), \tag{1}$$

is generated by the geometry of the graph of m and the dynamics of ball, $b(t)$. This reference trajectory (along with the induced velocity and acceleration signals) can then be directly passed to a robot joint controller. [3] In following the prescribed joint space trajectory, the robot's paddle pursues a trajectory periodically intersecting that of the ball. The impacts induced at these intersections result in the desired juggling behavior.

Central to this juggling strategy is a sensory system capable of "keeping it's eyes on the ball." We require that the vision system produce a 1 KHz signal containing estimates of the ball's spatial position and velocity (six measurements). Denote this "robot reference rate" by the symbol $\tau_r = 10^{-3} sec$. Two RS-170 CCD television cameras constitute the "eyes" of the juggling system and deliver a frame consisting of a pair of interlaced fields at 60 Hz, so that a new field of data is available every $\tau_f = 16.\overline{6} \cdot 10^{-3} sec$. The CYCLOPS vision system, described in [8, 4], provides the hardware platform upon which the data in these fields are used to form the input signal to the mirror law, (1). The remainder of this section describes how this is done.

2.2. Triangulation and Flight Models

We work with the simple projective stereo camera model,

$$p : I\!R^3 \rightarrow I\!R^4$$

that maps positions in affine 3-space to a pair of image plane projections in the standard manner. Knowledge of the cameras' relative positions and orientations together with knowledge of each camera's lens characteristics (at present

---

[3] In the case of a one degree of freedom arm we found that a simple PD controller worked quite effectively [2]. In the present setting, we have found it necessary to introduce a nonlinear inverse dynamics based controller [10]. The high performance properties of this controller notwithstanding, our present success in achieving a spatial two-juggle has required some additional "smoothing" of the output of the mirror law as described in [6].

we model only the focal length, and image center) permits the selection of a pseudo-inverse or "triangulation function",

$$p^\dagger : IR^4 \to IR^3,$$

such that $p^\dagger \circ p = id_{IR^3}$. We have discussed our calibration procedure and choice of pseudo-inverse at length in previous publications [9, 8].

For simplicity, we have chosen to model the ball's flight dynamics as a point mass under the influence of gravity. A position-time-sampled measurement of this dynamical system will be described by the discrete dynamics,

$$
\begin{aligned}
w_{j+1} &= F^s(w_j) \overset{\triangle}{=} A_s w_j + a_s; \\
A_s &\overset{\triangle}{=} \begin{bmatrix} I & sI \\ 0 & I \end{bmatrix}; \quad a_s \overset{\triangle}{=} \begin{bmatrix} \frac{1}{2}s^2\tilde{a} \\ s\tilde{a} \end{bmatrix} \\
b_j &= Cw_j; \quad C = [I, 0],
\end{aligned}
\tag{2}
$$

where s denotes the sampling period, $\tilde{a}$ is the gravitational acceleration vector, and $w_j \in IR^6$.

2.3. Signal Processing: A Linear Observer

Following Andersson's experience in real-time visual servoing [1] we apply a first order moment computation to a small window of a threshold-sampled (thus, binary valued) image of each field. Thresholding is the only "early vision" strategy required in a visually structured environment, and we presently illuminate white ping-pong balls with halogen lamps while putting black matte cloth cowling on the robot, floor, and curtaining off any background scene. Thus, the "world" as seen by the cameras contains only one or more white balls against a black background. When only one white ball is presented, the camera system reports a pair of pixel addresses, $v \in IR^4$, containing the centroid of the single illuminated region seen by each camera. A field from each camera is acquired with period τ_f, and centroid computations over a small subwindow of 30 by 40 pixels follow during the next τ_f seconds. Finally triangulation is performed to extract the ball's spatial position from the centroid measurements.

A linear observer can recover from these discrete time measurements an estimate of the full state (positions and velocities). As described above, the window operates on pixel data that is at least one field old,

$$p_k = F^{-\tau_f}(w_k),$$

to produce a centroid. We use p_k as an "extra" state variable to denote this delayed image of the ball's state. Denote by W_k the function that takes a white ball against a black background into a pair of thresholded image plane regions and then into a pair of first order moments at the kth field. The data from the triangulator may now be written as

$$\bar{b}_k = p^\dagger \circ W_k \circ p(Cp_k).
\tag{3}$$

Thus, the observer operates on the delayed data,

$$\hat{p}_{k+1} = F^{\tau_f}(\hat{p}_k) - G(C\hat{p}_k - \bar{b}_k), \tag{4}$$

where the gain matrix, $G \in I\!R^{6 \times 3}$, is chosen so that $A_{\tau_f} + GC$ is asymptotically stable — that is, if the true delayed data, Cp_n, were available then it would be guaranteed that $\hat{p}_k \to p_k$.

In principle, one might choose an optimal set of gains, G^*, resulting from an infinite horizon quadratic cost functional, or an optimal sequence of gains, $\{G_i^*\}_{i=0}^k$, resulting from a k-stage horizon quadratic cost functional (probably a better choice in the present context), according to the standard Kalman filtering methodology. Of course, this presumes rather strong assumptions and a significant amount of à priori statistical information about the nature of disturbances in both the free flight model (2) as well as in the production of $\bar{b}$ from $\hat{d}$ via the moment generation process. To date we have obtained sufficiently good results with a common sense choice of gains G that recourse to optimal filtering seems more artificial than helpful.

We provide our juggling algorithm with an appropriately extrapolated and interpolated version of these estimates as follows. The known latency is corrected by the prediction,

$$\hat{w}_k = F^{\tau_f + \iota_k}(\hat{p}_k),$$

where ι_k denotes the time required by the centroid computation at the kth field. Subsequently, the mirror law is passed the next entry in the sequence,

$$F^{i\tau_r}(\hat{w}_k), (i = 1, ..., \tau_f - \iota_{k+1})$$

until the next estimate, $\hat{p}_{k+1}$ is ready.

2.4. Sensing Issues Arising From the Two-Juggle

Initial attempts to implement the spatial two-juggle (simultaneously batting of two balls), demonstrated that the simple system described above required a number of modifications. Central to juggling two balls is the ability for the sensor system to withstand transient losses of data for one or both cameras. These events arise from balls passing outside the field of view of one or both cameras and from balls passing arbitrarily close together in the image planes of the cameras.

For reasons reported earlier [7] we have chosen to rely on the observer's prediction of the balls flight across such events, rather than resorting to more complicated algorithms or fundamentally different strategies. Expanding the moment computations performed on a window to include the zeroth and second order moments affords a reliable measure of the "ballness" of an object, and allows for selective rejection of images with unreliable first order moments (caused either by having either two or no balls present in the window)[4]. Un-

---

[4] As reported in [7] the potential for skipping images and dealing with prolonged out-of-frame events has necessitated that the output of the observer be used to locate the windows in successive images, as opposed to simply centering the windows on the previous measurements. Thus the observer is being used both to drive the juggling algorithm and locate the subwindows used for centroid computations mentioned above.

fortunately, a loss of data from one camera precludes triangulation and, thus, update of the ball's state by means other than simple integration of the flight model (2). This amounts to setting $G = 0$ in (4) whenever there is an occlusion event.

3. Dynamic Triangulation

Underlying our new estimation technique is the simple idea of augmenting the standard (linear) Newtonian flight model, $\ddot{b} = \tilde{a}$, with a nonlinear "output map," $v = p(b)$, where p denotes the camera transformation introduced in Section 2.1. We now present this procedure, a substitute for (4), that promotes the use of each camera's data when available.

In Section 3.1, we will note a useful algebraic property of rational linear transformations that relates image plane errors to cartesian errors through a scaled version of the camera jacobian. In Section 3.2 we introduce a "toy" first order observer theory for purposes of illustrating the utility of the previous algebraic fact. The convergence properties of this version of the procedure are simple to establish. Of course, the physical plant of interest is characterized by second order dynamics, and in Section 3.3 we offer an extension of the simple first order procedure to this setting. Arguments establishing the favorable convergence properties of this algorithm are presently in progress.

3.1. Camera Model Properties

Recall that the stereo camera transformation, p, is formed by stacking together the perspective projections due to the two individual cameras,

$$v \stackrel{\triangle}{=} p(b) = \left[\begin{array}{c} \frac{f_1}{\Pi_3\,{}^l H_0\, b} \Pi_{1,2}\,{}^l H_0\, b \\ \frac{f_2}{\Pi_3\,{}^2 H_0\, b} \Pi_{1,2}\,{}^r H_0\, b \end{array} \right]. \tag{5}$$

Here, ${}^i H_0$ is the the homogeneous-matrix representation of the base frame in the ith camera's frame, f_i is the ith camera's focal length, Π_i denotes projection onto the ith component, and $\Pi_{1,2} \stackrel{\triangle}{=} \left[\begin{array}{c} \Pi_1 \\ \Pi_2 \end{array} \right]$.

In this section we note that

$$\begin{aligned} p(\hat{b}) - p(b) &= \Lambda(\hat{b}, b) P(\hat{b}) \left(\hat{b} - b \right) \qquad &(6) \\ &= \Lambda(b, \hat{b}) P(b) \left(\hat{b} - b \right), \end{aligned}$$

where $P(b)$ is the jacobian of p evaluated at b and

$$\Lambda(\hat{b}, b) \stackrel{\triangle}{=} \left[\begin{array}{cc} \frac{\Pi_3\,{}^l H_0\, \hat{b}}{\Pi_3\,{}^l H_0\, b} I_2 & 0 \\ 0 & \frac{\Pi_3\,{}^2 H_0\, \hat{b}}{\Pi_3\,{}^2 H_0\, b} I_2 \end{array} \right].$$

This fact emerges directly from computation, given b lying in the frame of reference of a camera with focal length f, we have

$$p(b) \stackrel{\triangle}{=} \frac{f}{b_3} \left[\begin{array}{c} b_1 \\ b_2 \end{array} \right]. \tag{7}$$

The Jacobian of this projection is then given by

$$P(b) = \frac{f}{b_3} \begin{bmatrix} 1 & 0 & -\frac{b_1}{b_3} \\ 0 & 1 & -\frac{b_2}{b_3} \end{bmatrix}. \tag{8}$$

Finally expanding the right hand side of the top row of (6) in these coordinates gives

$$f\frac{\hat{b}_3}{b_3} \left(\frac{\hat{b}_1}{\hat{b}_3} - \frac{b_1}{b_3} \right) \tag{9}$$

, similar results follow for the remainder of the rows, which confirms the original assertion.

Throughout the remainder of this paper we will assume that Λ is positive definite. Geometrically this requires that both b and $\hat{b}$ always lie on the same side (front/back) of all the cameras at all times. In practice this is not an unrealistic assumption since it merely requires that neither the actual object cross the singularity in p (5) nor that the initial errors in the observer system become so large as to cause the estimated object location to cross this same singularity. Formally, this assumption is expressed as

$$(\hat{b} - b)P^T(\hat{b})(\hat{v} - v) > 0. \tag{10}$$

3.2. A First Order Observer

Consider the dynamical system described by

$$\begin{aligned} \dot{b} &= Ab + u \\ v &= p(b). \end{aligned} \tag{11}$$

This system is "fully-measurable" – algebraic triangulation can fully reconstruct it's state b. But it seems helpful to underscore the utility of (6) by considering an observer for this system to both filter the resultant state estimates and allow for estimation during loss of measurements. The observer takes the form

$$\begin{aligned} \dot{\hat{b}} &= A\hat{b} + u - KP^T(\hat{b})(\hat{v} - v) \\ \hat{v} &= p(\hat{b}), \end{aligned} \tag{12}$$

with $K \in I\!R^{3 \times 4}$. Forming the error dynamics for $\tilde{v} \overset{\Delta}{=} \hat{v} - v$ we have

$$\dot{\tilde{b}} = A\tilde{v} - KP^T(\hat{b})(\hat{v} - v). \tag{13}$$

Making use of (6) allows us to substitute for $(\hat{v} - v)$, and results in

$$\dot{\tilde{b}} = \left(A - KP^T(\hat{b})\Lambda(\hat{b}, b)P(\hat{b}) \right) \tilde{b}. \tag{14}$$

From (10) we know $P^T(\hat{b})\Lambda(\hat{b}, b)P(\hat{b})$ is positive definite from which it follows that there exists a K such that[5] $\lim_{t \to \infty} \tilde{b} = 0$.

---

[5]Note, due to the time-varying nature of (14), the choice of stabilizing K is necessarily influenced by the initial conditions. In practice a reasonable bound could be placed on the initial errors such that a suitably large fixed K might readily be chosen.

3.3. An Observer for Mechanical Systems with Linear Dynamics

In contrast to the completely measurable system presented above, let us now reconsider the system, $\dot{b} = \tilde{a}$, written more generally as

$$
\begin{aligned}
\dot{b}_1 &= b_2 \\
\dot{b}_2 &= A_1 b_1 + A_2 b_2 + u \\
v &= p(b_1),
\end{aligned}
\tag{15}
$$

where b_1 and b_2 represent the position and velocity of the object respectively. The associated observer now takes the form

$$
\begin{aligned}
\dot{\hat{b}}_1 &= \hat{b}_2 - \Gamma_1 P^T(\hat{b}_1)(\hat{v} - v) \\
\dot{\hat{b}}_2 &= A_1 \hat{b}_1 + A_2 \hat{b}_2 - \Gamma_2 P^T(\hat{b}_1)(\hat{v} - v) \\
\hat{v} &= p(\hat{b}_1),
\end{aligned}
\tag{16}
$$

with gain matrices Γ_1 and Γ_2 free to be chosen. Proceeding as above we take differences to determine the error dynamics

$$
\begin{aligned}
\dot{\tilde{b}}_1 &= \tilde{b}_1 - \Gamma_1 P^T(\hat{b}_1)(\hat{v} - v) \\
\dot{\tilde{b}}_2 &= A_1 \tilde{b}_1 + A_2 \tilde{b}_2 - \Gamma_2 P^T(\hat{b}_1)(\hat{v} - v),
\end{aligned}
\tag{17}
$$

which simplifies to

$$
\begin{aligned}
\dot{\tilde{b}}_1 &= \left(I - \Gamma_1 P^T(\hat{b}_1)\Lambda(\hat{b}_1, b_1)P(\hat{b}_1) \right) \tilde{b}_1 \\
\dot{\tilde{b}}_2 &= \left(A_1 - \Gamma_2 P^T(\hat{b}_1)\Lambda(\hat{b}_1, b_1)P(\hat{b}_1) \right) \tilde{b}_1 + A_2 \tilde{b}_2.
\end{aligned}
\tag{18}
$$

A proof of convergence for this system is in progress.

4. Implementation

Although the analysis of the previous section is at best in its infancy we have forged ahead with a number of simulations and a functional implementation of the class of observer described here. As usual, the real world departs from the assumptions underlying these models in certain important regards. What follows is a brief discussion of the differences between the previous section and the actual system, along with both experimental and simulation results demonstrating the utility and pitfalls for this type of observer.

4.1. Choice of Observer Gains

Having no immediate insight at the outset concerning choice of the gain matrices Γ_1 and Γ_2 in (16), we chose to use the same gains as for the linear observers. Poor convergence in our first simulations demonstrated that this simple choice was inadequate. The primary cause for this effect was that the spatial dependence of $P(\hat{b})$ leads to widely differing effective gains depending on the ball's location in space. We were able to successfully compensate for this by making use of non-linear gain matrices of the form

$$
\Gamma = \Gamma_0 \left(P^T(\hat{b})P(\hat{b}) \right)^{-1}.
$$

290

This essentially amounts to performing local triangulation (i.e. $\Gamma P^T(\hat{b})$ is the linear approximation to $p^\dagger$ at $\hat{b}$), and dramatically improved the convergence behavior of the observer.

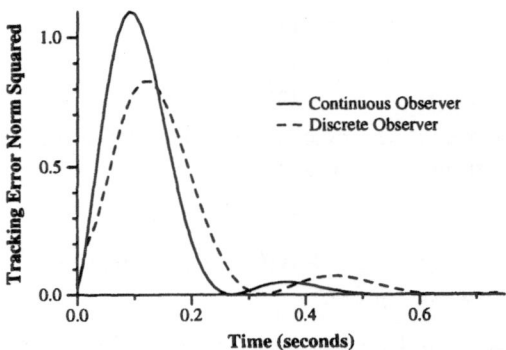

Figure 2. Simulation: Convergence of the continuous and discrete time observers for small initial error.

4.2. Time Sampled Implementation for Second-Order Systems

Real cameras are rarely continuous devices – they generally take snapshots of the world at a fixed sampling rate, in our case, 60 Hz. Since the observed system's motion is significant relative to this rate (near impact, the ball often travels in excess of 10 cm between successive images), sampling considerations cannot be ignored. For the observer of Section 2.3 (with explicit triangulation) implementation with sampling presents no problem since the dynamical system we are observing is linear. Traditional discrete time systems theory affords a reliable observer (4). However no analogous theory is available for our new nonlinear dynamic triangulator.

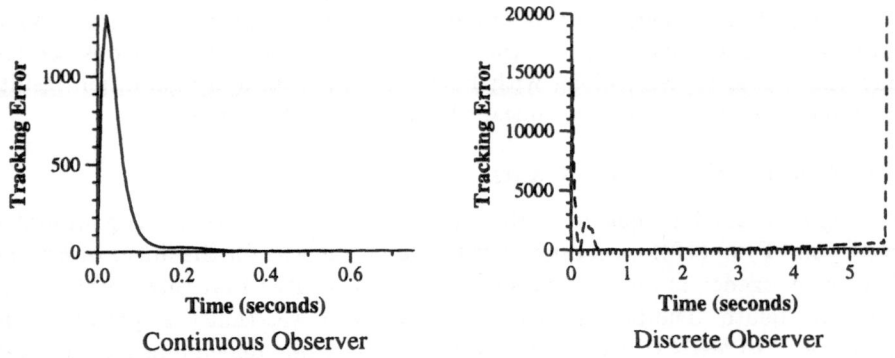

Figure 3. Simulation: Convergence of the continuous and discrete time observers for larger initial error.

In the absence of any theory we have numerically studied both the continuous and discrete time systems. Figures 2 and 3 demonstrate how a change in the initial conditions can result in instability for the discrete system, while the continuous version remains well behaved. Figure 2 depicts a case where the discrete and continuous system demonstrate comparable behavior for identical gains and small initial errors, Figure 3 demonstrates that the same systems can display markedly different behavior for different initial conditions. In this particular example the continuous system converges reasonably quickly, while the discrete version initially behaves reasonably well, then slowly begins to fail until 5.5 seconds, when it "explodes".

4.3. The Dynamic-Triangulator and Juggling

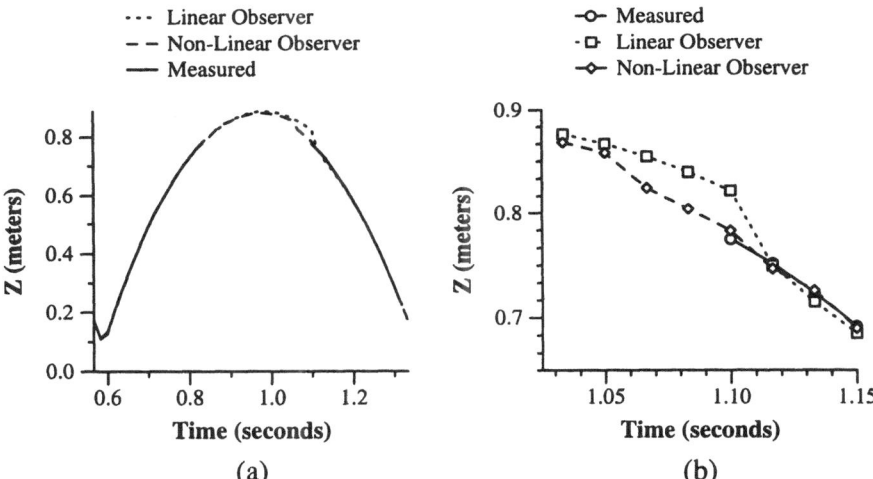

Figure 4. Experimental Data: Triangulated ball height and estimated ball height from both observers during recovery from a typical out of frame event.

Although we are forced to implement the discrete time version of this observer for the juggling system, we are fortunate in that it is relatively easy to control the initial errors. In particular the observer is started only after a ball comes into the view of both cameras. This allows us to triangulate at start-up and correctly initialize the position estimate. Since the balls are always manually presented, we can safely initialize the velocity estimate to zero.

The primary benefit of this observer scheme is its ability to make use of data from a camera even if there is no data from the other camera. This has ramifications for the larger juggling system, since a ball will often pass outside the field of view of one or both cameras for periods of time in excess of 0.25 sec. Figure 4 demonstrates the difference between this observer and and the triangulator/linear-observer system in just such a situation. Figure 4(a) shows the overall flight of the ball as estimated by both observers, and measured by the triangulator (absence of the solid line indicates that the ball was out of frame). In this example the ball travels out of frame for approximately 0.2 sec.

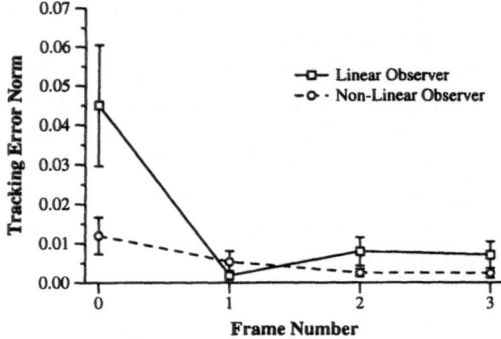

Figure 5. Experimental Data: Mean and standard deviation for the spatial observer errors immediately after recovery from out of frame, averaged over 102 events.

As can be seen in Figure 4(b) (a blowup of the ball returning into the field of view) the dynamical triangulator is capable of updating its estimate while the triangulator/observer pair are forced to simply predict the trajectory (note the differing behavior from 1.05 to 1.10 seconds). Significant reduction in tracking error then results as the ball reappears in both camera's fields of view at 1.10 seconds. This anecdotal picture is confirmed by experimental statistics. Figure 5 shows the mean and standard deviation of the norm squared tracking errors (position only) for the first four frames after recovery from an out of frame event for 102 typical events.

5. Conclusion

In this working paper we have

- discussed certain algebraic properties of camera maps that suggest a new class of state estimators for cartesian dynamical systems

- explored the nature of these estimators through various simulations

- successfully implemented a particular estimator for this class in our laboratory juggling robot.

Note that we have not proposed an "image-plane observer" as discussed in [5], since our estimates reside in *cartesian* space not the image planes. Yet, as an immediate benefit, we need no longer throw away data from both cameras when one of them is occluded. In the longer term, these ideas may provide a framework for turning more general problems in sensor fusion into nonlinear observer problems.

The price we pay for these advantages is manifest at once in

- the need for a new observer theory for linear systems with projective output maps to aid in choice of gain settings and provide convergence guarantees;

- the greater practical risks of implementing in discrete time a nonlinear continuous-time dynamical system (for which we cannot foresee the development of any principled sampling theory).

Further study — theoretical, simulation, and experimental — seems needed before a verdict on the attractiveness of this scheme can be fairly rendered.

References

[1] R. L. Andersson. *A Robot Ping-Pong Player: Experiment in Real-Time Intelligent Control*. MIT, Cambridge,MA, 1988.

[2] M. Bühler, D. E. Koditschek, and P. J. Kindlmann. Planning and control of a juggling robot. *International Journal of Robotics Research*, (to appear), 1993 .

[3] M. Bühler, D. E. Koditschek, and P.J. Kindlmann. A family of robot control strategies for intermittent dynamical environments. *IEEE Control Systems Magazine*, 10:16–22, Feb 1990.

[4] M. Bühler, N. Vlamis, C. J. Taylor, and A. Ganz. The cyclops vision system. In *Proc. North American Transputer Users Group Meeting*, Salt Lake City, UT, APR 1989.

[5] Bijoy k. Ghosh, Mrdjan Jankovic, and Y. T. Wu. Some problems in perspective system theory and its application to machine vision. In *Int. Conf. on Intelligent Robots and Systems*, pages 139–146, July 1992.

[6] A. A. Rizzi and D. E. Koditschek. Further progress in robot juggling: The spatial two-juggle. In *IEEE Int. Conf. Robt. Aut.*, pages 3:919–924, May 1993.

[7] A. A. Rizzi and D. E. Koditschek. Toward the control of attention in a dynamically dexterous robot. submitted for presentation at the IEEE/RSJ Int. Conf. on Intelligent Robots and Systems, July 1993.

[8] Alfred Rizzi and Daniel E. Koditschek. Preliminary experiments in robot juggling. In *Proc. Int. Symp. on Experimental Robotics*, Toulouse, France, June 1991. MIT Press.

[9] Alfred A. Rizzi and D. E. Koditschek. Progress in spatial robot juggling. In *IEEE Int. Conf. Robt. Aut.*, pages 775–780, Nice, France, May 1992.

[10] Louis L. Whitcomb, Alfred Rizzi, and Daniel E. Koditschek. Comparative experiments with a new adaptive controller for robot arms. *IEEE Transactions on Robotics and Automation*, 9(1):59–70, 1993.

Section 6
Robot Design

It is well known that mechanical design parameters have crucial effects on overall dynamic robot performance. The papers in this section discuss the use of parallel mechanisms or the modular structure, and nonlinear characteristics of harmonic drives.

Hayward describes a prototype shoulder joint whose structure is designed so as to make full use of the advantage offered by parallel linkage mechanisms such as light weight, high mobility, and favorable structural characteristics.

The paper by Schütte, Moritz, Neumann, and Wittler is on a modular revolute-joint manipulator TEMPO developed through the complete design cycle from modelling over identification up to realization of the control aided by a mechatronic design environment. The paper points out that a modular system is very flexible and reliable concerning the kinematic structure, the modelling of the entire system, and the controller design.

Kircanski, Goldenberg, and Jia describe a new phenomenon observed in the behavior of harmonic drives named 'torque-transmission paradox' related to the inability of transmitting the input motor torque to the output link. A simple soft-windup model in addition to the nonlinear stiffness and hysteresis model is also introduced.

Design of a Hydraulic Robot Shoulder Based on a Combinatorial Mechanism

Vincent Hayward

Center for Intelligent Machines, McGill University
3480 University Street
Montréal, Québec, Canada, H3A 2A7

Abstract— In previous papers, I have argued that while parallel mechanisms are well known for their favorable structural properties, their utility is generally limited by an inherently small workspace. I have also argued that proper use of actuator redundancy can simultaneously increase the workspace, remove singularities, and dramatically improve overall kinematic, structural, and actuator performance, while keeping the complexity low. This paper discusses a prototype shoulder joint more appropriately described as a combinatorial mechanism which exhibits the features. Additional benefits in terms of modularity, self-calibration, reliability, self-test, and degraded modes of operation are briefly discussed in the conclusion.

1. Introduction

While a great deal pf research activity is devoted to the design of robot manipulator wrists (see [18]) and articulated hands, the design of other joints or groups of joints is not less challenging. A robot manipulator, like most technological creations: automobiles, airplanes, excavators, etc., is an integrated machine. Weakness of any of its constituents: actuators, sensors, structural properties, materials, kinematic properties, dynamic properties, control techniques, and many more factors, will affect dramatically the performance of the complete machine [5].

Note that the great majority of existing manipulator arms, man-made or found in Nature, can be viewed as the assembly of three joint groups: the shoulder group which provides for the overall orientation or the pointing direction of the manipulator, the elbow group which provides for radial extension or reach, and the wrist group which provides for the final orientation of the effector. Of course, there are exceptions to this rule: some crustacean or insect limbs, for example, are not so clearly organized. This rule nevertheless appears to be well enforced for all larger animals.

Among man-made manipulators in use in the manufacturing industry (welding, painting, assembly, part handling, deburring, polishing, etc.), the author has seldom seen an exception to this rule. In the application of robotics for intervention in remote or hazardous locations, this rule seems to be followed even more meticulously: for undersea, outerspace, and other applications, among many existing or proposed designs that the author has surveyed, the only exception found are the "variable truss manipulators" dealt with in the next paragraph.

In research and development laboratories, there are numerous exceptions to the suggested rule. Although it would be lengthy to discuss them all, it may be noticed that the greatest majority of these exceptions either fall in the category of trunks, spines, snakes, tentacles, and other **segmented designs** (for example see [3]), or in the category of platform manipulators (see [17] for example). These categories often overlap (see [15] and [4] for example).

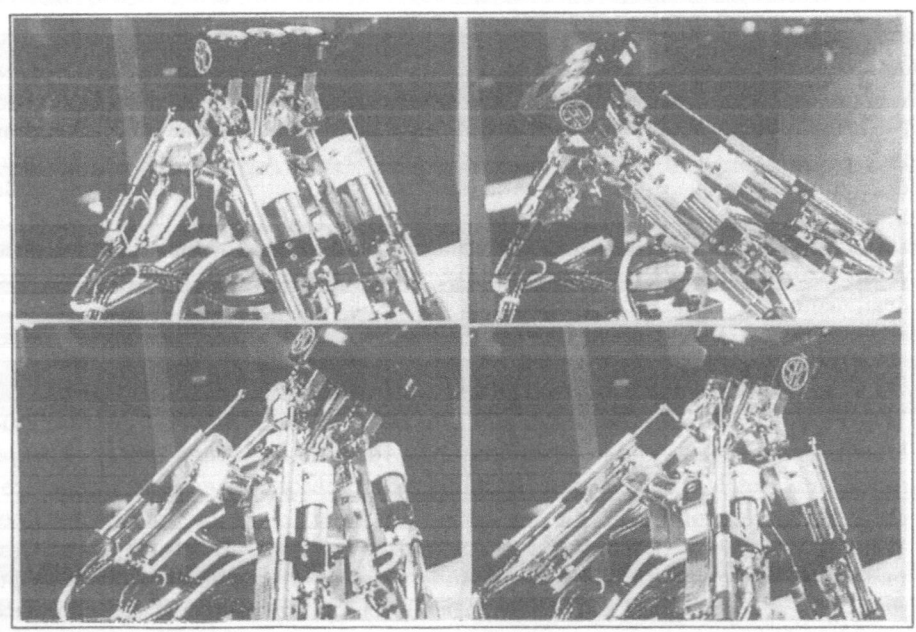

Figure 1.

Following what has just been said, some liberties with the conventional terminology are taken: the word 'joint' will refer to a group of joints, actuated or passive, instrumented or not, which contribute toward a common function: orientation or extension, as in the 'shoulder joint' or the 'wrist joint'. In this very brief survey, a large number of examples could not be commented on. Since the focus is on hydraulic actuation, the reader is referred to three recent designs of hydraulic manipulators [12, 19, 21] for purposes of comparison. All three examples have a serial linkage architecture, although in the first two cases, larger versions of these manipulators have one or two proximal revolute joints replaced by four-bar 'crank-piston' joints.

Here, the design of a shoulder joint of unconventional architecture is described and the characteristics of a prototype discussed. This paper is a sequel to [7]. Please see Fig. 1 for a view of the prototype.

2. Design

The major characteristic of a manipulator shoulder joint is that it works with the **worst possible mechanical advantage**, and yet its bulk and weight must be minimized, although this last requirement is obviously less critical than it is for distal joints. In the applications of robots in hazardous environments, this requirement is nevertheless crucial because a manipulator is typically **not** a grounded structure, but instead, is supported by a gross positioning device: a vehicle, a boom, or a crane.[1] In [7], a promising architecture to achieve light weight, high mobility and favorable structural characteristics is discussed and represented in Fig. 2.

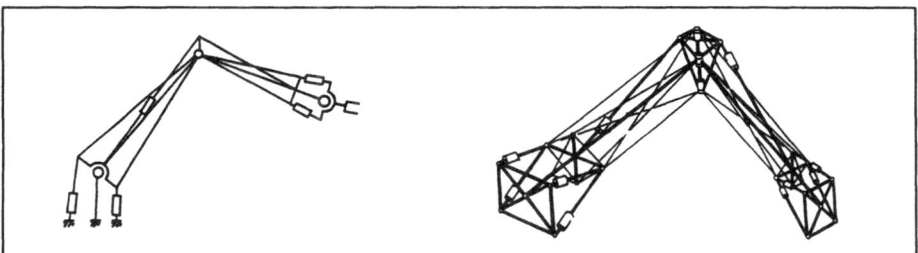

Figure 2. On the right, a spatial version of the planar case shown on the right is diagrammed.

The kinematic/structural concept is shown on Fig 3. According to the prevailing terminology, such a mechanism is termed actuator redundant. The author much prefers the term *combinatorial* because this adjective describes more accurately the underlying concept. No actuator is 'redundant': all are used and well. Their combined utility can easily be appreciated by observing the dramatic loss of performance of the entire system when one actuator in removed (much more than 25%, anyway we look at it). Interestingly, the loss of an actuator does not leave the device completely crippled, in fact, a notable level of functionality is preserved, albeit in a greatly reduced controlled workspace (by about an order of magnitude).

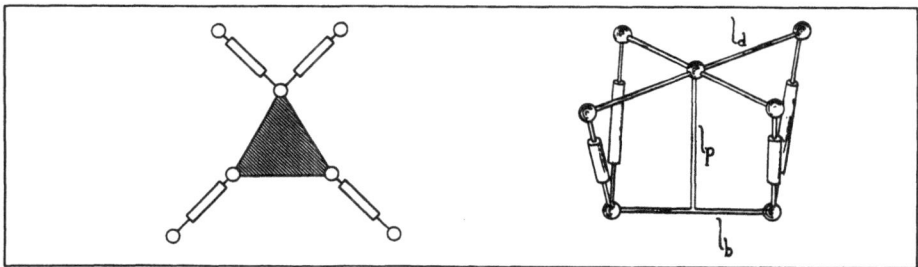

Figure 3. Combinatorial spherical mechanism, analogous to the planar version shown on the left

---

[1]This paragraph applies to human arms as well!

2.1. Kinematic Concept

The operation of this architecture is best understood by viewing it as a combination of four piston-crank systems working cooperatively in a differential fashion, a spatial version of a two piston-one-crank mechanism, so to speak. In this case, four topological regions can readily be distinguished, labelled by the relative velocities of the two pistons, as seen on the Fig. 4.

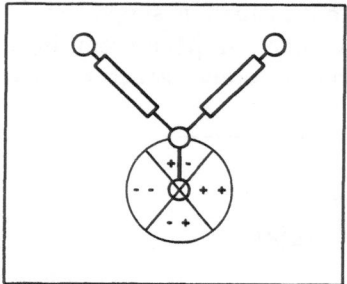

Figure 4. The four topological regions of the simplest *combinatorial* mechanism.

For the more general case discussed in this paper, in the vicinity of the central position, tilting motions occur when the pistons attached to opposite edges of the square platform move with velocities whose signs are different, while swiveling motions occur when pistons in opposite corners have velocities with differing signs as illustrated in Fig. 5.

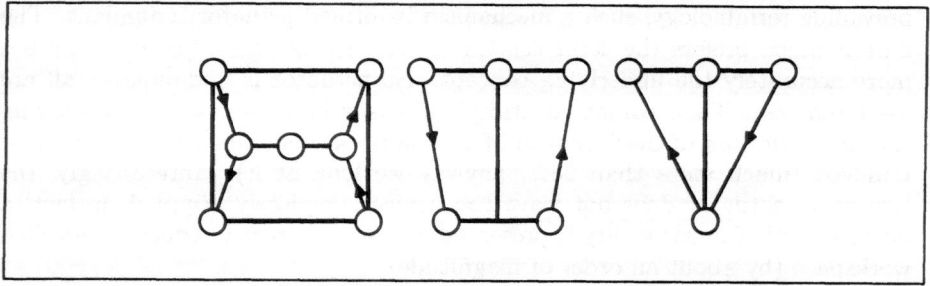

Figure 5. Velocity sign combinations.

Motions with all four equals velocity signs are kinematically prevented since the platform is constrained to a spherical motion. These sign relationships are captured by the signs of each entry in the Jacobian matrix of the mechanism's kinematic map [8]. Thus, there is an upper bound of 2^{12} of such topological regions, but at this point there is no proof that they all exist. The transition from one region to another is illustrated in the Fig. 6 during a swiveling motion with no tilt.

2.2. Expected Performance

The study in [13] indicates that the theoretical optimal workspace free of singularities (all parts with zero thickness) is as large as 180° of tilting motion

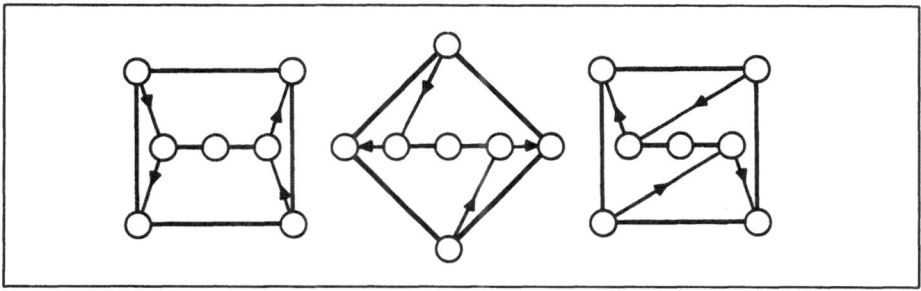

Figure 6. The platform swivels, two of the actuators' contributions to the swiveling torque vanish while the other two's reach their maximum mechanical advantage. A similar situation occurs for all the topological region transitions, which explains the exceptional extent of the working region having high kinematic conditioning.

in both directions and 270° of swivel. Of course, this ideal performance must be compromised for kinematic performance, structural properties, and space to lodge bearings and actuators. What has actually been achieved in the first prototype is about 90° of tilting motion in both directions and 180° of swivel. These figures are expected to be improved substantially in a future prototype.

The large workspace increase makes it possible to take full advantage of the efficiency of parallel linkages as discussed in [9]. In effect, during most motions of the joint all four actuators move, each contributing mechanical power combined at the output link, because the actuator efforts sum and they all move. The worst case occurs when two actuator velocities vanish, in which case these actuators may only play a structural role. The best case occurs when all four actuators move at the same velocity and exert the same effort.

In a competing serial arrangement, during movements where the joint is kinematically well conditioned, only one actuator is producing mechanical power, the others have low velocity and uselessly dissipate power to support their neighbors, while the strength of the joint is limited by the weakest link. The worst case occurs when the joint is near a singularity when some actuators move at high speed exerting little effort, thus operating at low efficiency. Moreover, in a serial arrangement the stiffness of the group is limited by the stiffness of the weakest joint, whereas in the parallel case, their stiffness add.

As a result of the previous observations, excellent strength-to-weight ratio and power-to-weight ratio are expected over a large workspace, made possible by 'actuator redundancy'. The moving mass lumped at the output link is about 1 Kg at a lever arm of about 0.1 m, resulting in a moment of inertia of the order 0.01 Kg.m^2. Considering that the joint can produce a torque of 200 N.m (with a low supply pressure of 3.3 MPa-500 psi), flat within a bandwidth of at least 100 Hz in isometric conditions around any of the three rotation axes, its performance can be quite remarkable. We are now in position to examine the tradeoffs which were confronted to come up with a practical shoulder joint.

2.3. Dimensioning

The investigations carried out to determine the range of parameters for a useful device resulted in several important observations reported in [13]. It was found that conventional kinematic indices could in fact be quite misleading in the search for a useful device if physical limitations were not taken into account. Another important and related observation is the **low sensitivity** of kinematic performance as a function of the design parameters, except at **scattered** trouble spots. This was typified by the need to use log scales to concisely summarize results. The upshot of the study is that kinematic design for manipulators is better described as the avoidance of debilitating conditions rather than targeting for sharp optima.

When $\mathbf{J}$ is the manipulator Jacobian, $\sigma_1, \sigma_2, \sigma_3$ are its singular values, $k(\mathbf{J}) = \sigma_1/\sigma_3$ is its condition number, three such indices were used in a hierarchical design method: global conditioning, actuator forces minimization, and global gradient index. The second and third indices were introduced for the first time:

$$D_g = \frac{\int_W D_l\, dw}{\int_W dw} = \frac{\int_W 1/k(\mathbf{J})\, dw}{\int_W dw} \tag{1}$$

$$F_g = \frac{\int_W F_l\, dw}{\int_W dw} = \frac{\int_W 1/\sigma_3\, dw}{\int_W dw} \tag{2}$$

$$GD_g = \max_W GD_l = \max_W \|\nabla D_l\| \tag{3}$$

Referring to Fig. 3, it was found that the general case mechanism could be made isotropic with a conditioning ideally flat for the values $l_b = l_p = 2l_d$ (units are irrelevant: since it is a spherical mechanism, its properties are scale invariant). In practice, for a shoulder mechanism, there is no reason why it should be isotropic. Actually, it should have greater strength around the axis working against gravity. The scale of the device is directly related to the stroke of the actuators.

Due to the vast number of actual design parameters, a trial and error was used to precisely determine the final dimensions of the device: first with the construction of a series of physical models of increasing fidelity and then checking with the indices listed above that performance was indeed acceptable. Most importantly, the decision tree was rooted in the actuator choice (see section 2.5). At the first level, many decisions were dictated by the necessity of piston-type actuators to have their shortest length exceed their maximal stroke. From this observation, attaching the cylinders by their ends would necessarily lead to poorly conditioned mechanisms. To overcome this difficulty, the cylinders must protrude and be placed in cradles, thus ruling out use the square platform (Fig. 3) as the output link. The final dimensioning decisions where made by considering avoidance of self collisions and structural strength (see section 2.6). The entire decision process resembled searching for a Nash equilibrium (conflicting objectives) in game theoretic optimization problem [20].

2.4. Reduction of Passive Joints Count

One fundamental disadvantage of parallel mechanisms (particularly spatial mechanisms) is the need for numerous passive joints. In the case studied in this paper, a straightforward implementation of the diagram shown on Fig. 3— replacing each spherical joint by a three axis gimbal (each with two forks) and providing for torsion decoupling bearings for linear displacement sensors piggy backed on the pistons—leads to a large number of bearings (45). Unless great care is exercised with respect to the strength and precision of all of these joints, wear, backlash, or failure (not mentioning high cost) is to be expected, defeating the supposed advantages of parallel mechanisms. This problem is further compounded by the lack of room usually available to lodge these bearings.

Inoue *et al.* [11] designed an iso-static platform based on three pantograph (5-bar) mechanisms which displays advantages over the piston driven McCallion-Truong design [14]. While this design reduces the number of passive joints by having a pair of pistons share two revolute joints in a double Hooke joint arrangement, further saving can be obtained by noticing that a single revolute joint is needed to orient a plane—the plane containing the pantograph. This fact was also noticed by Dunlop et al. [6], who further pointed out that this would save torsion decoupling bearings for linear sensors mounted on screw actuators. The combination of these ideas to the case of piston actuators leads to a simple structure illustrated Fig. 7. A structurally sound implementation of this chain can be realized with 25 bearings, which saves almost half the number of bearings as compared to the straightforward design.

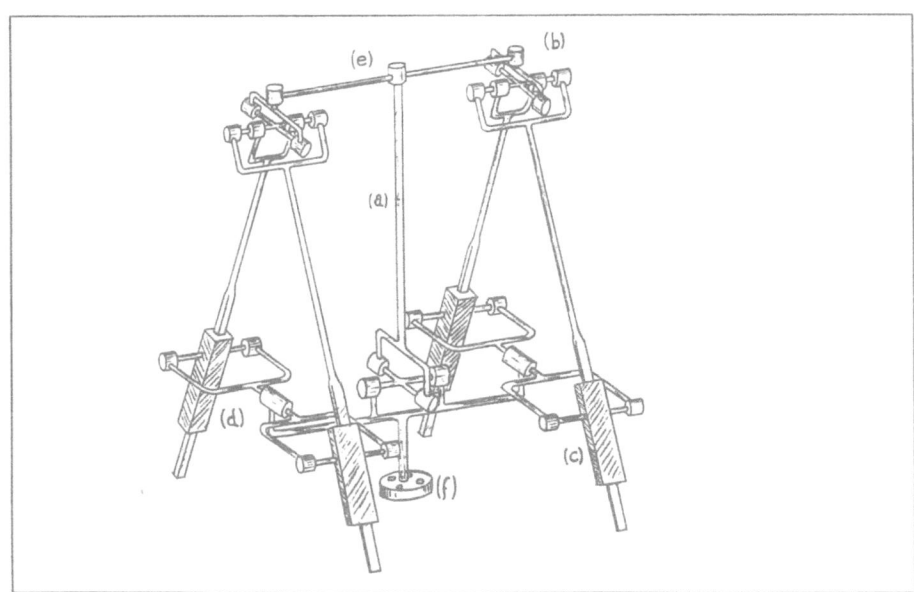

Figure 7. Complete kinematic chain.

The bearings, although supporting large loads, could be made very small and yet have little backlash thanks to a technique developed at the Center for Engineering Design at the University of Utah.

Finally, yet another benefit of actuator redundancy contributes to ensure the elimination of the residual backlash. In any given orientation of the mechanism, internal stresses can be actively created, independently from the external load, thus bias-loading all the mechanism's bearings.

2.5. Actuators

It is suggested here that the actuator analysis problem in robotics might be split into four items:

- **Energy Storage.** Under this heading, one considers the form under which the energy needed to actuate the robotic device is stored. In a manufacturing application, this question of is small importance since it can reasonably be assumed that an unlimited supply of energy is available. In other applications of robotics, this item can play a major role, for example for planetary explorations vehicles or untethered submarine rovers, in which case a capacious energy storage must be part of the robot itself.

- **Energy Transport.** One considers the method by which energy is transported from storage to a final stage, where it must be available in mechanical form. In most cases, this function is performed in multiple steps. For example in an electric industrial robot, energy is transported in electrical form, converted into mechanical form in the manipulator's motors and then, most often, once again mechanically transported to the joints via gears, shafts, belts, chains, cables, tendons, hydraulic or pneumatic conduits, or other mechanical energy transmission techniques. A 'direct drive' robot is characterized by the absence of mechanical transmission of energy, except by the structure of the manipulator itself (hence most hydraulic robots are direct drive).

- **Energy Throttling.** One considers here the method whereby energy is throttled to ensure control over the robot. Energy throttling is hard to achieve mechanically (clutches do that inefficiently). Among a large number of possibilities, electric energy throttling is usually the method of choice, even in a multi-step process as for electromagnetic servovalves. Regardless of the type of energy being throttled, there are three basic methods. In the first method, constant power is delivered by the supply, and the throttling mechanism arranges for a variable amount of the effort to be diverted to the load, usually by creating an effort imbalance. An electric version of this principle leads to 'class A' amplifiers, a hydraulic version of it leads to the jet pipe hydraulic valve. In the second method, the power delivered by the supply varies with that delivered to the load ('class B') limiting the loss in the throttling mechanism. But in this case the throttling mechanism must have large bandwidth because the power must be switched from throttling to quiescence, and vice versa, at each reversal of the direction of the energy flow. In the third method, in some sense a limiting case of the previous case, power is switched, and switching timing is used to throttle power. Then it must have even higher bandwidth to approximate ideal switching. This topic is not often studied in robotics, although it is quite a thorny problem more often than generally assumed (example: variable reluctance electric actuators).

- **Energy Conversion.** Under this last item, one must examine which principle is applied to convert a transportable form of energy into mechanical energy. The reader is referred to [10] for a recent survey.

Most actuator systems accomplish transport, throttling and conversion in multiple steps. Hydraulic techniques have been selected for use in this shoulder joint prototype enabling direct application in areas in which new manipulator designs are in need, in particular telerobotics. In this area, often, the levels of performance are attainable only with hydraulic actuation. Also the recently available force-controlled ASI integrated hydraulic actuators offered a path toward rapid creation of a high performing prototype. In a number of telerobotic applications, there is need for human-like performance in terms of bulk, precision, strength, reach and dexterity. Again, only hydraulic actuators can be made compact enough to even consider approaching the level human performance. Thus, with this in mind, the four items listed above can be discussed.[2]

Because of the large amount of power required by hydraulic actuators, the energy storage can only be utility-supplied, or be stored chemically in conjunction with a first conversion stage provided by a turbine or an internal combustion engine. This is the method used in many 'pre-robotic' devices such as cranes, excavators, forestry or mining equipment. In submarine applications, high levels of power may be made available through umbilical cables. With hydraulics, the transport function is accomplished with hydraulic lines which must be run from the source of pressure to the actuators. Lines have their own dynamical properties which are actually best described as distributed parameters systems. For this reason, their dynamics are not simple and they can account for significant losses in energy and performance. Hence, compact high-bandwidth valves that can be co-located with the actuators themselves are quite advantageous from a control view point.

With this technique, excellent performance using local compensation can be obtained, even with long and ill-modelled lines. The valve itself is only a part of the energy throttling mechanism since it is driven by an electromagnetic motor, itself driven by an electronic amplifier. In the case of the ASI actuators it was found that the electromagnetic energy conversion mechanism was in fact a limiting factor in the performance of the complete actuator. The last stage of the energy throttling mechanism is based on forcing the fluid through orifices whose areas can be controlled and the last stage of energy conversion is accomplished simply by letting the fluid pressure act differentially on the

---

[2]It a worthwhile to reflect on Nature's solution to these four problems.

sides of pistons or vanes. Details on the properties and compensation control of these actuators can be found in [1].

2.6. Structural Design and Collision Avoidance

The structural design was carried out while attempting to take full advantage of the opportunities offered by parallel linkage mechanisms. For example, pistons, as opposed to rotary vane actuators, can be used as structural members and thus economically contribute to strength. Piston actuators have in general numerous structural advantages over rotary actuator while being simpler to manufacture and maintain, and they have greater efficiency.

The principal structural parts are listed by order of decreasing design difficulty: (a) Central stem (1 unit); (b) Double Hooke joints (2 identical units); (c) Actuator cradles (4 identical units); (d) Double actuator forks (2 identical units); (e) Output flange (1 unit); (f) Ground link (1 unit); As it can be seen, the number of parts is quite reasonable are they can be made to have rather simple shapes. The loads types and magnitudes in each bearing was examined, and correspondingly sized.

In the current state of prototype, no systematic structural optimization was carried out, using finite element methods for example, but that could be done in the future. It is encouraging to see that the first sizing estimates used in this first prototype already produced quite a light and compact system. Certain deficiencies were noticed and will have to be corrected in future prototypes.

The main culprit in the reduction of the practical workspace is the possibility of self collisions well within the singularity-free work region. This topic is not studied in any systematic fashion because of the great difficulty in its formalization. General guidelines have been followed to alleviate this problem. Ideally all parts should be as small as possible, but of course, this approach has its limits.

Except for the output flange which is made of aluminum, the material selected for the construction of most of the structure is stainless steel to optimize the strength to size ratio of its parts.

3. Experimental Results

The shoulder joint was tested in the laboratory using a L-shaped member designed to approximate the construction of a complete arm. It is made of two steel tubes welded at a right angle with lengths 0.3 m each. It is bolted to the output flange shoulder such that the axis of first tube coincides with its axis of symmetry. This member is quite rigid and has a mass of 1.6 Kg. An additional load mass of 1.4 Kg is affixed to the end of the second tube. The moments of inertia of this composite load about the joint center of rotation are approximately 0.62 Kg.m^2, 0.48 Kg.m^2, and 0.15 Kg.m^2.

In an attempt to decouple the basic performance of the design from that of the various control algorithms currently under investigation, the simplest feedback controller is used. It is a proportional position feedback controller about each actuator, augmented by a lag term for improved steady-state accuracy.

The controller is implemented digitally at a servo rate of 1 kHz. We have also implemented a force regulator of similar design.

We executed a small step motion about the axis of highest inertia, in the middle of the joint range. In this condition, the shoulder is capable of producing accelerations of 130 m/s^2 at the tip where the wrist joint is to be located, and a maximum tip velocity of 0.45 m/s was observed. The maximum static force produced at the wrist varies from a worst case of 200 N to a best case of 400 N, according to direction of the exerted force, at the center of the shoulder joint range. The force control bandwidth is observed to be about about 100Hz. Position resolution could not be measured with the available equipment.

The basic low velocity and force resolution performance is not a function of the design since it can be qualified of being "direct drive" (there are no motion transmission elements other than simple linkages interposed between the actuators and the load). Low velocity performance and force resolution depend completely on the actuators, sensors and local feedback control. To date, experiments in this area have been limited solely by the resolution of the 12 bits analog to digital converters used for digital control.

4. Conclusion

A manipulator shoulder joint design based on a combinatorial mechanism which addresses several issues of concern in telerobotics was described. A large workspace was achieved despite the use of parallel linkages, thereby offering an opportunity to take advantage of their possibilities in terms of structural strength and precision. The joint is efficient because any motion in any direction combines the contribution of at least 3 and almost everywhere of 4 actuators, where a serial joint could use only one actuator. They are also advantages from a control view point, in terms of impedance modulation and control robustness, when compared to a serial arrangement as described in [2].

As a result of the preceding items, a dramatic improvement in power-to-weight ratio has been achieved while construction was kept simple and the number of parts minimized, while low pressure hydraulics led to a significant torque output. Moreover, four identical piston type actuators, which are the simplest to manufacture and to maintain, were used exclusively.

This design provides a path toward the development of a highly reliable system since it has built-in redundancy. There is natural provision for self calibration: sensors, kinematics, and dynamics. Initial success in this direction has been achieved [16]. Self-testing is natural extension of self calibration: Work is under way to investigate the automatic diagnostic of partial structural, actuator, and sensor failures as well as backlash. Finally, there is built-in provision for degraded modes of operation with missing sensors, missing actuators and minor structural failures. Moreover backlash can be actively compensated.

At the time of this writing, an elbow joint using two actuators identical to those used in the shoulder is under construction. A compact wrist mechanism with four smaller actuators is also in the design stage. All these joints are based on design principles described in this paper and will result in a manipulator arm with high dexterity, high strength-to-weight ratio, powered by 6 replicated piston actuators and 4 smaller ones, also identical.

5. Acknowledgments

I am indebted to many people who contributed to this project in various ways: in particular Dr. Fraser Smith and David Barret from Sarcos Research Corp. for their first class engineering design and prototype realization, Prof. Stephen C. Jacobsen for his encouragements and valuable comments, Benoit Boulet and Ronald Kurtz, for their excellent work, Prof. Laeeque Daneshmend for his contribution to modeling and control of actuator systems, Chafye Nemri and Duncan Baird for their engineering support. Finally, I would like to thank Profs. I. W. Hunter and J. M. Hollerbach for their support within the project "High-Performance Manipulators" (C3) funded by IRIS, the Institute for Robotics and Intelligent Systems part of Canada's National Centers of Excellence program (NCE). Additional funding was provided by a team grant from FCAR, le Fond pour les Chercheurs et l'Aide à la Recherche, Québec, and an operating grant from NSERC, the National Science and Engineering Council of Canada.

References

[1] Boulet, B., Daneshmend, L. K., Hayward, V., Nemri, C. 1993. System Identification and Modelling of a High Performance Hydraulic Actuator. *Experimental Robotics 2*, Chatila, R., Hirzinger, G. (Eds.), *Lecture Notes in Control and Information Sciences*, Springer Verlag.

[2] Boulet, B., Hayward, V. 1993. Robust control of a robot joint with hydraulic actuator redundancy. Part I: Theory application and controller design. Part II: Practical variations and experimentation. TR-CIM-93-6 and TR-CIM-93-7. Technical Report, *McGill Center for Intelligent Machines*. McGill University, Montréal Canada.

[3] Burdick, J. W. and Chirikjan. Hyper-redundant robots: kinematics and experiments. in Preprints of 6th International Symposium on Robotics Research, Oct. 2-5, Hidden Valley, PA.

[4] Charentus, S., Renaud, M. 1990. Modelling and Control of a Modular, Redundant Robot Manipulator. in *Experimental Robotics 1*, Hayward, V., Khatib, O. (Eds.), Lecture Notes in Control and Information Sciences 139. Springer Verlag, pp. 508–527.

[5] Dietrich, J. Hirzinger, G., Gombert, B., Schott, J. 1990. On a unified concept for a new generation of light-weight robots. in *Experimental Robotics 1*, Hayward, V., Khatib, O. (Eds.), Lecture Notes in Control and Information Sciences 139. Springer Verlag, pp. 244–286.

[6] Dunlop, G. R., Johnson, G. R., Afzulpurkar. 1991. Joint design for extended range Stewart platform applications. *Proc. Eight World Congress on the Theory of Machines and Mechanisms*. Prague, Czechoslovakia. Vol. 2. pp. 449–451.

[7] Hayward V. 1993. Borrowing some design ideas from biological manipulators to design an artificial one. In *Robots and Biological Systems, NATO Series*, P. Dario, P. Aebisher, and G. Sandini, (Eds.), Springer Verlag. pp. 139–151.

[8] Hayward, V. Kurtz, R. 1991. Modeling of a parallel wrist mechanism with actuator redundancy. In *Advances in Robot Kinematics*. Stifter, S., Lenarcic (eds.). Springer-Verlag. pp. 444–456.

[9] Hirose, S., and Sato, M. 1989. Coupled drive of the multi-DOF robot. in *Proc. IEEE Int. Conf. on Robotics and Automation*, pp. 1610-1616.

[10] Hollerbach, J. M., Hunter, I. W., Ballantyne, J. 1991. A comparative analysis of actuator technologies for robotics. In *Robotics Review 2*, Khatib, O., Craig, J., Lozano-Perez, T. (eds.), MIT Press.

[11] Inoue, H., Tsukasa, Y., Fukuizumi, T. 1986. Parallel manipulator. In *The Third International Symposium on Robotics Research*. Faugeras, O. E., Giralt, G. (Eds.). MIT Press. pp. 321–327.

[12] Jacobsen, S. C., Smith, F. M., Backman, D. K., and Iversen, E. K. 1991. High performance, high dexterity, force reflective teleoperator II. in *Proc. ANS Topical Meeting on Robotics and Remote Systems*.

[13] Kurtz, R., Hayward, V. 1992. Multi-goal optimization of a parallel mechanism with actuator redundancy. IEEE Transactions on Robotics and Automation. Vol. 8, No. 5. pp 644–651.

[14] McCallion, H., Truong, P. D. 1979. The analysis of a six degree of freedom work station for mechanized assembly. Proc. *5th World Congress on the Theory of Machines and Mechanisms*, pp 611–616.

[15] Morecki, A., Malczyk, G. 1987. Mathematical Model of a flexible manipulator of the elephant's-trunk-type. in *Proc. Sixth CISM-IFToMM Symposium on Theory and Practice of Robot and Manipulators*, Morecki, A., Bianchi, G., Kedzior, A. (Eds.), Hermes, Paris.

[16] Navhi, A., Hollerbach, J. M., Hayward, V. Calibration of a parallel robot using multiple kinematic loops. Submitted to 1994 IEEE Int. Conf. on Robotics and Automation.

[17] Pierrot, F., Dauchez, P., Fournier, A. 1991. Hexa a fast six-dof fully parallel robot. Proc. *5th Int. Conf. on Advanced Robotics*, Pisa, Italy. IEEE Press.

[18] Rosheim, M. E., 1989. *Robot wrist actuators*, John Wiley & Sons.

[19] Regan, B. 1991. ATLAS-8F Advanced Bilateral Manipulator System. In Proc. of ROV-91 Conference, Marine Technology Society. Washington DC.

[20] Vincent, T. L. 1983. Game theory as a design tool. *J. of Mechanisms, Transmissions, and Automation in Design, Trans. of ASME*, Vol. 105, No. 1, pp. 165–170.

[21] Yoshinada, Hiroshi, Yamazaki, T., Suwa, T., Naruse, T. 1992. Design and control of a manipulator system driven by seawater hydraulic actuator. *Proc. Second Int. Symposium on Measurement and Control in Robotics (ISMCR)*, Tsukuba Science City, Japan, pp. 359–364.

Practical Realization of Mechatronics in Robotics

Schütte, H., Moritz, W., Neumann, R., Wittler, G.

Mechatronic Laboratory Paderborn, MLaP
(Prof. Dr.-Ing. J. Lückel), University of Paderborn, Pohlweg 55
D-33098 Paderborn, Germany

Abstract - Design of a very *fast* and *precise modular* revolute-joint *manipulator* is shown in detail. The complete design cycle from *modelling* over *identification* up to *realization* of the control is aided by a *mechatronic design environment* which is based on a unified model description in state-space. Special emphasis lies on the generation of a reliable entire model of this complex system and on the controller design. The derivation of a *symbolic model* of the robot including *elasticities* in gears and bearings is the basis of the design of a complex controller including a *disturbance estimator*. *Measurements* show the efficiency of the proposed proceeding.

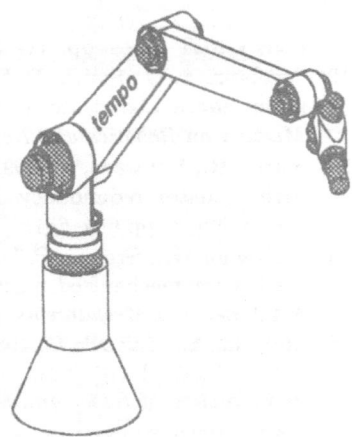

Figure 1. Robotic System
TEMPO

2. Introduction

In order to prove the efficiency of the design methods and software tools developed at the Mechatronic Laboratory Paderborn (MLaP) for the design of *mechatronic systems*, a six-axis revolute-joint manipulator was designed and built within a very short term.

One of the aims was to create a very fast manipulator of high precision, especially during the entire motion along a trajectory, as a very broad basis of experimental verification of complex robust linear and nonlinear control schemes; and all this not only as a simplified laboratory system but as a system of industrial standard, concerning all its components. Another point is that we want a modular system (see Fig. 2) because such a system is very flexible concerning the kinematic structure that are to be built and it can support the modelling of the entire system because of its clear mechanical structure. The mechatronic side of this

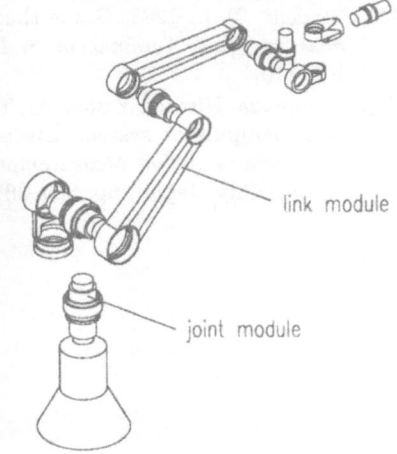

Figure 2. Modular Structure

matter is that in spite of the modular structure it can be shown that a good adaptation of the mechanical parts to the electronic resp. software parts leads to very high performance. Consideration of the joint and its control as an integrated system result in several modifications of the mechanics because of the requirements coming from the controller design.

As a side effect of the development of the system TEMPO, we hope to reduce the big gap between industrial robots and the actual results in robotics research; we also hope that it is possible to open new fields of application for precise and fast robot systems, such as laser cutting or laser welding.

3. Mechatronic Function Module

The basic mechanical element is a scalable joint module (see Fig. 3) made up of a gear, a brushless DC motor, a permanent magnet brake, contact-free end position switches, and the angle measuring devices.

The main part is the Harmonic Drive gearbox that, in a very compact form, integrates the gear and a bearing which can directly be used as a bearing for the links of the robot. It has a gear ratio of 1:121 and ensures a backlash-free transmission of the moments. The main disadvantage of this gear type is its high viscous and Coulomb friction, which will be compensated for by means of a disturbance estimator.

A special feature of this joint is its high-precision angle measurement system for the motor and additionally for the load side of the gear, each with a resulting resolution of 720000 inc./(loadside rev.). This additional measurement allows e. g. the detection of the gear torque, an active vibration damping and disturbance estimation and under the assumption of rigid links, the exact calculation of the end-effector position.

Furthermore, a high ratio of motor torque to motor side inertia leads to a high acceleration capability.

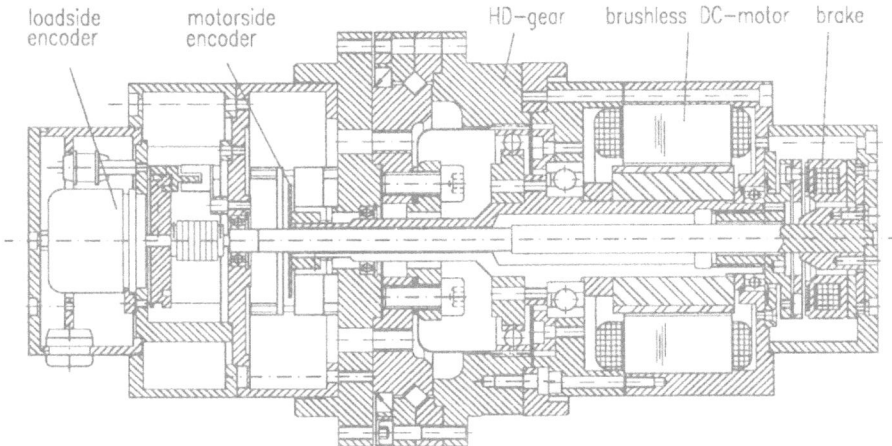

Figure 3. Scalable Joint Module

This joint element is a so-called mechatronic function module because in connection with some software and hardware components it forms an autonomous system for several applications, not only for robots but also for systems where a fast and precise positioning is required, such as in machine tools.

4. Mechanical Model for an Elastic Six-axis Robot

The dynamical equations for multibody systems (MBS) are generated by a MAPLE[1]-based program developed at the MLaP. The following analytical facts are used for *rigid body systems* but it is also possible to deal with systems with distributed parameters via the Ritz formalism.

$$M(q)\ddot{q} + P\dot{q} + Cq + h(q,\dot{q}) = f$$

$$M(q) = J^T(q)M_{diag}J(q), \quad q \in R^{(2n+b)}, \quad n = nb\,of\,joints$$

$$J(q) = [...,J_{Ti}^T, J_{Ri}^T,...]^T, \quad J_{Ti} = \frac{\partial v_i^{(i)}}{\partial \dot{q}^T}, \quad J_{Ri} = \frac{\partial \omega_i^{(i)}}{\partial \dot{q}^T}, \quad i = 1(1)2n$$

$$M_{diag} = [...,m_i, m_i, m_i, Jx_i, Jy_i, Jz_i,...]$$

$$h(q,\dot{q}) = \left(A - \frac{1}{2}A^T\right)\dot{q}, \quad \text{with } A = \frac{\partial(M(q)\dot{q})}{\partial q^T}$$

$$v_i^{(i)} = \dot{r}_i^{(i)} + \tilde{\omega}_i^{(i)}r_i^{(i)}, \quad \tilde{\omega}_i^{(i)} = S_{Ii}^T\dot{S}_{Ii}$$

Attention should be paid to the fact that only the jacobian matrix of the system has to be calculated and thus it is possible to avoid the redundant operations of the Lagrange formalism. The given representation of the centrifugal and Coriolis forces is proved to be very efficient for symbolic computations.

For this MBS program a simple user interface (see Fig. 4) for revolute joint manipulators with the following convention is created :

- relative minimal coordinates
- tree-structured systems
- coordinate systems according to the Denavit-Hartenberg convention
- elasticities in the gears and the bearings

```
# MAPLE input file for generating a DSL-model of the TEMPO
# revolute joint robot with joint elasticity
# author : h. schuette                    date: june 1993

datnam        := 'rob_j3';
nb_of_links   := 3;
kreisel       := true;
DH_para       := matrix(nb_of_links,3,[[ Pi/2,    0, lA1],
                                        [   0,  lA2,   0],
                                        [   0,  lA3,   0],
                                        [-Pi/2, lA4,   0],
                                        [ Pi/2,   0,   0],
                                        [   0,    0, lA5]]):

Sr[2] := motor_gear_link(q1, q2,  q13,q14, [   0,    0,JM1z],  mM1,
                                  [   0,    0,JA1z],  mA1,  cd1,ue1,1);
Sr[3] := motor_gear_link(q3, q4,  q15,q16, [   0,    0,JM2z],  mM2,
                                  [JA2x,JA2y,JA2z],  mA2,  cd2,ue2,2);
Sr[4] := motor_gear_link(q5, q6,  0,q17,   [   0,    0,JM3z],  mM3,
                                  [JA3x,JA3y,JA3z],  mA3,  cd3,ue3,3);
Sr[5] := motor_gear_link(q7, q8,  0,q18,   [   0,    0,JM4z],  mM4,
                                  [JA4x,JA4y,JA4z],  mA4,  cd4,ue4,4);
Sr[6] := motor_gear_link(q9,q10,  0,  0,   [   0,    0,JM5z],  mM5,
                                  [JA5x,JA5y,JA5z],  mA5,  cd5,ue5,5);
Sr[7] := motor_gear_link(q11,q12, 0,  0,   [   0,    0,JM6z],  mM6,
                                  [JA6x,JA6y,JA6z],  mA6,  cd6,ue6,6);
Sr[8] := EE_mass([JEx,JEy,JEz], mE, 7);

save_kin();
read 'calc_jacobian';
read 'make_dsl';
```

Figure 4. Input of the MBS Program

This user interface will also those enable to get the differential equations who have not any experience with the methodology.

The state-space representation of the system (necessary for coupling and a homogeneous representation) is obtained by a symbolic Cholesky decomposition of a pattern matrix of the mass matrix M. In this way the extremely high expenditure for the symbolic inversion of the mass matrix as used in the equation

$$\dot{x} = \begin{bmatrix} 0 & I \\ -M^{-1}C & -M^{-1}P \end{bmatrix} x + \begin{bmatrix} 0 \\ M^{-1}(f-h) \end{bmatrix}$$

can be avoided.

Simplifications for reducing the extension of the equations play a crucial role in the symbolic computations, not only for the efficiency of the numerical evaluations but also for an optional analytical processing that may follow (e. g. for decoupling purposes). A considerable reduction of the terms can be achieved by applying the trigonometric identities and addition theorems. Furthermore, an algorithm was designed and implemented for a subsequent and systematic abbreviation of terms that depend only on physical parameters; an extended function for setting parentheses is used.

The modelling of the system TEMPO starts out from a system with six axes

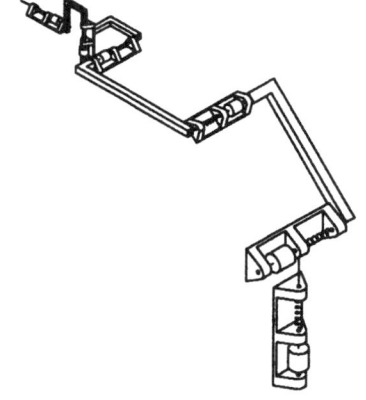

Figure 5. Physical Model of the TEMPO

having gear elasticities, but during the identification process described in the next chapter it turned out that the model has to be augmented by six degrees of freedom corresponding to the elasticities in the bearings resp. flanges and arms.

With the input file (Fig. 4) the MBS program generates a model with eighteen degrees of freedom in DSL syntax (Dynamic System Language, see [5]) which is similar to the model of a three-axis robot shown in the appendix. Such a model can directly be used for nonlinear simulation or for frequency response calculations.

5. Identification

Identification, i. e. in this context determination of the physical parameters of a mechanical system, such as masses, stiffnesses or dampings, is done by means of frequency measurements. With a MATLAB[2]-based program, several measured transfer functions will be adapted simultaneously to frequency responses calculated with the help of the mathematical model. According to the varying loadside inertia, the transfer paths are measured in different positions of the robot.

$$\varepsilon_{magk} = \frac{1}{n}\left\{ \sum_{i=1}^{n} [20\log(|G_{Gk}(j\omega_i)|) - 20\log(|G_{Mk}(j\omega_i)|)]^2 w_{ki}\right\}$$

$$\varepsilon_{phak} = \frac{1}{n}\left\{ \sum_{i=1}^{n}\left[20\log\left(1 + |\angle G_{Gk}(j\omega_i) - \angle G_{Mk}(j\omega_i)|\frac{\pi}{180}\right)\right]^2 w_{ki}\right\}$$

$$\varepsilon_k = \varepsilon_{magk} + \varepsilon_{phak}, \qquad\qquad k = 1(1)m$$

An optimizer keeps varying the physical parameters of the system until a vectorial error criterion $\underline{\varepsilon} = [\varepsilon_1,...,\varepsilon_m]$ containing the errors of all transfer paths reaches a minimum.

The error criterion shown above plays an important part in the optimization because its choice determines the course of the optimization and the adaptation quality. The index k denotes the transfer path in question and n the number of measured frequency points. The measured resp. computed magnitudes G_M and G_G are weighted with the central differences of the logarithm of the frequency points w_{ki}. This avoids an over-emphasis on higher frequency ranges. Moreover, the weightings take into account the measured coherence to soften the influence of faulty or inaccurate measurements.

An addition of amplitude and phase error leads to redundancies that have a positive effect on the optimization process.

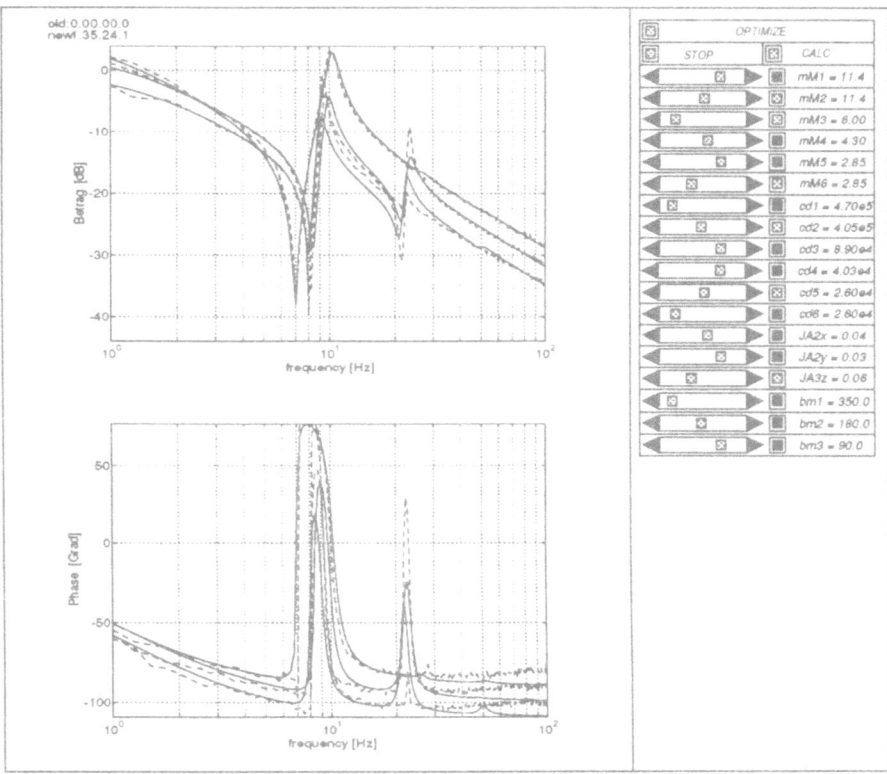

Figure 6. Identification via Frequency Measurements (- - - measured,
—— calculated)

Summing up, it can be said that this criterion yields a good adaptation in
the frequency range, i. e. the amplitude in dB and the phase in degrees.
A graphical interface to the software tool (Fig. 6) allows the user to vary
individual parameters with the help of sliders. After some modifications
of parameters it is possible to calculate a frequency response and the
corresponding errors. This procedure may help the user to find appropriate
starting values or to decide whether a parameter has or has not an influence on the optimization. Then the parameters in question can be set
active or passive by means of the buttons. Lower and upper bounds for the
parameters ensure that the parameters stay physically meaningful during
the optimization process.

6. Modelling with DSL

For the final controller design process it is necessary to obtain a reliable
entire DSL-model not only for the mechanical part but also for amplifiers,
controllers, estimators, models for measurement devices, measurement
filters and for the Coulomb friction. The Dynamic System Language
mentioned in chapter 4 can be used for modelling and coupling of such
systems in a modular and hierarchical manner.

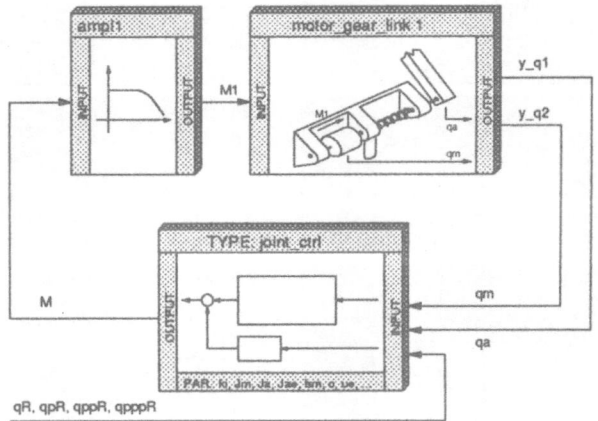

Figure 7.Coupling of Joint, Controller, Amplifier

For one joint of the robot TEMPO, this coupling is exemplified in Fig. 7, 8, and 9.

A BASIC SYSTEM includes a set of linear or nonlinear STATE EQUATIONs and AUXILIARs and provide a defined interface with PARAMETERs, INPUTs and OUTPUTs (see appendix). If a special system, such as the controller in Fig. 8 or an amplifier, is used several times only varying in its parameters, it can be defined as a BASIC SYSTEM TYPE.

```
BASIC SYSTEM TYPE joint_ctrl
 (PARAMETER: k1, k2, k3, k4, Jm,Ja,Jae, bm, c, ue, fd, v1,v2;
  INPUT    : qm "rad", qa "rad",
             qR "rad", qpR "rad/s", qppR "rad/s^2",
             qpppR "rad/s^3", fg    "Nm";
  OUTPUT   : M "V") IS
STATE    : x_qm, x_qa;
AUXILIAR : kpv, kvv, kav, kapv, Td, Jg;
  Td   := 1/(8*atan(1)*fd);
  Jg   := Jm*ue**2 + Jae;

  -- feedforwards --------------------------------
  kpv  := k1 + k3;
  kvv  := k2 + k4 + bm;
  kav  := Jm*ue**2 + Jae - Td*(k2+k4) + k1*Ja/c;
  kapv := bm*Ja/c + k2*Ja/c + (k2+k4)*Td**2 +
          v1*(Jae*(Jm*ue**2) - Ja*(Jm*ue**2) +
          (Jae*Jae - Jae*Ja)) + v2*(bm*Jae - bm*Ja);
STATE_EQUATION :
  x_qm':= -1/Td * x_qm + 1/Td * qm;
  x_qa':= -1/Td * x_qa + 1/Td * qa;
OUTPUT_EQUATION :
  M    :=(-(k1  * qm + k2  * x_qm' + k3  * qa + k4 * x_qa')
         + kpv * qR + kvv * qpR + kav * qppR + kapv * qpppR
         + k1  * fg*ue/c)/ue;
END joint_ctrl;
```

Figure 8. Joint Controller Module

Such a BASIC SYSTEM TYPE can then be instantiated with a certain set of parameters and coupled to the other subsystems in a so-called COUPLED SYSTEM. The coupling is done by means of names in dot notation.

```
COUPLED SYSTEM TEMPO_MCS (PARAMETER :  ...
                         INPUT     :  ...
                         OUTPUT    :  ...     ) IS
SUBSYSTEM : TEMPO, exc_sm, ampl;

  -- some BASIC SYSTEM or BASIC SYSTEM TYPE

  SYSTEM joint_ctrl1 IS joint_ctrl (k1  =>   6.16861E+05,
                                    k2  =>   5.73270E+03,
                                    k3  =>  -6.16861E+05,
                                    k4  =>   1.76020E+03,
                                    Jm  =>   0.00288015,
                                    Ja  =>      50.1,
                                    Jae =>      50.1,
                                    bm  =>     153.7,
                                    c   =>     4.2e5,
                                    ue  =>     121.0,
                                    fd  =>     200.0);
PARAMETER_CONDITION :  ...
COUPLE_CONDITION    :
  -- excitation ----------------------------------
  joint_ctrl1.qR    := exc_sm1.qR;
  joint_ctrl1.qpR   := exc_sm1.qpR;
  joint_ctrl1.qppR  := exc_sm1.qppR;
  joint_ctrl1.qpppR := exc_sm1.qpppR;

  -- ctrl j1 --------------------------------------
  joint_ctrl1.qm    := TEMPO.y_q1;
  joint_ctrl1.qa    := TEMPO.y_q2;

  ampl1.u := joint_ctrl1.M (+ dist_esti1.M_dist);

  TEMPO.M1 := ampl1.y (- frict_m1.Mf_m);

INPUT_CONDITION    :  ...
OUTPUT_CONDITION   :  ...
END TEMPO;
```

Figure 9. Instantiation and Coupling of a Joint Controller

A COUPLED SYSTEM can contain several BASIC SYSTEMs and it can by itself contain COUPLED SYSTEMs which have INPUTs and OUTPUTs. With the help of this mechanism a modular and hierarchical structuring of large systems is supported by DSL.

For the system TEMPO an integral model of order 72 consisting of six amplifiers, six controllers, six estimators, and the mechanical model of the TEMPO can be obtained. For special analysis, additional models of friction and motor nonlinearities are taken into account.

This DSL description is the basis of several CACE tools described in e. g. [2] and [6], developed during the last ten years at the MLaP. Especially it is possible to get an exact linearized entire model. With this linearized model resp. the DSL description, linear analysis (e. g. calculation of frequency response and eigenvalues, linear simulation) and synthesis (e. g. Riccati design) can be performed. Furthermore, a multi-objective parameter optimization can be done with a tool called LINEX. LINEX can be used for the controller design (see chapter 7) or e. g. for the optimization of free construction parameters. A convenient tool named SIMEX [4] for nonlinear simulation of the realistic complex model is available for the verification of the designed controller (see Fig. 13).

7. Controller Design

The main requirement on the integrated system TEMPO was a high accuracy of less than 0.5 mm along the entire cartesian trajectory in connection with a very high velocity and acceleration of about 3 m/s and 15 m/s² (EE-mass ‹ 10 kg, work-space radius = 2 m). The designed controllers must have realizable coefficients and should work with a sampling time of about 1 kHz.

The problems to be overcome were the elasticities in gears and bearing and the high viscous and Coulomb friction in the gears. Therefore the gear vibrations will be actively damped. Disturbance estimators similar to the estimator proposed in [3] will be used to compensate for the time-varying friction moments at the motorside.

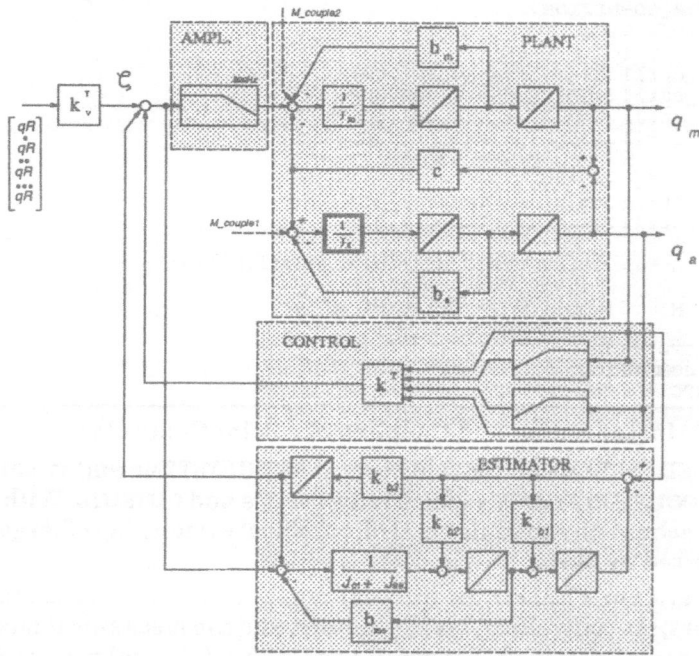

Figure 10. Model for Calculation of Feedforward Gains

Optimization of the feedback gains starts out from a predefined controller structure represented in Fig. 10. The loadside and motorside angle measurements and the corresponding velocities are fed back with constant gains. The velocities are obtained with the help of differentiation filters of about 200 Hz.

A linear disturbance estimator based on a rigid mass model is used for disturbance estimation and compensation. It has been proved that a simple integrator as disturbance model yields a very good behaviour. For the estimator a reusable BASIC SYSTEM TYPE is created that can be instantiated for every joint with certain time constants from which the resulting gains are calculated in the DSL model of the estimator by means of pole placement. The chosen time constants are of about $1/(2*\pi*60)$ sec.

The feedforwards (see also Fig. 8), i. e. here the coefficients of the reference position, velocity, acceleration and the derivation of the acceleration, are calculated according to a simplified plant shown in Fig. 10. In spite of omitting the coupling moments in the model, a very good result for the feedforward compensation can be achieved. The coupling moments and the gravity forces are separately calculated and compensated for. Their influence on the feedforwards are taken into account in such a way that stationary accuracy of the loadside is obtained.

$$G(s) \quad = \quad \frac{Q_a}{\zeta} = \frac{b_m s^m + \ldots + b_3 s^3 + b_2 s^2 + b_1 s + b_0}{a_n s^n + \ldots + a_3 s^3 + a_2 s^2 + a_1 s + a_0}$$

$$b_i \quad = \quad b_i(k_x, k_b, phys.para.plant\&estim., T_d, k_v)$$

$$a_i \quad = \quad a_i(k_x, k_b, phys.para.plant\&estim., T_d)$$

$$\zeta = [k_{vl}] Q_R, Q_R = \frac{1}{s} \quad \Rightarrow \lim_{s \to 0} s(1 - G(s)[k_{vl}])\frac{1}{s} = 0 \quad \Rightarrow k_{vl}$$

$$\zeta = [k_{vl} + k_{v2}s] Q_R, Q_R = \frac{1}{s^2} \quad \Rightarrow \lim_{s \to 0} s(1 - G(s)[k_{vl} + k_{v2}s])\frac{1}{s^2} = 0 \quad \Rightarrow k_{v2}$$

$$\ldots \quad \Rightarrow \quad k_{v3}, k_{v4}$$

Special emphasis is put on the fact that the gear elasticities, the amplifier dynamics, the differentiation filters, and varying load side inertia are taken into account. This means especially that the entire mass-matrix of the system is additionally to the linear parts of the control (order 30) are evaluated in every control cycle (1kHz). The diagonal elements are used for the feedforwards, and the nondiagonal elements, as mentioned above, are utilized for decoupling.

The necessary calculations are performed by means of a MAPLE program according to the formulas given above. They reflect the well-known fact that the coefficients of numerator and denominator of the transfer function of the closed control loop including the feedforwards should be equal up to the highest possible order. In order to control the link position the transfer function from the input ζ to the loadside position is considered.

Through the proposed method much smaller control errors can be achieved than by using the computed torque method, because stationary accuracy is obtained up to third order of differentiation of the load side coordinate, all with a model having elasticities in the gears.

The stability of the nonlinear system is ensured by the multi-objective parameter optimization tool LINEX, which is able to minimize several RMS values simultaneously. Due to the nonlinearities of the system a multi-model approach is used.

320

For this purpose the entire model is linearized in different positions to take into account the varying loadside inertia and a parallel model with the desired dynamics (edge frequency of 12 Hz, $\zeta = 0.8$) is created. Then the differences in the RMS values of the output of the parallel model and the outputs of the linearized models are minimized according to Fig. 11.

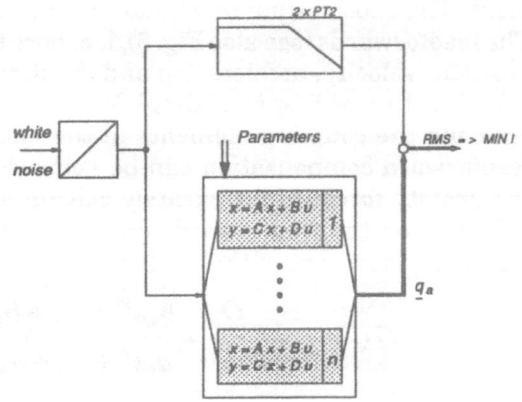

Figure 11. Optimization of Feedback Gains

An additional damping of poles which have e. g. a small residuum can be achieved by a pole area defined by the user. The optimizer tries to move all poles into this area during minimisation of the error criterion.

In this case, starting with a stable motor controller, 24 parameters are minimized step wise at a model of order 72.

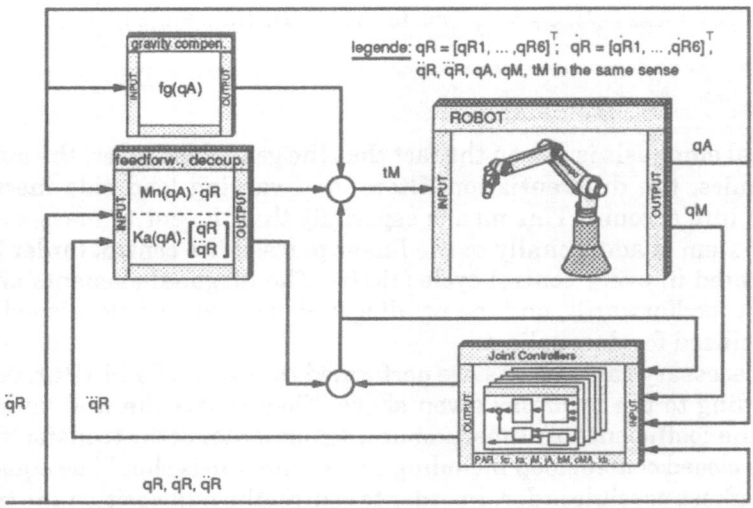

Figure 12. Coupled Entired System

8. Simulations and Measurements

Fig. 13 displays the simulation results for the second axis. Only this axis is rotated by 30 degrees with a maximal acceleration of 5.0 rad/s², a velocity of 0.6 rad/s. The upper part shows the motor, the link, and the reference velocity. In the lower part the control errors of the motor and the link can be seen. Due to the fact that a loadside position control was designed the link error is smaller than 0.5 mrad and during the phases with constant

velocity stationary accuracy is obtained. The error of the motor is an effect
of the gear deflection resulting from the gravity forces, which vary during
the movement.

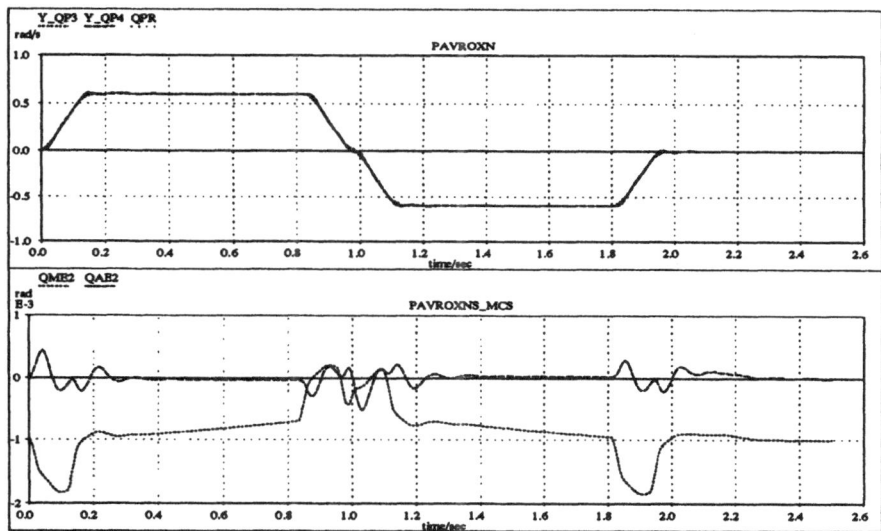

Figure 13. Simulation Results : Controller Error for the Second Axis

The next figure shows a comparison of simulated data on the left and
measured data on the right-hand side for the first axis. In the middle part
a very good correspondence between the simulated and the measured
control error can be seen.

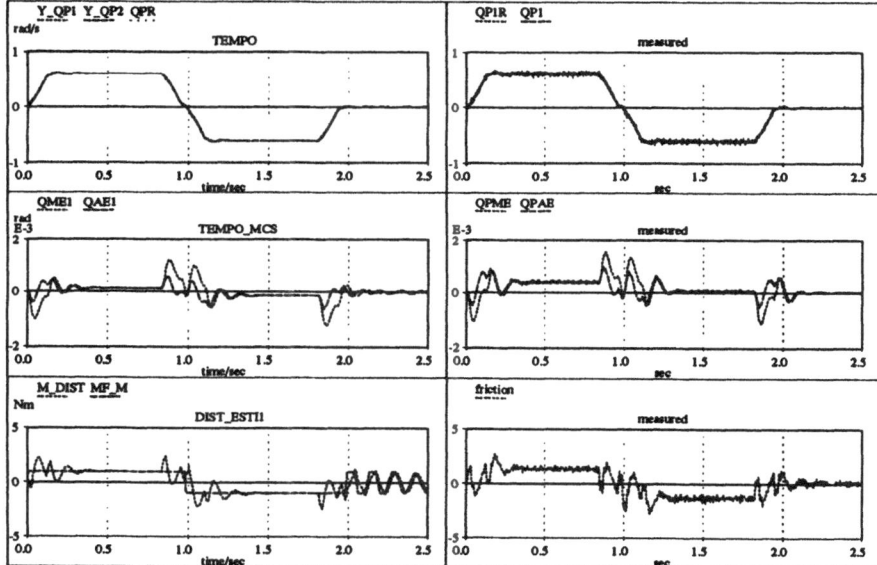

Figure 14. Comparison: Simulation and Measurement

In the lowest part the simulated and the measured friction is shown. The block-shaped line on the left-hand side shows the friction applied to the simulation model and the dashed line is the estimated friction which shows a very good conformity with the friction estimated on the real system. The transient behaviour during the acceleration phases results from the fact that during these phases the rigid mass model of the estimator does not match the mechanical system, which can approximately be regarded as a two-mass system.

9. Controller Implementation

The controller (of order 30 with 12 inputs and up to 10 outputs) is implemented on a transputer-based multi-processor hardware. A so-called **Transputer Interface Board (TIP)** developed at the MLaB (see[10]) can carry up to four transputers and is connected to commercial AD/DA converters, incremental encoders and digital I/O boards via a so-called Peripheral High Speed (PHS) bus. All components mentioned can reside e. g. in an IBM-compatible PC.

For this hardware a code generation tool (GENOCC, see [10]) is available which transforms the linear part of the controller that is given in DSL into OCCAM code. This code can after compilation and linking with a so-called **Peripheral Modul (PM)** and several modules for the feedforward control, directly be used to control the robot. In consideration of offsets and scales the PM connects the controller inputs and outputs with the hardware via names. So it is possible to test new controllers within a few minutes.

For online and offline path planning, i. e. calculation of the reference values, software packages are developed, especially for cartesian path calculation, inverse transformation, and a fine interpolation. A speciality of the path planning tool is that not only the reference position but also the velocity and the acceleration are analytically calculated.

A comfortable graphical user interface on the host computer allows the downloading of motion programs (e. g. straight lines, splines, optimized velocity profiles) and a teach-in programming. A very important feature for the control engineer is that an online monitoring is possible, i. e. that all controller inputs and outputs and e. g.control errors can directly be visualized (see Fig. 14).

10. Conclusions

With the help of the realization of a new robot system it could be shown that the mechatronic idea, i. e. the integration of mechanics, electrics/electronics and computer science, leads to systems with high performance, flexibility and modularity. From our point of view the integrational factor seems to be system resp. control theory, which has an influence not only on the computational parts of the system but also on the mechanical design. In this paper a complete design cycle and the corresponding software tools are outlined in an exemplary way. For the generation of the symbolic dynamical equations of elastic robots a user-friendly tool was introduced. After several simplifications of the equations a state-space model can be

obtained. Identification of physical parameters has been proved to be effective in the frequency domain. The augmentation of the mechanical model by models of amplifiers, controllers, estimators, and measurement filters is supported by a unified model description in state space.
The augmented nonlinear model of order 72 is used for the design of a feedback control via a multi-model approach. The feedforward parts of the controller are calculated analytically and take into account, among other things, the nonlinearities of the plant (varying inertias, coupling moments), the elasticities of the gears, and the dynamics of the estimator. The reliability of the model and the justification of the introduced controller structure are verified by the high accuracy of the controllers and the good correspondence between measurements and simulations, that were achieved with these controllers.

11. Acknowledgement

This project was sponsored by the Ministry of Science and Research of North Rhine-Westphalia, Germany.

12. References

[1] H. Henrichfreise. Aktive Schwingungsdämpfung an einem elastischen Knickarmroboter. Dissertation Universität-GH-Paderborn, 1989; Fortschritte der Robotik 1, Vieweg, 1989.

[2] R. Kasper, J. Lückel, K.-P. Jäker, J. Schröer. CACE Tool for Multi-Input, Multi-Output Systems Using a New Vector Optimization Method. Int. J. Control 51, 5(1990), pp. 963-993.

[3] R. Neumann, W. Moritz. Observer-Based Joint Controller Design of a Robot for Space Operation. Proceedings of the 8th CISM-IFToMM Symposium on Theory and Practice of Robot and Manipulation Ro.man.sy '90. Krakau, Poland, 2.-6.7.1990, pp. 496-507.

[4] U. Lefarth. Modelling and simulation of mechatronic systems. Sensor '93. 6. Internationale Fachmesse mit Kongreß für Sensorik & Systemtechnik. Nürnberg, Deutschland, 11.-14. Oktober 1993.

[5] J. Schröer. A Short Description of a Model Compiler/Interpreter for Supporting Simulation and Optimization of Nonlinear and Linearized Dynamic Systems. CADCS 91. 5th IFAC/IMACS Symposium on Computer Aided Design in Control Systems. Swansea, Wales, 1991.

[6] K.-P. Jäker, P. Klingebiel, U. Lefarth, J. Lückel, J. Richert, R. Rutz. Tool Integration by way of a Computer-Aided Mechatronic Laboratory (CAMeL), in: CADCS 91. 5th IFAC/IMACS Symposium on Computer Aided Design in Control Systems, Swansea, Wales, 15.- 17. 07. 1991.

[7] W. Moritz, R. Neumann, H. Schütte. Control of Elastic Robots Using Mechatronic Tools. Harmonic Drive International Symposium 1991, Hotaka, Nagano, Japan, 23. - 24. 5. 1991.

[8] R. Neumann, W. Moritz. Modelling and Robust Joint Controller Design of an Elastic Robot for Space Operation, in: Proceedings of the 2nd German-Polish Workshop on "Dynamical Problems in Mechanical Systems", Paderborn, 10. - 17. 03. 1991.

[9] H. Schütte, W. Moritz, R. Neumann. Analytical Calculation of the Feedforwards up to Their Second Derivatives and Realization of an Optimal Spatial Spline Trajectory for a 6-DOF Robot, in: Preprints of the IFAC Symposium on Robot Control SYROCO '91, Wien, Österreich, 16. - 18. 9. 1991.

[10] U. Honekamp, T. Gaedtke. Modulkonzept zur Echtzeitverarbeitung, Transputernetze in der Mechatronik und Regelungstechnik. Elektronik 6/1993, pp. 95-101.

13. Appendix

```
-- SYSTEMNAME          : rob_j3
-- AUTHOR              : h. schuette
-- DATE                : 15.07.93
-- VERSION             : 1.0

BASIC SYSTEM rob_j3 (
    PARAMETER : g       :=    9.81        "kgm/s^2",
                ue1     :=  -121.0        "-",
                        :
                JM1z    :=  0.00288015    kgm^2",
                        :
                JA1z    :=  0.210         "kgm^2",
                JA2x    :=  0.0345        "m^4",
                        :
                mM3     :=  11.55         "kg",
                        :
                1A2     :=  0.750         "m",
                1A2s    :=  0.324         "m",
                        :
                cd1     :=  420500.0      "Nm/rad",
                cd2     :=  405910.0      "Nm/rad",
                cd3     :=   87111.0      "Nm/rad",
                bm1     :=  153.70        "Nm/(rad/s)",
                bm2     :=   97.90        "Nm/(rad/s)",
                bm3     :=   65.10        "Nm/(rad/s)",
    INPUT       : M1 "Nm", ... M3 "Nm",
    OUTPUT      : y_q1 "rad",  y_qp1 "rad/s", ... y_qp6 "rad/s") IS

STATE : q1 :=0.0 "rad", qp1 :=0.0 "rad/s",.. qp6 :=0.0 "rad/s";

AUXILIAR :
-- abbreviations ----------------------------------------------
s4    := sin(q4);     c4   := cos(q4);
s6    := sin(q6);     c6   := cos(q6);
s46   := sin(q4+q6);  c46  := cos(q4+q6);

-- constants -------------------------------------------------
c_1   := -1A2s*mA2*g-1A2*mM3*g   ;
c_2   := -1A2*mA3*g   ;
c_3   := -mA3*g*lA3s   ;
c_4   := ue1**2*JM1z   ;
c_5   := (mA3+mM3)*1A2**2-JA2x+JA2y+1A2s**2*mA2   ;
c_6   := 2*mA3*1A3s*lA2   ;
c_7   := mA3*1A3s**2+JA3y-JA3x   ;
c_8   := JA1z+JA3x+JA2x   ;
c_9   := ue2**2*JM2z   ;
c_10:=(mA3+mM3)*1A2**2+1A2s**2*mA2+JA2z+JA3z+JM3z+mA3*1A3s**2;
c_11  := ue3*JM3z   ;
c_12  := ue3**2*JM3z   ;
c_13  := mA3*1A3s*lA2   ;
c_14  := mA3*1A3s**2+JA3z   ;
c_15  := -2*c_5   ;
c_16  := -2*c_7   ;

-- massmatrix -----------------------------------------------
m1_1 :=   c_4   ;
m2_2 :=   c_8+c_5*c4**2+c_7*c46**2+c_6*c4*c46   ;
m3_3 :=   c_9   ;
m4_4 :=   c_10+c_6*c6   ;
m4_5 :=   c_11   ;
m5_5 :=   c_12   ;
m4_6 :=   c_14+c_13*c6   ;
m6_6 :=   c_14   ;

-- CHOLESKY decomposition M = R*R' (R*b=f and R'*qpp= b) ------
R1_1 :=   m1_1**(1/2)   ;
R2_2 :=   m2_2**(1/2)   ;
R3_3 :=   m3_3**(1/2)   ;
```

```
    R4_4  :=    m4_4**(1/2)    ;
    R5_4  :=    m4_5/R4_4    ;
    R5_5  :=    (m5_5-R5_4**2)**(1/2)    ;
    R6_4  :=    m4_6/R4_4    ;
    R6_5  :=    -R5_4*R6_4/R5_5    ;
    R6_6  :=    (m6_6-R6_4**2-R6_5**2)**(1/2)    ;

-- stiffnessvector ------- dampingvector --------------------
    fQ_1  :=  cd1*q1-cd1*q2;    fQ_1  :=  b1*qp1-b1*qp2 + bm1*qp1;
    fQ_2  :=  -cd1*q1+cd1*q2;    fQ_2  :=  -b1*qp1+b1*qp2    ;
    fQ_3  :=  cd2*q3-cd2*q4;    fQ_3  :=  b2*qp3-b2*qp4 + bm2*qp3;
    fQ_4  :=  -cd2*q3+cd2*q4;    fQ_4  :=  -b2*qp3+b2*qp4    ;
    fQ_5  :=  cd3*q5-cd3*q6;    fQ_5  :=  b3*qp5-b3*qp6 + bm3*qp5;
    fQ_6  :=  -cd3*q5+cd3*q6;    fQ_6  :=  -b3*qp5+b3*qp6    ;

-- centrifugal and coriolis forces ------------------------
a2_4  := c_16*s46*c46*qp2+(c_15*s4*c4+(-s4*c46-c4*s46)*c_6)*qp2;
a2_6  := c_16*s46*c46*qp2-qp2*c_6*c4*s46    ;
a4_6  := -qp4*c_6*s6-c_13*s6*qp6    ;
a6_6  := -qp4*c_13*s6    ;

    fh_2  := a2_4*qp4+a2_6*qp6    ;
    fh_4  := -1/2*a2_4*qp2+a4_6*qp6    ;
    fh_6  := -1/2*a2_6*qp2-1/2*a4_6*qp4+1/2*a6_6*qp6    ;

-- generalized forces ---------------------------------------
    fgenkr_1    :=    ue1*M1    ;
    fgenkr_3    :=    ue2*M2    ;
    fgenkr_4    :=    c_2*c6*c46+c_1*c4+c_3*c46+c_2*s46*s6    ;
    fgenkr_5    :=    ue3*M3    ;
    fgenkr_6    :=    c_3*c46    ;

-- right side -----------------------------------------------
    f_1 := fgenkr_1 - fQ_1 - fP_1 - fh_1;
        :
    f_6 := fgenkr_6 - fQ_6 - fP_6 - fh_6;

-- forward recursion ----------------------------------------
    z_1 :=    f_1/R1_1    ;
    z_2 :=    f_2/R2_2    ;
    z_3 :=    f_3/R3_3    ;
    z_4 :=    f_4/R4_4    ;
    z_5 :=    (f_5-R5_4*z_4)/R5_5    ;
    z_6 :=    (f_6-R6_4*z_4-R6_5*z_5)/R6_6    ;

-- backward recursion ---------------------------------------
    x_6 :=    z_6/R6_6    ;
    x_5 :=    (z_5-R6_5*x_6)/R5_5    ;
    x_4 :=    (z_4-R5_4*x_5-R6_4*x_6)/R4_4    ;
    x_3 :=    z_3/R3_3    ;
    x_2 :=    z_2/R2_2    ;
    x_1 :=    z_1/R1_1    ;

STATE_EQUATION :
    q1'  := qp1;  qp1' := x_1;  ...  q6'  := qp6;  qp6' := x_6;

OUTPUT_EQUATION :
    y_q1  := q1;  y_qp1 := qp1;  ...  y_q6  := q6;  y_qp6 := qp6;
END rob_j3;
```

DSL Model : Three-Axis Revolute-Joint Robot with Gear Elasticities

[1] MAPLE © by Waterloo Maple Publishing
     Univ. of Waterloo, Waterloo, Ontario, Canada N2L 3G1
[2] MATLAB® is registered trademark of Math Works, Inc.

An Experimental Study of Nonlinear Stiffness, Hysteresis, and Friction Effects in Robot Joints with Harmonic Drives and Torque Sensors

Kircanski N., Goldenberg A.A., and Jia, S.

Department of Mechanical Engineering
University of Toronto
Toronto, Canada

Abstract— Despite widespread industrial applications of harmonic drives, the source of some elastokinetic phenomena causing internal instability has not been fully addressed thus far. This paper describes a new phenomenon named "torque-transmission paradox" related to the inability of transmitting the input motor torque to the output link. Also, we describe experiments and mathematical models related to other physical phenomena, such as nonlinear stiffness, hysteresis and soft-windup. The goal of our modeling strategy was not in developing very precise and possibly complicated model, but to distill an appropriate model that can be easily used by control engineers to improve joint behavior. To visualize the developed model, equivalent mechanical and electrical schemes of the joint are introduced. Finally, a simple and reliable estimation procedure has been established not only for obtaining the parameters, but also for justifying the integrity of the proposed model.

1. Introduction

Understanding inherent properties of robot drive systems is of paramount importance for achieving smooth motion at low speed, as well as compliant behavior in contact situations. A crucial component of a robot joint is the gear that must embody many features such as zero-backlash, low-weight, low friction losses, compact size, and high torque capacity. Developed over 30 years ago, harmonic drives have been used primarily in aerospace and military applications before becoming widely accepted for industrial applications as well. They have been implemented extensively, over the last decade, in robotics due to their unique and inherent properties, first of all virtually zero-backlash.

In spite of its widespread acceptance in industry, the harmonic drive has not attracted complementary attention of researchers, with the exception of those in USSR [1]. Most of the available literature covers particular aspects of harmonic drives, like kinematics and geometry of tooth engagement [2], kinematic error analysis, harmonic analysis of vibrations [3], etc.

An attempt to explain the regular occurrence of periodic position errors can be found in [4]. The error is attributed to the geometry of the wave generator. A different approach based on parametric curves to model flexspline deformation reported in [5] captures the first harmonic of the experimentally measured gear error. Nye and Kraml found that the first nine harmonics in gear error can be attributed to the lack of concentricity, misalignments, out-of-roundness and the tooth placement errors, [3]. Their experiments with "S" tooth profile demonstrated well-behaved gear error free of higher harmonics. Apparently, the controversy about the source and genesis of harmonic drive errors has not been resolved thus far.

The transmission compliance mechanism is another controversial issue regarding the primary source of flexibility. Hashimoto at al., [6], studied the deformations of the flexspline under torque-loaded conditions using finite-element stress analysis. Some other researchers claim that the flexspline deformation is not the dominant source of elasticity, but the gear-tooth engagement-zone which determines the overall compliance of a harmonic drive as well as its torque capacity [7]. Finally, the wave-generator deformation is considered as essential and dominant by Margulis, et al., [1]. Clearly, a reliable finite-element analysis of all components of a harmonic drive combined with experimental verification is still missing to resolve this problem.

Numerous contributions to intuitive understanding and analytical description of harmonic-drive-based robot joints have been reported by Tuttle and Seering, [8, 9]. Their experimental observations show that the velocity response to step commands in motor current are not only contaminated by serious vibration, but also by unpredictable jumps. The velocity response observations were used to guide the development of a series of models of increasing complexity to describe the harmonic-drive behavior. The most complex model involves kinematic error constraints, a nonlinear spring between the flexspline base and the teeth, and the gear-tooth interface with frictional losses.

The preceding discussion shows that despite the 30 years history of harmonic drive technology some physical phenomena have been discovered only recently. In this paper we describe a new phenomenon [11], we came across upon the experimentation with the newly designed IRIS multi-manipulator setup [12]. We assigned the name "torque-transmission paradox" to this phenomenon. In contrast to the unrestrained motion experiments described in [8], most of our experiments relate to the drive system behavior in contact with a stiff environment. When the joint is unrestrained, we found that the intrinsic joint dynamics including nonlinear stiffness, hysteresis and some friction terms are covered-up and almost not observable due to the fact that the operating point can not move easily across the whole range of possible states. It happened that IRIS Facility joints do not exhibit the turbulent behavior described by Tuttle and Seering in [8, 9]. The torque transmission paradox is described in Section 2.

In this paper, a simple soft-windup model in addition to the nonlinear stiffness and hysteresis model is introduced. A moderate complexity of the model makes it applicable for control purpose. The model distinguishes two subsystems: a fast one with high-frequency modes reflecting the wave-generator and rotor dynamics, and a slow one influenced by the output link inertia properties. These two subsystems are coupled through the elastic deformation of the

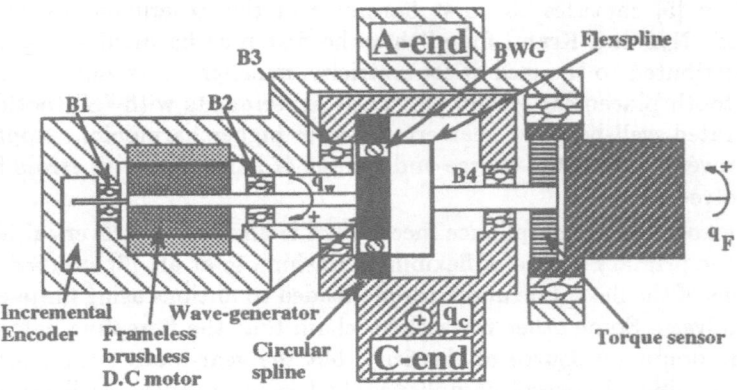

Figure 1. Joint structure

flexspline. In Section 4, equivalent mechanical and electrical representations of a harmonic-drive-based joint are introduced. Section 5 presents a systematic procedure based on bias-compensating least squares estimation algorithm (Ref. [13]).

2. Experimental Setup

This section provides a short description of the experimental setup - IRIS Facility - developed at the Robotics and Automation Laboratory at the University of Toronto [12]. The facility is a versatile, reconfigurable and expandable setup composed of several robot arms that can be easily disassembled and reassembled to assume multitude of configurations.

The constituent element of the system is the *joint module* that has its own input and output link. Each module is equipped with a brushless DC motor coupled with harmonic drive gear, and instrumented with a position and torque sensor (Fig. 1). The components of the harmonic drive: wave-generator (WG), flexspline (FS) and circular spline (CS) are placed concentrically around the rotor axis. The torque sensor is designed specifically for IRIS joints to have minimum weight and much higher stiffness coefficient than that of the harmonic drive itself. The torque sensor electronics including the instrumental amplifier and signal conditioner is placed close to the torque sensor inside the joint. An external optical encoder can be mounted to measure output joint angle.

3. Experimental results

The most promising way to highlight the intrinsic joint phenomena and to distill an appropriate modeling strategy, including an automated parameter estimation procedure, appears to be through a series of experiments, which incorporate both restrained and unrestrained joint motion. It will be shown that the experiments that assume the output link to be in full contact with an environment of known stiffness are much more revealing in comparison to that

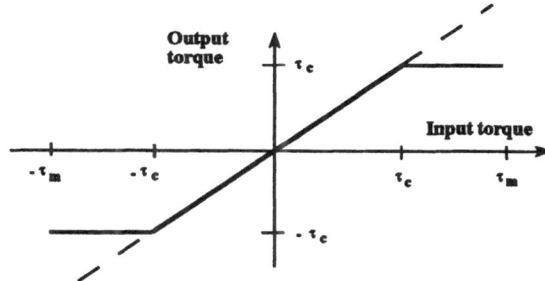

Figure 2. Torque transmission paradox - TTP

allowing the unrestrained motion of the link.

3.1. Torque-transmission paradox

We assume that the output link is in full contact with a stiff environment under a constant driving torque generated at the motor side. The motor torque is opposed by the action of a constant force on the output link tip. The input and output torque must have the same value, except for some small amount due to Coulomb friction inside of the joint.

Theoretically, the input-output torque characteristic is a straight line in the range $(-\tau_m, \tau_m)$, where τ_m is the maximum torque that can be delivered by the given DC motor and amplifier. Nevertheless, the measured input-output torque characteristics saturates in the region $\tau_c \leq |\tau| \leq \tau_m$, where τ_c stands for a "critical torque" which is in the range 0 to τ_m, (Fig. 2). Upon reaching the critical torque, the input torque is being lost, which is unexplainable from the static condition of torque equivalence. The existence of this paradox is experimentally verified by measuring the motor current, force at the load-cell, and the torque of the joint torque sensor, (Fig. 3). In this experiment, the motor current is increased very slowly towards the saturation of the motor current-mode amplifier. Notice that the torque measured by the joint torque sensor and the external load cell are nearly equal and saturate at about 12 Nm, while the input torque extends up to nearly 15 Nm and than slowly decreases and saturates at about 13.3 Nm as a consequence of current limiting by the amplifier.

The explanation of this paradox is as follows: Upon reaching a certain value of deflection, the flexspline stiffness increases giving rise to small vibrations neither visible by a position sensor nor the external load cell. The oscillations are measurable only by the highly sensitive joint torque sensor (see the corresponding signal in Fig. 3). The energy dissipation increases in these regions of resonance and hinders the increase of joint torque. In conclusion, the virtually static system is actually converted into a dynamic system upon reaching the critical torque, or, equivalently, the critical value of motor current.

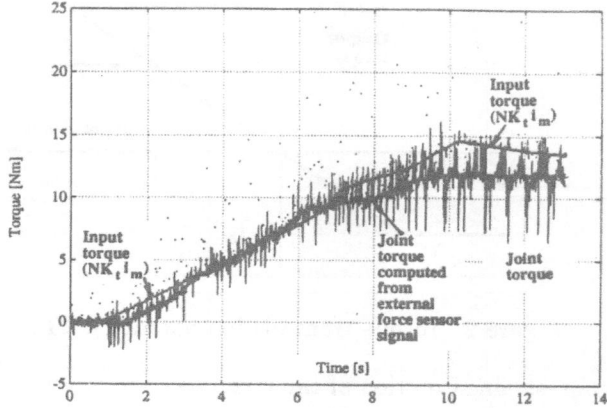

Figure 3. Experimental evidence of TTP

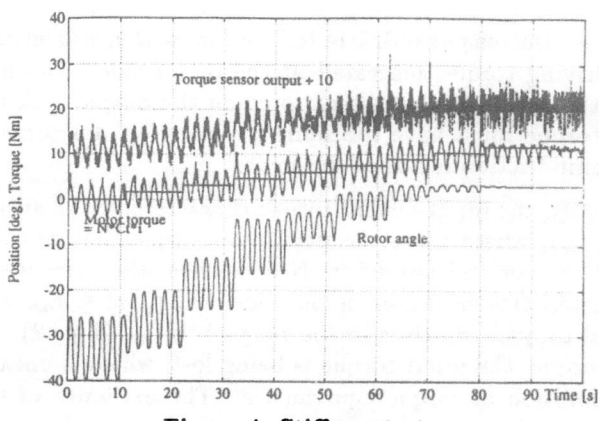

Figure 4. Stiffness test

3.2. Soft-windup, nonlinear stiffness and friction characteristics

There are three physical phenomena related to the behavior of the flexspline: nonlinear stiffness, quasi-backlash due to soft windup effect, and hysteresis ([8, 10]). All these phenomena can be captured by the use of the same experimental setup as the one described above.

Suppose that the output link faces a very stiff environment and that the motor current follows a low-frequency sinusoidal function with different mean values and amplitudes (Fig. 4). The rotor follows the input signal due to rotational compliance of the joint. The joint torque sensor response can be used to measure the torque-torsion coefficient around an operating point determined by the motor torque mean value. The joint stiffness coefficient as a function of applied torque is shown in Fig. 5. The stiffness is evidently lower at lower torques as a consequence of the soft windup effect.

We used the same experiment to observe the hysteresis effect due to Coulomb friction. Changing the motor current slowly to follow a sinusoidal function with

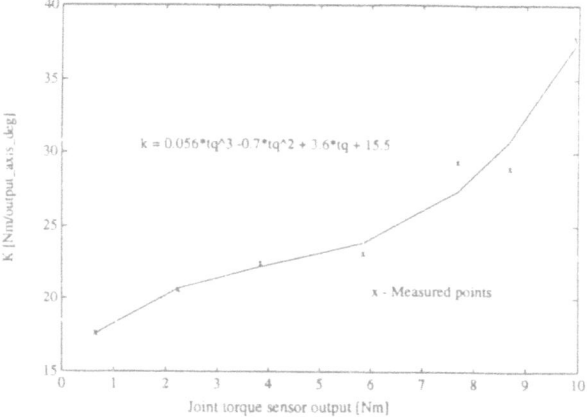

Figure 5. Joint stiffness vs. torque

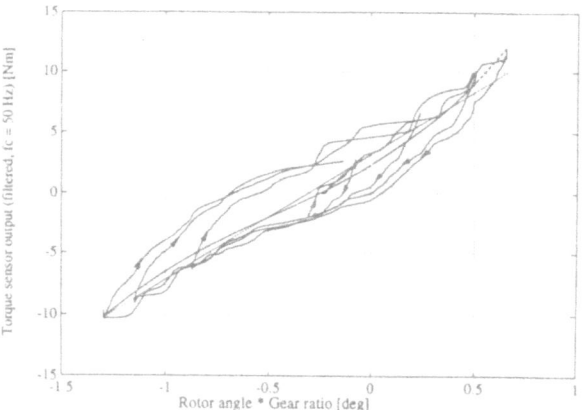

Figure 6. Hysteresis effect

zero mean value and variable magnitude, and recording the torque sensor output as a function of angular displacement, we obtained the diagram shown in Fig. 6. We see that after applying a certain torque pulse at the motor side, the torsional angle will increase, but will not return back to the initial value. This is the well-known hysteresis effect that can be fully characterized by the diagram shown in Fig. 6. The Coulomb friction term can be easily obtained from the hysteresis plot since it is equal to the difference between the input motor torque and the torque sensor output. Fig. 7 shows the Coulomb friction torque as a function of torque sensor output. As expected, the friction torque depends mostly upon the direction of rotation, and is nearly independent of the velocity.

3.3. Frequency response

With harmonic-drive joints the frequency response experiments can be very dangerous because of internal resonance effects that can generate unpredictable jumps in velocity (as those reported by Tuttle, [8, 9]), and possibly damage the

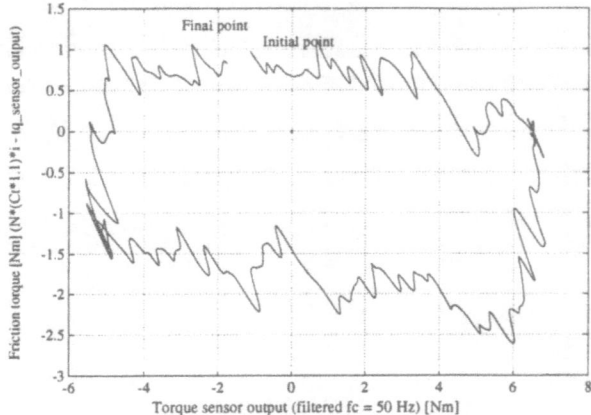

Figure 7. Coulomb friction

gear teeth. In our experiments the sinusoidal joint torque signal was corrupted by extremely sharp pulses of high magnitudes due to the elastic coupling between high-frequency "internal" and low-frequency "external" dynamics. In contrast to the pathological instability effects when unrestrained, the behavior of the joint in full contact with a stiff environment is completely predictable under sinusoidal input regardless of the frequency and amplitude of the excitation. While measuring the joint torque signal, a sinusoidal motor current with a sweeping frequency ranging from 0 to 300 Hz was applied. We obtained the resonant peak at about 30 Hz, while the slope of the magnitude characteristics was 40 db/decade. Apparently, the joint behaves as a second order system having two complex conjugate poles.

4. Mathematical model

The goal of this work is to derive a general model of minimum complexity to allow control engineers to use the model not only for evaluation of a control strategy but also as a basis for control synthesis. The experimental observations provided in the previous section will be used for model development and verification.

4.1. Model equations

The basic components of the IRIS joint (Fig. 1) are DC motor, harmonic-drive gear and the torque sensor. The entire system including all these components can be divided into four subsystems: motor-amplifier subsystem, wave-generator subsystem, flexspline subsystem and circular-spline subsystem. The last three subsystems are coupled through the harmonic-drive.

4.1.1. Motor-amplifier subsystem

The frameless DC motor in partnership with the PWM current-mode amplifier can be modelled as a "torque source" with a limiting capability due to the action of back e.m.f. and the current-limiter of the amplifier. The relation between the control output u, motor current i_m and rotor velocity $\dot{q}_w$ can be established easily as

$$i_m = k_s u, \qquad |i_m| \leq \min(I_s, (U_s - K_b|\dot{q}_w|)/r_m) \qquad (1)$$
$$\tau_m = K_t i_m \qquad (2)$$

where k_s is the amplifier gain, I_s is the upper-bound of the current defined by the amplifier electronics, U_s is the maximum output voltage of the amplifier, K_b is the back e.m.f. constant of the motor, r_m is the motor resistance, and K_t is the motor torque constant.

4.1.2. Wave-generator subsystem

The DC motor applies the torque τ_m to the input inertia which is composed of the motor armature and the harmonic-drive wave-generator. The torque driving the wave generator which is ellipsoidal in shape and surrounded by a ball bearing will be considered as the subsystem output. The following model involves Coulomb and viscous friction at the bearings **B1**, **B2**, and the wave-generator ball bearing **Bwg** (Fig. 1):

$$\tau_m = J_{mw}\ddot{q}_w + \tilde{B}_m(\dot{q}_w) + \tilde{B}_w(\dot{q}_w + \dot{q}_f) + \tau_w \qquad (3)$$

where q_f is the flexspline angular velocity with the reference direction opposite to the wave-generator velocity, J_{mw} is the combined inertia of the rotor, shaft and the wave-generator, $\tilde{B}_w(\dot{q}_w)$ is the total friction torque generated at bearings **B1** and **B2**, $\tilde{B}_w(\dot{q}_w + \dot{q}_f)$ is the friction term of the wave-generator bearing **Bwg**, and τ_w is the torque exerted on the wave-generator.

The friction term $\tilde{B}(\dot{q})$ involves kinetic (Coulomb), viscous and static friction:

$$\tilde{B}(\dot{q}) = B\dot{q} + (B^+, B^-)sgn(\dot{q}) \qquad (4)$$

where B is viscous coefficient, B^+ and B^- are positive-direction and negative-direction Coulomb friction torques. The last term in (4) is equal to B^+ for $\dot{q} > 0$, while it is equal B^- for $\dot{q} < 0$. This expression is consistent with the experimental results reported in the previous section (Fig. 7).

If the flexspline is fixed to ground (input link) the velocity $\dot{q}_f$ is negligible comparing to $\dot{q}_w$. If the flexspline is free to move, $\dot{q}_f = \dot{q}_w/N$ is still much lower than $\dot{q}_w$ since the typical value of the gear ratio N is 50, 100, etc. Taking this into account and the analysis above, the wave-generator subsystem model becomes:

$$\tau_m = J_{mw}\ddot{q}_w + \tilde{B}_{mw}(\dot{q}_w) + \tau_w \qquad (5)$$
$$\tilde{B}_{mw}(\dot{q}_w) = B_{mw}\dot{q}_w + (B_{mw}^+, B_{mw}^-)sgn(\dot{q}_w) \qquad (6)$$

4.1.3. Flexspline subsystem

The experimental observations reveal three physical phenomena related to the behavior of the flexspline: nonlinear stiffness, hysteresis and quasi-backlash due

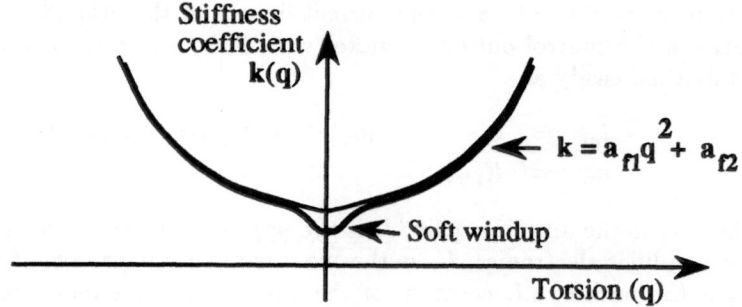

Figure 8. Stiffness coefficient

to soft windup effect. We found that the following simple model adequately reflects the real behavior of the subsystem (see Figs. 6, 7):

$$\tau_f = \tilde{k}_f(q) + \tilde{B}_f(\dot{q}) \tag{7}$$

where τ_f is the torque transmitted to the flexspline, q is the flexspline torsion, and $\tilde{B}_f$ is the friction operator defined analogously to that given in (4). The term $\tilde{B}_f(\dot{q})$ is introduced to match the hysteresis curve shown in Fig. 6.

A careful analysis of the experimental data summarized in Fig. 5 reveals that the stiffness coefficient of the harmonic drive, $k(q)$, can be modeled as a second-order polynomial in torsion q, except for the region close to zero, where the soft windup effectively decreases the joint stiffness (Fig. 8):

$$k(q) = a_{f1}q^2 + a_{f2} + K_{sw}(q) \tag{8}$$

where a_{f1} and a_{f2} are coefficients, and $K_{sw}(q)$ is the soft windup correction factor that can be modeled as a saddle-shaped function:

$$K_{sw}(q) = -k_{sw}e^{-a_{sw}q^2} \tag{9}$$

with k_{sw} and a_{sw} being the parameters which can be determined from the experimental plot in Fig. 5. Consequently, the first term on the right side of equation (7) can be presented as

$$\tilde{k}_f(q) = a_{f1}q^3 + a_{f2}q + Ksw(q)q \tag{10}$$

The torsion q is defined as $q = q_f - q_\tau$, where q_f is the angular displacement of the flexspline at the tooth engagement cross-section, while q_τ is the angular displacement of the torque sensor attached to the flexspline base (Fig. 1). The compact form of the flexspline model is thus

$$\tau_f = \tilde{k}_f(q_f - q_\tau) + \tilde{B}_f(\dot{q}_f - \dot{q}_\tau) \tag{11}$$

The torque sensor acts as a pure spring with the stiffness coefficient k_τ much larger than that of the harmonic drive. The torque τ_f is measured by the torque sensor, i.e.

$$\tau_f = \tau = k_\tau(q_\tau - q_F) \tag{12}$$

where q_F is the angular displacement of the F-end port of the joint (1). This angular displacement is zero if the F-end is grounded. The stiffness coefficient k_τ is usually determined by using before integrating the sensor with other joint components.

4.1.4. Circular-spline subsystem

The output link to the joint can be attached either to the flexspline or to the circular spline. In the first case the circular spline port is grounded, while in the second case the flexspline is connected to the input link. Therefore, one of the following two equations will be used to describe C-end/F-end link dynamics:

$$\tau_c = J_c \ddot{q}_c + \tilde{B}_c(\dot{q}_c) + \tau_{Cl} \tag{13}$$

$$\tau = J_F \ddot{q}_F + \tilde{B}_F(\dot{q}_F) + \tau_{Fl} \tag{14}$$

where q_c is the C-end angular displacement, τ_{Cl} is the C-end load torque, while τ_c is the driving torque transmitted to the circular spline at the engagement zone cross-section. The parameters J_c and those in $\tilde{B}_c$ are inertia and friction parameters at the C-end. The variables and parameters on the right hand side of (14) are defined analogously.

4.1.5. Harmonic-drive

Having developed the dynamic equations of all subsystems, the entire system model can be integrated provided that the harmonic-drive constitutive relationships are known. The harmonic drive consists of 3 parts: (1) the wave-generator - a thin raced ball bearing assembly fitted into an elliptical plug, (2) the flexspline - a thin cylinder cup with external teeth on slightly smaller pitch diameter than the circular spline, fitted over and elastically deflected by the wave-generator, and (3) the circular spline - a rigid ring with internal teeth, engaging the teeth of the flexspline across the major axis of the wave-generator. The flexspline is usually having two fewer teeth than the circular spline, and thus each full turn of the wave-generator moves the flexspline two teeth in the opposite direction relative to the circular spline. This directly implies the kinematic constraint

$$q_w = N q_f + (N+1) q_c \tag{15}$$

where N is the gear ratio defined as $N = N_t/2$, where N_t is the number of teeth on the flexspline outer circumference. It is noteworthy that there is only one kinematic constraint relating coordinates at the three ports. For this reason one of the two output ports is often fixed to the ground. The second important observation is that the gear error is not represented by Eq. 15. One way to introduce this error is given in [8, 9]. The gear error can be compensated electronically by modifying the motor current at the lowest control level in such a way to effect an error cancellation, [3]. For this reason, we shall not focus on this topic.

Applying the power conservation law on the 3-port device described by kinematic constraint (15), or more precisely on its derivative form:

$$\dot{q}_w = N \dot{q}_f + (N+1) \dot{q}_c \tag{16}$$

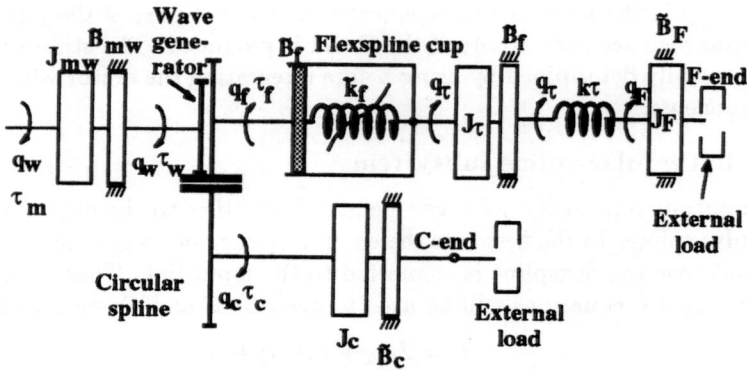

Figure 9. Equivalent mechanical scheme

yields the input/output torque relationships (see Ref. [8, 9])

$$\tau_f = N\tau_w \qquad (17)$$

$$\tau_c = (N+1)\tau_w \qquad (18)$$

For building an intuitive understanding of the harmonic-drive operation it is necessary to consider tooth-level geometry and interaction forces, [2, 8, 9]. We studied the distribution of forces interacting between the gear teeth and established a framework for understanding the torque transmission relations without introducing simplifications in the system. This part of our research is presented in [11].

4.2. Equivalent mechanical and electrical schemes

The models introduced in the previous section can be integrated into a single model describing kinematic and dynamic behaviour of the robot joint. The following subsystem models have been introduced: (i) PWM current-mode amplifier (1); (ii) DC motor (2); (iii) wave-generator subsystem (5) and (6); (iv) flexspline subsystem (11), (12) and (14); (v) circular spline subsystem (13); and (vi) harmonic drive constraints (15) - (18). All these equations constitute the joint model. An illustrative presentation of these equations is given in Fig. 9 which represents an equivalent mechanical scheme of the joint. Notice that the harmonic drive is represented by an ideal 3-port device, while all physical phenomena like nonlinear stiffness, hysteresis, and friction effects are extracted from the harmonic drive and attributed to the components outside this idealized transmission device. This is a common practice both in mechanical and electrical engineering.

The mechanical scheme in Fig. 9 can now be easily transformed into an equivalent electrical scheme, Fig. 10. It is noteworthy that there is a full analogy between a harmonic drive and a voltage transformer with two secondary windings. Both are 3-port devices, exhibit hysteresis and have internal losses. Furthermore, both can be represented by ideal substitutes, while all losses can

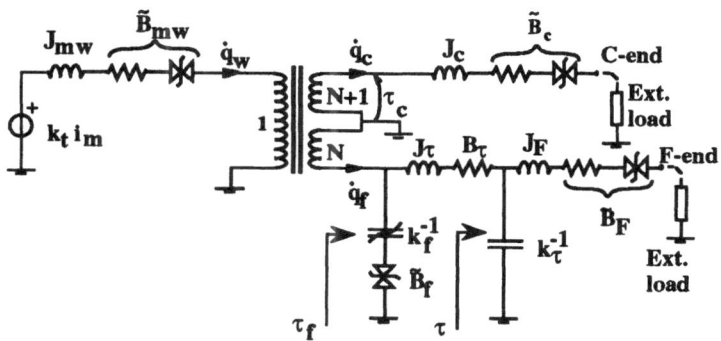

Figure 10. Equivalent electrical scheme

be attributed to external elements. Similarly, the Coulomb friction can be represented by a couple of Zener diodes, while the nonlinear spring corresponds to a nonlinear capacitor.

In both schemes shown in Figs. 9, and 10 the F- and C-ends are independently loaded by external loads. In practice, either F- or C-end is grounded. This simplifies both the equations and equivalent schemes. For example, suppose that the F-end is fixed to the input link. In this case, in addition to the PWM amplifier and DC motor equations, the model reduces to the following three equations:

$$\tau_m \;=\; J_{mw}\ddot{q}_w + \tilde{B}_{mw}(\dot{q}_w) + \frac{1}{N}\tau \tag{19}$$

$$\tau \;=\; \frac{N}{N+1}g_c(\ddot{q}_c, \dot{q}_c, q_c, x) \tag{20}$$

$$\tau \;=\; \tilde{k}_f(\frac{1}{N}q_w - N'q_c) + \tilde{B}_f(\frac{1}{N}\dot{q}_w - N'\dot{q}_c) \tag{21}$$

where $g_c(\ddot{q}_c, \dot{q}_c, q_c, x)$ stands for C-end dynamics, and $N' = (N+1)/N$. These equations will be used as a basis for the parameter estimation procedure described in the next section.

5. Parameter estimation and model evaluation

It is well known that the estimation of unknown physical parameters of a system based on a presumed system model and a set of experimental data, is the most reliable way to obtain an accurate information about both the model validity and the parameters' values. An unstable estimation process indicates that the assumed model can not describe the system behavior. Even if the model perfectly matches the system behavior, the estimates may have bias due to the influence of sensor noise. For this reason, the model output has to be compared to the real-system output under the conditions different from those that have been used for parameter estimation. We eliminated the bias by the use of a "bias-compensating" least-squares algorithm, [13].

The experiments described in Section 3 indicate that the dynamic behavior of the joint highly depends upon the operating conditions. A physical phenomenon dominating under one set of conditions may be literally invisible upon changing the operating point. For example, the flexspline hysteresis was a dominant effect only when the joint was stuck against an obstacle. Otherwise, it was much less visible. For this reason, instead of estimating all parameters at once, it is much more convenient to divide the estimation process into three independent procedures corresponding to the three subsystems mentioned above.

5.1. Motor - wave-generator parameters

By observing the model of the wave-generator subsystem (5), (6), it is clear that the Coulomb friction parameters B_{mw}^+, and B_{mw}^- can be obtained from the experimental plot in Fig. 7. In this experiment the motor current was varied very slowly in the range -2 to +2 amps while keeping the output link fixed. The variations of q_w due to elasticity were small, so that the magnitudes of $\ddot{q}_w$a and $\dot{q}_w$ were negligible. In this case the model (5) reduces to

$$\tau_m - (1/N)\tau = (B_{mw}^+, B_{mw}^-)sgn(\dot{q}_w) \qquad (22)$$

implying that the torque $\tau_m - (1/N)\tau$ is equal to the Coulomb friction torque. Since both τ_m and τ are measurable, the unknown constants on the right side of (22) can be accurately estimated. From the experimental data (Fig. 7) we got $B_{mw}^+ = 0.014Nm$, and $B_{mw}^- = -0.028Nm$.

In the second step we introduced a composite sinusoidal excitation $u(t)$ covering the joint bandwidth uniformly up to 5 Hz, and observed the unrestrained motion of the joint. We used the model (5) in the form

$$\tau_e = J_{mw}\ddot{q}_w + B_{mw}(\dot{q}_w) \qquad (23)$$

to estimate the unknown inertia and viscous friction. The torque $\tau_e = \tau_m - (1/N)\tau - (B_{mw}^+, B_{mw}^-)sgn(\dot{q}_w)$ was computed on-line from the motor current, joint torque sensor signal, and motor velocity. By applying the bias-compensating least-squares method, we got the following estimates: $J_{mw} = 8.010^{-6}Kgm^2$, and $B_{mw} = 6.410^{-4}Nm/(rad/sec)$.

To validate the modeling assumption, as well as the estimated parameters, we replaced the original trajectory with a modified one and performed the same experiment once again. The comparison of the real and simulated responses (Fig. 11) shows that the estimates are very well tuned to the real parameters.

5.2. Output link parameters

The estimation of the output link parameters is tremendously simplified by the introduction of the joint torque sensor. The joint torque measurement makes the decomposition of the overall system possible. The equations (20) and (21) can be regarded as virtually independent due to availability of $\tau(t)$.

Similarly to the previous case, we used a composite sinusoidal excitation to cover the joint bandwidth as uniformly as possible. If the output link is moving

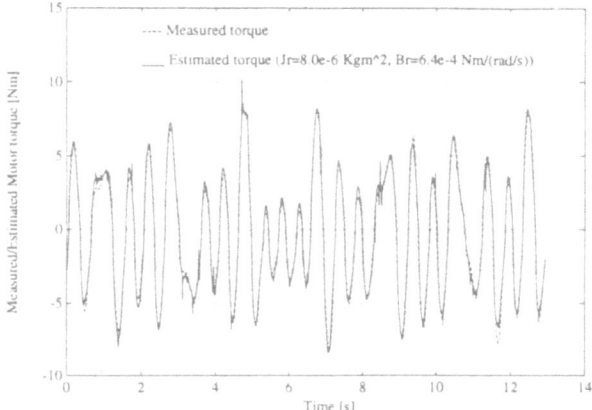

Figure 11. Measured/estimated motor torques

in the horizontal plane, the model (21) becomes very simple:

$$\tau = \frac{N}{N+1}(J_c\ddot{q}_c + B_c\dot{q}_c) \tag{24}$$

The unknown parameters estimated as $J_c = 0.013 Kgm^2$ and $B_c \approx 0$ were verified by comparing the measured and estimated torque sensor output.

5.3. Flexspline parameters

To distinguish the flexspline subsystem behavior from both the input (wave-generator) subsystem and the output (circular-spline) subsystem, the torque sensor signal is used again. Note that the flexspline model (21) is a quasi-algebraic model encompassing the nonlinear stiffness and the hysteresis. If the output link is constrained, $q_c = 0$, the equation (21) can be accurately approximated as follows:

$$\tau = a_{f1}q^3 + a_{f2}q - k_{sw}e^{-a_{sw}q^2}q + (B_f^+, B_f^-)sgn(\dot{q}) \tag{25}$$

where $q = (1/N)q_w$ represents the flexspline torsion. Given a direction of motion, $sgn()$ becomes either 1 or -1, and the torque-torsion relation becomes purely algebraic. The parameters a_{f1}, a_{f2}, k_{sw}, and a_{sw} can be obtained by the use of the stiffness diagram in Fig. 5, while the parameters B_f^+, and B_f^- are easier to obtain from the hysteresis plot in Fig. 6. For IRIS joint we obtained the following values in SI units: $a_{f1} = 6.61, a_{f2} = 5.27, k_{sw} = 7.8, a_{sw} = 25.0, B_f^+ = 2.0$, and $B_f^- = -2.0$. Similar results are obtained for the case when the C-end is grounded.

6. Conclusions

Based on experimental and theoretical studies, we have developed a systematic way to capture and rationalize the behavior of a harmonic-drive-based robot joint. In addition, the paper reveals a new phenomenon related to intrinsic joint instability, termed as "torque-transmission paradox". We found that the restrained motion experiments were more convenient for examining the joint phenomena than that with free motion. Simple models for the soft-windup, hysteresis and friction were established and used as constituents of the entire joint model described by three basic nonlinear differential equations modeling high-frequency, low-frequency and coupling dynamics of rotor - wave-generator, circular-spline and flexspline, respectively. The great significance of the joint torque sensor has been highlighted from the standpoint of observability and controllability of joint behavior. At the end, a systematic parameter estimation procedure has been proposed not only to gather the parameters, but also to support the validity of the developed model.

References

[1] Margulis, M.V., and D.P. Volkov, "Calculation of the Torsional Rigidity of a Harmonic Power Drive with a Disc Generator,", Soviet Engineering Research, Vol. 7, No. 6, pp. 17-19, 1987.

[2] Paul D. Oei, Transfer of Electrical Power Across a Rotating Joint Using Harmonic Drive Principles, M.Sc. in Engineering thesis, Univ. of Pennsylvania, 1991.

[3] Nye, T.W., R. P. Kraml, "Harmonic Drive Gear Error: Characterization and Compensation for Precision Pointing and Tracking", Proc. of the 25th Aerospace Mechanisms Symp., pp 237-252, 1991.

[4] Hsia, L.M., "The Analysis and Design of Harmonic Gear Drives," Proc. 1988 IEEE Int. Conf. Sys., Man, Cybern., Vol. 1, pp 616-619, 1988.

[5] Ahmadian, M., "Kinematic Modeling and Positional Error Analysis of Flexible Gears," Clemson Univ. Research Report under grant no. RI-A-86-8, 1987.

[6] Hashimoto M., Y. Kiyosawa, and R.P. Paul, "A Torque Sensing Technique for Robots with Harmonic Drives,", IEEE Trans. on Robotics and Automation, Vol. 9, No. 1, pp 108-116, Feb. 1993.

[7] Kiyosawa, Y., Sasahara, M., and S. Ishikawa, "Performance of a Strain Wave Gearing Using a New Tooth Profile," Proc. of the 1989 Int. Power Transmission and Gearing Conf., ASME, pp 607-612, 1989.

[8] Tuttle, T.D., Understanding and Modeling the Behavior of a Harmonic Drive Gear Transmission, MIT Masters Thesis, Technical Report No. 1365, 1992.

[9] Tuttle, T.D., and W. Seering, "Modeling a Harmonic Drive Gear Transmission", Proc. of the 1993 Int. Conf. on R&A, pp 624-629, 1993.

[10] Good, N.C., L.M. Sweet and K.L. Strobe, "Dynamic Models for Control System Design of Integrated Robots and Drive Systems," ASME J. Dy. Sys., Meas. & Control, Vol. 107, pp 53-59, 1985.

[11] Kircanski N., IRIS Grasping & Manipulation Facility - Performance Evaluation, Modeling and Identification, RAL Int. Report, Department of Mechanical Engineering, Univ. of Toronto, Feb. 1993.

[12] Hui R., N. Kircanski, A. Goldenberg, C. Zhou, P. Kuzan, J. Wiercienski, D. Gershon, P. Sinha, "Design of the IRIS Facility - A Modular, Reconfigurable and Expandable Robot Test Bed," Proc. 1993 IEEE Int. Conf. R&A, Vol. 3, pp 155-160, May 2-7, Atlanta, 1993.

[13] Jia S., Model Identification of a Reconfigurable Robot Manipulator, M.A.Sc. Thesis, Department of Mechanical Engineering, Univ. of Toronto, 1993.

Section 7
Teleoperation

An essential problem associated with a master-slave teleoperation system is the way of translating visual or force information detected by sensors from the slave side to the human operator. Some of the papers in this section emphasize that tactile display of high-frequency force information can enhance performance in teleoperated manipulation tasks. The others discuss the issues of control and motion coordination that arise in operating teleoperation systems.

The paper by Howe and Kontarinis present direct manipulation experiments that demonstrate the importance of high-frequency vibration information in manipulation. This result suggests that sensing and display of high-frequency information can improve performance in teleoperation.

Sato, Ichikawa, Mitsuishi, and Hatamura propose a novel configuration of a teleoperation system suitable for micro object handling. The system consists of the upward CRT display monitor and a force feedback pencil, and enables intuitive teleoperation because of the sensitive bilateral micro operation capabilities and the coincidence of operating and force sensing coordinates with monitoring coordinates.

Anderson presents an algorithm for generating virtual forces in a bilateral teleoperator system. The virtual forces are generated from a world model approximated with convex polyhedral primitives by applying Gilbert's polyhedra distance algorithm.

The paper by Ganeshan, Ellis, and Lederman addresses the problem of length perception in a teleoperation context, using the sense of touch alone. A novel sensor to perceive the length of an object is designed based on the theory of thin plates, and two different models of length perception are proposed through parallel robotic and human psychophysical studies.

Yokokohji and Yoshikawa introduce a new framework of human-robot cooperating systems that consists of six operation modes for cooperating between manual operation and autonomous functions. They also point out the importance of the sequence of mode change to achieve intuitive and smooth teleoperation.

Salcudean and Wong present a novel teleoperation system developed by equipping a conventional robot with a magnetically levitated fine-motion wrist and using an identical wrist as a master. The issues of control of such a fine master and a coarse-fine slave system are discussed from the viewpoint of 'transparency' and stability.

High-Frequency Force Information in Teleoperated Manipulation

Robert D. Howe and Dimitri A. Kontarinis

Division of Applied Sciences
Harvard University
Cambridge, MA 02138 USA

Abstract— This paper presents a preliminary investigation of techniques for sensing and display of high frequency force information in telemanipulation. For a human teleoperator, force variations above 50-100 Hz are unresolved vibrations, permitting the use of simple means for relaying this information from the remote slave manipulator. Vibrations convey important information about manipulation events such as contact and slip, and about object properties such as friction and surface texture. In this study, a sequence of experiments demonstrates that sensing of vibrations can be important in manipulation and that a vibratory display can convey this information in teleoperation. Using a tapping task, the experiments confirm that: (1) humans make use of vibrations; (2) a vibration display can convey appropriate high-frequency information; and (3) vibratory sensing and display can be used with a teleoperated hand system to improve task performance. Finally, to demonstrate the use of high frequency information in an inspection and maintenance task, subjects used a teleoperated hand system to identify a worn ball bearing set.

1. Introduction

We are studying the fundamental question of bandwidth requirements for precision manipulation by human and robot hands. Precision manipulation tasks require accurate control of the finger tips, but little is known about the range of frequencies that must be sensed for good performance. In addition to the relatively low-frequency forces and motions usually considered in theoretical analysis of manipulation, many tasks also produce high-frequency vibrations. Events which generate these vibrations include contact between fingers and object surfaces, contact between grasped objects and surfaces in the environment, and slip. This paper presents preliminary experiments which demonstrate that tactile display of vibratory phenomena can enhance performance in teleoperated manipulation tasks.

Most research on haptic feedback in teleoperation has focused on force reflection, with bandwidths typically limited to less than 10 Hz (Sheridan 1992). Work has recently appeared on tactile display of shape in teleoperation and virtual environments (Hasser et al. 1993, Kontarinis and Howe 1993), as well as sensory substitution aids for the blind such as the Optacon (Bliss 1970). These devices often use vibrating pin elements to stimulate the skin. In each of these cases the vibratory stimulus is designed to provide information about another physical parameter such as shape, optical intensity, or force, and no attempt was made to relay information about vibrations that occur as part of a task. One system that is designed to portray vibratory information for virtual environments has been developed by Minsky et al. (1990). This joy-stick based device simulates the vibrations and mechanical interactions produced by stroking a stylus over various surface textures and features.

Physiological studies show that humans can perceive vibrations higher than 1 kHz in frequency and smaller than 1 micrometer in amplitude. These studies also suggest that high frequency vibration information is important for sequential coordination of activities during task execution (Vallbo and Johansson 1984, Johansson and Westling 1987). In particular, these studies show that the Pacinian nerve endings, the mechanoreceptors principally responsible for detection of phenomena above 50-100 Hz, cannot localize vibratory stimulus on the skin. This leads to the prediction that a simple vibratory display can take the place of more elaborate and expensive spatially resolved force feedback for high frequency information.

Our previous work on force feedback bandwidth requirements supports this prediction (Howe and Kontarinis 1992). In this study subjects performed a close-tolerance assembly task using the force reflecting teleoperated hand system described in section 3.1 below. The force feedback signal from the slave manipulator to the master manipulator was low-pass filtered to remove frequency components above a variable cutoff frequency. Task performance did not significantly improve when the bandwidth was increased above 8 Hz. This suggests that, at least for tasks that are limited by the ability to control resultant forces, force reflection bandwidth may be restricted to relatively low frequencies.

In this paper we explore sensing and display of high frequency information, consisting of vibrations above 50-100 Hz. The work reported here is a sequence of four experiments. Since performance in the assembly task in the above study appears to be governed by the ability to control resultant forces, we chose a tapping or probing task where subjects use a stylus to make contact with a target surface. Subjects are constrained to minimize contact forces, so vibrations are essential to successful completion of the task. First, we show that humans use vibrations in this task; when perception of task-generated vibrations are masked by extraneous vibrations, performance decreases markedly. The second experiment confirms that a vibration display (i.e., a device for producing vibrations on the operator's finger tip) can effectively convey vibratory information. In this experiment perception of vibrations is difficult because subjects use a massive stylus which reduces the amplitude of the vibrations transmitted from the stylus tip. When a vibratory display in the stylus is used to signal the subject that contact has occurred, performance improves. The third experiment verifies that vibratory sensing and display can be used with a teleoperated hand system to improve performance in this task. Finally, to demonstrate the use of high frequency information in an inspection and maintenance task, subjects used the teleoperated hand system to identify a worn ball bearing set. As in the preceding experiment, the addition of vibration sensing and display significantly improved performance.

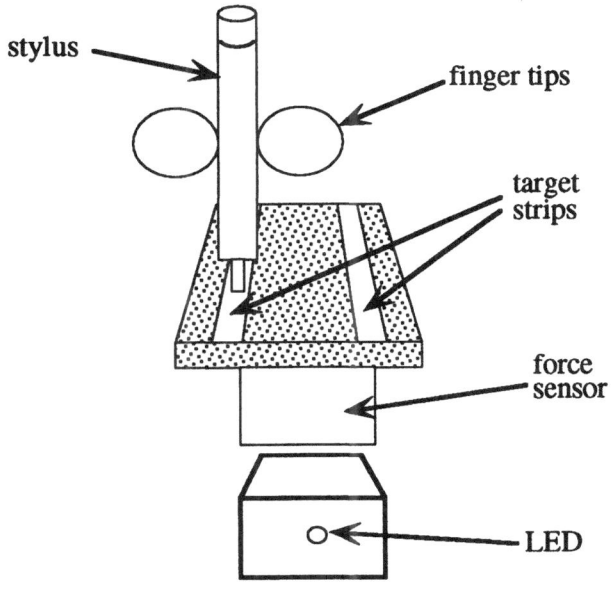

Figure 1. Tapping task experimental set up.

2. Direct manipulation experiments

Our hypothesis is that high-frequency tactile information is used by humans in manipulation, and that a high-frequency display can convey this information to the operator of a teleoperated system. We began with direct manipulation experiments in which subjects handled instrument objects bare-handed. For the experiments reported in this paper we chose a remote contact task since this type of task places a premium on high-frequency information. Subjects were asked to alternately touch two target strips with the tip of a hand-held stylus while minimizing contact force, as when probing a delicate surface (Figure 1). This task requires detection of vibrations to indicate that the contact has occurred before large forces have developed. The surface on which the strips were marked rested on a force sensor which measured the contact force between the stylus and the strip. Every time a subject exceeded a force threshold a light-emitting diode (LED) would light indicating to the subjects that they had pressed too hard. During a brief training period subjects performed the task until they felt comfortable. After the training period subjects were asked to alternately tap the strips trying not to exceed the force threshold that would light the LED. Scoring was based on the number of taps that did not exceed the threshold in a 30 sec test period. During the execution of the task, subjects wore headphones playing white noise to mask any audio feedback produced during the task.

2.1. Experiment 1: Use of vibratory information

The first series of experiments involved tapping using a lightweight stylus approximately 150 mm long, which subjects held in a writing grasp as they would a pencil. The low mass (39 grams) of this stylus permitted good perception of the small vibrations that indicated the earliest stages of contact, and subjects were able to execute a large number of successful taps in the test periods. To confirm that this performance was facilitated by high-frequency information, a solenoid mounted inside the stylus was actuated continuously in some test periods. The high frequencies generated by the solenoid were designed to mask the vibrations generated by contact, preventing subjects from using this type of information. As anticipated, performance was far worse when the masking vibrations were present. Subjects made only one-sixth the number of taps compared with the unmasked case. This confirms that sensing of high frequency vibrations is important for successful task execution.

2.2. Experiment 2: Vibratory display

The next experiment was designed to determine whether a high-frequency display could be used to provide vibratory information. In this experiment subjects used a stylus of approximately the same size but far higher mass (245 grams) to perform the same task. The mass of this stylus acted as a low pass filter which decreased the amplitude of the vibrations reaching the subject's fingers. This made it far more difficult to perceive the initial contact vibrations, and performance was significantly worse. To examine the utility of a vibratory display, in half of the test periods a solenoid in this stylus produced a small burst of vibrations when the tip of the stylus made electrical contact with the target strips. (Electrical contact occurred at the first stages of contact, before the force had risen significantly.) In test periods when the tactile display was active, subjects' scores were an average of 1.6 times higher than periods without the display, confirming that the display improved performance.

3. Teleoperated manipulation experiments

3.1. Teleoperated dextrous hand system

Having shown that a vibratory display can be useful in direct manipulation experiments, we proceeded to experiments using the teleoperated hand system developed in our laboratory (Howe 1992, Howe and Kontarinis 1992). Most work on dextrous teleoperated hands has focused on creating lightweight master manipulators with many degrees of freedom for end-of-arm mounting (Jau 1992, Burdea et al. 1992). This presents an extremely difficult design problem since it requires placing many sensors and actuators around the human operator's hand at the end of the arm. Proposed solutions often entail the use of cables or gears, which makes it difficult to accurately control the small forces and

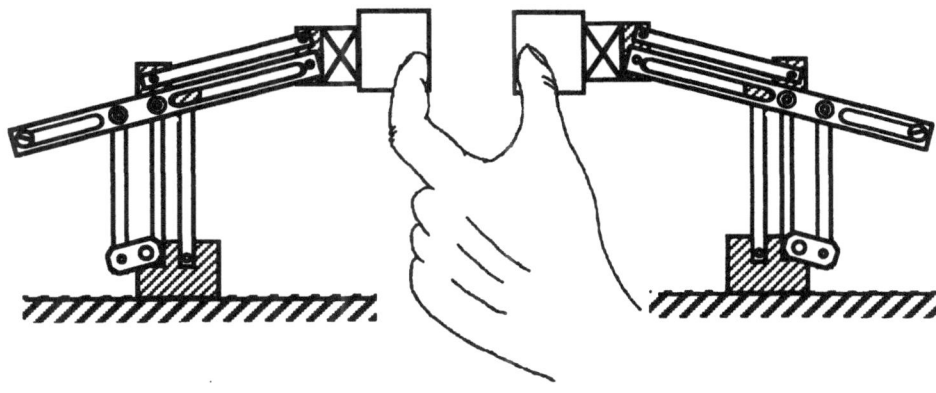

Figure 2. Master manipulator with operator's hand. The slave manipulator uses identical mechanisms with round finger tips containing skin acceleration vibration sensors.

displacements that are essential for effective tactile sensing and display. In contrast, our system trades a limitation on the number of joints for a clean and simple mechanical design. The system has high bandwidth and large dynamic range, which permits accurate control of contact forces and small motions.

The system is designed to execute tasks that humans usually accomplish with a precision pinch grasp between the thumb and index finger. For most tasks the operator's wrist rests on the table top and the operator makes contact with the master only at the tips of the fingers (Figure 2). Master and slave manipulators are identical two-fingered hands, with two degrees of freedom in each finger, so finger tip position or force can be controlled within the vertical plane. The controller uses a conventional bilateral force reflection control scheme, so the slave follows the motions of the master and the master applies the forces measured at the slave finger tips to the operator's fingers. Position bandwidth of the slave manipulator is at least 18 Hz for 4 mm displacements, and force bandwidth of the master is at least 70 Hz for a 0.5 N step against an operator's finger. Further details of the manipulator system design and performance are presented in (Howe 1992, Howe and Kontarinis 1992).

These experiments required both high-frequency sensing at the remote manipulator and high-frequency display to the operator. A skin acceleration sensor (Howe and Cutkosky 1989) was mounted on a remote slave manipulator finger tip to detect vibrations produced during task execution (Figure 3). These signals triggered the tactile display on the master manipulator. This tactile display used a small solenoid to bring a metal plunger into contact with a metal frame, producing impacts with substantial high frequency content. This frame was mounted on the fingers of the master manipulator so that the operator could clearly feel the vibrations produced.

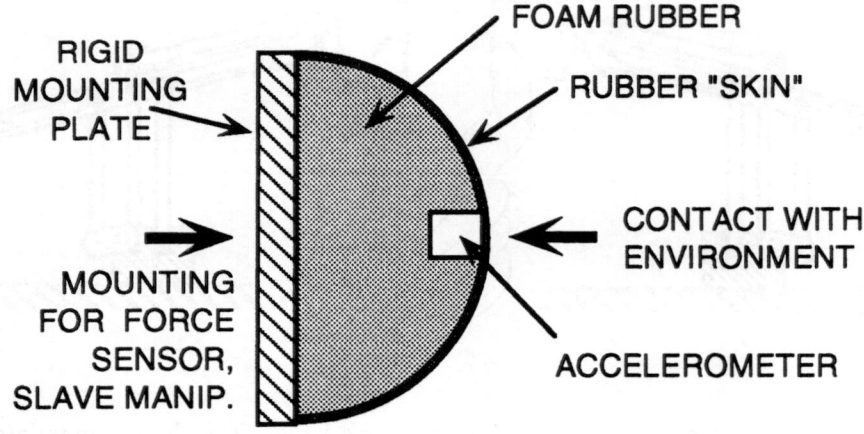

Figure 3. Cross section of slave finger tip with skin acceleration sensor.

3.2. Experiment 3: Telemanipulation with the vibration display

Subjects performed the same tapping task as in the direct manipulation experiments, using a lightweight stylus gripped by the slave manipulator. To isolate only high-frequency vibrations the skin acceleration signal was filtered through a bandpass filter with corner frequencies at 300 Hz and 1 kHz. Each time a preset acceleration threshold was exceeded the display was triggered, producing a 25 ms wide vibration pulse on the master finger tip. The time delay between the time the skin acceleration sensor signal exceeded the threshold and the rise of the pulse was 15 ms. Most subjects were able to use the high-frequency information conveyed by the display with little training. Subjects' performance improved by a factor of 1.8 when the display was active.

3.3. Experiment 4: Ball bearing inspection task

As a final demonstration of the utility of high-frequency sensing and display in teleoperated manipulation, we used the system in a illustrative inspection task, the identification of a worn ball bearing set. For this experiment, the binary high frequency "impact" display used above was replaced with an analog voice coil display mounted on the master manipulator. This improved display recreated the same pattern of vibration at the operator's finger tips that the skin acceleration sensor measured at the slave finger tips. Subjects were presented with a pair of ball bearings sets, one of which was new while the other was worn and produced vibrations when rotated (Figure 4). Subjects used one finger of the hand to rotate each bearing and then made a forced choice as to which one was bad. Between trails the spatial order of the bearings could be altered without the knowledge of the subject.

Subjects performed the task under four feedback conditions: visual feedback only, visual and force feedback, visual and vibratory feedback, and visual,

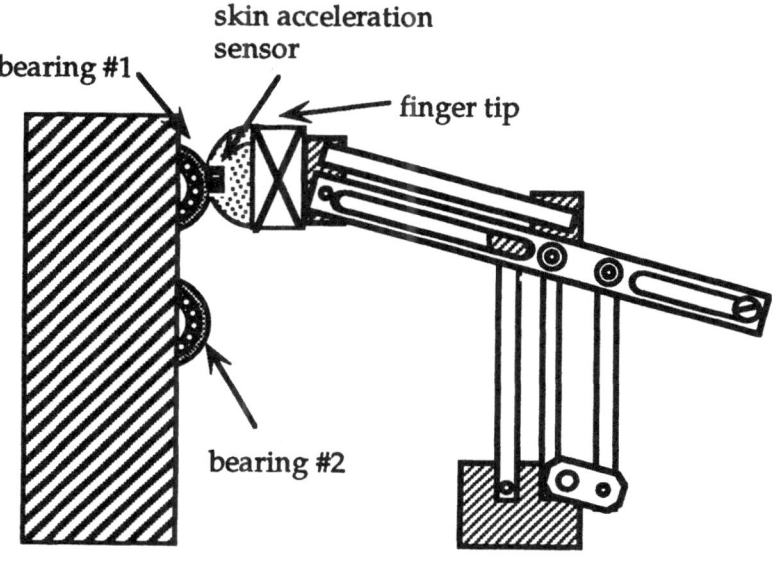

Figure 4. Experimental setup of the ball bearing inspection task.

vibratory, and force feedback. With visual feedback only, the correct response rate was no better than chance (50%), indicating that it was impossible to detect the bad bearing visually. The use of force feedback raised the correct response rate to 80%, while with vibratory feedback subjects were able to select the bad bearing every time, either with or without force feedback. Aside from the obvious conclusion regarding the effectiveness of the vibration display, several related observations are pertinent. Although the vibration display was clearly the best way to detect the bearing quality, force feedback was also important. Without force feedback, subjects fumbled a great deal in attempting to rotate the bearings. The result was longer completion times and excessive contact forces between the finger tips and the bearing sets. Thus effectiveness of the vibration display is enhanced by force feedback. Similarly, subjects reported that use of force feedback without the vibration display in this and other tasks felt "dead," and that the vibration display enhanced the subjective quality of the manipulation system.

4. Discussion

These results show that vibrations can be important in task execution, and that they can be effectively conveyed by a simple sensing and display system. The importance of high-frequency information varies greatly among tasks. Tasks involving only low speeds and continuous motion will produce few high-frequency vibrations, while rapid motions and contact with hard surfaces will produce copious vibrations. For these initial studies we selected tasks where perception of vibrations is essential for task completion in order to confirm the utility of our system. However, studies of human and robotic manipulation show that vibrations can play an important role even in tasks where vibrations are not consciously perceived (Johansson and Westling 1987, Howe et al. 1989).

Complex manipulation tasks consist of a sequence of phases or subtasks separated by discrete events. Events such as slip and the making and breaking of contact indicate the status of the task as it proceeds and allow for the sequential coordination of activities that will lead to the completion of the task. For example, consider a simple grasp-and-lift task. Before contact, the fingers approach the object. During this phase the positions of the fingers are controlled. Contact of the finger tips with the object initiates a change in the control strategy, so that the forces exerted by the fingers will firmly grasp and lift the object. Lift off (break of contact as the object is lifted from the table) is another event that indicates that the object has been successfully grasped and subsequent phases of the task can be performed. For a given manipulation task the spatiotemporal range of physical parameters (forces, accelerations, etc.) depends not only on the nature of the overall task but also on the events which delineate the phases of the task. It appears that many of these events may be best detected through vibratory signals.

The high-frequency display may serve to extend the dynamic range of a manipulation system at minimal cost. For conventional force reflection, a force sensor is used to measure the force applied by the slave to the environment or the object under manipulation, and a master manipulator is used to display this force to the operator. This works well at low frequencies, but high bandwidth sensing and display is difficult and expensive, if it is possible at all. Instead, these experiments suggest that high frequency information may be effectively conveyed by a specialized sensor (e.g. a skin acceleration sensor) and a specialized display (a vibrator) that have good response at high frequencies and no response at low frequencies. This approach effectively divides force feedback into separate displays of low and high bandwidth, depending on the sensory information conveyed. The specifications for such distinct sensors and displays are better defined and easier to meet than those of a single wide bandwidth display. For certain applications it may be possible to use the vibratory sensing and display system without traditional force feedback, resulting in a vastly less expensive system.

This work on high-frequency feedback is part of a larger effort in our laboratory to develop devices for sensing and display of a range of tactile sensation. One device we have developed is a tactile shape display (Kontarinis and Howe 1993). The current version of the shape display consists of a 3 x 3 array of pin elements that are raised against the operator's finger tip to approximate the shape sensed at the slave manipulator's finger tip with a tactile array sensor. The actuators are shape memory alloy (SMA) wires. Forces of 1.2 N per tactor (to withstand force reflection) and displacements of 3 mm have been achieved. A 6 x 4 array is under construction with the same resolution and smaller overall dimensions. The low bandwidth and hysteretic response of SMA wires are the major disadvantage of this type of actuation. However, a feedforward control law and air cooling significantly improve the performance of the SMA wires, resulting in step rise and fall times of about 65 ms. We believe that with further development it may be possible to use this display to convey both shape and high-frequency vibrations by superimposing the vibratory signal on the low frequency shape information.

These experiments represent a first attempt to define the role of high-frequency sensing and display. The display device used here has not been optimized, and a number of design issues remain to be resolved. Nonetheless, these experiments demonstrate that high-frequency vibration information can play an important role in manipulation. Humans make use of this information, and sensing and display of high-frequency information can improve performance in teleoperation. Fundamental questions still remain about the range of frequencies generated in particular manipulation tasks. Further study of the human tactile system may help to determine parameter ranges such as useful frequencies to sense and display, and ways of combining this modality with other tactile displays.

5. References

J. C. Bliss, M. H. Katcher, C. H. Rogers, and R. P. Shepard, "Optical-to-Tactile Image Conversion for the blind," *IEEE Trans. on Man-Machine Systems*, Vol. MMS-11, No. 1, pp. 58 - 65 ,1970.

G. Burdea, J. Zhuang, E. Roskos, D. Silver, and N. Langrana, "A portable dextrous hand master with force feedback," *Presence*, Vol.1, No.1, pp. 18-28, 1992.

C. J. Hasser and J. M. Weisenberger, "Preliminary evaluation of a shape memory alloy tactile feedback display," in H. Kazerooni, J. E. Colgate, and B. D. Adelstein, eds., *Advances in Robotics, Mechatronics, and Haptic Interfaces 1993*, DSC-Vol. 49, pp. 73-80, ASME Winter Annual Meeting, New Orleans, LA, Nov. 28-Dec. 3, 1993.

R. D. Howe, "A Force-Reflecting Teleoperated Hand System for the Study of Tactile Sensing in Precision Manipulation," in *Proc. 1992 IEEE International Conference on Robotics and Automation*, Nice, France, pp. 1321-1326, May 1992.

R. D. Howe and M. R. Cutkosky, "Sensing skin acceleration for texture and slip perception," in *Proc. 1989 IEEE International Conference on Robotics and Automation*, Scottsdale, AZ, May 1989.

R. D. Howe and D. Kontarinis, "Task performance with a dextrous teleoperated hand system," in H. Das, ed., *Telemanipulator Technology Conference*, Proc. SPIE vol. 1833, Boston, MA, Nov. 15-16, 1992, pp. 199-207.

R. D. Howe, N. Popp, P. Akella, I. Kao, and M. Cutkosky., "Grasping, manipulation, and control with tactile sensing," In *Proc. 1990 IEEE International Conference on Robotics and Automation*, pp. 12581263, Cincinati, Ohio, May 1990.

B. M. Jau, "Man-equivalent telepresence through four fingered human-like hand system," *Proc. 1992 IEEE Intl. Conf. on Robotics and Automation*, Nice, France, May 1992.

R. S. Johansson and G. Westling, "Signals in tactile afferents from the fingers eliciting adaptive motor responses during precision grip," *Experimental Brain Research*, Vol . 66, pp. 141-154, 1987.

D. Kontarinis and R. D. Howe, "Display of high frequency tactile information to teleoperators," in W. Kim, ed., *Telemanipulator Technology and Space Robotics Conference*, Proc. SPIE vol. 2057, Boston, MA, Sept. 7-9, 1993, in press.

D. A. Kontarinis and R. D. Howe, "Tactile Display of Contact Shape in Dextrous Manipulation," in H. Kazerooni, J. E. Colgate, and B. D. Adelstein, eds., *Advances in Robotics, Mechatronics, and Haptic Interfaces 1993*, DSC-Vol. 49, pp. 81-88, ASME Winter Annual Meeting, New Orleans, LA, Nov. 28-Dec. 3, 1993.

M. Minsky, M. Ouh-young, O. Steele, F.P.J. Brooks and M. Behensky, "Feeling and Seeing: Issues in force display," In *Proc. 1990 Symp. on interactive 3D Graphics*, pp. 235- 243, 1990.

T. B. Sheridan, *Telerobotics, Automation, and Human Supervisory Control*, The MIT Press, 1992.

A. B. Vallbo and R. S. Johansson, "Properties of cutaneous mechanoreceptors in the human hand related to touch sensation," *Human Neurobiology*, Vol. 3, pp. 3-14, 1984.

Micro Teleoperation System Concentrating Visual and Force Information at Operator's Hand

Tomomasa SATO*, Junri ICHIKAWA**,
Mamoru MITSUISHI**, Yotaro HATAMURA**

*Research Center of Advanced Science and Technology,
The University of Tokyo
**Faculty of Engineering, The University of Tokyo

Abstract— This paper proposes a novel configuration of a teleoperation system suitable for micro object handling. In the system, a CRT display monitor is placed beneath a table with its face up keeping both surface on the same plain. An Operator applies necessary operations to the image of target object on the monitor by utilizing a pencil-shaped master manipulator with force feedback function. This configuration derives from features of manual micro tasks. The system has three characteristic features: firstly the system enables intuitive teleoperation because it realizes the coincidence of operating coordinates and force sensing coordinates with monitoring coordinates, secondly the pencil-shaped master manipulator enables sensitive bilateral micro operation because of its lightness of moving components and its small inertia, thirdly the operator can command with a wide dynamic range of force feedback because they holds the master manipulator in a pen-holder grip fixing side of their hands during operation, i.e., the operator can feel fedback force simultaneously from both pressure/slip sensory receptor on their fingertips and muscle stretch sensory receptor on fingers during operation.

1. Introduction

Micromachines have been attracting growing interests. Semiconductor technology such as silicon process has developed to fabricate very small objects which has simple mechanical functions. However, because of the difficulty to fabricate complex three-dimensional small objects or parts of micromachines by this technology, assembling and handling technology of the small objects are essential to complete micromachines with complex functions. Furthermore, there are much needs to test or tune the completed micromachines. There are also keen needs for checking or repairing LSI circuit pattern, and executing such micro surgery as inner ear surgery or blood vessel suture. These micrometer scaled tasks must be realized by a telerobot system which couples the human operator and micrometer scaled robot world.

Although a master-slave teleoperation system which translates operational commands from the operator in human world to micrometer scaled world with shrinkage in scale, and translates visual or force information detected by sensors from micrometer scaled world to human operator with magnifying transformation may be able to perform micro teleoperation, the system suffers from following problems of human interface devices and those of human force perception.

As for the human interface devices problems, a monitoring point does not coincide with an operating point in conventional teleoperation systems, i.e.. a display monitor is placed in front of the operator while input/output devices are placed around the operator's hand. Therefore it is necessary for the operator to change their monitoring point (focus of attention) very frequently to verify whether their actions applied to the device works well or not. Moreover in order to apply desired robot motion the operator must do mental transformation from desired direction of robot actions on a display monitor to the direction to which the operator should move their hand and arms[1]. Also conventional bilateral master manipulators don't have quick response because of their large inertia.

As for the human force perception problems, conventional master manipulators appeal force only to force sensory receptor which measures the angle of finger or arm joint as muscle stretch information. But in micro object handling, the operator senses resistant force not only by this sensory receptor but also tactile sensory receptor which measures the pressure applied to the operator's fingertips simultaneously. Utilizing both receptors enables responsive operation with wider dynamic range of force. It may avoid breaking or flipping away the target objects.

In this paper, we propose a novel configuration of a micro teleoperation system which utilizes an upward CRT display monitor and a pencil-shaped master manipulator with force feedback function called force-feedback pencil. It is based on the consideration of the characteristic features in manual micro object handling, and is capable of mitigating both the previously mentioned inherent problems in conventional teleoperation systems.

2. Features of manual micro operation

To construct a micro teleoperation system, new concepts are needed which are different from conventional one in usual macro scale. There are such problems and characteristic features about micro object handling as: 1)problems derived from smallness of target objects, and 2)posture of the operator in execution of manual micro tasks. The followings are the details of them and we will show needed functions to construct a human interface system for micro teleoperation.

2.1. Micrometer sized object effects

1. **Smallness of target objects**

   The smallness of target objects impels to be watched by magnifying devices. It causes the problems on human interface how to show the magnified visual information in small world to an operator in human world without losing intuition of micro task execution.

2. **Characteristics of small objects in small world**

One of the accidents which often occurs unexpectedly in execution of micro tasks is flipping the target objects away from the field of view of a microscope. This derives from that the operator tends to apply much larger force to the objects because the operator incorrectly estimates force. The fact is that the influence of inertia of the small objects relatively decreases more than expected, because the weight of the object decreases in proportion to the third power of length reduction. Another accidents are the damage of the objects. The stress caused by outer force is much larger than expected because the cross section area of the objects decreases in proportion to the second power of the length reduction. To avoid these accidents, it is necessary to realize sensitive bilateral control with high response time.

2.2. Posture of manual micro operation

In execution of manual micro tasks, the operator poses as shown in figure 1(a). The followings are reasons and effects of such posture.

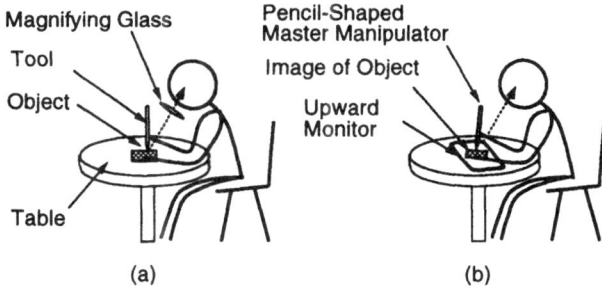

Figure 1. Posture in execution of manual micro tasks

1. **The operator places the target objects between his arms and watch them from above**

   This placement enables easy to watch the target objects. Magnified monitoring can be easily obtained by accessing operator's face to the object whenever the operator wants to detect details, even if the operator is using magnifying glasses (zooming function). Furthermore, the posture enables the operator to approach tools to the objects from any desired direction using his both arms (wide range of approach direction).

2. **The operator fixes the side of his or her hands**

   Generally, micro tasks are executed on a table. Fixing the side of the operator's hands to the table realizes very sensitive positioning and force control because the kinematic chain have much shorter length than the chain without fixing (short kinematic chain length) as shown in figure 2. The kinematic chains starts from the ground, runs through the table, the operator's hand, finger, and the tool and ends up with the ground. For example, the operator can write precise characters or signs because the pose enables fine positioning and force control. This is the advantage of this posture.

3. **Employment of fingertip pressure/slip and finger muscle stretching sensory information**

Finger force sensing includes tactile sensory receptor and stretch sensory receptor. The tactile perception receptor is equipped on fingertips and gets such three main force informations as contact, pressure/force and slip[2]. It senses small value of force applied to the fingertips with high sensitivity and quick response time[3]. On the other hand, the stretch perception receptor is equipped on muscle of fingers or arms. It senses larger value of force with rougher sensitivity than tactile perceptive receptor. The operator can execute delicate manual micro tasks by simultaneously integrating both sensory informations because the integration realizes very sensitive human force sensing with wide range of sensing (sensor fusion).

4. **Operating coordinates and force coordinates coincide with visual coordinates**

In execution of manual micro tasks, force sensing point and watching point concentrates within an operating area. In ideal micro teleoperation, operating and force sensing point also should be placed on the display monitor on which the watching point locates because the concentration helps verification of the operator's commands. Furthermore, in execution of manual micro tasks, operating coordinates, force sensing coordinates and watching coordinates also coincides with each other (three-dimensional coincidence of visual, force sensing and operating coordinates). It also helps intuitive verification.

3. Fundamental configuration of micro teleoperation system

Based on the requirements in former section, micro teleoperation system is proposed as shown in figure 1(b). The system consists of the upward monitor and the force-feedback pencil. Followings are the characteristic features of the system.

3.1. Upward display monitor

The CRT display monitor is placed beneath a table with its face up. This enables the operator to look down on the monitor, and to place an image of objects between the arms. Setting up a "superimposing" master manipulator realizes coincidence of operating coordinates and force information coordinates with visual information coordinates which are actually located above the surface of the monitor. This makes the micro teleoperation intuitive because of the concentration of these information within the area where the image of the object is monitored. Furthermore the direction the operator looks down on the monitor coincides with the direction in a case of looking into an ordinary stereo microscope. The coincidence releases the operator from the need of spatial transformation from actual robot coordinates to the displayed robot coordinates on the monitor.

3.2. Force-Feedback Pencil

In the system, the force-feedback pencil is utilized as bilateral master manipulator. Followings are the description of problems on conventional human interface devices and comparison with the force-feedback pencil in terms of human interface.

3.2.1. Problems on conventional human interface devices

Although such interface devices as a mouse, data gloves, joysticks, and a multi-link master manipulator have been known and used as human interface devices, each of them has problems in some cases. Mouse is most widely used input device using operator's arm and hand motion. It is naturally suitable for two-dimensional position data input, but poor at three-dimensional position data input and force feedback which is caused by collision between robot and object to the operator. Furthermore, it forces the operator to transform monitoring coordinates to operating coordinates because of the incompatibility of two co-ordinates, the one the table surface on which the operator move the mouse around, the other the monitor surface on which an icon is displayed. Data glove is the device which can command position and force by operator's finger movement, some of them with force feedback function[4]. However it is not so convenient device as the mouse because it needs efforts to wear or take off. Joystick is widely used device as position control units. However it is suitable for two-dimensional positioning. Three-dimensional positioning with the joystick needs some skill to command synthesized three-dimensional motion, even if it has three degrees of freedom which are x-axis, y-axis and rotation around z-axis. Multi-link master manipulators which has bilateral control function are capable of three-dimensional positional commanding[5], but it is difficult to operate with high accuracy and highly responsive force feedback because of the long kinematic chain as shown in figure 2(a), and furthermore the heavy weight of itself and counter balance weight.

3.2.2. Features of force-Feedback Pencil

In this paper, a force-feedback pencil is proposed as a master manipulator which solves these problems of human interface devices stated above. Followings are features of the force-feedback pencil. 1) The pencil is handled by the operator with pen-holder grip, and realizes force feedback by its moving and force sensing lead. 2) The pencil realizes the highly responsive force feedback because of its smaller weight and inertia than the conventional multi-link master manipulators. 3) The pencil realizes the high accuracy and highly responsive force feedback because of the shortening of length of kinematic chain by fixing side of the operator's hand.

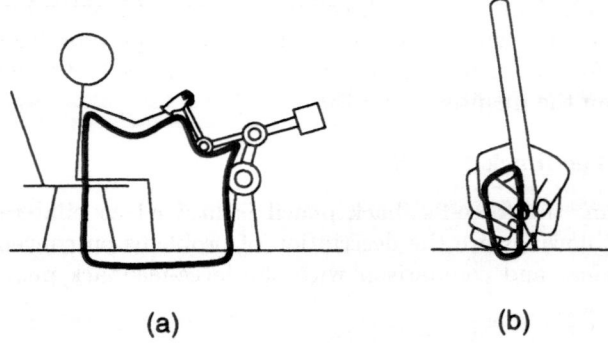

(a)   (b)

Figure 2. Conventional Master Manipulator and The Force-Feedback Pencil

4. Experimental system of micro teleoperation

Setting up the proposed configuration of micro teleoperation system are shown in figure 3. The experimental systems consists of the following components: 1)the robot, 2)the human interface, and 3)the computers and controllers.

4.1. Robot

The robot consists of a microscope and two robot arms as shown in figure 4.

The microscope is a optical stereo microscope which has magnification ratio from 45 to 384 with 39mm working distance. The visual information can be obtained by a CCD camera attached to the microscope.

Robot arms consists of a right arm and a left arm. The left arm to which a work table is equipped with two rotational and three translational degrees of freedom[6]. The right arm to which a tool is attached has two rotational degrees of freedom. One of the features of the robot is that the tip of the tool does not shift by its motion because all rotational axis intersects at one point, i.e., the center of the view field of the microscope. In the system, arms can approach from any direction.

A tool is exchangeable for various tasks, a three-axis force sensor can also be attached to. The three-axis force sensor has parallel plates structure and strain gauges. The sensor load and sensitivity are 1.4N and 0.92μst./mN (4 gauges) respectively.

4.2. Human interface

Human interface consists of the table, the CRT display monitor, the force-feedback pencil, a touch-sensor panel and CCD cameras.

The CRT display monitor is placed beneath the table. The monitor is 17 inch sized and the touch-sensor panel is installed on its surface which can sense contact with the side of operator's hand and/or tip of the pencil. On the surface of the touch-sensor panel, a conductive thin film which senses touch is pasted. Its surface senses touch of conductive objects like human fingers by measuring the change of the minute current in the film which caused by the

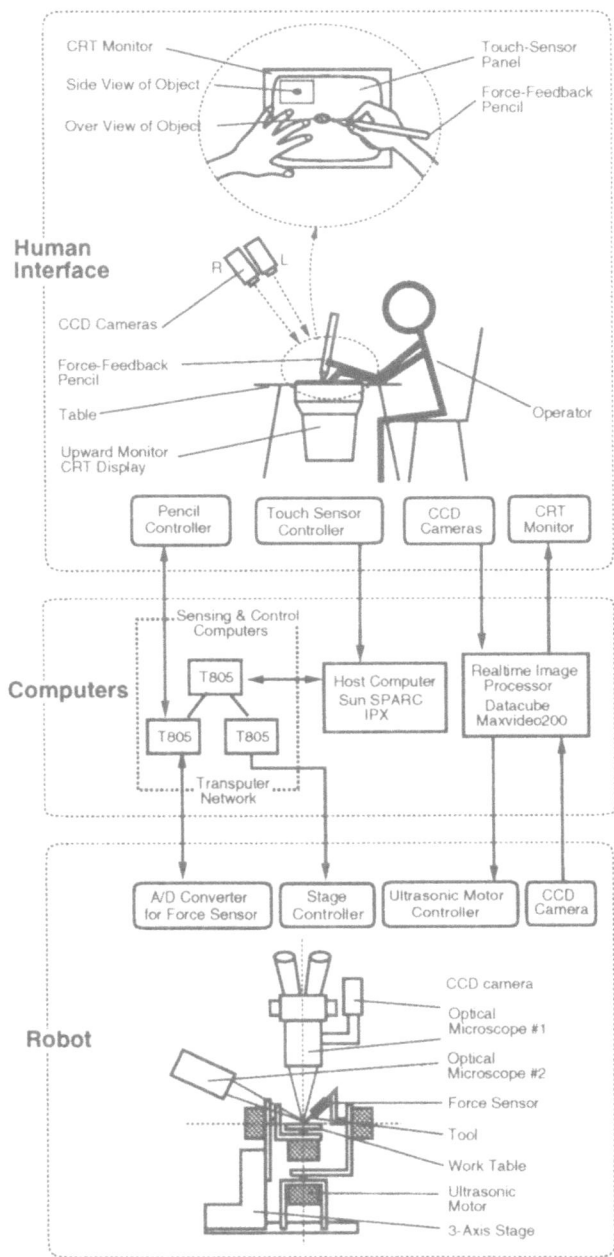

Figure 3. Block Diagram of the Micro Teleoperation System

touch. Available informations from it is sense of touch or untouch as well as two-dimensional touched position on the surface.

Figure 4. Outer View of Robot

The force-feedback pencil is shown in figure 5. Its radius, length and the weight is 19mm, 210mm, 82g, respectively. Inside of it a single axis force sensor and a linear actuator are installed. The linear actuator consists of a DC motor and a ball screw which transform rotational motion to linear motion of the lead. Force feedback is realized by the linearly moving lead of the pencil.

The optical encoder is also installed to measure the rotation of the motor. As for the sensor load, maximum force the pencil generates and stroke of the lead movement are 5N, 3.8N and 16mm respectively.

The two LED's are attached to the pencil to measure three-dimensional position by tracking these LED's using visual sensors.

The CCD cameras are placed over the table to monitor the LED's. Three-dimensional position of a point is identified using two camera information. Thus the three-dimensional position and two-dimensional angle of pencil is monitored in video rate by tracking two LED's simultaneously.

The teleoperation method is as follows: the operator sits at the table and looks down on the display monitor, applies necessary commands using the pencil in the operator's right hand or drag an image of the target objects by the operator's left hand.

In the system, as described in former section, translational commands causes a motion of the table because the translational degrees of freedom are assigned to the left arm manipulator. This configuration is derived from the necessity that the tip of the tool and the handled object should be always adjusted to be within the view field of the microscope. But it causes the incompatibility between operating information and visual information, i.e., even if the operator commands the tool to shift right, the monitor displays that the

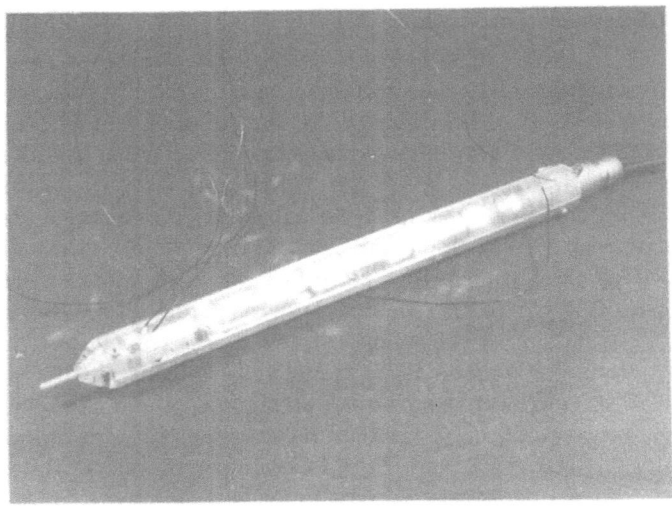

Figure 5. Outer View of Force-Feedback Pencil

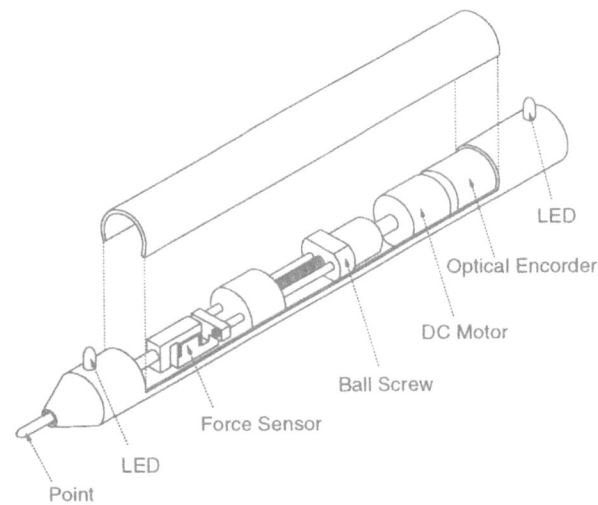

LED

Optical Encorder

DC Motor

Ball Screw

Force Sensor

LED

Point

Figure 6. Schematic Diagram of the Force-Feedback Pencil

table shifts left. It interferes the verification of the operator's commands. In the system, intuitive translational commanding is realized by the method of dragging the touch-sensor panel using operator's left hand as if the operator directly draws the image of the object.

4.3. Computers and controllers

The schematic diagram of the micro teleoperation system is shown in figure 3. Computers consist of workstations (SUN SPARC IPX and MaxTD), transputers (three T805s) and two image processors (Max Video 200). There are three main information flows as follows: 1)The output of visual information from CCD camera on the microscope is sent to image processors and displayed on the upward monitor. 2)The output of CCD cameras over the table is sent to image processors and transformed to rotational position of the pencil. This rotational position is fed to Ultrasonic Motor controllers to rotate right arm manipulator. The output of the touch-sensor panel is transformed to translational position commands and sent to transputers to control three-axis stage. 3)The output of the force sensors installed in the pencil and the robot are amplified by the strain amplifiers and sent to the transputers. One of a transputer executes bilateral control task using both force information from the pencil and the robot as an independent task. Force feedback information is D/A converted and sent to the pencil to rotate the DC motor.

5. Experimental results

5.1. Force-feedback pencil

The followings are the fundamental results of characteristics of realized force-feedback pencil. Figure 7 shows step response from 0mm to 5mm displacement. The rise time is about 50ms. Figure 8 shows gain frequency response of sign wave inputs. It is the result of amplitude response when the output of 0.5Hz is set to be 0dB.

5.2. Micro teleoperation experiments

With the system, impact feeling and rugged feeling are successfully obtained by giving impact force to the robot sensor and executing such tasks as drawing the tip of the manipulator on sandpapers or sponges. Also the following micro teleoperational tasks are successfully executed as pushing a component of micro air turbine which is about 450μm in diameter to desired place, suturing an artificial blood vessel, producing micro models of Japanese tiny doll "kokeshi". Figure 9 shows a photograph of a "kokeshi." The "kokeshi" is actually carved out of wood and painted to imitate a girl. With the system, a micro "kokeshi" is carved out of a red colored lead of mechanical pencil. The micro "kokeshi" is 200μm in diameter and its face has eyes and mouse. Figure 10 and figure 11 shows a photograph of the micro "kokeshi" and scene of tele-carving the micro "kokeshi" using the proposed system. These micro-scaled tasks includes complicated three-dimensional operation such as suturing, carving, sweeping, or writing.

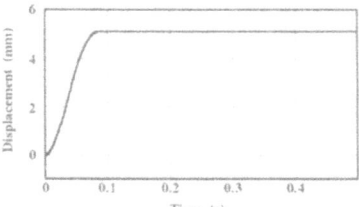

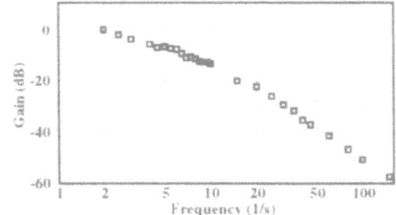

Figure 7. Step Response   Figure 8. Gain Frequency Response

Figure 9. A photograph of Tiny   Figure 10. A photograph of
doll "Kokeshi"   Micro "Kokeshi"

6. Conclusions

We proposed a novel configuration of a teleoperation system suitable for execution of micro tasks. The system consists of the upward CRT display monitor and an force-feedback pencil. The Operator applies necessary operation to the image of target object on the monitor by utilizing the force-feedback pencil. The characteristics of the system can be summarized as follows: (1)The system enables intuitive teleoperation because it realizes the coincidence of operating and force sensing coordinates with monitoring coordinates. (2)The pencil-shaped master manipulator enables sensitive bilateral micro operation because of the lightness and small inertia of its moving components (3)Handling by pen-holder grip enables the operator to command with force feedback which has wide dynamic range, because the operator can feel fedback force by both pressure perception receptor of fingertips and muscle stretch perception receptor of fingers during operation.

With the system such micro teleoperation tasks are successfully executed as pushing a parts of micro air turbine and suturing a blood vessel. These execution reveal that the system configuration are reasonable.

A more sensitive multi-axis force sensor with high stiffness will be further developed to avoid flipping the objects away. The technique to realize multi-

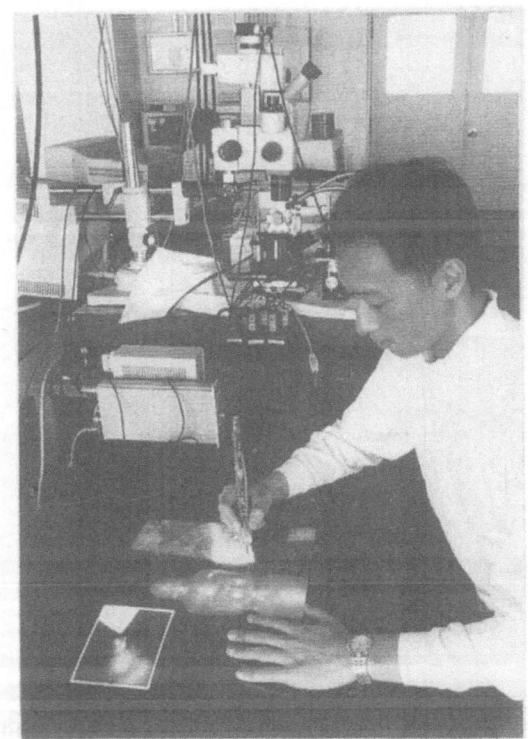

Figure 11. Scene of Tele-Carving a Micro "Kokeshi" by the System

axis force feedback without enlarging and weighting the pencil is also being implemented.

Acknowledgements

We would like to acknowledge to Mr.Yoshinobu Kotani for the contribution in development of the force-feedback pencil. This research is performed under support of grant-in-aid for scientific research of Ministry of Education.

References

[1] Osuga,S., "Human Interface", Ohmsha Inc., 1992, in Japanese.

[2] Caldwell,D.,Gosney,C., "Multi-Modal Tactile Sensing and Feedback(Tele-Taction) for Enhanced Tele-Manipulator Control," *Proceedings IEEE/RSJ International conference on Intelligent Robots and Systems* Vol.3,pp.1487-1494,Jul 1993.

[3] Takahashi,K., "Neuro Circuit and Living Body Control", Biological Science Lec-

ture, Vol.6, Mar 1976, in Japanese.

[4] Buss,M.,and Hashimoto,H., "Information and Power Flow During Skill Acquisition for the Intelligent Assisting System - IAS," *Proceedings IEEE/RSJ International conference on Intelligent Robots and Systems* Vol.1,pp.25-32,Jul 1993.

[5] Mitsuishi,M.,Warisawa,S.,Hatamura,Y.,Nagao,T.,and Kramer,B., "Human-Friendly Operating System for Hyper-Environments," *Journal of Robotics and Mechatronics* Vol.4 No.2,pp.167-178,Jan 1992.

[6] Sato,T.,Koyano,K.,Nakao,M.,and Hatamura,Y., "Novel Manipulator for Micro Object Handling as Interface Between Micro and Human Worlds," *Proceedings IEEE/RSJ International conference on Intelligent Robots and Systems* Vol.3,pp.1674-1681,Jul 1993.

[7] Hatamura,Y.,and Morishita,H., "Direct Coupling System Between Nanometer World and Human World," *Proceedings IEEE Micro Electro Mechanical Systems,* pp.203-208, 1990.

[8] Hunter,I.,Lafontaine,S.,Nielsen,P.,Hunter,P.,and Hollerbach,J., "Manipulation and Dynamic Mechanical Testing of Microscopic Objects Using a Tele-Micro-Robot System," *IEEE Control Systems Magazine,* pp.3-9,Feb 1990.

[9] Farry,K and Walker,I., "Myoelectric Teleoperation of a Complex Robotic Hand," *Proceedings IEEE International Conference on Robotics And Automation,* Vol.3,pp.502-509,Jul 1993.

[10] Hudgins,B.,Parker,P.,Scott,N., "A New Strategy for Multifunction Myoelectric Control," *IEEE Transitions on Biomedical Engineering,* Vol.40,No.1,pp.82-94,Jan 1993.

[11] M,Hirose., "Virtual Reality," Industrial Library, 1993. in Japanese.

Teleoperation with Virtual Force Feedback

Robert J. Anderson

Sandia National Laboratories
Intelligent Machine Systems Division
Albuquerque, NM[1]

Abstract — In this paper we describe an algorithm for generating virtual forces in a bilateral teleoperator system. The virtual forces are generated from a world model and are used to provide real-time obstacle avoidance and guidance capabilities. The algorithm requires that the slave's tool and every object in the environment be decomposed into convex polyhedral primitives. Intrusion distance and extraction vectors are then derived at every time step by applying Gilbert's polyhedra distance algorithm, which has been adapted for the task. This information is then used to determine the compression and location of nonlinear virtual spring-dampers whose total force is summed and applied to the manipulator/teleoperator system. Experimental results validate the whole approach, showing that it is possible to compute the algorithm and generate realistic, useful psuedo forces for a bilateral teleoperator system using standard VME bus hardware.

1. Introduction

Sandia National Laboratories has been developing telerobotic technology for environmental restoration, waste management and agile manufacturing. Our goal is to achieve the optimal blend of human and robot behaviors for a given task. One of the prerequisites for intelligent teleoperation is the ability to convey world model information to the operator. This is useful for preventing misguided human operator commands (e.g., knocking out the windows in a glove box, or breaching a waste tank container's walls) for intelligently guiding redundant robots through cluttered environments (e.g., for decommissioning nuclear facilities) or for creating entirely virtual worlds for operator training and task planning. In this paper we describe the development of a real-time, high-resolution 6 DOF, obstacle avoidance representation and avoidance capability for telerobotic systems, using a priori world model information to generate force barriers. The algorithm not only prevents inadvertent collisions, but generates intuitive contact information which the operator can feel using a force-reflecting master.

Sandia's search for flexible, modular teleoperator control systems has lead to the development of SMART (Sequential Modular Architecture for Robotics and Teleoperation) which is currently being used at multiple sites inside and outside of Sandia [1]. SMART is a real-time telerobotic control architecture which allows the

[1] This work was performed at Sandia National Laboratories and supported by the U.S. Department of Energy under contract DE-AC04-76DP00789.

user to construct a telerobot system using independent modules describing input devices (e.g., space ball, force reflecting masters), manipulators (e.g., PUMA, Schilling Titan) sensors (proximity sensors, force sensors). Each module represents a one-port or a two-port network element which can either perturb the force or the velocity of the teleoperator system. This paper describes the development of a module which will generate large forces whenever the manipulator tool comes into close proximity to a modeled object in the environment. Additional modules can be combined with this module to achieve force reflection, force compliance or obstacle avoidance behaviors for the manipulator. By attaching a force-reflecting master, the operator can feel virtual forces [2].

2. Approach

Using the network based passivity philosophy of SMART we decided to generate boundary functions around obstacles using non-linear springs and dampers, where the spring/damper combination would provide zero force disturbance outside a given threshold region, and would ramp up to a large force at the surface of the object. Figure 1 shows a diagram to illustrate this approach. As the gripper approaches the corner workpiece, computer generated spring-dampers push away the gripper from points of nearest contact. The method of using nonlinear springs is similar in concept to the potential field approach [3], but substantially different in implementation. Here, only object geometry is embedded and no attraction functions exist. The springs provide a barrier function over a very limited region, and serve only to repel. The influence function is zero over the vast majority of the workspace.

In order to calculate the forces from a number of nonlinear springs we developed an algorithm which would compute the distances and the points of nearest contact in real-time. The first step in the development of this algorithm is data representation. We considered three candidates, memory map, C-space and convex polyhedra primitives. Memory map techniques represent a number of methods by which the Cartesian space of the manipulator has been segmented into a 3-D array, and geometric data (e.g. occupancy or object potential) is stored at each location. These methods, although offering great speed, do not readily deal with the tool's geometry, and cannot be readily queried for surface normal information. C-space methods [4], we felt could deal with tool geometry, but would be difficult to apply to systems having 5 or more DOF. Both C-space and Cartesian memory-map approaches we felt would have resolution problems and would not be easily derivable from our polyhedra based graphical representation. Thus, by default, we decided to decompose the environment into convex polyhedra.

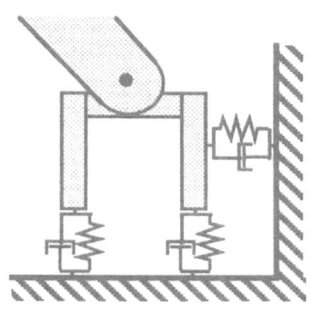

Figure 1. Use of spring-dampers for simulating contact.

368

Consider the gripper shown in Figure 2a. It can be decomposed into convex polyhedra in many ways. We prefer to choose a decomposition consisting of the minimum number of sections using the maximum amount of overlap. We want the minimum number because it reduces the computation required, and we want maximal overlap to reduce the likelihood of driving the system to a zero-potential at the interface of two polyhedra. Such a decomposition is shown in Fig 2b. The overlap areas are shown in grey.

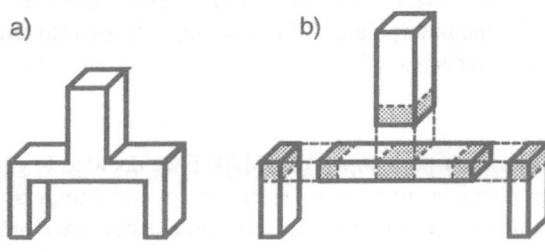

Figure 2. Convex object decomposed into primitives.
a) original object; b) decomposed object with shading
showing the overlap.

Once every object of interest in the workspace has been decomposed into convex polyhedra, and the manipulator tool has been similarly decomposed, we need to determine when object pairs (i.e., one object from the set of gripper polyhedra, and one object from the set of environment polyhedra) come into close proximity. Furthermore, this needs to be done at a high sampling rate in order to achieve realistic stiffness and a minimal region of perturbation. After studying various possible approaches [5, 6], we decided to implement an algorithm based on Gilbert's algorithm [7] using a few enhancements. The basic algorithm computes the unique distance and the not necessarily unique vector between the convex hulls of two sets of points in Cartesian space. The convex hull of a set of points generates a polyhedron. The algorithm has two main drawbacks. It won't return anything if the two obstacles are in collision, and if two objects have parallel edges or sides the algorithm will only return a single vector with no information about the type of contact. The next two subsections explains how we deal with these drawbacks.

2.1 Dealing with object collisions

The first problem with Gilbert's algorithm is that it gives no informative answer when objects are in collision. It comes back immediately telling when two polytopes overlap, but does not tell which direction the polytopes should move to best extricate themselves. For our real time architecture this was unacceptable. Because our real-time system can only generate a boundary function spring with a finite maximum stiffness (infinite proportional gain is difficult to implement in a sampled data system), we cannot guarantee a priori, that a barrier won't be breached. Thus actual collision of polytopes is always possible. In some cases collision is desirable as we try to simulate membranes and other soft tissue which can naturally be penetrated.

The simplest solution is to drive the objects apart in the opposite direction from the initial collision direction. This solution is unacceptable for sustained collision however, since movement along some other direction might provide quicker disentanglement. To explain this scenario, consider Figure 3, which shows a small rectangle as it moves into a box from the left side. In Fig 3a the polytopes first collide and the extraction vector should oppose the motion. In Fig 3b however the rectangle has intruded substantially into the rectangle, and the extraction vector should be at some angle. Finally, by Fig 3c the rectangle should be extracted by directing a force upwards.

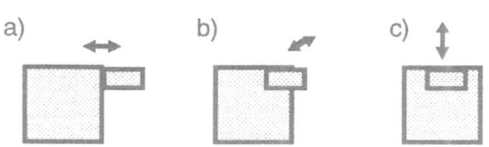

Figure 3. Determining extraction vectors for objects in collision a) Objects just collided; b) Objects after substantial intrusion; c) Objects after full intrusion.

Our solution to this problem is to use reduced primitives. For each convex object primitive we generate a linked list of sub-polytopes such that each sub-polytope is fully contained and centered inside a delta window of its parent. Each sub-polytope has its own embedded sub-polytope until finally the original polytope is reduced to a point. Figure 4 gives several examples. For example a cube can be reduced to any number of smaller cubes, and finally to a point. A long square box can be reduced to a line then to a point. An arbitrary box can be reduced to a plane, then a line, and then finally to a point. A cylinder can be reduced to a line and then to a point.

Using the reduced polytope concept, if two convex objects are found to be in collision then the same test is made using smaller and smaller sub-polytopes until no collision condition exists. The nearest distance vector between the sub-polytopes is used as the extraction vector for the parent polytopes, and the "distance" is derived from the computed sub-polytope distance and the combined delta functions. The beauty of this approach is that the same algorithm can be used, and that time spent in computing overlapping polytopes can be recovered since the sub-polytopes typically contain fewer vertices and are thus faster to compute.

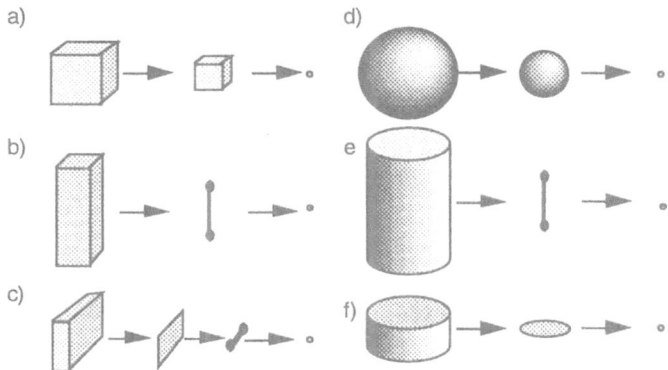

Figure 4. Convex object reductions: a) cube; b) long square box;
c) generic box; d) sphere; e) cylinder; f) disk.

370

Furthermore, the distance function between objects can be made continuous and monotonically non-increasing (although not monotonically decreasing) as a function of object position. This is important since the total spring force is a direct function of position and should not jump in magnitude for an infinitesimally small change of the object's position.

The following two-dimensional example clarifies the issue. Consider the triangle moving toward the rectangle in Fig. 5. The thick lines show the polytope used for making the distance and extraction vector calculations. In samples a) through d) the extraction vector will point to the right. In sample e) it will point straight up, and in samples f) thru h) it will point to the left. The grey area on the triangle shows the "equidistance" region on the triangle. When the nearest distance vector points to this region on the triangle, the distance function stays constant. This is illustrated in Figure 6, which gives the distance function for the example shown in Figure 5. The sub-polytopes method allows us to use Gilbert's algorithm without significant modification and allows us to achieve a suitable extraction vector whenever two objects collide.

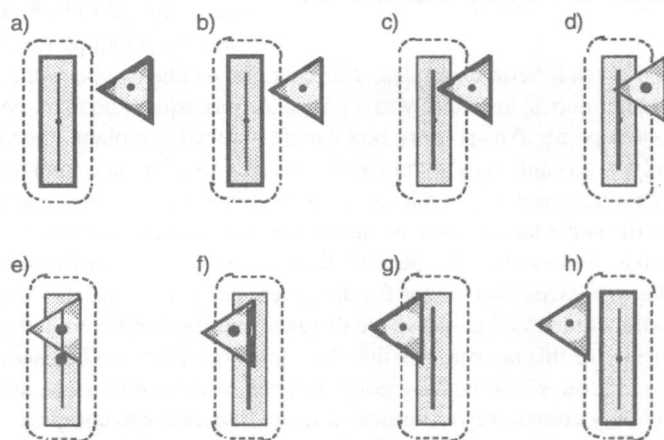

Figure 5. Example showing use of sub-polytopes to determine extrication vectors and distances:a) no collision; b) within delta window; c) Using rectangle line sub-polytope; d) using triangle point sub-polytope; e) Using both point sub-polytopes; f) Using line sub-polytope; g) Using-full triangle polytope; h) Using both full polytopes.

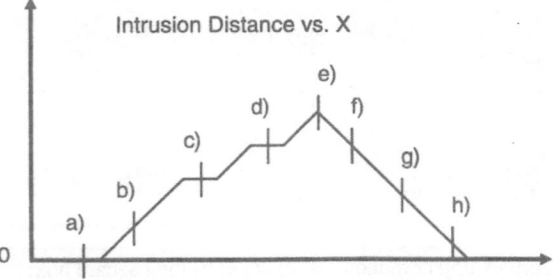

Figure 6. Distance function for rectangle/triangle collision path example of Figure 5.

2.2. Determining the type of contact

Rigid three dimensional convex polytopes can interact in a finite number of ways. Corners can collide with surfaces, edges or other corners. Edges can align in parallel or skewed. A summary of edge contact conditions is given in [8, 9], and numerous examples are shown in Figure 7.

Gilbert's algorithm does not distinguish between the type of contact, it just returns a unique nearest distance, and a not necessarily unique direction vector and interaction point. Thus instantaneously there is no way to distinguish between point-to-point, point-to-edge, edge-to-edge, point-to-surface, edge-to-surface and surface-to-surface contacts. Furthermore, if surfaces and edges are parallel the algorithm typically returns a vertex rather than a midpoint.

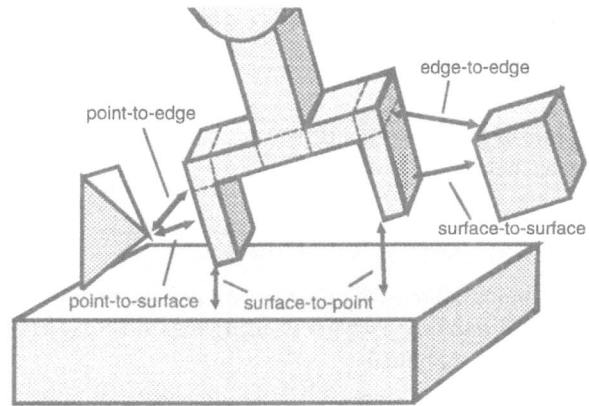

Figure 7. Surface Contact types.

This originally caused concern since we thought an interaction point at the middle of a parallel surface made the most sense for representing parallel contact. Later we conjectured that the instantaneous determination of contact type and the exact location of nearest contact vectors was inconsequential. What mattered was how the system reacts to instantaneous rotations. If small shifts in orientation are opposed in two directions there is surface contact, if opposed in one there is edge contact, if opposed in no directions there is point contact. In essence, the type of contact did not need to be determined instantaneously, but could be determined over time by gauging the net effect of reaction forces.

Consider the parallel plate interaction shown in Figure 8. The distance algorithm might return any of the four points (A, B, C or D) as being the nearest point. The algorithm presented in this paper will generate a virtual force at this point, which due to the impedance algorithm implemented for the manipulator will result in that vertex rotating away from the surface. This will cause another vertex to become the "nearest" vertex, and the algorithm will next rotate that corner away. This will continue indefinitely. The extrusion vector will race all across the surface of

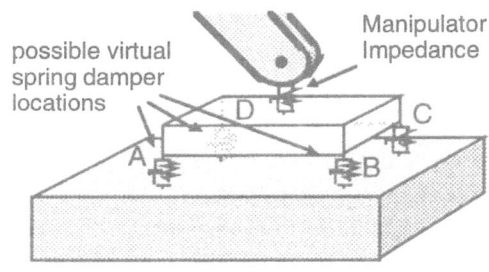

Figure 8: Parallel plate interaction.

the tool opposing any infinitesimal infraction, and the plates should remain in parallel

The question to be answered through experimental work was would it work? Could an algorithm which utilized rapid switching of virtual spring contact points achieve smooth tool-environment interactions through some type of averaging phenomena, or would objects chatter wildly whenever brought near parallel? Could the algorithm be computed fast enough using commercial hardware to make it viable for building virtual force fields in a telerobotic system? Was determination of contact type unecessary for developing obstacle avoidance algorithms?

3. Algorithm Summary

Before we describe the results of the experiments, we summarize the algorithm used for generating virtual forces below:

Virtual Force Algorithm

I. Decompose tool and the environment into overlapping convex object primitives.
    A. Divide environment and tool into convex objects, using maximal overlap.
    B. Approximate convex objects by polytopes.
    C. For each convex object primitive determine a list of embedded sub-polytopes ending in a point.

II. Determine object interactions.
    A. Use bounding boxes to eliminate object pairs which are definitely outside a delta region.
    B. Apply Gilbert algorithm (using the last computed distance vectors as a starting point) to all remaining pairs of objects.
    C. For any objects pairs found to be in collision use the reduced polyhedra until an intrusion distance, an interaction point, and an extrusion vector can be determined.

III. Determine force on tool.
    A. For each pair within a delta distance, compute the force (f_i) as a function of intrusion distance based on non-linear spring/damper (spring constant changes from 0 at object interaction boundary to maximum value at contact point).
    B. Apply repulsive force element along the extrusion vector at the interaction point.
    C. Determine net force (f_{TOOL}) and moment (N_{TOOL}) $f_{TOOL} = \Sigma f_i$
    on tool by summing up repulsive forces (f_i), and
    computing sum of cross products of the distances $N_{TOOL} = \Sigma d_i \times f_i$
    to the point of nearest contact (d_i) .

IV. Compute motion of manipulator based on impedance law.
    A. Add force from world model interaction to forces from other inputs and sensors.
    B. Compute delta change of position based on local dynamics and impedance.
    C. Go back to II.

---

4. Experimental Setup

The algorithm was implemented in C using Gilbert's code as a starting point. Gilbert's algorithm utilizes varying amounts of iteration, requiring irregular amounts of computation time. For instance it typically takes more time to compute extraction vectors for objects near parallel. Because each iteration of the algorithm had to be computed in a finite time step the following approach was taken. Every time stop the Gilbert algorithm was run on as many convex object pairs as could be computed. If all pairs couldn't be computed in the time step then old vector/distance data would be used for that pair during that time sample.

The module representing obstacle avoidance was incorporated into the SMART architecture as the OBSTACLE module. SMART is a multi-processor based real-time telerobot control architecure which runs under VxWorks in a VME-Bus environment. The OBSTACLE module itself was run on a Mercury MC860 Intel I860 based attached processor. Update rates of 400 Hz were achieved for a tool consisting of 4 convex objects, and an environment consisting of 8 convex objects.

The SMART module connection diagram for this experiment is shown in Figure 9, and uses the following modules: A torque arm for 1 DOF force reflection, a MULTIPLEX module for coupling a 1 DOF device to 6 DOF and scaling forces and velocities, a SPACEBALL module for enabling 6 DOF unilateral motion from a Dimension 6 Force Ball, a RELIEF module to enable compliance along arbitrary DOF, a KBB2 module to provide filtering and reduce wave reflections, the OBSTACLE module implementing the algorithm described here, a PUMA_KIN module for mapping the PUMA's kinematics from world space to joint space, a LIMITS module for imposing joint limit restrictions, a VISUAL module for displaying the system in a 3-D modeling environment (Deneb's IGRIP) and a PUMA_JOINTS module to connect to the PUMA 560 hardware using a VAL II controller and Unimation's SLAVE software.

The system was run using the two input devices and with various compliance modes. The simulated gripper was driven into environment objects at various angles. At all times the resulting motion of the gripper was smooth and intuitive. No chattering was ever observed for any type of contact. In addition the virtual forces provided more than obstacle avoidance capabilities, they generated constraint forces which could be utilized to achieve task objectives. For example, tools could be aligned with virtual surfaces and slid into virtual corners.

5. Conclusions

The experiments described in this work demonstrated that it is possible to use a very simple algorithm with no knowledge of interaction types, to achieve smooth, intuitive, object interactions for a complex gripper tool working in a complex environment. Although requiring substantial computation every sample instant, the computation can be achieved using relatively inexpensive off the shelf computer hardware.

374

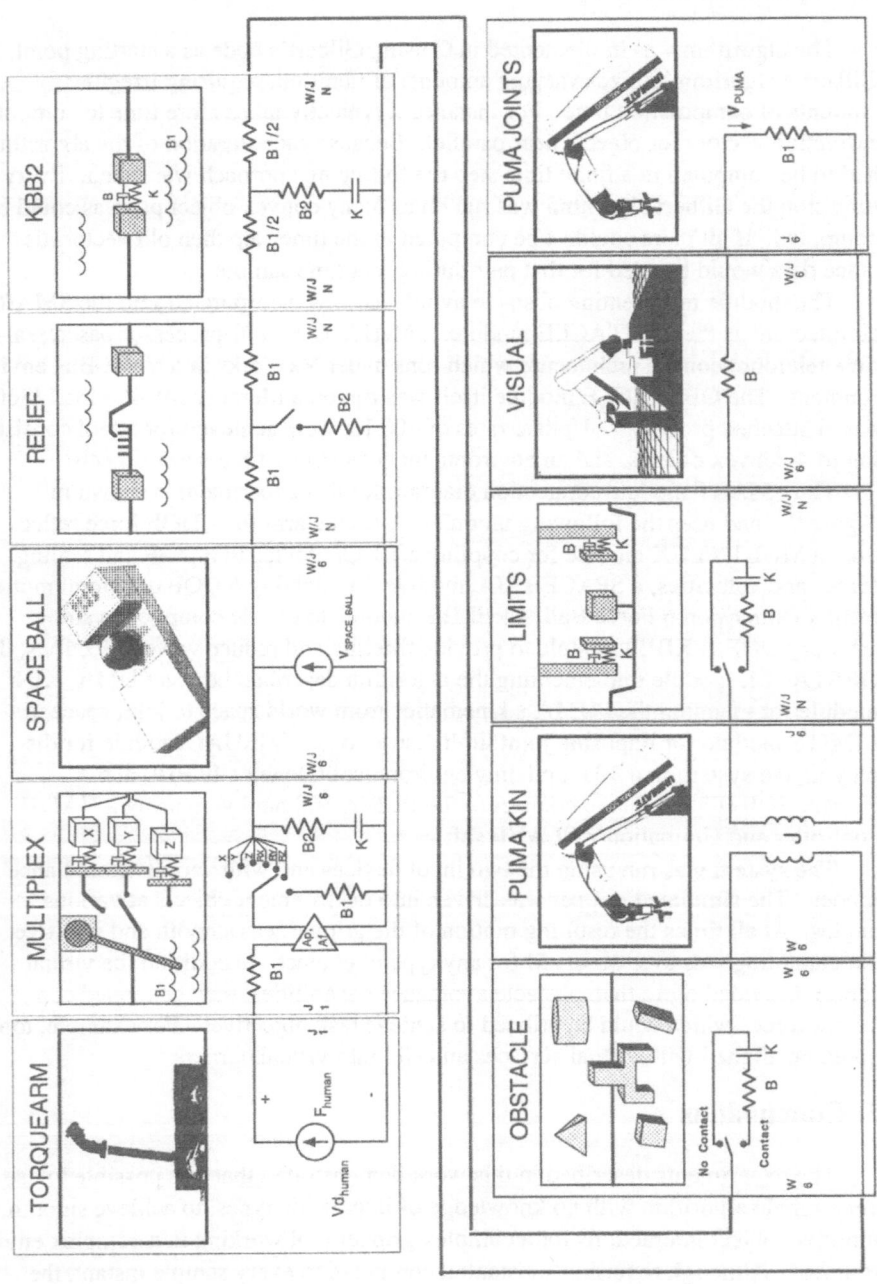

Figure 9: SMART Modules used for experimental validation.

The SMART module deriving from this work (i.e., the OBSTACLE module) is now a core module of the SMART architecture and virtual forces generated from this module can be used in addition to the existing sensor feedback and simulated dynamics, to control any type of robot. In particular, it is being applied to obstacle avoidance in our underground storage tank laboratory and for task planning in our waste processing operations lab, and is providing virtual force information for a kinesthetic virtual reality laboratory.

Currently, we are extending the virtual force concept to the entire arm, rather than just the tool. In the case of the PUMA this requires modeling the arm and forearm as two addition rigid body polyhedra moving with respect to the same set of fixed polyhedra in the environment. The net virtual force signal will then be mapped back into joint space using the appropriate Jacobians to achieve the desired behavior.

References

[1] R. J. Anderson, "SMART: A Modular Architecture for Robotics and Teleoperation", *Proceedings of the Fourth ISRAM*, 1992, Santa Fe, NM, pp 467-474.

[2] W. S. Kim, "Developments of New Force Reflecting Control Schemes and an Application to a Teleoperation Training Simulator," *1992 IEEE Int. Conf. on R&A*, pp 261-266, Nice.

[3] O., Khatib, , "Real-time Obstacle Avoidance for Manipulators and Mobile Robots", *The International Journal of Robotics Research*, Vol 5, No 1, Spring 1986.

[4] T. Wikman, and W. S. Newman, "A Fast On-line Collision Avoidance Method for a Kinematically Redundant Manipulator Based on Reflex Control", *1992 IEEE Int. Conf. on R&A*, pp 1412-1419, Nice.

[5] J. E. Bobrow, "A Direct Minimization Approach for Obtaining the Distance Convex Polyhedra", *The International Journal of Robotics Research*, Vol. 8, No. 3, 1989, 65-76.

[6] M. C. Lin, and Canny J. F., "A Fast Algorithm for Incremental Distance Calculation", *Proc. IEEE Int. Conf. on Robotics and Automation*, 1991, pp. 1008-1014.

[7] E. G. Gilbert, D. W. Johnson, and S. S. Keerthi, "A Fast Procedure for Computing the Distance Between Complex Objects in Three-Dimensional Space," *IEEE Journal of R&A*, vol. 4, no. 2, pp193-203, April 1988.

[8] S. Hirai and K. Iwata, "Recognition of Contact State Based on Geometric Model", *1992 IEEE Int. Conf. on R&A*, pp 1507-1518, Nice.

[9] M.T. Mason, "Compliance and Force Control for Computer Controlled Manipulators", *IEEE Trans on Systems, Man, and Cybernetics*, vol. 11, pp. 418-432, 1981.

Teleoperated Sensing by Grasping:
Robot Sensors and Human Psychophysics

S.R. Ganeshan[*]   R.E. Ellis[*†]   S.J. Lederman[‡*]

[*] Department of Computing and Information Science
[†] Department of Mechanical Engineering
[‡] Department of Psychology
Queen's University, Kingston, Ontario, Canada K7L 3N6

Abstract— An interesting problem for a robot that is manipulating an object in the environment, either directly or by using a tool, would be to perceive the length of the object and/or tool. This paper addresses the problem of length perception in a teleoperation context, using the sense of touch alone. To establish a relation between the forces acting on the robot gripper and the length of the gripped object, a rigorous mechanical analysis of the gripper-object system is performed. With regard to robotic length perception, a novel sensor that derives directly from the results obtained from the mechanical analysis is proposed. The mechanical analysis is also applied to the question of how humans perceive the length of a rod held stationary in the hand, and tested experimentally in a sequence of psychophysical investigations. The implication of this study for design of teleoperated systems under human force-reflection control is discussed.

1. Introduction

Teleoperated tasks often involve manipulation of objects in the environment, either directly or through tool use. In haptic manipulations involving unconstrained grasping of tools, we ask whether the human operator (or an autonomous robot system) can determine the characteristics of a tool simply by holding it. Rods may be viewed as one of the simplest tools a human operator might be asked to control remotely. In this paper, we address the task of determining the length of a uniform rod by grasping it.

We approach this problem in a series of four steps:
1. define the required task that needs to be accomplished
2. study the physical and other constraints that apply
3. determine a system that satisfies both the task and constraints
4. evaluate system performance.

We have chosen the task-driven model for teleoperation, rather than the technology-driven approach that attempts to determine the usefulness of a system developed from an existing equipment base, because a task-driven model better addresses the human-interface issues that arise in teleoperation.

2. Haptic perception of length by grasping

For a remote robot controlled by a human operator, accurate length perception may be important in situations where the object is manipulated in space-constrained environments: if the dimensions of the object are known, then the object can be manipulated (reoriented, relocated, etc.) without causing damage to other surrounding objects or to the robot itself. Perceiving length may also be required in situations where it is important to know the location at which the object is grasped, or in tasks that require further manipulation based on the knowledge of length (e.g., assembly of components of a space-station structure).

In the haptically teleoperated task, an object is grasped by a remote robot at the slave end. The forces are presented to the hand of a human operator, who controls the task at the master end. Thus, to fully understand the problem of haptic length perception within the teleoperation context, it is necessary to:

1. understand the nature of the interaction of the remote robot with the environment, particularly the effects of the environment on the robot due to the interaction, and

2. determine the human's ability to accomplish the task of estimating tool length when provided with appropriate haptic feedback.

The first goal lies within the domain of *machine perception* or *robotics*, while the second goal within the domain of *experimental psychology*. Thus, what links the seemingly diverse fields of experimental psychology and robotics is the more general goal of *telerobotics* (or more generally, teleoperation) whereby a human operator must control the activities of a robot in a remote workspace.

We approach this problem by first performing a force analysis based on the general principles of mechanics, devising a suitable system (for the remote robot) that fulfils all the mechanical constraints specified in the analysis, extending the results of the force analysis to the human haptic subsystem, and finally evaluating the performance of both systems.

2.1 Force analysis of the gripper-object system with respect to length perception

In order to perceive the length of a gripped object, it is necessary to begin by establishing the relation between the forces acting on the gripper and the length of the object. The term 'gripper' refers either to the end-effector of the remote robot equipped with force-torque sensors, or to the hand of the human operator.

In the force analysis of the gripper-object system, we make the following assumptions:

1) the gripped object is made of homogeneous material such that any element cut from the body possesses the same specific properties of the body.

2) the gripped object is uniform, i.e., it has the same geometric properties throughout its length.

3) the angle or orientation of the object with respect to the ground plane is assumed to be known.

4) the gripper-object system is at rest and in a state of static equilibrium where there is no macroscopic motion. This assumption excludes any situation where damage or permanent deformations may occur to either the object or the gripper due to excessive forces and torques.

When the object is gripped at one extreme end, the force, **F**, and torque, **T**, acting on the gripper are:

$$F = \rho_A L g \tag{1}$$

$$T = \frac{1}{2} \rho_A L^2 g \cos(\theta) \tag{2}$$

where, ρ_A is the mass per unit length of the object, L is the length of the object, g is the gravitational acceleration, and θ is the orientation of the object with respect to the ground plane. When the object is gripped at an intermediate location, the torque is given by:

$$T = -\frac{1}{2} \rho_A [(L-x)^2 - x^2] g \cos(\theta) \tag{3}$$

where x is the distance of the grip location from one end of the object. The gravitational acceleration g and the orientation of the object θ with respect to the ground are known quantities, the force **F** and torque **T** are measured quantities. The quantities ρ_A, L and x must be deduced.

The force analysis reveals that the length of the object becomes indeterminate when the object is held perpendicular to the ground plane, or at an intermediate location. To overcome the latter problem, it is proposed that two force-torque measurements be obtained by gripping the object at two independent locations simultaneously. If the two force-torque measurements are physically close to each other, the system can be easily split into two independent subsystems, and the length of the grasped object can be determined individually for both sides by using equations (1) and (2).

3. Robotic perception of length

The results of the force analysis implied the need for further sensor development since the existing gripper and tactile technologies were inadequate. For references and reviews of current technology in tactile sensing, see Ellis, et. al. [4], Regtien [11] and Nicholls & Lee [9]. To provide adequate sensing in the gripper, we developed a novel sensor based on thin-plate deformation.

3.1 Design of robotic sensor

The proposed sensor employs detection and use of force information to predict the length (and mass) of the gripped object. For reviews of the detection of force information, the interested reader is referred to Brock [2], Bicchi, et. al. [1], and Okada [10].

The sensor consists of two sets of plate arrangements. Each plate arrangement constitutes two thin plates arranged such that they directly face each other, i.e., one plate is mounted on the top inner surface of the gripper while the other is mounted on the inner bottom. Each plate is a plane structure whose thickness is very small when compared to its other two dimensions, and bound geometrically by straight lines. For ease of design and analysis, only square plates were considered.

The boundary conditions prescribed for the sensor plates were two *fixed* supports at two opposite edges, with the other two edges *simply supported* (Fig. 1). Simply supported edges offer adequate sensitivity to the sensor plate and also make it relatively easy to obtain analytical solutions, whereas fixed supports offer ruggedness to the design. Since the analysis is relatively straight-forward when the structure is symmetrical, the boundary conditions for the opposite edges were chosen to be the same.

With the object gripped by the proposed sensor arrangement, the predicted loading pattern on each sensor plate is essentially a non-uniform one (Fig. 2). The sensor plates may be subjected to loads that are present on the entire or partial surface of the plate, or they may be line loads. Here, only the case in which the sensor plates are loaded on the entire surface is considered. For a more detailed discussion of different loading patterns, the reader is referred to Ellis, Ganeshan & Lederman [4].

For a square plate with sides of length 'b', the sides $x = 0$ and $x = b$ simply supported and the sides $y = \pm b/2$ fixed, the loading function (assuming that the load varies in the x-direction only) may be represented by the following series (Timoshenko & Krieger [14]):

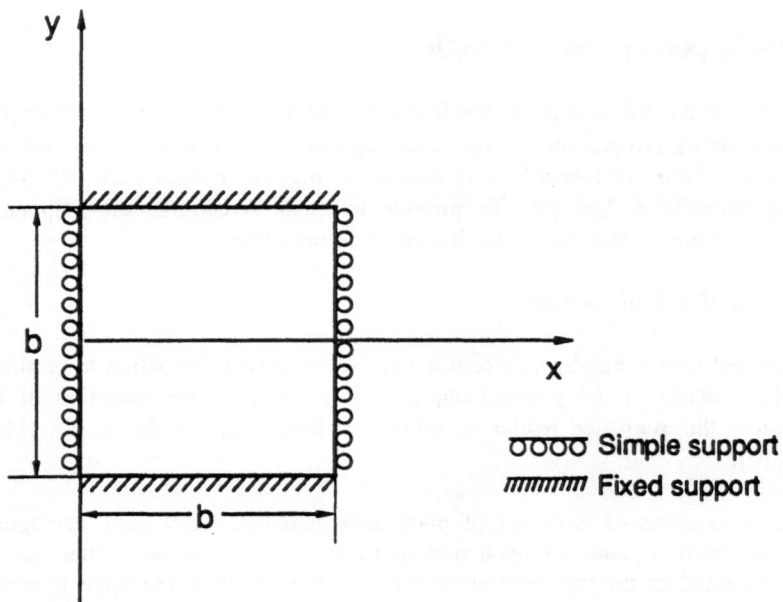

Figure 1 Boundary Conditions for the Proposed Sensor Plate

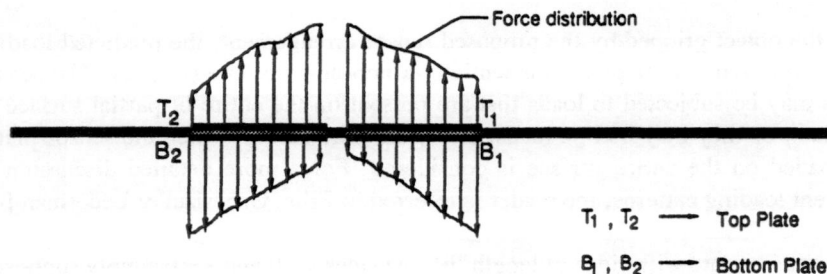

Figure 2 Predicted Loading Pattern on Sensor Plates

$$F_z = \sum_{i=1}^{\infty} F_i \sin\left[\frac{i\pi x}{b}\right] \qquad (4)$$

where, F_z is the lateral load acting on the surface of the plate. The strain ϵ_x (x,y) at a particular location on the plate is then given by:

$$\epsilon_x(x,y) = \frac{1}{2}\left[\frac{h\pi^2}{b^2}\right]\sum_{i=1}^{\infty}\left[\frac{b^4 F_i}{i^4\pi^4 D} + A_i\cosh\left(\frac{i\pi y}{b}\right) + B_i\left(\frac{i\pi y}{b}\right)\sinh\left(\frac{i\pi y}{b}\right)\right]\sin\left(\frac{i\pi x}{b}\right)i^2 \qquad (5)$$

where,

ϵ_x is the strain developed in the x-direction   of the plate
h   is the thickness of the plate
b   is the width of the plate
D   is the flexural rigidity of the plate
A_i, B_i are the coefficients

$$A_i = -\frac{b^4 F_i}{i^4\pi^4 D}\left[\frac{1}{\cosh(\alpha_i)} + \beta_i\alpha_i\tanh(\alpha_i)\right]$$

$$B_i = -\frac{b^4 F_i\beta_i}{i^4\pi^4 D}$$

$$\alpha_i = \left(\frac{i\pi}{2}\right)$$

$$\beta_i = \frac{\tanh(\alpha_i)}{\sinh(\alpha_i) + \alpha_i\cosh(\alpha_i) - \alpha_i\tanh(\alpha_i)\sinh(\alpha_i)}$$

If the material and geometric properties of the plate are known, the strain ϵ_x can be determined to a desired level of accuracy by choosing the number of terms required to represent the load equation. For most loading conditions, only the first few terms need to be considered. For four terms, an accuracy of 98% has been reported for the triangular loading pattern, which is similar to the loading pattern prevalent in the present situation (Timoshenko & Krieger [14]). Rearranging the terms, we obtain:

$$\epsilon = Cf$$

where C is a constant representing the compliance of the plate. Since, C is a square matrix, given that its determinant is not equal to zero, the force set **f** can be expressed as

$$f = C^{-1}\epsilon$$

3.2 Experimental design and methodology

In this section, the mechanical design and the experimental method used to evaluate the proposed sensor are presented. A strip of low carbon stainless steel was bent at right angles at a chosen distance from both ends. The flat portion was then used as the sensing surface, and the bent portions were used to simulate the fixed boundaries of the sensor plate. For the simple supports, aluminum-channel sections with razor-sharp edges were used. The channel section and the plate arrangement were then screwed to a wooden base block that was attached to a bench vise.

Four electrical strain gauges were used for measuring the strain on each plate. These gauges were mounted on the bottom surface of the plate along the centre line between the simple supports, and the locations (x and y coordinates with respect to the origin) of these gauges were recorded. In order to account for errors in strain gauge installation, such as misalignment, lack of proper bonding, etc., the gauge readings were calibrated against a known value of strain.

Four different objects were used in the experiment. The width of all four objects was the same as that of the plate. Two of the objects were of oak wood, one was of pine and the other was of aluminum. Each of these objects was gripped at an intermediate location between its ends. The total mass for each of the objects, and the lengths extending on the two sides of the plate sensor are shown in Table 1.

When the object is gripped by both the left and right set of plate sensors, each set predicts the overhang length plus the length of the gripped portion of the object. Four different sets of readings were obtained for each rod from different trials, and the mean was used for the length estimates. The adapted strain-gauge calibration method could only provide for force calibration. In order to provide for torque calibration, a least-squares quadratic fit of the data was obtained and the estimated lengths were computed using the empirically derived equation:

$$CL = 0.1357 \times 10^{-03} (PL)^2 + 0.6852 (PL) + 51.78$$

where, CL is the length obtained after the above correction is applied, and PL corresponds to the length perceived by the plate sensor.

3.3 Results

The length perception results are shown in Table 2, where the mean perceived length is the mean of the four different trials. After the torque correction was applied, the estimated length yielded an average error of 3.82% (standard deviation = 2.17). The average error in mass perception was found to be 1.6 % (standard deviation = 1.14).

Thus, rod length can be estimated to within 6% accuracy using the proposed plate sensor; but by downsizing the strain gauge lengths, we expect that measurement error can be further reduced. The results obtained for the 60 mm length were found to be spurious as they deviated greatly from the average results; hence they were discarded as a statistical outlier. The strain developed for the 60mm-length rod was in the low range of 1 - 5 μstrain, and the installed strain gauges were not sensitive enough to record them accurately.

4. Human perception of haptically derived length

Earlier in the paper, we argued that in order to fully understand the problem of length perception within the teleoperation context, it is not only necessary to study the remote robotic system, but to also investigate the human haptic system. A necessary and essential step in this investigation is to determine the human's ability to accomplish the task of estimating object length by grasping under conditions of restricted exploration. Here, restricted exploration refers to the situation where the objects are held stationary in the hand (no translatory or rotary motion).

The problem of haptic length perception has been approached from various viewpoints by a number of researchers (Jastrow [7], Teghtsoonian and Teghtsoonian [13]). However, a number of more recent experiments directed at the perception of length in humans are based on an entirely different approach. These studies involve wielding or hefting, or just holding the stimulus still in the hand (Carello, et. al. [3], Solomon & Turvey [12]).

4.1 Length Perception - Psychophysical or Cognitive?

When a rod is held in the hand, stresses and strains are developed there due to the effects of the action of forces. These direct physical inputs cause the deformation of the relevant tissues of the haptic system, from which the metric information regarding the forces and torques is transduced. Based on the force analysis presented earlier, we propose that the perceived length of unseen objects held stationary in the hand depends on both the force applied by the rod and on the torque subtended by the rod at the grip (Force-Torque model).

It is important to note that in certain situations, as in the case where the rod is held perpendicular to the ground plane or gripped at its centre of mass, torque is unavailable; in certain other situations, as in the case where the rod is held at different orientations with respect to the ground plane, the weight may remain constant while the torque varies. It is hypothesized that the weight of the rod and the torque subtended at the grip will both contribute to the perception of the length of unseen objects, their strengths depending on the extent to which each source is available.

In the pilot studies that were conducted prior to the actual formal experiment, subjects asked to provide psychological reports after the completion of the experiments reported that they first perceived weight, and then used these estimates to perceive length. These reports suggested that length estimates are probably derived from the corresponding weight percepts. Since the cue related to weight is present in every situation, irrespective of the grasp location or the orientation of the rod with respect to the ground plane, this raises an important question for human length perception: "Is perception of length affected by the perception of weight?". One possible hypothesis is that length percepts are associated with the corresponding weight percepts. This approach to the problem of length perception is based on a more cognitive approach to perception that hypothesizes that perception of a certain quality of the object is based on the computations performed on other perceptions (Percept-Percept model). Thus, we approach the problem of human length perception from two theoretical viewpoints - the psychophysical approach that is based on the force/torque analysis, and the cognitive approach that is based on the theory that one percept is derived from another percept.

4.1.1 Experimental Methodology

In this human psychophysical study, subjects assigned numerical estimates in proportion to the perceived length of different statically grasped rods. In the experiment, the effects of different stimulus parameters were systematically assessed, including material composition of the rods, mass of the rod and orientation of the rod.

Subjects were presented with visually occluded rods of widely varying masses to obtain a broad range of values for the force and torque. Rods of three different lengths and three different masses were used. This design allows us to study the effect of the mass of the rods on perceived length, when actual length remains a constant; in other words, it allows us to see if different rods having the same length but different mass (due to different densities) are *perceived* as having the same length. To decorrelate the weight of the rods from the torque subtended at the grip, each rod was also presented in three orientations - vertical, horizontal or at 45 degrees to the ground plane. In addition to the length estimates, subjects also responded with weight estimates. The lengths, masses, forces and torques for each of the nine different stimulus rods, each held at three different orientations, are as shown in Table 3.

4.1.2 Results

The length estimates and weight estimates were initially transformed logarithmically (base 10) as human psychophysical functions are usually best described by a power function. Length estimates increased with increasing length and decreasing orientation (vertical, oblique and horizontal). They also varied consistently with material increasing in the following order: wood, plastic and aluminum. In other words, rods of the same length but different materials were perceived to be of different lengths: the wooden rods were perceived to be the shortest, while the aluminum rods were perceived to be the longest. It was also found that the effect of rod orientation was smaller for shorter, wooden rods. The linear equation obtained for the psychophysical model that predicts $\log_{10}$ perceived length from $\log_{10}$ actual weight and torque (since simple correlations with perceived length were a little higher when these parameters were used) is as follows:

$$PL = 1.23 \log \text{Force} + 0.26 \log \text{Torque} + 0.26$$

where PL is the perceived length.

Weight estimates similarly varied with material, in the order wood, plastic, and aluminum. Overall, the weight estimates varied in strikingly similar fashion to the length estimates ($r = 0.984$). Also, both length and weight magnitude estimates were more strongly related to the logarithmically transformed values of weight ($r = 0.93$ and 0.95, respectively) than to torque ($r = 0.74$ and 0.75, respectively).

To evaluate the percept-percept model, a multiple regression was first performed in which log weight estimates were predicted from log weight and torque:

$$PWT = 1.23 \log \text{Force} + 0.26 \log \text{Torque} + 0.26$$

where PWT represents the weight estimates. The R^2 value was found to be 0.965. The phenomenological reports of the subjects also suggested a dependency of length estimates on perceived weight. Eleven out of fifteen subjects volunteered that they based their judgements of length on the estimated weight of the object, while the other four suggested that they were influenced by the twisting effect felt at the grip.

The results of the experiment suggest that both models - the force-torque model based on the psychophysical approach and the percept-percept model, can predict the length estimates exceptionally well, although the latter model may be marginally better.

4.2 Pitting the Force-Torque Model against the Percept-Percept Model

To differentiate between the relative effectiveness of the force-torque model and the percept-percept model, another experiment was conducted to effectively assess the role of the weight percept over and above the forces and torques in judging rod length. The technique we used involves inducing a static version of the haptic size/weight illusion in the subject. When subjects were presented with two objects of the same weight but different sizes, Ellis & Lederman [5] have reported that the larger one when hefted felt lighter than the smaller one. This illusion has been termed the haptic size/weight illusion. In order to pit the force-torque model against the percept-percept model, the subject was asked to statically grasp rods whose diameters (size) varied, while the physical length, weight and torque subtended at the grip remained unchanged (Ellis & Lederman [5]). The percept-percept model would predict that subjects should estimate the narrower diameter rod to weigh more than the larger diameter rod, and consequently, should further estimate its length to be longer. The force-torque model would predict no such differences as the force and torque at the grip remain unchanged.

4.2.1 Experimental Methodology

Blindfolded subjects were presented with rods that were held in a steady, horizontal position relative to the ground plane. One group assigned numerical estimates in proportion to the perceived length of the grasped rods, while another group made only weight estimates. Three pairs of rods were used in the experiment, with pair lengths of 22.9 cm (8"), 27.9 cm (11") and 35.6 cm (14"). The outer diameters of the small and large rod of each pair were 1.75 cm (11/16") and 76.3 cm (3"), respectively. The masses for each pair of rod lengths were 384, 525, and 666 g, respectively. The torques subtended at the grip by each pair were 431.3, 719.8, and 1162.9 N-m respectively.

4.2.2 Results

Subjects from the weight estimation group judged the narrower diameter rod to weigh more than the wider diameter rod. In the other group, subjects judged the narrower rod to be longer than the wider rod. The simple correlation coefficient between the perceived length and perceived estimates was higher (0.99) than with log force or log torque. The subjects' estimates of length were almost perfectly correlated with the weight estimates. Thus, we conclude that the percept-percept model explains the estimation of the perceived length of statically held rods best.

5. Discussion

Haptically teleoperated tasks are usually based on the contact paradigm, for example, inserting a peg in a hole. Another haptic manipulation task arises in the unconstrained grasping of tools, where we ask whether the human operator (or an autonomous robot system) can determine characteristics of a tool simply by holding it. This paper addresses a particular problem in machine and human perception concerned with acquiring the necessary information to determine the length of a gripped object by static contact. We believe the current task also poses interesting questions for a robot that must manipulate objects in the environment, whether directly or through the use of a tool. With regard to machine perception, we address the "forward" problem, namely how can a machine perceive the length of a statically gripped rod? In contrast, the question posed of human perception addresses the "inverse" problem: what environmental properties determine or contribute to length perception of statically held rods?

A mechanical analysis of the system consisting of the gripper and the object was first performed using the general principles of statics. Here, the term "gripper" refers to either the hand or the end-effector of a robot. The results obtained from the mechanical analysis revealed that the length of a gripped object is indeterminate when the object is held perpendicular to the ground plane or when it is gripped at a single intermediate location. To overcome the latter limitation with regard to robotic perception, we proposed that two simultaneous and independent force-torque measurements be obtained.

With respect to robotic length perception, this result implied that existing gripper and tactile technologies were not adequate, indicating the need for further sensor development. We proposed a unique and novel sensor design based on the theory of thin plates, using strain-gauge technology. Some of the advantages of this plate sensor are as listed below:

1. The sensor is designed to establish direct contact with the environment. Since forces and torques are transduced directly from the parameters of contact interaction between the sensor and object, measurement error is minimized.

2. The design provides economical and simple alternatives for transducing forces and torques.

3. The proposed sensor is sophisticated in design, requiring few modifications for directly mounting on the inner sides of the robot gripper.

4. An experimental evaluation of the proposed sensor suggests that it can be effectively employed to perceive the length of an object that is statically gripped at an arbitrary location.

5. The sensor can be made physically sturdy to withstand rough handling.

The results of the static analysis were also extended to the study of human length perception. Based on the analysis of forces and torques, two different models of length perception were proposed. The psychophysical model uses the forces and torques acting on the hand to predict the length estimates directly in a single processing step, while the percept-percept model uses the forces and torques first to predict the perceived weight, which is then used to predict perceived length.

The first experiment focused on investigating the effectiveness of both the force-torque and percept-percept models in predicting rod lengths. The results indicated that perceived length could be predicted by both models remarkably well, although perhaps the percept-percept model based on weight percepts was marginally better. The direct assessment of perceived weight was achieved in the second experiment by successfully inducing a static version of the haptic size-weight illusion. This technique allowed us to maintain the values of the mechanical variables constant, while altering the perceived weights of rods of equal length and weight but different diameters. As predicted, the manipulation produced a corresponding change in the length estimates: the narrow rod felt both heavier and longer than the correspondingly wider one. The phenomenological reports obtained from subjects also suggested that the length estimates could be derived from the estimates of weight. The weight percept proved to be a single variable that varied in remarkably similar fashion to the length percept for the rods of different materials across all three orientations (in the first experiment) and for rods of all sizes (in the second experiment). The second experiment pitted the force-torque model against the percept-percept model by inducing a haptic size/weight illusion: the results indicated that the perceived weight estimates could account for 98.4% of the variance in the length estimates. Overall, the percept-percept model can account for length estimates of rods varying in material, orientation, length and diameter.

The results from these parallel robotic and human psychophysical studies further suggest that the highly accurate forces and torques sensed by the robot may need to be appropriately altered when delivered as inputs to the operator's hand to counteract the biases documented in human length perception. Otherwise, the operator may perceive the rod length to change when the robot gripper alternates between tools of different materials, and when it alters tool orientation. Hence, providing length information directly in the form of forces and torques at the wrist and elbow of the human operator - without compensating for the effects of relevant perceived parameters - is likely to lead to considerable perceived distortion of a rod that is grasped by a remote teleoperated robot.

Table 1 :   Mass and length of objects used to experimentally evaluate the plate sensor

Material	Total Mass	Length (left)	Length (right)
Pine	229.2	60	1060
Aluminum	238.6	225	120
Oak	447.1	510	350
Oak	796.0	895	650

Table 2 :   Results of length perception from plate sensor

Actual Length (mm)	Mean perceived length (mm)	Mean Corrected length (mm)	Mean % Error	Standard Deviation (mm)
120	102.5	123	+2.5	1.14
225	223.5	211	-6.22	7.21
350	433.8	374	+6.86	4.34
510	582.3	497	-2.55	1.35
650	781.0	670	+3.08	2.22
895	957.0	853	-4.69	2.06
1060	1199.8	1069	+0.85	3.70

Table 3 :   Length, mass and values of the relevant mechanical variables pertaining to objects used in the human experiment

Material	Length (mm)	Total Mass (kg)	Force (N)	Torque (vert) (N-mm)	Torque (45°) (N-mm)	Torque (hor) (N-mm)
Pine	305	0.124	1.216	0	76.0	107.5
Acrylic	305	0.171	1.678	0	144.4	204.2
Aluminum	305	0.304	2.982	0	338.0	478.0
Pine	457	0.144	1.413	0	143.2	202.5
Acrylic	457	0.224	2.197	0	301.8	426.8
Aluminum	457	0.418	4.101	0	686.5	970.8
Pine	610	0.169	1.658	0	244.3	345.5
Acrylic	610	0.274	2.688	0	508.2	718.8
Aluminum	610	0.535	5.248	0	1164.2	1646.5

390

References

[1] Bicchi, A., Salisbury, J.K. & Brock, D.L., Contact Sensing from Force Measurements, *The International Journal of Robotics Research*, 12(3):249-262, 1993.

[2] Brock, D.L., Enhancing the Dexterity of a Robot Hand using Controlled Slip, *Proceedings of the IEEE International Conference on Robotics and Automation*, 249-251, 1988.

[3] Carello, C., Fitzpatrick, P., Domaniewicz, I., Chan, T. C. & Turvey, M. T., Effortful Touch with Minimal Movement, *Journal of Experimental Psychology: Human Perception and Performance*, 18:290-302, 1992.

[4] Ellis, R. E., Ganeshan, S., & Lederman, S. J., Sensor Design based on the Theory of Thin Plates, *Robotica* (in press)

[5] Ellis, R. R., & Lederman, S. J., The Role of Haptic versus visual volume cues in the size-weight illusion, *Perception and Psychophysics*, 53:315-324, 1993

[6] Ganeshan, S., Robotic and Haptic Perception of Length of Statically Held Objects, *MSc. Thesis*, Queen's University, Kingston, Ontario, Canada, 1992

[7] Jastrow, J., Perception of Space by Disparate Senses, *Mind*, 11:539-554, 1886

[8] Lederman, S. J., Ganeshan, S., & Ellis, R. E., "Effortful Touch with Minimum Movement" Revisited, *Journal of Experimental Psychology: Human Perception and Performance* (submitted)

[9] Nicholls, H.R. & Lee, M.H., A Survey of Robot Tactile Sensing Technology, *The International Journal of Robotics Research*, 8(3):3-30, 1989.

[10] Okada, T., A New Tactile Sensor Design based on Suspension Shells, *Dextrous Robot Hands*, Springer-Verlag, 267-285, 1990.

[11] Regtien, P.P.L., Tactile Imaging, *Sensors and Actuators*, 31:83-89, 1992.

[12] Solomon, Y., & Turvey, M. T., Haptically perceiving the distances reachable with hand-held objects, *Journal of Experimental Psychology: Human Perception and Performance*, 14:404-427, 1988

[13] Teghtsoonian, M., & Teghtsoonian, R., Seen and Felt Length, *Psychonomic Science*, 3:465-466, 1965

[14] Timoshenko, S.P., & Woinowsky-Krieger, S., *Theory of Plates and Shells*. McGraw-Hill: New York, 1970

Task Sharing and Intervention in Human-Robot Cooperating Systems

Yasuyoshi YOKOKOHJI and Tsuneo YOSHIKAWA

Department of Mechanical Engineering
Kyoto University
Kyoto 606, Japan

Abstract— In a human-robot cooperating system, a given task can be performed cooperatively between the human operator and the robot, or can be shared among them. The operator can also leave a repetitive or monotonous task to the robot and intervene it if necessary. In this paper, a new framework of task sharing and intervention in human-robot cooperating systems is proposed. A prototype of human-robot cooperating system is shown. Experimental results show the effectiveness of the proposed framework.

1. Introduction

In a human-robot cooperating system, a given task can be performed cooperatively between the human operator and the robot, or can be shared among them[1][2][6][10]. The operator can also leave repetitive or monotonous tasks to the robot and intervene them if necessary.

There are mainly two types of human-robot cooperating systems such as (i) direct cooperating type and (ii) indirect cooperating type. In the direct type cooperating system, the operator and the robot are connected physically and directly. Typical example is the power amplifier by Kazerooni[8]. Fujisawa et al.[5] also proposed a man-robot system where the human operator can grasp the tip of a manipulator mounted on a vehicle as shown in Fig.1(a). On the contrary, in the indirect type cooperating system, the operator cooperates with

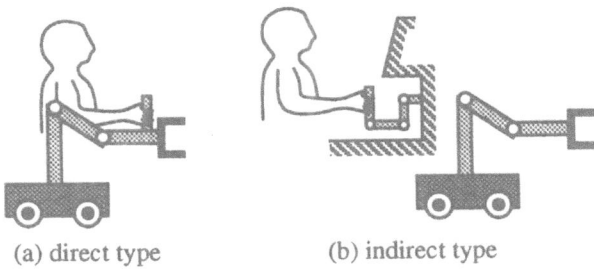

(a) direct type   (b) indirect type

Figure 1. Direct type and indirect type of cooperations

the robot indirectly through a control device such as joy-stick or master arm as shown in Fig.1(b). Typical example is, of course, a master-slave system.

In the direct cooperating systems, the operator could intervene the system intuitively while observing the scene of the task and feeling the reaction forces. However, this type contains danger for the operator in nature, especially when the robot has powerful actuators. On the other hand, the indirect cooperating systems could be applied to wider area than the direct type, including hazardous environments and micro environments. Moreover, since the operator is not physically connected to the robot (i.e. the motion of the operator need not be the same as that of the robot), the operator can intervene the task in various ways. Certainly, the indirect type has a drawback in the sense that the information of the task (e.g. visual information and force information) may not be so rich compared to the case of direct type. However, it would be possible to overcome this drawback by introducing a 3D display or improving the performance of bilateral control.

Considering the above points, we can conclude that the indirect type has more potential than the direct type. Note that the indirect cooperation systems cover wider cases than the conventional teleoperation systems, including the case where the operator is just beside the robot. In this paper, we focus on the indirect cooperating systems and propose a new framework of task sharing and intervention between the operator and robot. Experimental results are also shown using a prototype cooperating system.

2. Classification of Indirect Cooperating Systems

Let us classify the indirect type human-robot cooperating systems as shown in Fig.2. In manual control (Fig.2(a)), the computer is used just for data transmission and coordinate transformations between the operator site and the task site.

Indirect cooperating systems can be categorized in two types such as traded

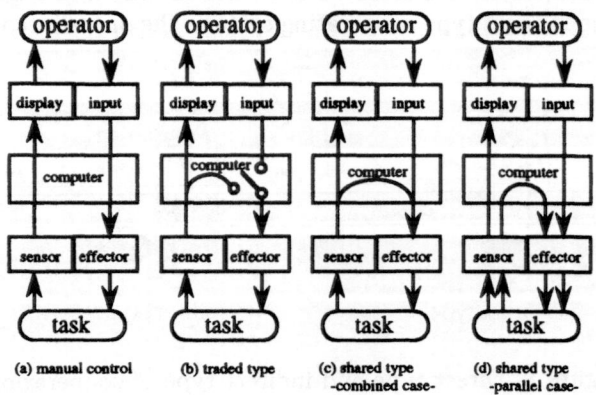

Figure 2. Classification of indirect cooperating systems

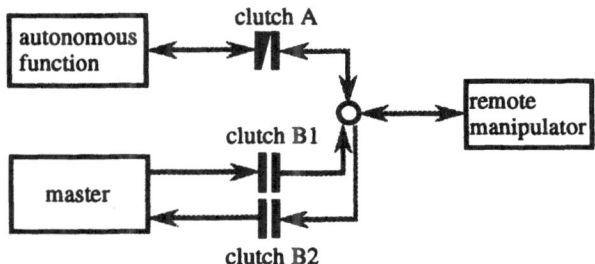

Figure 3. Conceptual sketch of cooperating system

type and shared type[12]. In traded type (Fig.2(b)), manual mode and autonomous mode are switched in serial. Conventional supervisory control[4] is in this category.

In the shared type, both manual operation by the human operator and autonomous functions are merged or run in parallel. This shared type has two cases such as combined case and parallel case. In the combined case (Fig.2(c)), control inputs both from the manual operation and from the autonomous functions are merged. This case could occur when the autonomous function modifies the command by the operator or when the operator modifies the motion generated by the autonomous function. Obstacle avoidance[2][9] is in this category.

In parallel case (Fig.2(d)), the given task is shared among the manual control and autonomous control in different components of the task coordinates. For example, in order to carry a cup filled with water, the orientation components can be controlled autonomously so that the cup orientation keeps horizontal, whereas the position of the cup can be commanded by the operator. This example is well known as the software jig[11]. Note that the parallel case is available only for the multiple DOF systems.

In the next section, we discuss a simple one DOF system. We assume that there is no time delay for data transmission between the operator and the robot. Extension into the multiple DOF case will be discussed in section 5.

3. Conceptual Model of Human-Robot Cooperating Systems

3.1. Operation modes

Fig.3 shows a conceptual sketch of the human-robot cooperating system where three clutches are implemented. Conway et al.[3] proposed the concept of time clutch and position clutch for teleautonomous systems at a distance with time delay. Three clutches in Fig.3 are used for changing the operation modes. Clutch A connects or disconnects the contribution of the autonomous function module into the data transmission line between the input device of operator's motion (master arm) and the robot (remote manipulator). This clutch continuously connects the line so that the contribution of autonomous module can be increased or decreased continuously. Clutches B1 and B2 connect or disconnect the lines between the master and the remote manipulator. These are discrete on/off type clutches and each one functions as a one-way clutch. Among the several on/off combinations of these three clutches, we chose six meaningful combinations from them and set the following six operation modes.

(i) Bilateral mode (mode B)

Mode B is a bilateral control mode without any aid of autonomous functions. This mode corresponds to the case when only clutch A is disconnected in Fig.3. A serious problem of indirect type is the difficulty for the operator to get the precise information of the task. The authors have proposed an advanced bilateral control scheme of master-slave manipulators where the dynamics of both the master and slave arms are taken into account[13]. By introducing this scheme, the total system behaves like a virtual rigid bar connecting the operator and the task environment as shown in Fig.4 (neglect the virtual operator in this mode). With this scheme, the maneuverability becomes equivalent to or better than that of the direct type. Mode B is used for teaching some motions for autonomous functions or recovering from errors caused by the autonomous function.

(ii) Free mode (mode F)

In free mode, all the clutches are disconnected. This mode is used when the operator wants to change the relative position between the master arm and the manipulator.

(iii) Unilateral mode (mode U)

In unilateral mode, only clutch B1 is connected and the master is used just like a joy-stick without force reflection.

(iv) Autonomy aided bilateral mode (mode AB)

In this mode, an autonomous function is added to mode B. All clutches in Fig.3 are connected. The motion command from the autonomous function is mixed into the command from the operator. Therefore, the operator can intervene the task performed by the autonomous function and simultaneously he/she can get the feedback information from the remote task. This mode is intuitive for the operator because the master and slave sides are still connected by a virtual rigid bar and an additional input from the autonomous function applies on this bar as if another person grabs the bar as shown in Fig.4.

(v) Autonomous mode (mode A)

In this mode, the remote manipulator becomes a full autonomous robot where

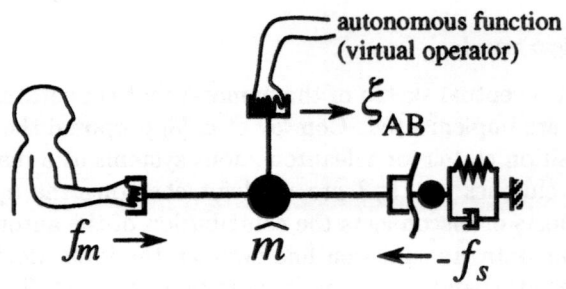

Figure 4. Equivalent model of mode AB

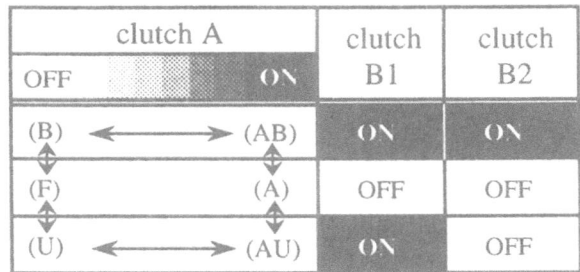

Figure 5. Diagram of clutch connection (ON)/ disconnection (OFF)

the operator cannot intervene the task but can just monitor the task. In Fig.3, only clutches A is connected. Master arm motion becomes free and the operator can move the master arm to a convenient posture before intervening the task in other modes.

(vi) Autonomy aided unilateral mode (mode AU)

This mode corresponds to the case when only clutch B2 is disconnected in Fig.3. The operator can intervene the task without feedback information. In this mode, the master arm is used just like a joy-stick as in mode U.

3.2. Mode change sequence

The sequence of mode change among the six modes will be fixed as shown in Fig.5, that is, when clutch A is disconnected, the sequence will be fixed as (B)↔(F)↔(U) and when clutch A is connected, the sequence will be fixed as (AB)↔(A)↔(AU). By connecting or disconnecting clutch A continuously, one can change the mode continuously between the corresponding two modes: (B)↔(AB) and (U)↔(AU). Note that one cannot go to mode A immediately from mode F, since it would be dangerous to start the autonomous module in mode A where the operator cannot intervene or correct the output of the module.

Fig.5 means that one cannot skip the intermediate modes like (B)↔(A) or (AB)↔(AU). For example, suppose that the mode has been changed from mode AB directly to mode AU. In this case, the master arm will suddenly be free or fixed so as to be a joystick in mode AU and the operator may input an unintended command thorough the master arm in mode AU. This problem can be avoided by inserting mode A between modes AB and AU. Furthermore, if the mode goes back from AU directly to AB, the master arm will immediately be connected to the slave arm and the operator cannot adjust the relative position between the master and slave. Changing modes from A directly to B, or from AU directly to B, etc. has the similar problems.

3.3. Control schemes

In this subsection, we discuss control inputs for master and slave arms which realize each operation mode mentioned above.

The dynamics of master arm and slave arm is given by the following equations:

$$\tau_m + f_m = m_m \ddot{x}_m + b_m \dot{x}_m \tag{1}$$

$$\tau_s - f_s = m_s \ddot{x}_s + b_s \dot{x}_s \tag{2}$$

where x_m and x_s denote the displacements of the master and slave arms respectively. m_m and m_s denote masses of master and slave arms, and b_m and b_s are viscous coefficients. f_m denotes the force that the operator applies to the master arm and f_s denotes the force that the slave arm applies to the object. Actuator driving forces of master and slave arms are represented by τ_m and τ_s respectively.

(i)Modes B and AB

In mode B, control schemes proposed by the authors[13] is used. In mode AB, control input from the autonomous function is mixed with the bilateral mode.

$$\tau_m = m_m[(1 - \frac{\widehat{m}}{2m_m})\ddot{x}_{ms} + k_1(\dot{x}_{ms} - \dot{x}_m) + k_2(x_{ms} - x_m)]$$
$$+ b_m \dot{x}_m - f_{ms} + \frac{\alpha}{2}\xi_{AB} \tag{3}$$

$$\tau_s = m_s[(1 - \frac{\widehat{m}}{2m_s})\ddot{x}_{ms} + k_1(\dot{x}_{ms} - \dot{x}_s) + k_2(x_{ms} - x_s)]$$
$$+ b_s \dot{x}_s + f_{ms} + \frac{\alpha}{2}\xi_{AB} \tag{4}$$

where $x_{ms} \triangleq \frac{x_m + x_s}{2}$, $f_{ms} \triangleq \frac{f_m + f_s}{2}$. The control input by the autonomous function in mode AB is ξ_{AB}, and $\alpha(0 \le \alpha \le 1)$ denotes the coefficient representing the state of clutch A. In mode B, $\alpha = 0$. Note that this control input ξ_{AB} is added not only to the input τ_s but also to the input τ_m so that the operator can feel the reaction force caused by his/her own input only.

Substituting (3) and (4) into (1) and (2), we get

$$\ddot{e} + k_1 \dot{e} + k_2 e = 0 \tag{5}$$

$$\widehat{m} \ddot{x}_{ms} = f_m - f_s + \alpha \xi_{AB} \tag{6}$$

where $e \triangleq x_m - x_s$. As shown in Fig.4, by applying the above two equations, the total system behaves like a virtual rigid bar connecting the operator and remote environment. Mass of this rigid bar is $\widehat{m}$. When we set $\widehat{m} = 0$ in mode B, the operator can maneuver the system as if he/she were manipulating the remote object directly. Continuously increasing α from 0 to 1, a new input ξ_{AB} is added to the virtual rigid bar connecting between master and slave sides as shown in Fig.4.

(ii) Modes F and A

The slave arm motion becomes free or locked in mode F. In mode A the slave arm is controlled by the autonomous function only.

$$\tau_m = 0 \tag{7}$$

$$\tau_s = \xi_A \tag{8}$$

where ξ_A denote the autonomous function in mode A. It should be noted that changing between mode AB and mode A may cause discontinuity of τ_s between (4) and (8). Therefore it is necessary to change the control input continuously between (4) and (8).

(iii) Modes U and AU

In these modes, the master arm motion becomes free or fixed according to the task. For example, in the case when the master arm is free, the control schemes are given by

$$\tau_m = 0 \tag{9}$$

$$\tau_s = \eta_{op} + \alpha\xi_{AU} \tag{10}$$

where η_{op} denotes the operator's command through the master arm in mode AU.

3.4. Examples of autonomous functions

We show two examples of autonomous functions.

(i) Force control

The following scheme is an example of force control case.

$$\xi_{AB} = f_d + K_f \int_0^t \{f_d - (f_s - f_m)\}d\tau \tag{11}$$

$$\xi_A = f_d + K_f \int_0^t (f_d - f_s)d\tau \tag{12}$$

$$\xi_{AU} = f_d + K_f \int_0^t f_d d\tau \tag{13}$$

where f_d is the desired force. The desired value f_d is determined in advance or can be set by the actual value of f_m taught in mode B. In modes U and AU, the master arm is fixed and the control input from the operator is given by

$$\eta_{op} = f_m + K_f \int_0^t (f_m - f_s)d\tau \tag{14}$$

(ii) Position control

The following dynamic control scheme can be applied for tracking the desired trajectory x_d.

$$\xi_{AB} = \widehat{m}\ddot{x}_d + K_v(\dot{x}_d - \dot{x}_{ms}) + K_p(x_d - x_{ms}) \tag{15}$$

$$\xi_A = m_s\ddot{x}_d + K_v(\dot{x}_d - \dot{x}_s) + K_p(x_d - x_s) \tag{16}$$

$$\xi_{AU} = m_s\ddot{x}_d + K_v\dot{x}_d + K_p x_d \tag{17}$$

In modes U and AU, the master arm is used as a joy-stick and the control input from the operator becomes

$$\eta_{op} = m_s\ddot{x}_m + K_v(\dot{x}_m - \dot{x}_s) + K_p(x_m - x_s) \tag{18}$$

398

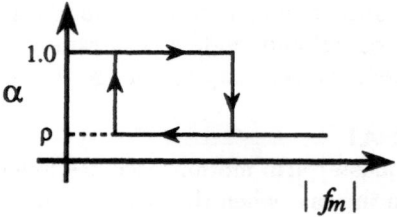

Figure 6. Changing α in mode AB

Feedback gains K_v and K_p in (15) through (18) should be large enough in order to track the trajectory precisely. In mode AB, however, the intervention by the operator can be regarded as a disturbance for (15), and large feedback gains may not accept the modification by the operator. Therefore, the autonomous function should detect the intention of the operator and accept his/her modification. This problem is discussed in the next section.

4. Understanding the intentions of human operator

In the human-robot cooperating systems, the priority should belong to the human operator. In this sense, understanding the operator's intentions is an important function of the system. We considered following two cases.

(i) **Reconnection between the master and remote manipulators**
We should be careful about changing the mode from A to AB or from F to B where master arm is connected again to the remote arm. When the position difference[1] between the master and remote arms is within a certain range (e.g. the relative distance is within 5cm), we can interpret that the operator intentionally tried to adjust the master arm position to that of the remote arm and the controller can correct the position error gradually into zero. On the contrary, when the position deference is larger than that range, we can interpret that he/she intentionally changes the relative position between master and remote arms and the controller can set the current position difference as a new position offset.

(ii) **Detecting the intervention of the operator**
When the given task is position control based, feedback gains of the autonomous module should be large enough to follow the given trajectory precisely. In mode AB, however, the operator's intervention could be regarded as a disturbance for this module, and the module with large feedback gains may not accept the modification by the operator. Therefore, the autonomous function should detect the operator's intention and accept his/her modification. For example, as shown in Fig.6, when the system detects the force applied by the operator f_m beyond a certain level, clutch A can gradually reduce the coefficient α into a sufficiently low value $\rho(< 1)$ and restrain the contribution of the autonomous module so that the operator can intervene the task as if the virtual operator in Fig.4 relaxes the force.

---

[1]Of course, the position difference means the difference among the two end-effector position vectors with respect to the corresponding base coordinates.

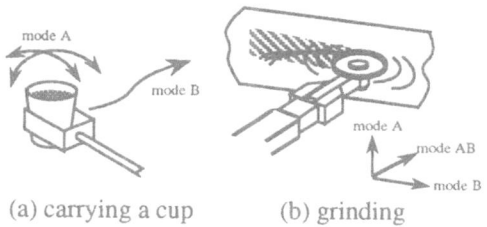

mode A

mode B

mode A

mode AB

mode B

(a) carrying a cup   (b) grinding

Figure 7. Examples of multiple DOF case

5. Task Sharing in the Task Coordinates

5.1. Extension into multiple DOF case

Now, let us discuss the extension into the multiple DOF case. Hayati et al.[7] discussed a shared control system architecture based on the hybrid position/force control. However, they did not consider changing the operation modes. When a task is given, we can set an appropriate task frame. By applying the control scheme for one DOF systems along with each axis of the coordinate frame, we can construct a human-robot cooperating system in multiple DOF.

For example, consider the task to carry a cup filled with water in three dimensional space as shown in Fig.7(a). In this case, two orientational components can be in mode A whereas remaining one orientational component and three translational components can be in mode B resulting in the parallel case. When the operator has to intervene the orientation control of the cup, the operation mode of the orientation components can be changed into AB or AU, resulting in the combined case. Fig.7(b) shows another example (grinding).

However, in three dimensional space, we need six switches for changing the mode in each coordinate axis and the operator could be confused with these six switches. This problem remains for future work.

5.2. Control schemes in multiple DOF

In this subsection, we extend the control schemes for the multiple DOF case. The dynamics of master and slave arms are given by the following equations:

$$\boldsymbol{\tau}_m = \boldsymbol{M}_m(\boldsymbol{\theta}_m)\ddot{\boldsymbol{\theta}}_m + \boldsymbol{h}_m(\boldsymbol{\theta}_m, \dot{\boldsymbol{\theta}}_m) - \boldsymbol{J}_m^T(\boldsymbol{\theta}_m)\boldsymbol{f}_m \qquad (19)$$

$$\boldsymbol{\tau}_s = \boldsymbol{M}_s(\boldsymbol{\theta}_s)\ddot{\boldsymbol{\theta}}_s + \boldsymbol{h}_s(\boldsymbol{\theta}_s, \dot{\boldsymbol{\theta}}_s) + \boldsymbol{J}_s^T(\boldsymbol{\theta}_s)\boldsymbol{f}_s \qquad (20)$$

where $\boldsymbol{\tau}_m$ and $\boldsymbol{\tau}_s$ denote joint driving force vectors of master and slave, $\boldsymbol{\theta}_m$ and $\boldsymbol{\theta}_s$ are joint vectors, $\boldsymbol{M}_m(\boldsymbol{\theta}_m)$ and $\boldsymbol{M}_s(\boldsymbol{\theta}_s)$ are inertia matrices, $\boldsymbol{J}_m(\boldsymbol{\theta}_m)$ and $\boldsymbol{J}_s(\boldsymbol{\theta}_s)$ are Jacobian matrices, $\boldsymbol{h}_m$ and $\boldsymbol{h}_s$ are Coriolis and centrifugal forces. Force vectors $\boldsymbol{f}_m$ and $\boldsymbol{f}_s$ correspond to f_m and f_s in the one DOF case.

In order to implement the control scheme for one-DOF systems along with each coordinate axis, we can linearize and decouple the system as follows:

$$\boldsymbol{u}_m = \widehat{\boldsymbol{M}}_m\ddot{\boldsymbol{x}}_m - \boldsymbol{f}_m \qquad (21)$$

$$\boldsymbol{u}_s = \widehat{\boldsymbol{M}}_s\ddot{\boldsymbol{x}}_s + \boldsymbol{f}_s \qquad (22)$$

where x_m and x_s are the end-effector position/orientation vectors of master and slave, $\widehat{M}_m$ and $\widehat{M}_s$ are constant diagonal inertia matrices, and u_m and u_s are new input vectors.

If we apply the following control inputs to the original systems (19) and (20), we get eqs.(21) and (22).

$$\tau_m = J_m^T[G_m \widehat{M}_m^{-1} u_m + (G_m \widehat{M}_m^{-1} - I)f_m] + h_m - M_m J_m^{-1} \dot{J}_m \dot{\theta}_m \quad (23)$$

$$\tau_s = J_s^T[G_s \widehat{M}_s^{-1} u_s - (G_s \widehat{M}_s^{-1} - I)f_s] + h_s - M_s J_s^{-1} \dot{J}_s \dot{\theta}_s \quad (24)$$

where $G_m = J_m^{-T} M_m J_m^{-1}$, $G_s = J_s^{-T} M_s J_s^{-1}$.

6. Experiment

6.1. Experimental system

We confirmed the validity of the proposed operation modes and the mode change sequence by experiments.

Fig.8 shows the overview of experimental system. The experimental system consists of two DD arms for the master and slave arms, and a 32-bit personal computer. Two DD-arms have same configurations with 3 DOF (link length: l_1=0.25[m], l_2=0.3[m]). The arm tip can change its position and orientation in the horizontal plane. In the experiment, the last joint (wrist) was left free. The operator can change the operation modes by three toggle switches located beside the operator.

6.2. Experimental tasks

We have already confirmed the validity of the proposed method in a simple one DOF case[14]. We then confirm the validity of task sharing and intervention in 2-DOF case by the following two experiments.

(i) Experiment 1

A circle (5[cm] radius) was drawn on the table as shown in Fig.9(a). The given task is to track this circle by the tip of slave arm. By teaching the position of the center of the circle and its radius, we can implement the autonomous

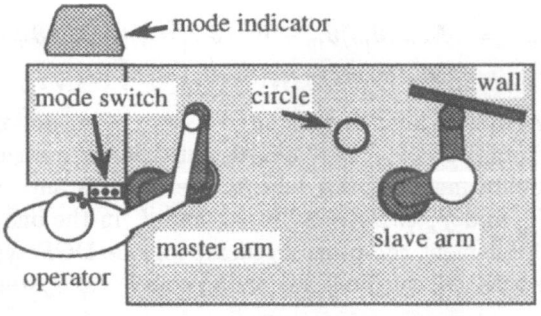

Figure 8. Overview of experimental system

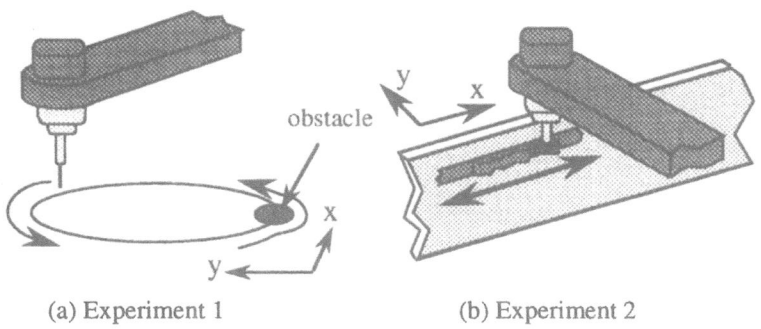

(a) Experiment 1   (b) Experiment 2

Figure 9. Tasks in experiments 1 and 2

function that can track the circle (5[sec/round]). We supposed an unexpected obstacle on the circle where the operator has to intervene the task to avoid it.
(ii) Experiment 2
The manipulator should go and back along with the surface of a wall exerting constant force. We suppose a grinding task as shown in Fig.9(b). By teaching the location of the wall and the amount of desired constant force, this task can be performed autonomously. We assume an unexpected region where more large force must be applied, supposing that this region has rough surface and requires a intervention to apply additional force by the operator.

6.3. Selecting the operation modes

The prototype of the mode selection device is shown in Fig.10. This device is located just beside the operator so that the operator can select the mode while he/she is maneuvering the master arm. Three toggle switches are implemented; the left switch is for selecting the frame axis; the middle switch is for selecting the mode among (B), (F) and (U) or among (AB), (A) and (AU); the right switch is for adding/removing the autonomous function. Since the right switch is on/off type, the operator can just add or remove the autonomous function discontinuously, while the system can adjust the contribution of the module continuously when it detects the operator's intervention.

A CRT monitor is located in front of the operator as the mode indicator. Fig.10 shows the case of experiment 2, where it indicates that x-axis (tangential direction) is in mode AB while y-axis (normal direction) is in mode U. The arrow in the left side indicates the selected axis. The selected axis can be changed by clicking the left switch. Clicking the right switch, the mode menu bar will be changed between (AB)(A)(AU) and (B)(F)(U). For example, in Fig.10, an autonomous function is added in the x-axis while no autonomous function is introduced in y-axis. In the case of experiment 1, only one mode bar is displayed because the operation modes in x and y axes are always the same. We are not satisfied with this first prototype. It would be better if we could mount these switches onto the gripper of master arm.

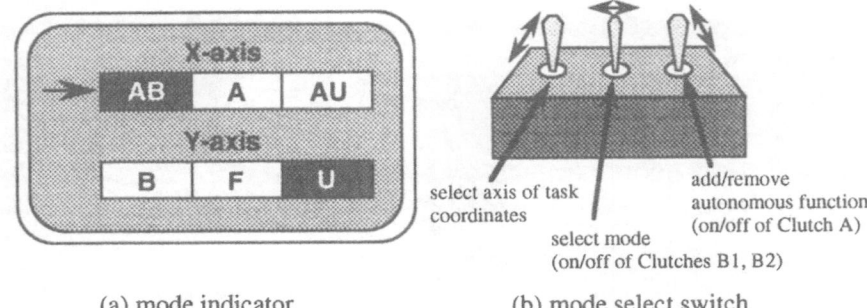

(a) mode indicator   (b) mode select switch

Figure 10. Select switch and mode indicator

6.4. Experimental results

Figs.11 and 12 show the experimental results.

(i) Experiment 1

Diagonal inertia matrices in eqs.(21) and (22) were set to $\widehat{M}_m = \widehat{M}_s =$ diag$(10, 9)$[kg]. Feedback gains k_1 and k_2, and intervening mass $\hat{m}$ in eqs.(3) and (4) were set to 24.0[1/s], 100.0[1/s^2], 14.0[kg] in X-axis, and to 26.7[1/s], 111.1[1/s^2], 12.6[kg] in Y-axis, respectively. Feedback gains K_v and K_p in eqs.(15) through (18) were set to 150[Ns/m] and 600[N/m].

In experiment 1, as shown in Fig.11, the operation mode was started from B and the operator tried to track the circle by conventional manual operation. After changing the mode into AB, the autonomous function started to track the circle, and the operator did not need to track the circle by himself. The operator intervened the task at 10[sec] to avoid the obstacle. Note that the position responses of master and slave are equal in this mode. When the mode had been changed into A, the master arm became free. In mode AU, the master arm was used as a joy-stick and the operator intervened again in this mode at about 25[sec]. Going through mode A again, the master arm was connected to the slave arm in mode AB at about 33[sec]. In this case, the relative position error between the master and slave was large enough, and the system regarded this error as new offset. At 45[sec], the operator tried to connect the master arm to the slave arm again. At this time, the relative error was so small that the system automatically corrected the error. Finally the mode went back to B and the experiment was finished.

(ii) Experiment 2

Diagonal inertia matrices were set to $\widehat{M}_m = \widehat{M}_s =$ diag$(12, 14)$[kg]. Feedback gains k_1 and k_2, and intervening mass $\hat{m}$ in eqs.(3) and (4) were set to 20.0[1/s], 83.3[1/s^2], 14.4[kg] in X-axis, and to 17.1[1/s], 71.4[1/s^2], 16.8[kg] in Y-axis, respectively. In X-axis direction, go-and-back motion control module were implemented, whereas a force control module was implemented in Y-axis direction. Feedback gains K_v and K_p in eqs.(15) through (18) were set to 150[Ns/m] and 600[N/m] for X-axis, and force feedback gain K_f in eqs.(11) through (13) was set to 0.6 for Y-axis.

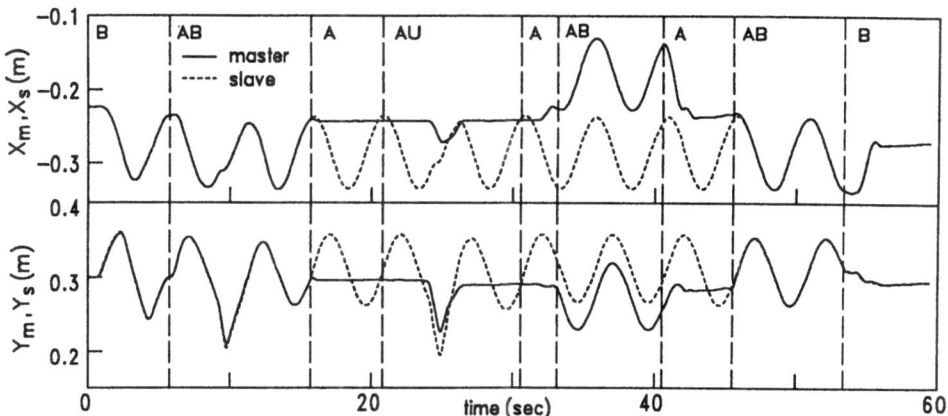

Figure 11. Results of experiment 1: End-effector position responses of X-axis and Y-axis components in the task frame

In Fig.12, the upper graph shows the position responses of master and slave in X-axis direction (tangential direction of the wall surface), while the lower graph shows the force responses of the master and slave in Y-axis direction (normal direction of the wall). Starting from mode B, the operator performed the grinding task by exerting about 1.5[kgf] force against the wall. Then the mode was changed into AB only in the X-axis component, and the go-and-back motion became automatic. Then the mode in X-axis changed into A, and the master arm became free only along with X-direction. After selecting Y-axis, the mode changed into AB, where the system automatically exerts constant force (about 1kgf) and the operator did not need to exert the constant force. In this mode, the operator intervened the task by applying additional force at about 31[sec] when the slave arm came into the rough region, and the slave arm force was increased. In mode AU, the operator intervened the task again at about 49[sec]. Selecting X-axis again, the master arm was connected to the slave in mode AB with new offset. Finally both components went back to mode B, and the experiment was completed.

Note that in modes B and AB in both experiments, the positions of the master arm and slave arm are almost the same as if both arms are connected by a rigid bar. Accordingly, the operator can maneuver the system just like a direct cooperating type system.

In mode B of experiment 2, f_m and f_s in Y-axis direction should be identical in ideal sense; however, in Fig.12, f_s is much lower than f_m. We believe that this error can be corrected by calibrating the force sensors and actuators.

7. Conclusion

In this paper, we have proposed a new framework of human-robot cooperating systems where six operation modes for cooperating between manual operation

404

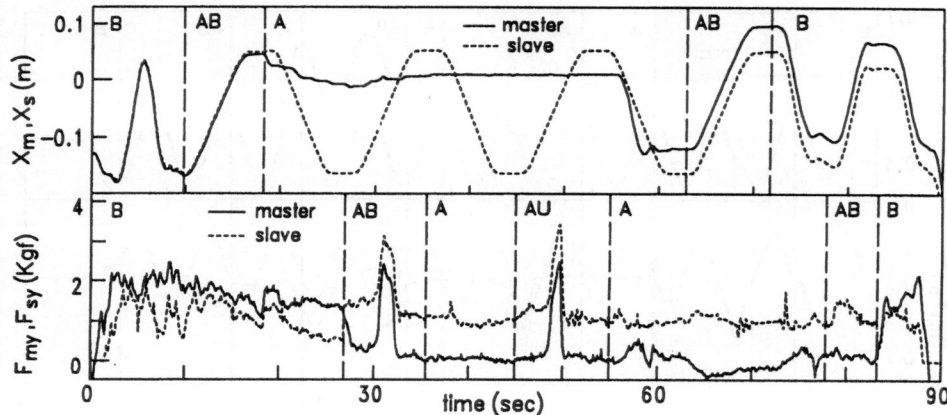

Figure 12. Results of experiment 2: Position response of X-axis component (tangential direction of the wall), force response of Y-axis component (normal direction of the wall) in the task frame

and autonomous functions are introduced. We have also specified the sequence of mode change so that the operator can change these operation modes intuitively and smoothly.

We have discussed one DOF systems, then extended it into the multiple DOF case. Control scheme for each operation mode has been shown. We have experimentally confirmed the validity of the proposed operation modes and the mode sequence by using a prototype manipulator system.

We have pointed out that the human-robot cooperating system should understand the operator's intention in various situations, and several function have been implemented in the experimental system.

Developing more easy-to-use device for mode selection, and experiments in more complex task with 3 DOF system are future works.

Acknowledgment

The authors would like to express their appreciation to Mr. Norio Hosotani and Mr. Hitoshi Hasunuma who are graduate students of Kyoto University, for their valuable help in the experiments.

This research work was supported by a Grant-in-Aid for Scientific Research of the Ministry of Education (No.05650258).

References

[1] P. Backes and K. Tso, "UMI: An Interactive Supervisory and Shared Control System for Telerobotics", In *Proceedings, 1990 IEEE International Conference on Robotics and Automation*, pp.1096–1101 (1990)

[2] P. T. Boissiere and R. W. Harrigan, "Telerobotic Operation of Conventional Robot Manipulators", In *Proceedings, 1988 IEEE International Conference on Robotics and Automation*, pp.576–583 (1988)

[3] L. Conway, R. Volz and M. Walker, "Teleautonomous Systems: Projecting and Coordinating Intelligent Action at a Distance", IEEE Trans. on Robotics and Automation, Vol.6, No.2, pp.146–158 (1990)

[4] W. R. Ferrell and T. B. Sheridan, "Supervisory Control of Remote Manipulation", IEEE Spectrum, Vol.4, No.10, pp.81–88 (1967)

[5] Y. Fujisawa et al., "Control of Manipulator/Vehicle for Man-Robot Cooperation Based on Human Intention", In *Proceedings, 1992 IEEE International Workshop on Robot and Human Communication*, pp.188–193, September 1–3, Tokyo (1992)

[6] W.Kim, B. Hannaford, and A. Bejczy, "Force-Reflection and Shared Compliant Control in Operating Telemanipulators with Time Delay", IEEE Trans. on Robotics and Automation, Vol.8, No.2, pp.176–185 (1991)

[7] S. Hayati and S. T. Venkataraman, "Design and Implementation of a Robot Control System with Traded and Shared Control Capability", In *Proceedings, 1989 IEEE International Conference on Robotics and Automation*, pp.1310–1315 (1989)

[8] H. Kazerooni, "Human-Robot Interaction via the Transfer of Power and Information Signals", IEEE Trans. of Systems, Man, and Cybernetics, Vol.20, No.2, pp.450–463 (1990)

[9] V. Lumelsky and E. Cheung, "Towards Safe Real-Time Robot Teleoperation: Automatic Whole-Sensitive Arm Collision Avoidance Frees the Operator for Global Control", In *Proceedings, 1991 IEEE International Conference on Robotics and Automation*, pp.797–802 (1991)

[10] N. P. Papanikolopoulos and P. K. Khosla, "Shared and Traded Telerobotic Visual Control", In *Proceedings, 1992 IEEE International Conference on Robotics and Automation*, pp.878–885 (1992)

[11] T. Sato and S. Hirai, "MEISTER: A Model Enhanced Intelligent and Skillful Teleoperation Robot System", Robotics Research -The Fourth International Symposium-, The MIT Press, pp.155–162 (1988)

[12] D. R. Yoerger and J. E. Slotine, "Supervisory Control Architecture for Underwater Teleoperation", In *Proceedings, 1987 IEEE International Conference on Robotics and Automation*, pp.2068–2073 (1987)

[13] Y. Yokokohji and T. Yoshikawa, "Bilateral Control of Master-Slave Manipulators for Ideal Kinesthetic Coupling –Formulation and Experiment–", In *Proceedings, 1992 IEEE International Conference on Robotics and Automation*, pp.849–858 (1992)

[14] Y. Yokokohji et al., "Operation Modes for Cooperating with Autonomous Functions in Intelligent Teleoperation Systems", In *Proceedings, 1993 IEEE International Conference on Robotics and Automation*, Vol.3, pp.510–515 (1993)

Coarse-Fine Motion Coordination and Control of a Teleoperation System with Magnetically Levitated Master and Wrist

S.E. Salcudean and N.M. Wong

Department of Electrical Engineering
University of British Columbia
Vancouver, BC, V6T 1Z4, Canada

Abstract— *A novel teleoperation system was developed by equipping a conventional robot with a magnetically levitated fine-motion wrist and using an identical wrist as a master [1].*

Here, the issues of control of such a fine-master, coarse-fine slave system are discussed. It is proposed that the master control the slave through a combination of position and rate control, with the conventional robot controlled in rate mode and its wrist in position mode. Kinesthetic feedback is achieved through wrist-level coordinated force control. Transparency is improved through feedforward of sensed hand forces to the master and environment forces to the slave. To maintain stability, the slave damping is controlled by the sensed environment forces. Experimental results are discussed.

1. Introduction

The "transparency" of conventional teleoperation systems is limited by the inadequacy of existing mechanical hardware, both at the master and at the slave level [2, 3]. Low frequency responses, high impedances and friction, the inability to exert accurate forces, backlash, etc., make the remote execution of delicate assembly tasks almost impossible. Since modelling errors put severe limitations on achievable performance [4], mechanical hardware shortcomings cannot be overcome by feedback control alone.

In order to overcome the limitations due to conventional design, a new teleoperation system based on parallel actuation and a "coarse-fine" approach to manipulation [5, 6] has been developed, built and tested by the authors at the University of British Columbia [1]. The teleoperation master employed is a six-degree-of-freedom (6-DOF) magnetically levitated (maglev) wrist with programmable compliance, while the slave manipulator is a conventional 6-DOF robot equipped with a maglev wrist identical to the master. In this paper the issues of motion coordination and bilateral control that arose in operating this novel teleoperation systems are discussed.

A hybrid position/rate control method is proposed for slave robot positioning. Based on experimental results, it is argued that the best implementation of this method is decoupling of the coarse and fine components of the slave

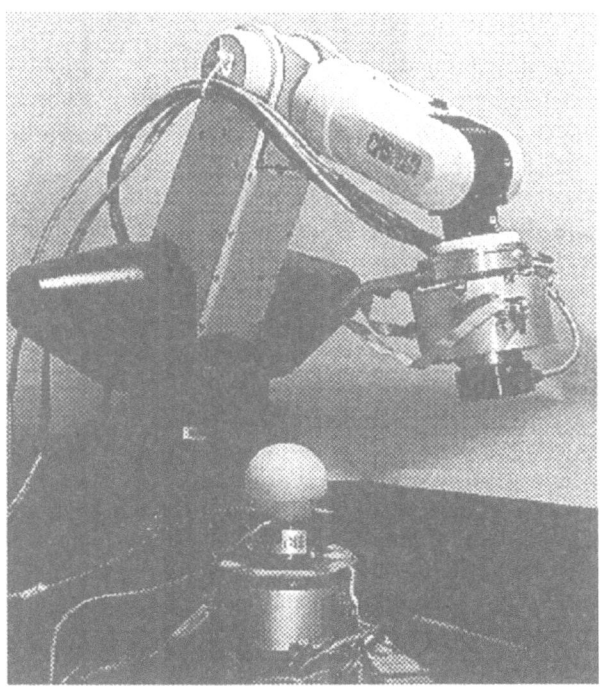

Figure 1. Photograph of the UBC Teleoperation System

manipulator motion by controlling the slave fine-motion wrist in position mode and the coarse-robot of the slave manipulator in rate mode.

Bilateral control is implemented at wrist level by using a coordinated force controller [7]. Transparency is improved by feeding forward sensed hand forces to the slave wrist and environment forces to the master wrist [8]. The tradeoff of transparency *vs* stability robustness is discussed and it is shown that some form of adaptation to environment stiffness is required. It is shown that environment force-dependent damping can be used to increase stability in contact tasks while keeping the master impedance low during free motion.

The paper is organized as follows: In Section 2, the design of the UBC maglev wrist and of the UBC teleoperation system is summarized. In Section 3, issues of control and motion coordination are discussed. Section 4 presents experimental results, while Section 5 presents conclusions and suggestions for future work.

2. Teleoperation System Design

This section will briefly describe the components of the UBC teleoperation system, as well as the computing subsystems and their interaction. Its photograph can be seen in Figure 1, which shows the conventional CRS A460 robot used in the system and the two maglev wrists. Each wrist is equipped with a force-torque sensor, enabling measurements of hand and environment forces.

2.1. UBC Maglev Wrist

This wrist, designed by Salcudean according to the principles outlined in [6], is an in-parallel actuated device in which six Lorentz forces are generated between two rigid elements - a stator and a "flotor", the lighter flotor being actively levitated.

An assembly sketch and a photograph of the UBC maglev wrist are shown in Figures 2 and 3, respectively, while its characteristics are summarized in Table 1. It can be seen that the Lorentz actuators are arranged in a star configuration with 120° symmetry. Each consists of a flat coil immersed in the magnetic field generated by four rectangular magnets whose flux is contained by two soft iron return plates. The optical sensor detecting the location of the flotor with respect to the stator consists of three light-emitting diodes (LEDs) projecting infrared light beams on the surfaces of three position-sensing diodes (PSDs) [6]. By comparison to the maglev wrists presented in [6, 9] which use coils arranged on the faces of a hexagonal cylindrical shell, the UBC wrist is substantially smaller, although it can produce the same forces and only slightly reduced torques,

Table 1. UBC MagLev Wrist Characteristics

Dimensions	Cylinder with $r = 66$ mm, $h = 110$ mm
Stator mass	2 Kg
Flotor mass	0.65 Kg
Payload (continuous)	2 Kg (along the z-axis)
Translation Range	± 4.5 mm from center
Rotation Range	$\pm 6°$ from center
Resolution	< 5 μm (trans.), < 10 μrad (rot.)
Force/Torque Freq. Response	>3 kHz
Closed-Loop Position Freq. Response	>30 Hz (trans.), >15 Hz (rot.)
Actuator force constant	2 N/A
Actuator gap	11.5 mm
Gap field	0.4 T
Flat coil thickness	2.5 mm
Coil resistance R_c	1.1 Ω (room temp.)
Coil inductance	0.3 mH
Power consumption (z-axis force)	$\frac{R_c}{12} f_z^2$ W
Max. continuous current	3 A / coil
Peak current	10 A / coil

2.2. Coarse Motion Manipulator

The coarse-motion robot used in the UBC teleoperation system is a CRS A460 elbow manipulator with an orthogonal spherical wrist. This robot is powered by DC servo motors with timing belt and/or harmonic drive transmissions. Similarly to the remote ALTER of PUMA robots, a robot controller state called REMOTE can be used for external computer control. While in this state,

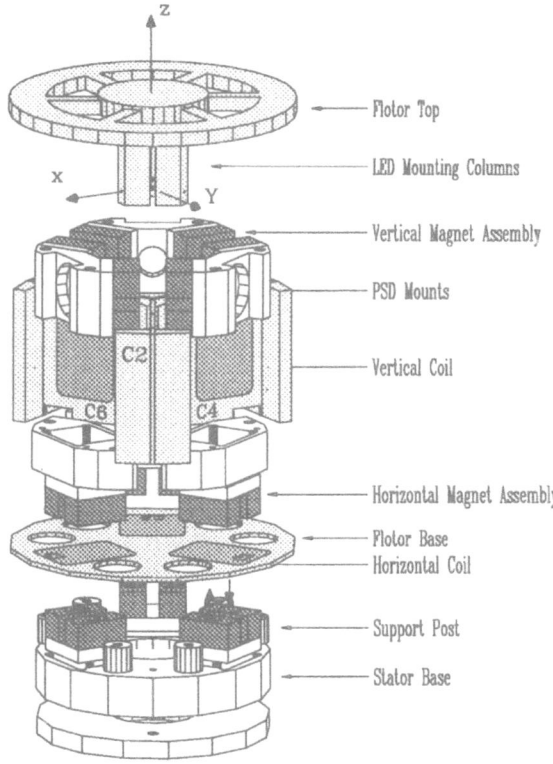

Figure 2. The Assembly Sketch of the Maglev Wrist

the robot controller reads and updates set-points from an internal data buffer at a fixed rate. The set-points can be specified in absolute or incremental, joint or cartesian modes, and are received from external hardware through serial or parallel ports. It was found that the use of cartesian data and serial communications severely limit the achievable REMOTE rate (the CRS Intel 8086-based controller is slow while the serial communication has substantial overhead). The incremental mode was selected because it is inherently safer and requires less data transmission.

2.3. Computing System

The real-time system employed for the coordinated control of the CRS robot, the maglev master and the maglev slave is shown in Figure 4. A DSP system using a TMSC30 chip does most of the calculations, while the PC host is used for communications, debugging and data collection.

The robot kinematics routines and the processing of strain-gauge data were ported to the DSP in order to gain flexibility and speed. The overall control

Figure 3. UBC Maglev Wrist/Hand Controller

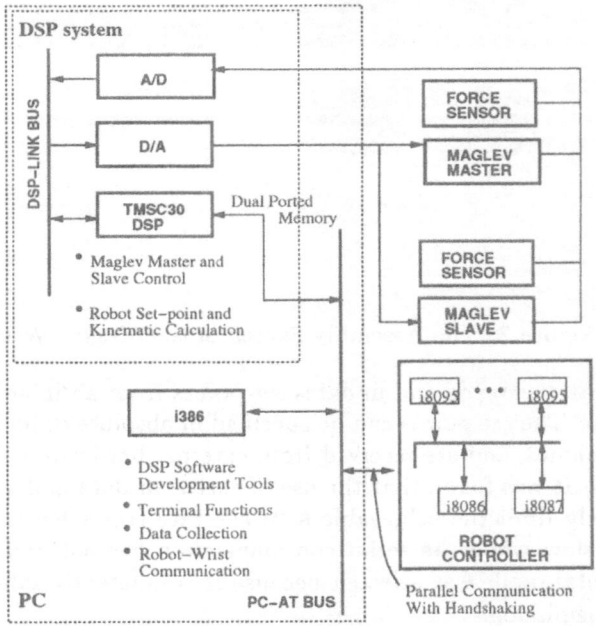

Figure 4. The Real-Time Computing System

sampling rate is roughly 380 Hz, while the exchange of location data and set-point with the CRS robot controller takes place at the slower 60 Hz. Details can be found in [10].

3. Motion Coordination and Control

Two issues must be dealt with. The first one, referred to as "motion coordination", is that of positioning of the slave manipulator[1], and is essentially a kinematic problem. The second one is that of bilateral control.

3.1. Motion Coordination

Because the master and the slave workspaces are so disparate, neither position control with scaling nor indexed position control [11] can be used to position the slave. The former would lead to poor resolution, the latter to almost continuous use of the index trigger. While the slave could be positioned by using rate control, *i.e.*, by commanding a slave velocity proportional to the master position, studies have shown that fine dextrous tasks are difficult to execute in this mode [12]. Furthermore, force feedback in rate control mode feels unnatural (mass is perceived as viscosity, viscosity as stiffness), and has poor stability margins [13].

3.1.1. Hybrid Position and Rate Control

Since rate control can solve the resolution problem encountered in position control, while position control can solve the spatial correspondence and force feedback problems encountered with rate control, a new hybrid position/rate control approach is proposed. The approach consists of dividing the master workspace into a central region designated for position control and a peripheral region designated for rate control, and can be interpreted by imagining that the slave manipulator is controlled in position mode about a local (slave) frame. When the master is about to exceed its workspace, the center of the local frame X_L is displaced or "dragged" towards the goal pointed at by the master, thus providing automatic indexing. This scheme is implemented as

$$X_{sd} = c_p\, x_m + X_L \tag{1}$$
$$\dot{X}_L = \begin{cases} f(\|x_m\|)\, x_m & \|x_m\| > r \\ c_v(X_s - X_L) & \text{otherwise,} \end{cases}$$

where x_m, X_{sd} are the master and desired slave manipulator absolute locations, X_s is the slave manipulator position, c_p is a position scaling factor, $f(\|x_m\|)$ is a positive increasing velocity scaling function, $\|\cdot\|$ is the max (or other) vector norm, and c_v is a small velocity constant. The goal is to map the master position workspace $\{x_m : \|x_m\| \le r\}$ into a scaled local position workspace $\{X_s : \|X_s - X_L\| \le c_p r\}$ about X_L. A little thought will reveal that the small centering motion of X_L caused by the term $c_v(X_s - X_L)$ is necessary to allow the slave manipulator to be positioned against a stiff obstacle with the teleoperation master flotor in its center [10]. This centering motion can barely be noticed by the operator.

---

[1]The redundant coarse-fine slave (robot *with* maglev wrist) will be referred to as the "slave manipulator".

412

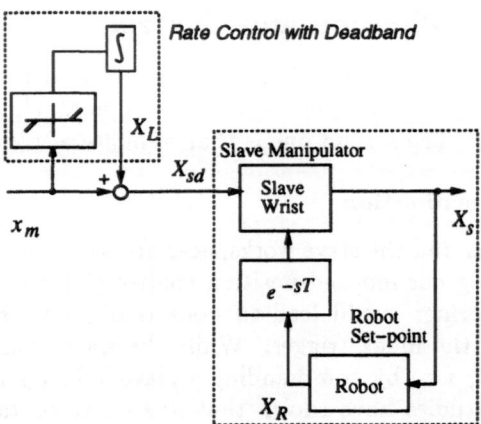

Figure 5. Hybrid Position/Rate Control (small centering motion not shown).

In the first attempt at implementing the scheme proposed above, the slave flotor was controlled as a rigid body in space, with the CRS robot programmed to track it continuously, as suggested in [6]. This is illustrated in Figure 5, where X_R is the absolute position of the CRS robot. Unfortunately, it was found that this control method leads to substantial step response overshoot, making positioning of the slave robot difficult. Simulations isolated the problem to the delay in communicating the robot position to the wrist controller [10]. As seen in Figure 5, this delay adds phase lag to the feedback loop controlling the slave and causes oscillatory behavior. In addition, the continuous tracking approach to coarse-fine coordination is inherently unsafe. A slave wrist power failure or payload overload would cause the robot to track the flotor until it would reach the boundary of its workspace.

3.1.2. Decoupled Coarse-Fine Coordination

To minimize the effect of communication delays, the robot and slave flotor should be controlled independently. Such decoupling of coarse and fine motions can be achieved by setting the master and slave flotors in kinematic correspondence relative to their stators, and by commanding the transport robot to track its wrist flotor only when the slave flotor is about to exceed its motion range. Although implemented and found to work well, the safety of this approach is also questionable. A better approach is to decouple the position and rate motions of the slave manipulator across the fine and coarse domains. This is achieved by controlling the transport robot in rate mode when the master is in its rate control region, while the two flotors are controlled in position mode relative to their stators. This amounts to letting the desired robot position X_{Rd} be equal to the local frame X_L and is implemented as follows:

$$x_{sd} = c_p x_m \tag{2}$$

$$\dot{X}_{Rd} = \begin{cases} f(\|x_m\|)\, x_m & \|x_m\| > r \\ c_v\, x_s & \text{otherwise,} \end{cases} \tag{3}$$

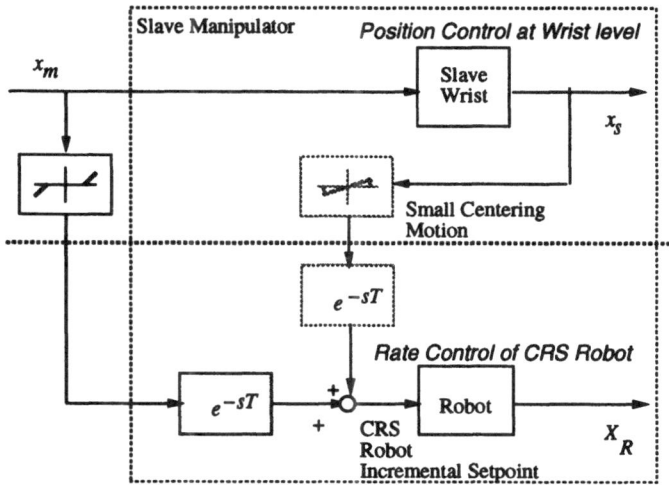

Figure 6. Decoupled Coarse-Fine Control

where x_s is the slave flotor position relative to the slave stator, x_{sd} is its setpoint, and X_{Rd} is the coarse robot setpoint. This method, illustrated in Figure 6, is safe and insensitive to robot motion delays. Its only disadvantage is that the position scaling c_p from master to slave cannot be larger than 1.

3.2. Wrist-Level Control

Because of the high ratio of the CRS robot harmonic drives, the dynamic interaction between the robot and the wrist attached to it can be neglected. It is also assumed, for now, that the master is within its position region so that the robot acceleration is negligible.

The flotors of the maglev wrists can be modelled as single rigid bodies. Through linearization and gravity feedforward, the differential equations describing the flotor motion can be transformed to decoupled double-integrator form [6], and justifies the use of single-axis mass models for the master and slave flotors:

$$ms^2 x_m = f_h + f_m = f_h^e - H(s)x_m + f_m \qquad (4)$$
$$ms^2 x_s = f_e + f_s = f_e^e - E(s)x_s + f_s, \qquad (5)$$

where m is the flotor mass (assumed, for simplicity, to be the same for master and slave), and f_m and f_s are the master and slave actuation forces. The hand and environment forces f_h and f_e can be thought as having active exogenous components f_h^e and f_e^e, respectively, and feedback components $-H(s)x_m$ and $-E(s)x_s$ dependent on the hand and environment impedances, respectively (see, for example [14]). Laplace transforms are used throughout.

Direct manipulation of the slave flotor when handled directly by the operator would correspond to

$$ms^2 x_s + H(s)x_s + E(s)x_s = f_h^e + f_e^e \qquad (6)$$

and would be stable in general. The goal of the teleoperation controller is to provide *transparency, i.e., stability* of the mapping from exogenous signals to positions and a *matching of the transfer functions* from f_h^e, f_e^e to x_m, x_s in (4,5) to that from f_h^e or f_e^e to x_s in (6), for all $H(s)$ and $E(s)$ encountered during direct operation. Precise definitions can be given for transparency in terms of worst-case transfer functions errors, but these are beyond the scope of this paper.

A bilateral controller based on a PID "coordinating force" f_c that emulates a stiff rigid-link between master and slave can be implemented as follows [15]:

$$f_c = (k_p + sk_v + \frac{1}{s}k_i)(x_m - x_s) \tag{7}$$

$$f_s = f_c \quad ; \quad f_m = -f_c .$$

The case in which the environment acts like a spring ($E(s) = k$), and the hand like a damper ($H(s) = bs$) will be considered since it is plausible and illustrative. As long as k_p and k_v are significantly larger than k and b, it can be shown that, although the coordinating torque controller does not provide transparency, it comes close. Indeed, the transfer function from external forces to master and slave positions emulate a mass-spring-damper system with the combined mass of the master and the slave. If k_p and k_v are not very large, performance deteriorates, as the master moves significantly more for the same applied force than it does in the case of direct manipulation. This can be the seen in Figure 7, which shows Bode plots from hand and environment forces to master and slave positions, as well as the "direct manipulation" (DM) transfer function. Because of the finite stiffness k_p, coordinated force controllers require

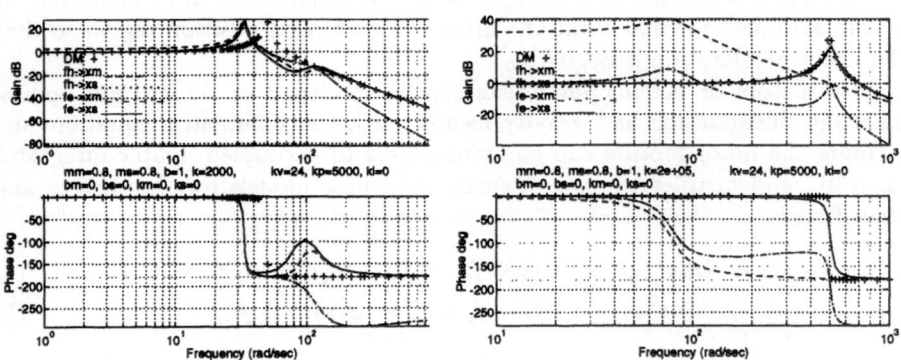

Figure 7. Coordinated force controller against compliant environment (left), stiff environment (right).

a position error $x_m - x_s$ to generate forces, thus causing a loss of kinematic correspondence between the master and the slave. When force sensors are available, this drawback can be alleviated by feedforward of the measured hand and environment forces:

$$f_s = f_c + f_h \tag{8}$$

$$f_m = -f_c + f_e .$$

It is easy to show that the transfer functions from forces to the position error $x_m - x_s$ becomes zero, and that the transfer functions from forces to positions exactly match that of direct manipulation. The price paid for such perfect transparency is the lack of stability robustness to errors in force sensing, delays, etc., especially against stiff environments. Indeed, for the stiffer environment shown in Figure 7, errors of only 5% in measured forces drive the teleoperation system unstable. The sensitivity to modelling error and measurement noise can be reduced by additional damping at the slave (or master and slave) when the slave is in contact with the environment. As seen in Figure 8, this additional damping causes a significantly lower frequency response relative to direct manipulation for soft environments or free motion. This is felt at the master as an increase in environment viscosity. In order to avoid trading-off free-motion performance for stability during contact tasks, the amount of damping in the system can be adapted to the sensed environment force. As an implementation

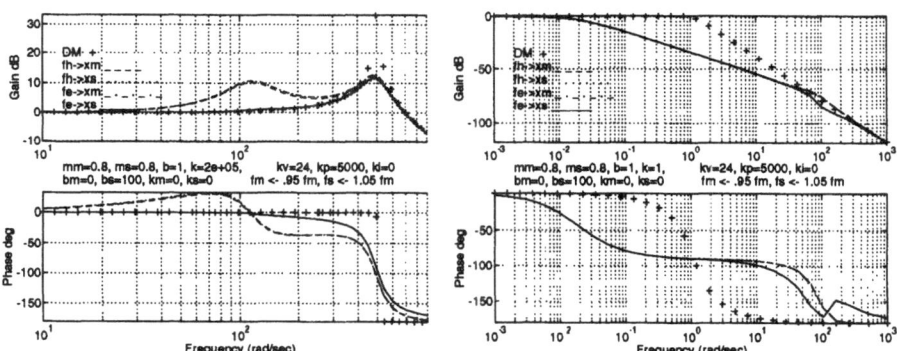

Figure 8. Coordinated force with force feedforward controller against compliant environment (left), stiff environment (right).

detail, it should be noted that since it is awkward to allow the master and slave flotors to drift about when the operator does not hold the master, PD centering terms are added to f_s and f_m (see Table 2 of Section 4).

3.3. Force Feedback during Rate Control

The motion of the transport robot adds an inertial force to (5) during free motion, and an environment-dependent force during contact tasks. Experiments show that, during free motion, the additional inertial term can be felt mainly at switch-over points between position and rate control. This can cause "move-stop-move" oscillations on the rate control boundary. In most contact tasks, the positioning robot is moving slowly and rate control does not affect reflected forces.

4. Experimental Results

The control and motion coordination approaches described in the previous section were successfully implemented. The ability to easily position the slave flotor via decoupled coarse-fine control was demonstrated by controlling the slave flotor with respect to the master stator, as well as with respect to the slave wrist stator (this corresponds to "gripper frame"). The most difficult aspect of slave robot positioning using the single-handed maglev master was found to be slave flotor orientation, which was found to take a bit of practice.

The kinesthetic feedback between master and slave was found to be excellent. Two side-effects of force-feedback were noticed. The first one is the previously mentioned "move-stop-move" oscillation at the onset of rate control. This oscillation can be easily controlled by stiffening the grip on the handle. The second one is the high force feedback felt when hitting a hard surface while the robot is still moving. This pushes the operator's hand back into position mode, which in turn pulls back the slave wrist. This counteracts the robot motion, which cannot stop instantly, and is a helpful feature in avoiding damaging objects upon contact.

Quantitative data will be presented below. The coordinating-force controller gains from (7) were chosen as a compromise between transparency and stability margin and are shown in Table 2. Unless specified otherwise, these gains are fixed, and the results presented are for the translational z-axis. Responses for other axes are similar.

Table 2. Controller Gains used in Experiments

Gains	Symbol	Value	unit
Proportional	k_p	5	N / mm
Velocity	k_v	0.024	N / (mm/s)
Integral	k_i	25*	N / (mm·s)
Centering spring	k_m, k_s	0.5	N / mm
and damping	b_m, b_s	0.005	N / (mm/s)

*There is a limit (6×10^{-3} $mm \cdot s$) on the integrator term in order to prevent windup. Thus the integrator effect is never greater than 0.15N.

4.1. Bilateral Free Motion Tracking

Wrist level position tracking while the control law (8) is applied is illustrated in Figure 9. Position tracking is good even at the highest frequency hand motion the operator could generate ($\simeq$ 7 Hz in these figures). The lack of sensed force feedforward (control law (7) instead of (8)) does not affect free-motion tracking substantially. Figure 10 illustrates the decoupled position and rate control, with the master having a rate control deadband from -3 mm to +3 mm. The slave flotor position x_s relative to its stator tracks the master flotor at all times. While the operator moves the master flotor within the rate

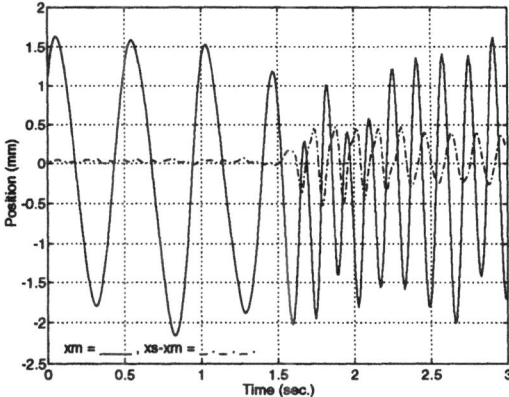

Figure 9. Wrist-level free-motion tracking - Master position and tracking error.

deadband ($t = 0$ to $t = 2$ sec.), the robot motion is small and due only to the slow centering term in (3). While the master flotor is outside its position deadband, ($t = 2$ to $t = 4$ sec.), the robot velocity tracks the master position. When the operator releases the master at $t = 4$ seconds, both the master and the slave flotors center themselves.

Backdriveability in rate mode is illustrated next, from $t = 5.2$ onwards, when the operator pulls the slave flotor back. The master tracks the slave flotor position and the robot tracks the master in rate control. The robot velocity can track the master position at frequencies up to 2-3 Hz.

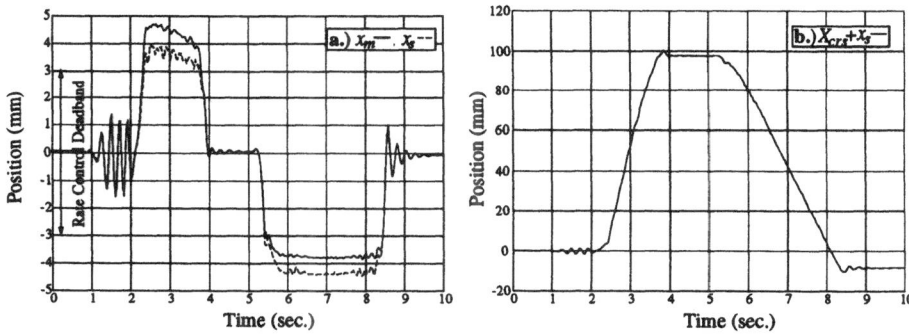

Figure 10. Hybrid position/rate control: a.) Flotor position tracking: x_m and x_s; b.) Slave system motion: $X_R + x_s$. The slave flotor position data, x_s, is measured from the stator center and is expressed in world coordinates.

4.2. Stability in contact with stiff environment

Figure 11 shows the master and slave positions and forces that occur when the operator brings the slave in contact with a solid steel steer attached to the table, while the wrists are in coordinated force control (7) (no feedforward is applied). A loss of kinematic correspondence between master and slave can be seen, although the force tracking is excellent.

When the hand and environment forces are fed forward (controller (8), the contact between the slave and the solid steer becomes unstable. By increasing

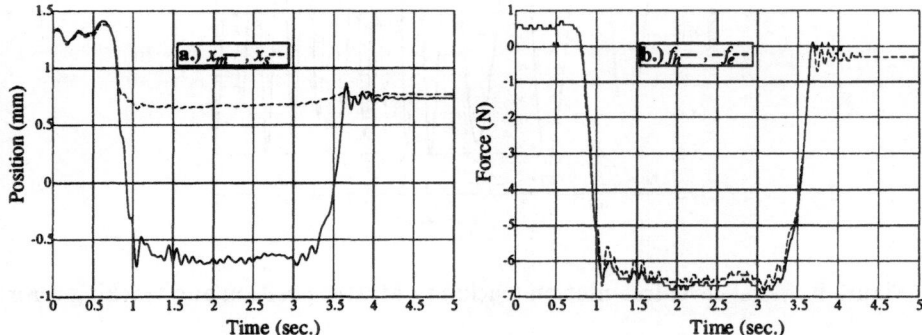

Figure 11. Contact with stiff environment with coordinated force control: a.) master and slave flotor positions x_m and x_s; b.) hand and environment forces: f_h and $-f_e$.

the damping term, b_s from 0.005 to 0.1 N/(mm/s), the hard contact instability disappears. The trade-off is that the operator feels the master flotor drag during free motion.

A simple way by which the viscous drag during free motion can be avoided is by adjusting the damping term according to the magnitude of the measured environment force. For instance, using minimum damping of $b_{min} = 0.0025$ N/(mm/s), and a scaling constant $k_b = 0.01$ (mm/s)$^{-1}$, and setting the slave wrist damping to

$$b_s = k_b \, |f_e| + b_{min}$$

removes instability problems when contacting stiff environments. Figure 12 shows the moment of contact ($t \approx 3$ seconds), and a larger force being applied at $t \approx 5.5$ seconds.

4.3. Exertion of forces against stiff environment.

The results obtained with the coordinated force controller are shown in the left-side plots of Figure 13. The operator moves the slave in rate control until the slave flotor hits the environment. The collision force pushes the master back into its position control region and the robot rate motion stops (except for the small centering motion tracking the slave flotor). After contact, the operator attempts to exert constant forces of 5N, 10N and 15N, while the sensed environment force is displayed on the PC monitor. Excellent force tracking is observed even though no force sensor is used in the controller. For high forces, the position error between master and slave builds up. As the operator intended to exert a higher force at $t = 12$ seconds, the master crosses into the rate region (±3mm), and the CRS robot moves towards the solid steer. A large feedback

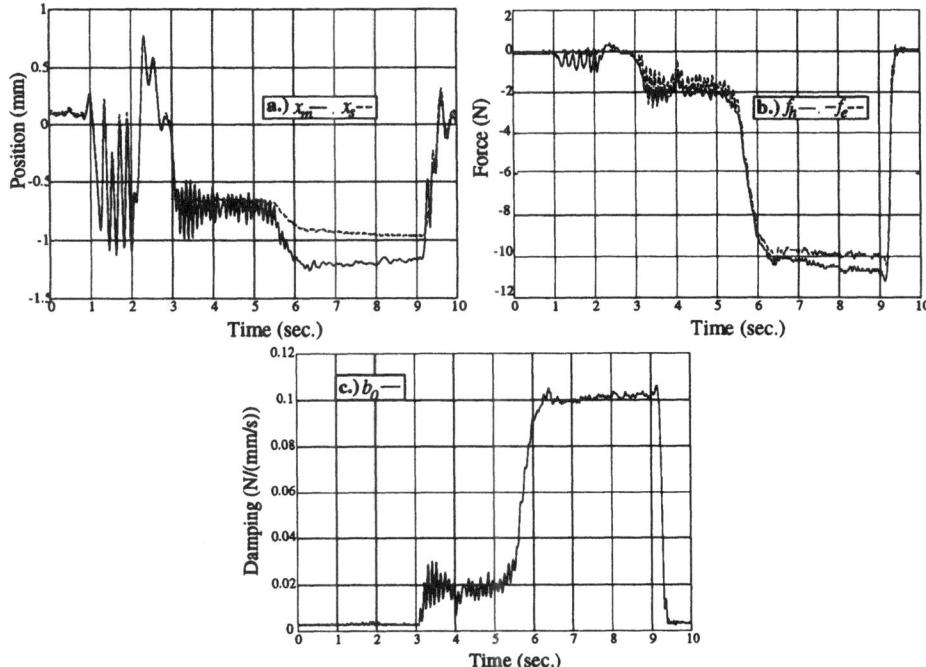

Figure 12. Contact with stiff environment during coordinated force control with force feedforward and variable damping: a.) Master and slave wrist positions: x_m and x_s; b.) Hand and environment forces: f_h and $-f_e$; c.) Damping factor, b_0.

force is generated to move the master back to the position region, following which the operator moves the robot away from the point of contact.

The right-side plots of Figure 13 show the results obtained with the coordinated force with force feedforward controller (the damping term b_s was 0.1 N/(mm/s) to maintain contact stability for this controller). Since the force sensor is used in this controller, there are no position errors needed to generate forces. The problem of deadband crossing encountered previously is eliminated and the operator has a better sense of the environment stiffness.

5. Conclusions and Future Work

This paper discussed the issues of motion coordination and control that arise in the operation of a novel fine-master, coarse-fine slave teleoperation system employing Lorentz magnetic levitation.

By combining position and rate control, an intuitive automatic indexing procedure was developed, allowing single-handed six-degree-of-freedom positioning of the slave. Feedforward of sensed hand and environment forces was proposed as a way to reduce the coordination errors between master and slave during bilateral control. It was also shown that both contact stability problems and

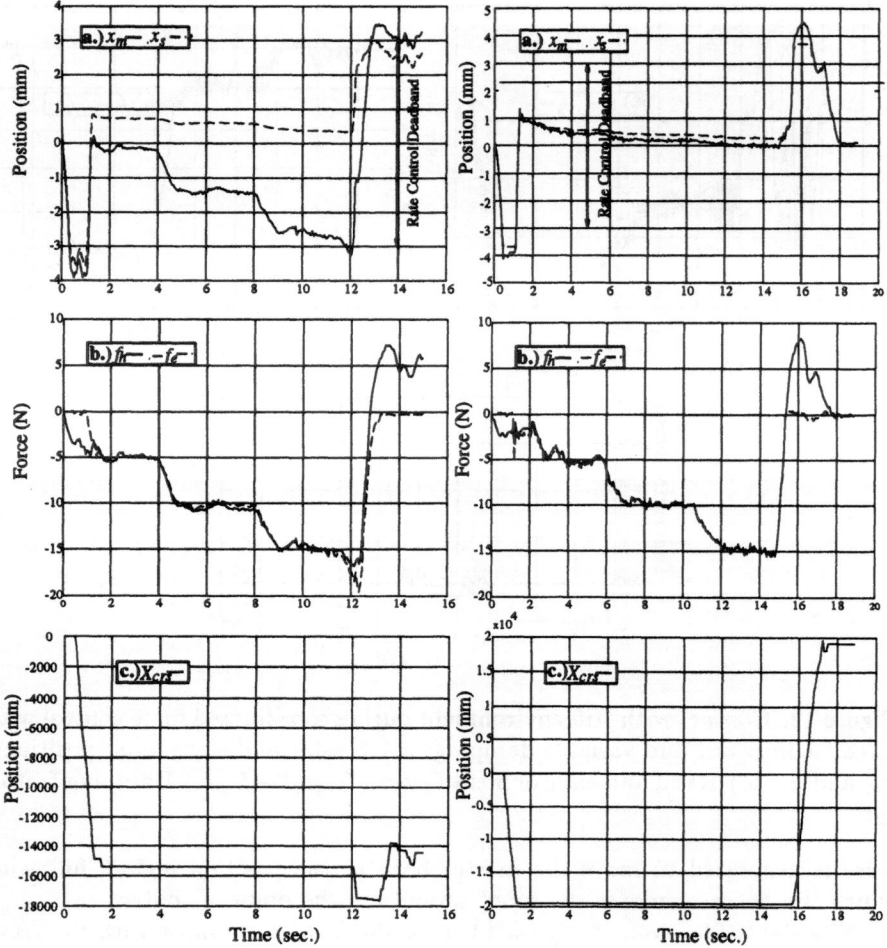

Figure 13. Exertion of forces under coordinated force control (left) and coordinated force control with force feedforward (right): a.) Master and slave wrist positions: x_m and x_s; b.) Hand and environment forces: f_h and $-f_e$; c.) CRS robot motion X_{CRS}.

free-motion drag can be avoided by applying additional environment-dependent damping at the slave.

Other methods of providing damping when in contact with stiff environments are being examined. One possibility is feedback of the measured environment force at the slave.

The issue of force feedback during slave rate control requires further research. Extensions of the teleoperation control scheme presented in Section 4 to more general master and slave models are also being pursued by using the parametrization of all stabilizing compensators and H^∞ or other optimization-based design methods [1]. On-line identification of inertial parameters of the master and slave flotors and their use in control algorithms is also being studied.

References

[1] S. Salcudean, N.M. Wong, and R.L. Hollis. A Force-Reflecting Teleoperation System with Magnetically Levitated Master and Wrist. In *Proc. IEEE Conf. Robotics Automat.*, Nice, France, May 10-15, 1992.

[2] J. Vertut and P. Coiffet. *Robot Technology, Vol. 3A: Teleoperations and Robotics: Evolution and Development.* Prentice-Hall Series on Robot Technology. Prentice-Hall, 1986.

[3] P. Fischer, R. Daniel, and K.V.Siva. Specification and Design of Input Devices for Teleoperation. In *Proc. IEEE Conf. Robotics Automat.*, pages 540–545, May 1990.

[4] J.C. Doyle and G. Stein. Multivariable feedback design: Concepts for a classical/modern synthesis. *IEEE Trans. on Automat. Contr.*, AC-26:4–16, 1981.

[5] R. H. Taylor, R. L. Hollis, and M. A. Lavin. Precise manipulation with endpoint sensing. *IBM J. Res. Develop.*, 2(4):363–376, July 1985.

[6] R.L. Hollis, S. Salcudean, and P.A. Allan. A Six Degree-of-Freedom Magnetically Levitated Variable Compliance Fine Motion Wrist: Design, Modelling and Control. *IEEE Trans. Robotics Automat.*, 7(3):320–332, June 1991.

[7] M. Handlykken and T. Turner. Control System Analysis and Synthesis for a Six Degree-of-Freedom Universal Force Reflecting Hand Controller. In *Proc. 19th IEEE Conf. Decision and Control*, December 1980.

[8] Y. Yokokohji and T. Yoshikawa. Bilateral Control of Master-Slave Manipulators for Ideal Kinesthetic Coupling. In *Proc. IEEE Conf. Robotics Automat.*, pages 849–858, Nice, France, May 10-15 1992.

[9] S.-R. Oh, R.L. Hollis, and S. Salcudean. Precision Assembly with A Magnetically Levitated Wrist. In *Proc. IEEE Conf. Robotics Automat.*, pages 127–134, Atlanta, USA, May 2-6, 1993.

[10] N.M. Wong. Implementation of a force-reflecting telerobotic system with magnetically levitated master and wrist. Master's thesis, University of British Columbia, December 1992.

[11] E.G. Johnsen and W.R. Corliss. *Human Factors Application in Teleoperator Design and Operation.* New York: Wiley-Interscience, 1971.

[12] W.S. Kim, F. Tendick, S.R. Ellis, and L.W. Stark. A comparison of position and rate control for telemanipulations with consideration of manipulator system dynamics. *IEEE J. Robotics Automat.*, RA-3(5):426–436, October 1987.

[13] N.R. Parker, S. Salcudean, and P.D. Lawrence. Application of Force Feedback to Heavy Duty Hydraulic Machines. In *Proc. IEEE Conf. Robotics Automat.*, pages 375–381, Atlanta, USA, May 2-6, 1993.

[14] B. Hannaford. A Design Framework for Teleoperators with Kinesthetic Feedback. *IEEE Trans. Robotics Automat.*, RA-5(4):426–434, August 1989.

[15] R.J. Anderson and M.W. Spong. Bilateral control of operators with time delay. *IEEE Trans. Automat. Contr.*, AC-34(5):494–501, May 1989.

Section 8
Mobile Robots

Although a large amount of results have been reported so far on mobile robot navigation, most of them are situated in known environments. However such an assumption of known environment rarely holds in practice; for example, it fails in a natural environment or a factory where environments are constantly changing. Autonomous navigation in unknown environments requires reliable and reactive strategies on perception, environment modelling, robot localization, motion planning, and execution.

Chatila, Fleury, Herrb, Lacroix, and Proust challenge a difficult problem of implementing all the capacities for autonomous navigation in natural environments. They emphasize the importance of the choice of adequate representations to model the complexity of a natural environment. The experimental testbed ADAM and the current implementation of the capacities for autonomous navigation are also described.

Zelinsky and Yuta present a practical approach to construct mobile robot navigation schemes that operate in unstructured indoor environments. They propose a unified approach that uses a combination of known and unknown environment path planning techniques. The validity of the proposed approaches is demonstrated by implementation of the navigation schemes on the autonomous robot Yamabico.

Cherif, Laugier, Milesi-Bellier, and Faverjon focus on motion planning for a mobile robot moving in a natural environment, especially on the generation of "executable and safe motions". The basic idea of the proposed method is to integrate geometric and physical models of the robot and the terrain in a two-stage trajectory planning process.

The paper by Krotkov and Simmons summarizes their approach to a six-legged autonomous walking robot Ambler with high mobility, high autonomy, and mission capabilities for planetary exploration. The Ambler integrated walking system consists of a number of distributed modules, the Task Controller for coordination of the distributed robot system, the Real-Time Controller for motion control, the perception modules for image processing, the planning modules for gaits, and the graphical user interface.

Rives, Pissard-Gibollet, and Kapellos discuss reactive approaches based on visual servoing applied to a mobile robot. The visual servo loop is closed with respect to the environment and allows to compensate the errors in the measurement or in the estimation of the robot position. To cope with the problem due to nonholonomic constraints, the developed robotic system uses a cart-like mobile robot carrying a two d.o.f. head with a camera.

Mondada, Franzi, and Ienne present a miniature robot Khepera for investigation in control algorithms such as the subsumption architecture, fuzzy logic, or artificial neural networks. Due to its small size, the programming environment, and the real time visualization tools, experiments for accurate validation of control algorithms can be carried out quickly and cost-effectively in a small working area.

The paper by Reister, Unseren, Baker, and Pin is the only one dealing with a task of the manipulator mounted on a mobile robot to follow an a priori unknown surface. It investigates the feasibility and requirements of simultaneous external-sensor-driven platform and redundant manipulator motions, and the interaction of the two subsystem's internal sensor-based position estimation.

Autonomous Navigation in Natural Environments

Raja Chatila   Sara Fleury   Matthieu Herrb
Simon Lacroix   Christophe Proust

LAAS-CNRS
7, avenue du Colonel-Roche
31077 Toulouse Cedex - France
{raja,sara,matthieu,simon,proust}@laas.fr

Abstract— This paper presents the approach, algorithms and processes we developed to perform autonomous navigation in a natural environment. After a description of the global approach, we discuss the characteristics of natural environment representations. Then the perception functions for terrain mapping and robot localization, as well as motion planning are described. Navigation strategies for selecting perception tasks and subgoals for motion planning are proposed. The current state of integration in the experimental testbed EDEN is finally presented. Results from this experiment illustrate the approach troughout the paper.

1. Introduction and Overview

A large amount of results exists today on mobile robot navigation, most of them related to indoor navigation, wherein obstacles are rather structured, and the terrain flat, or in some limited cases, includes slopes. Mobility in outdoors environments have been demonstrated for road following (e.g., the ALV [1], the NAVLAB [2]), and also in limited environment conditions (e.g., Robby on a dry river bed [3] [4]), or for robots using legged locomotion (e.g., the Ambler [5]).

In the "EDEN" experiment carried out at LAAS with the mobile robot ADAM[1] (figure 1), we aim to demonstrate a fully autonomous navigation in a natural environment, including perception, environment modelling, robot localization, and motion planning and execution on flat or uneven terrain.

The canonical task is "GO-TO (Landmark)" in an initially unknown environment that is gradually discovered by the robot. The landmark is an object known by its model, which can be recognised and localized on a video or laser image.

ADAM is equipped with a 3D scanning laser range finder, two orientable color cameras, and a 3-axis inertial platform. The motion controller also provides for odometry. On-board computing equipement is composed of two VME racks, one for locomotion and attitude control, and the other for the perceptual and decisional functions. All the functions developed for this experiment

---

[1] ADAM: "Advanced Demonstrator for Autonomy and Mobility" is property of FRAMATOME and MATRA MARCONI Space, currently lent to LAAS.

Figure 1. *The mobile robot Adam.*

are encapsulated in modules integrated together in a real-time environment for autonomous operation (section 5).

The laser range finder and the stereovision system are used for terrain modelling. During robot motion, a surveillance mode in which the laser scans a line two meters ahead can be activated.

The landmark position is given approximately, and is not necessarily in the field of view in the initial position of the robot. From the approximate given position of the landmark (e.g., 50 m, N-W.), goal coordinates are determined. The general approach that we have implemented is an incremental navigation that proceeds until the final goal is reached. Its main elements are the following, and will be detailed in the corresponding sections:

- In a natural terrain, there may be areas that are rather flat - cluttered or not by obstacles - and others which are uneven with respect to the locomotion capacities of the robot. Detailed modelling of the terrain is complex and resource consuming. Hence, we first use a simple algorithm to characterize the terrain into five classes: flat, with slope, uneven, obstacle and unknown (section 2.2). Only the uneven terrain regions will be modelled more accurately with an elevation map if necessary, i.e., if the robot has to move through such regions.

- A global navigation planner selects a subgoal within the known part of the environment, taking into account the nature of the regions to be crossed. The subgoal may be associated with a perception task to acquire new data.

- The adequate navigation mode is selected according to the nature of the traversed terrain: a reflex mode if the terrain is flat and free, a 2D planned motion if the terrain is flat and cluttered, or a 3D planned motion if it is uneven. (§3).

- After execution of the trajectory to the subgoal, the landmark is observed (if possible) and its position updated. A new terrain acquisition and classification is performed, and the procedure is iterated.

Navigation strategies concern motion strategies (definition of intermediate goals and navigation modes) and perception strategies (choice of the perception tasks to perform, point of view, etc.). The type of perception task necessary at a step of the mission execution is derived from an analysis of the result of the motion strategy: if an uneven area has to be crossed, it must then be interpolated, if the global goal to reach is in an unknown area, the model has to be completed, etc.

Terrain modelling mainly relies on laser data, but fusion with color stereo (currently under development) enables to interpret the nature of the terrain (§6).

The control system that executes the navigation procedure interprets the task and sequences the functional modules for its execution according to the context and specific modalities (§4). It selects the adequate sensors (vision, laser) according to the landmark description, the lighting conditions, etc. It also selects the motion planner that is adapted to the nature of the terrain to be traversed, and sets the conditions/reactions in case unexpected events are detected (e.g., obstacles). It sets the conditions for the achievement of the task, i.e., proximity and visibility of the landmark, and stops the robot near the landmark. The control system also verifies the coherence of the various functions (e.g., robot and target localization, motion execution).

We now detail these different issues.

2. Environment Representations

The complexity of a natural environment emphasizes the importance of the choice of adequate representations to model it: they must be easily manipulated, and also be well adapted to the main tasks involved during navigation. We present in this section a structural scheme in which are embedded all the different representations that coexist in the system, and briefly describe three important perception processes applied on 3D data provided by a laser range finder (LRF) or a correlation stereovision algorithm.

2.1. Types of Models and Relationships

The difficulty of modelling outdoor environments comes essentially from the fact that they are not intrinsically structured, as compared to indoor environments where simple geometric primitives match the reality. We have therefore favored the development of simple representations (polygonal maps, elevation maps...), easier to build and manage than those based on complex geometric primitives (linear or second degree surfaces, superquadrics...). Such more complex primitives can always be extracted from these basic representations if

428

necessary. In addition, semantic informations (e.g., the nature of some areas) can be easily included as labels in these representations.

The other characteristics of the representations are related to the robot sensors and task, namely navigation:

• The sensors are always imperfect: their data are uncomplete (lack of information concerning existing features) and not precise. They generate artefacts (information on non-existing features) and errors (wrong information concerning existing features). The same area when perceived again can therefore be differently represented. Hence environment representations must tolerate important **variations** [6].

• The environment is **incrementally discovered**: we must be able to manage local momentary representations, and merge them in a global description of the world[2].

During the navigation, environment representations are needed for three main functions:

• **Trajectory planning**: we consider three different navigation modes, depending on the nature of the areas to be crossed. The reflex navigation mode does not need any numeric information concerning the environment, as opposed to the other modes: the 2D planner [7] requires in our case a binary bitmap description, and the 3D planner [8] builds its own data structure on the basis of an elevation map on a regular cartesian grid.

• **Localization**: Localization based on environment features is always necessary, be it to correct odometry and inertial platform drifts. Specific representations of the environment are neccessary for this: a set of 3D points in the case of a correlation-based localization (iconic matching [9]), or a global map of detected landmarks (that must then be modeled, using particular geometric descriptions) in the case of a feature-based localization that we use [10].

• **Strategic decisions**: To perform the navigation strategies, a topological description of the perceived environment is necessary; we use an adjacency graph of labelled regions, on which classical heuristic search is performed [7]. As for the perception planning process, we use the bitmap description resulting from the terrain classification procedure (section 2.2).

To manage these different representations and maintain a global coherence, we developed a multi-layered heterogeneous model of the environment. In this model, the relationships between the various representations are of three different kinds (Figure 2):

• **Functional**: All the representations are not necessary at all times, but some are, such as the representation needed to perform strategic choices, in which the localization model must be included for instance (arrows labeled "1" in the figure);

• **Spatial**: There are areas of the perceived environment which may be more detailed to perform specific tasks such as 3D motion planning, localization, or object recognition[3]...(arrows labeled "2");

---

[2]A global representation is needed for an efficient navigation.
[3]for this last purpose, we are investigating the use of superquadrics

- **Constructive**: Some representations are determined on the basis of an other one (arrows labeled "3").

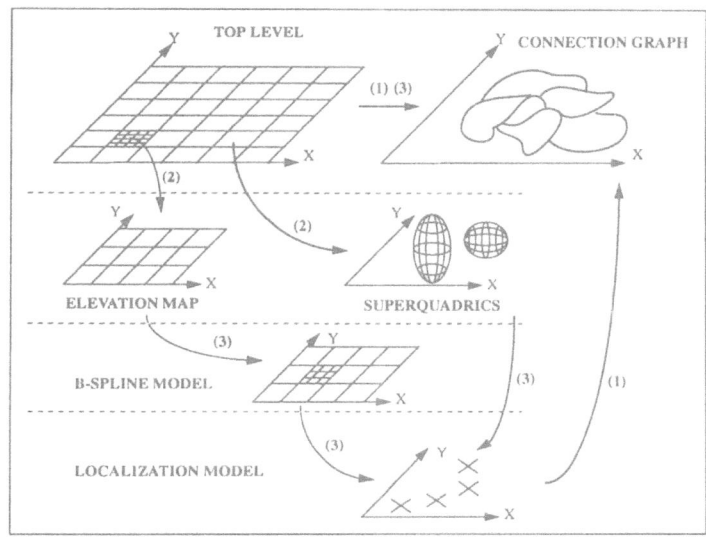

Figure 2. The representations used in the system

The top level of this heterogeneous model is a "bitmap" description of the environment, built upon the results of the fast terrain analysis algorithm (§2.2). The information encoded in every pixel of this bitmap is the terrain label and its confidence level, the estimated elevation, the identification of the region it belongs to. This structure is simple, adapted to the lack of geometrical structure of the environment, and flexible, in the sense that any supplementary information can easily be encoded in a pixel without reconfiguring the entire description and the algorithms that use it. Moreover, the techniques that allow to extract structured informations (regions, connexity...) from a bitmap are well known and easily implemented.

The main drawback of a bitmap representation is its memory occupancy, that rapidly becomes huge if it covers a large area. To cope with this, a "sliding bitmap" structure is used: only the area surrounding the robot, with a size limited by the sensor capacities and the local navigation needs, is described as a bitmap, whereas the remaining of the already perceived terrain is structured using a region merging algorithm (or any other classical image compression algorithm), that leads to a much more compact description.

2.2. Terrain Classification

Developing a method that allows to quickly determine the nature of the terrain is essential in our adaptative navigation approach: as explained in the introduction, this process is systematically applied on the 3D data produced by the sensors, and is the basis of the global model of the environment on which the strategic decisions are taken. Using this method, we aim at developing a "smarter" use of perception: time and resource consuming procedures, such as building a digital elevation map, are activated only when necessary, and are controlled in order to provide the useful information.

The principle of the classification method relies on a discretization of the perceived aera in "cells", that correspond to the projection on a virtual horizontal ground of a regular grid in the sensor frame (figure 3). If we observe the orthogonal projection of a set of 3D points along the vertical axis (figure 4), one can note that the density of the points corresponding to the flat ground is a decreasing function of their distance to the sensor, which directly depends on the scanning mode of the sensor. The obstacles correspond to a very high density of the projected points, whereas occluded areas have a very low density (theoretically null). Our discretization respects this property: all the cells covering flat areas contain a constant number of points, defined by the discretization rates, whereas the cells correponding to obstacles have a much greater number of points, and the ones correponding to "shadows" are empty.

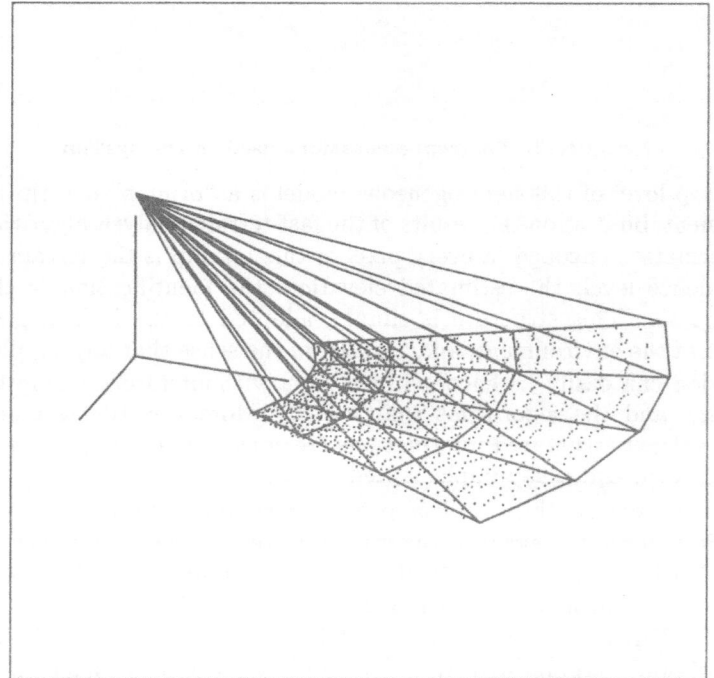

Figure 3. **Discretization of the pereived area**

This "cell density" information, along with other characteristics (mean altitude value, variance on the altitude, mean normal vector and variances on its coordinate) help to heuristically give a label to each cell as one of {*Flat, Slope, Uneven, Obstacle, Unknown*}.

The classification method has been run on a large number of different images and gives significant results. It is especially weakly affected by the sensor's noise

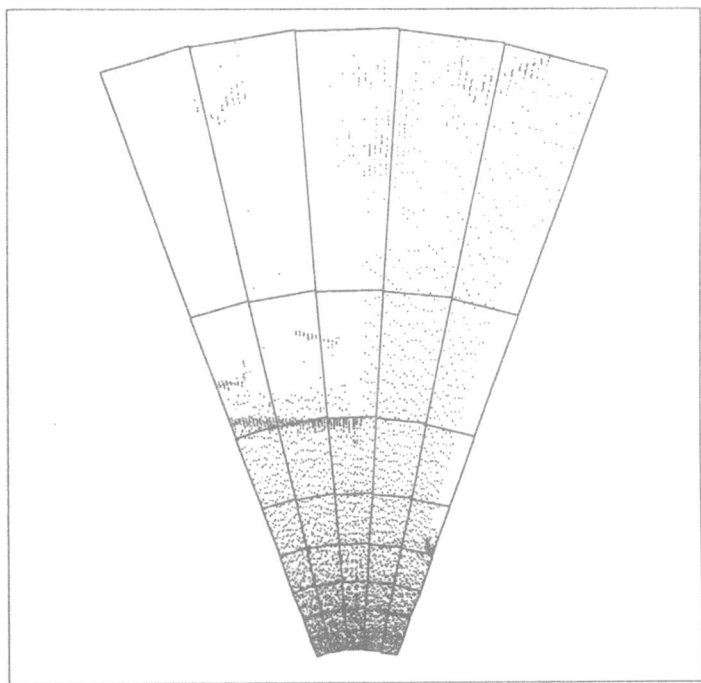

Figure 4. **Vertical projection of a 3D image**

(uncertainties and errors). It takes a reasonable time: the complexity of the procedure is $O(n)$, where n is the number of 3D points in the image. Figure 5 is a camera view of a scene composed of flat and very uneven areas. The corresponding classification result is illustrated in increasingly gray levels (from Unknown to Obstacle) in figure 6.

Fusion procedure The main problem raised by the fusion[4] of different perceptions is a possibly conflicting labelling of an area from a perception to another. These conflicts are due to the uncertainties of the sensor, whose behavior has therefore to be modeled to quantify the confidence on the data it returns.

The behavior of the logical sensor "terrain classifier" is a consequence of the range finder features whose accuracy on a 3D point coordinates is a decreasing function of the distance ρ. Hence the confidence on the label of each cell also decreases with its distance to the sensor. Another obvious observation is that this label confidence also depends on the label itself: for instance, a flat cell containing a few erroneous points can be labeled as an "uneven" one, whereas the probability that erroneous points perceived on an actually uneven zone lead to a "flat" label is very low. Figure shows a qualitative[5] estimate of the error probability $P(e)$ on a cell label.

---

[4] "fusion" here should be understood here as "aggregation", i.e., without position updating. Localization is performed by another system.

[5] The quantitative estimations are statistically determined.

Figure 5. **Video Image**

Fusion is a simple procedure: each cell resulting from the classification procedure is written in the bitmap using a polygon filling algorithm. When a pixel has already been perceived, the possible conflict with the new perception is solved by comparing the label confidence values. Figure 8 shows the fusion of 3 perceptions.

Many experiments have proved the robustness of this fusion method. Let's note that the confidence on the labels is also taken into account for the purpose of navigation strategies: it is essential that the artefacts (essentially badly labeled "obstacle" cells) due to the classification procedure do not mislead the robot's navigation.

3. Motion Planning

A reflex navigation mode enables to skip the modelling and motion planning steps. If the terrain to be traversed is flat and empty from obstacles, then the robot heads directly to its goal. If obstacles are detected by a proximity sensor (the surveillance mode of the LRF), the robot stops, and the nominal procedure of terrain modelling is started. The appropriate decision is taken after classification.

The nominal case is that motion planning is necessary. For this, we use two different motion planners according as the label of the regions to be crossed, is flat or uneven terrain.

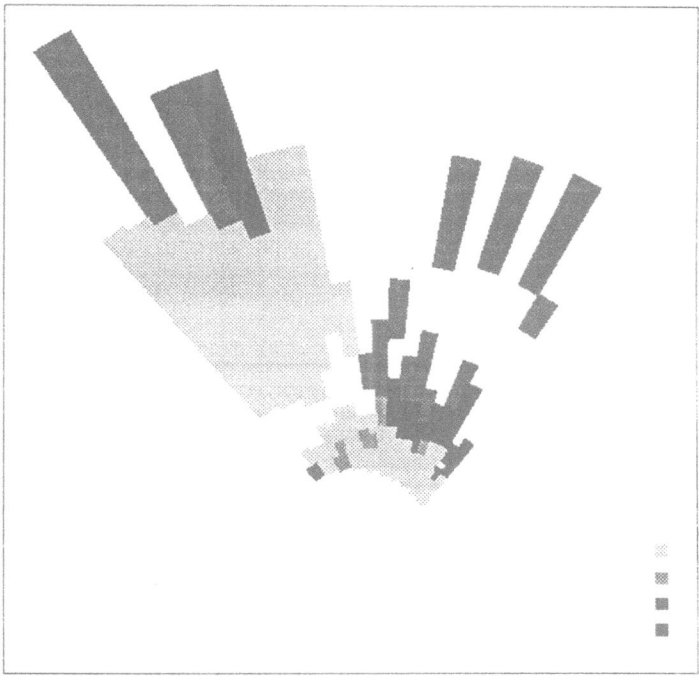

Figure 6. **Classification Result. From clear to dark: unknown, horizontal, flat with slope, uneven, obstacle)**

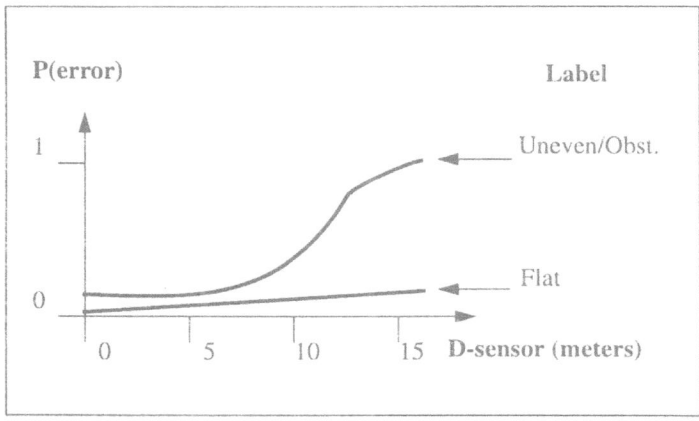

Figure 7. **Error probability on cell labeling**

3.1. Flat Terrain

The trajectory is searched with a simplified and fast method, based on bitmap and potential fields techniques. The robot is approximated by a circle, and its configuration space is two dimensional, corresponding to the robot's position

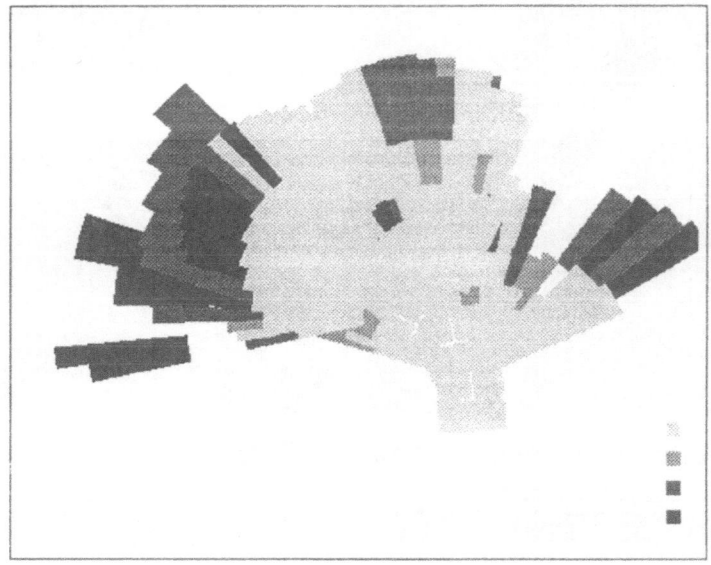

Figure 8. **Fusion of 3 perceptions**

in the horizontal plane. Path planning, detailed in [11], is done as follows :

1. A binary bitmap free/obstacle is built over the region to be crossed.

2. A distance propagation method similar to those presented in [12] produces a distance map and a discrete voronoï diagram.

3. This diagram is searched for a path reaching the goal.

4. A set of line segments is then extracted from this first path, and optimised in order to provide an executable trajectory, consisting of a minimum of straight lines separated by rotations.

Search time depends only on the bitmap discretization, and not on the complexity of the environement. This is very important in natural environments with an arbitrary high complexity. The path is obtained in at most 2.5 seconds on a 256 × 256 bitmap, on a Sparc 10 workstation.

3.2. Uneven Terrain

On uneven terrain, irregularities are important enough to alter the attitude and the motion of the robot. Here a more precise representation of the terrain in terms of an elevation map is necessary (with a step of 10 cm in our case). The path generated by the planner must verify the following constraints:

- The robot does not collide with the ground.

- The robot does not tip-over.

- the suspensions cannot be stretched beyond some limit length.

- The motion verifies the kinematic constraints of the vehicle.

A path planner, first presented in [8], generates incrementally a graph of configurations reached by a sequence of constant controls (i.e. rotations and straight lines, in the case of ADAM). The solution is obtained by searching this graph, using an A^* algorithm.

In the case of incremental exploration of the environment, an additional constraint must be taken into account: the existence of unknown areas on the terrain elevation map. Indeed, any terrain irregularity can hide part of the ground. The map generated contains blind areas, where elevation values are not known.

The path planner must avoid such unknown areas, when it is possible[6]. Otherwise, the planner must search the best way through unknown areas, and provide the best perception point of view on the way to the goal. This is performed with the following operations [11]:

- First, the unknown areas of the elevation map are filled by an interpolation operation, which provides a continuous elevation map.

- The planner can then search a way through unknown parts of the map. However, the avoidance of unknown areas is obtained by an adapted ponderation of the arc cost.

- In order to improve the heuristic guidance of the search, a new heuristic distance to the goal is built by bitmap techniques. A cost bitmap is first computed, including the difficulty of the terrain, and the proportion of unknown areas around the current patch. A potential propagation from the goal generates a distance information corresponding to the best way to the goal through terrain relief and unknown areas.

Once a trajectory reaching the goal is obtained, the first unknown area crossed by the robot is searched for. The trajectory is then truncated at the last point enabling to watch this area. The result is a motion stopped at the best point for a new perception.

Search time strongly depends on the difficulty of the terrain. The computation of the bitmap heuristic distance takes around 40 seconds for a 150×200 terrain map, and path search takes between 20 seconds to a few minutes, on Indigo R4000 Silicon Graphics workstations. Figure 9 shows a trajectory computed on a real terrain. The dark areas correspond to interpolated unknown terrain.

4. System Architecture and Control

4.1. Overview

The generic control architecture for the autonomous mobile robots developed at LAAS is organized into three levels [13, 14]. It is instanciated in the case of

---

[6]This is a caution constraint which can be more or less relaxed.

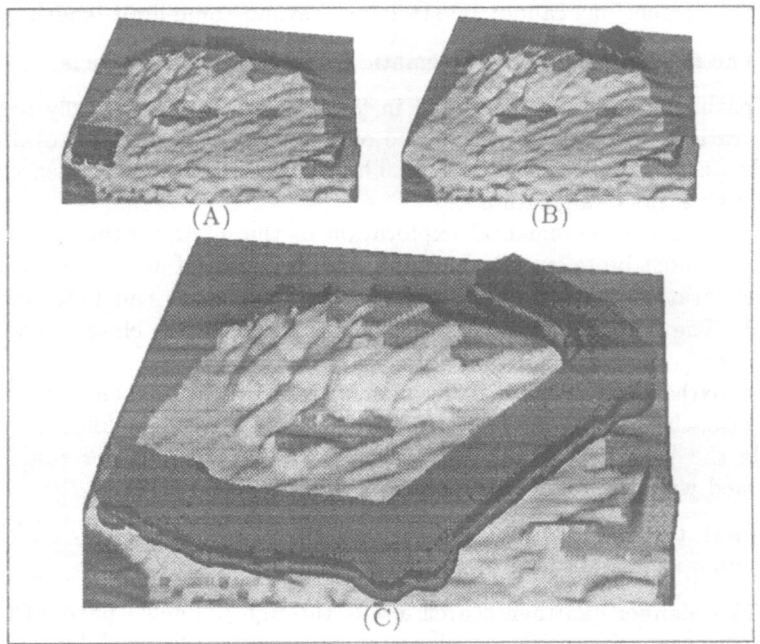

Figure 9. Start (A) and goal (B) configurations, and the trajectory (C) found in 30sec. CPU.

the EDEN experiment as shown in figure 10. The higher task planning level plans the mission specified by the operator in terms of tasks, with temporal constraints, interpretable by the robot and follows the execution of the mission. This level is not currently used in the experiment and will be embedded in a teleprogramming environment for experimenting the approach we propose for the development of intervention robots (e.g., for planet exploration) [15].

The on-board "Decisional Level" plans[7] the actions involved in the task - here navigation - and controls their real-time execution according to the context and the robot state to achieve them. The "Functional Level" embeds the on-board functions related to the control of sensors and processing of their data, and to trajectory planning and control of the effectors.

4.2. The Functional Level

The Functional Level includes the functions for acting (wheels, perception platform), sensing (laser, cameras, odometry and inertial platform) and for various data processing (feedback control, image processing, terrain representation, trajectory computation, ...).

To control robot functionalities and underlying resources, these functions

---

[7]We also use the word "refinement" instead of planning for this level.

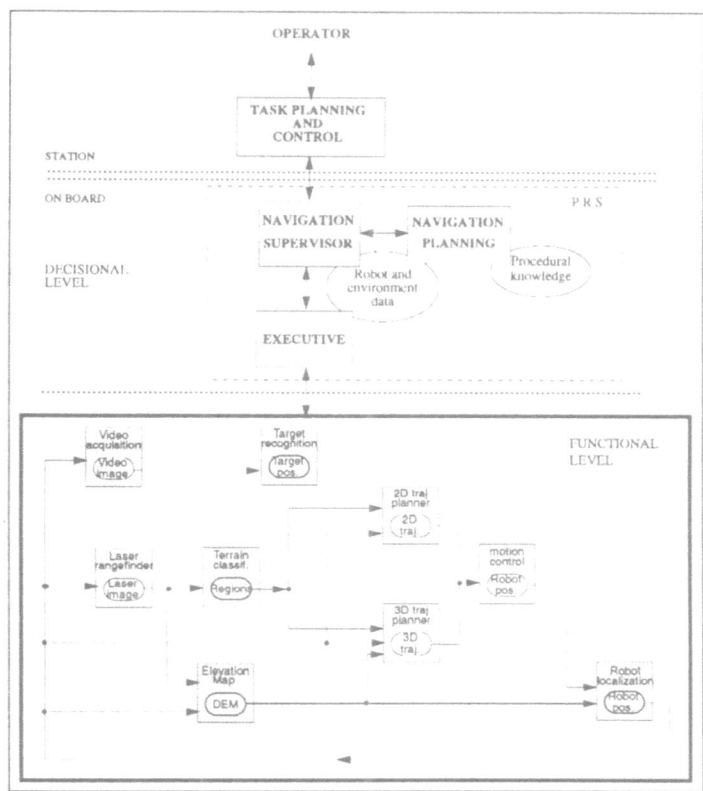

Figure 10. **Global control architecture. Connections between the modules at the functional level show data flow.**

are embedded into modules defined in a systematic and formal way, according to data or resources sharing. Thus, modules are servers which are called via a standard interface. This allows to combine or to redesign easily the functions.

Figure 10 shows the set of modules used for the experimentation and the data flow during the progress of a specific iteration. The connections are dynamically established by the decisional level according to the context. For instance the position of the robot is updated when the estimated deviation is higher than a fixed threshold; the visibility of the target is necessary to update its position.

Let's describe the steps executed during a nominal iteration:

1. A video image is acquired in the direction of the target (**Video acquisition module**).

2. The target is identified in a video image and localized in a global reference frame (**Target recognition module**). From the position of the target, the final goal (*e.g.* at a given distance in front of the target) is determined.

3. A 3D depth map is acquired (**Laser range-finder module**).

4. Using the laser image, the terrain is classified and the regions representation updated (**Terrain classification module**, §2.2). From this representation, the navigation planner in the decisional level selected sub-goals within the navigable regions.

5. If an uneven region is to be crossed, an corresponding elevation map is computed (**Digital Elevation Map module**).

6. The most suitable trajectory planner (2D or 3D) is used, according to the regions to be crossed, to find a trajectory inside the selected regions to reach the sub-goals (**2D trajectory planner and 3D trajectory planner modules**, §3).

7. The trajectory is executed with a feedback control loop on the position given by the odometry (**Motion Control module**).

8. The position of the robot is updated if needed necessary (**Robot Localisation module**).

4.3. The Decisional Level

This level includes the system that plans the navigation actions (navigation planner), and a supervisor that establishes the dependencies between modules in order to combine the functions according to the context and modalities at run-time. It also checks the conditions/reactions in case of external events. For example it will watch for obstacles when executing a trajectory. The "Executive" at the decisional level sends the requests to the functional level modules and checks the coherence of the various concurrent primitives. Each of these subsytems is described below. However, in our current implementation, the three entities of the decisional level have been simplified and merged together. They are implemented using PRS (Procedural reasoning System) [16].

4.3.1. The Supervisor

It receives a task to be executed and the associated events to watch for. The new task[8] is described in terms of actions to be carried out and modalities [14]. If the task is not directly executable, the supervisor calls the navigation planner that transforms it into the actions according to the knowledge on the environment, the execution modalities and robot state. The supervisor watches for events (obstacles, time-out, etc.). and reacts to them as planned and according to the dynamics of the situation and the state of the other running tasks. It sends to the Executive the different sequences of actions which correspond to the task. The supervisor sends back to the task planner (or the operator) the informations related to task (e.g., specific data) and the report about its execution.

---

[8]The task is in general not only related to navigation, and may include temporal constraints; however we focus here on navigation.

4.3.2. The Navigation Planner

In general, it is necessary to refine the task since the environment is roughly known at the moment of the task planning. The refinement planner uses procedures to carry out the task which is more or less decomposed in executable actions. Thus the purpose of the refinement planner is to produce a sequence of elementary actions, at run-time, on the basis of the current environment state, the modalities specified with the tasks, and the state of the robot. This is done in bounded time. The navigation procedure includes all the steps described above for the recognition of the target and landmarks, and the decision about the navigation modes and strategies.

4.3.3. The Executive

The executive launches the execution of actions by sending the related requests to the functional level modules. The parallelism of some sequences is taken into account: for instance the robot concurrently analyses the laser data and looks for the landmark, or moves toward a sub-goal while watching for obstacles. It sends back to the supervisor reports about the fulfilment of those basic actions. Futhermore the executive manages the access to ressources, the coherence of multiple requests at the syntactic level (for instance one cannot require concurrently the use of the laser to acquire data and the execution of a trajectory).

4.4. PRS: Procedural Reasoning System

To implement this control architecture we have used C-PRS [16]. It is an environment to develop and execute operational procedures (called "Knowledge Areas" : KA) which seemed well-designed to cope with the constraints of the control architecture. The representation is procedural which is convenient to define the different actions needed to carry out a task or to introduce some heuristic reasoning in case of non-nominal situations. The choice of the best procedure to solve a given sub-goal can be stated on the basis of the data corresponding to the context of execution or deduced by a meta-procedure that reasons about the best applicability of such procedures.

As an example, we describe the procedure "Goto Landmark loop" (Figure 11) which will loop until Adam has reached its goal. It achieves the goal (! (GOTO_LANDMARK)) which is posted by another KA ("Goto Cible"), the top level procedure that initializes the experiment. There is one interesting contruction in this procedure: the parallel execution of two goals in their own thread. It starts by checking if it has reached the terminal area. If it has, the procedure ends after achieving the goal. Otherwise, it posts 2 goals in parallel,[9] (! (UPDATE_GOAL_POSITION)) and (! (UPDATE_NAVIGATION_MAP)). The execution then proceeds with a join node, which means that whichever thread will complete first will have to wait the second one. From then, it is straight forward and loop back on the start node.

---

[9]The thick bottom of a node indicates that it is a "split"node. Similarly, a thick top of a node indicates a "join" of parallel threads.

440

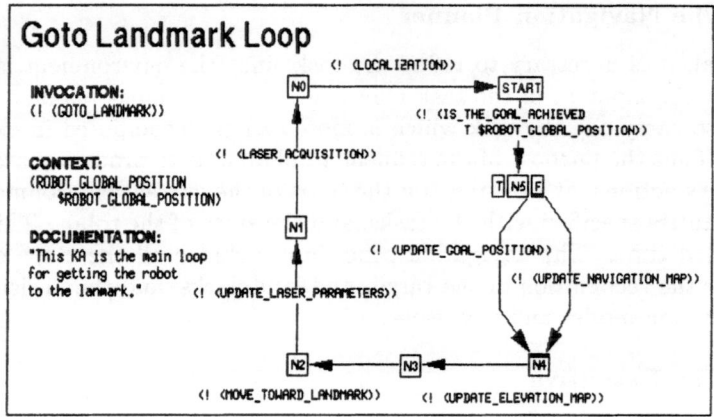

Figure 11. The main navigation procedure

5. Experimental Testbed and Integration

In this section we present our experimental testbed, the mobile robot ADAM and the current implementation of the elements described above. The experimental terrain is 12 by 50 meters with a maximum difference of level of 1.8 meter.

5.1. Hardware

The computing architecture is composed of two elements (figure 12): a fixed operating station (a Sun SparcStation 10-41) and two on-board VME racks connected to the operating station by an Ethernet link (figure 13). The first rack, includes of two 68030 CPUs and various I/O boards, is dedicated to the GNC (Guidance, Navigation and Control: motion execution and control, and localization with odometry and the inertial platform). They were provided with the associated software by Matra Marconi Space and Framatome with the Robot.

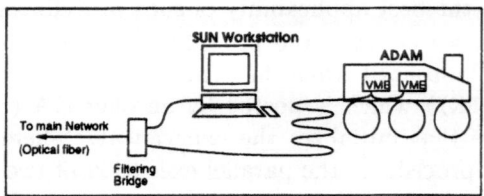

Figure 12. Network set-up

The second rack is composed of two 68040 CPUS, three Datacube boards and some I/O. It's dedicated to sensing activities: video image acquisition, laser range finder command and acquisition, local processing of data.

Both racks run under the real time operating system VxWorks.

During the experiments, most of the "high level" computing were done on the operating workstation to take benefit of its better debugging environment

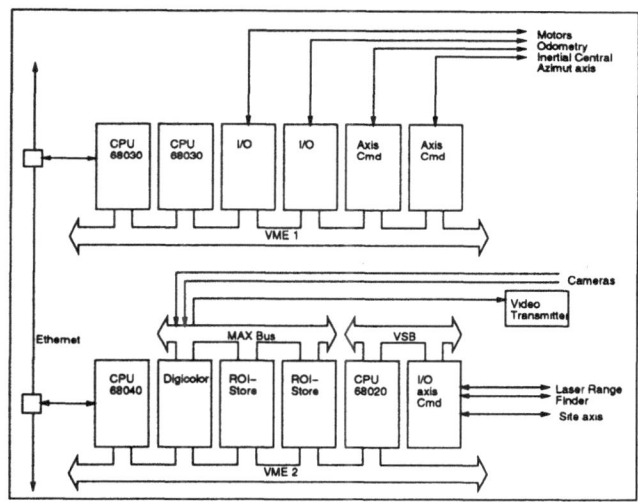

Figure 13. **On-board computing architecture of Adam**

and of the pre-existence of most of the software under Unix. However, we have the possibility to embark all the software in a near future: most of our Unix software already run under VxWorks. It is also possible to use an on-board Sparc CPU under Sun-OS.

5.2. Software Environment

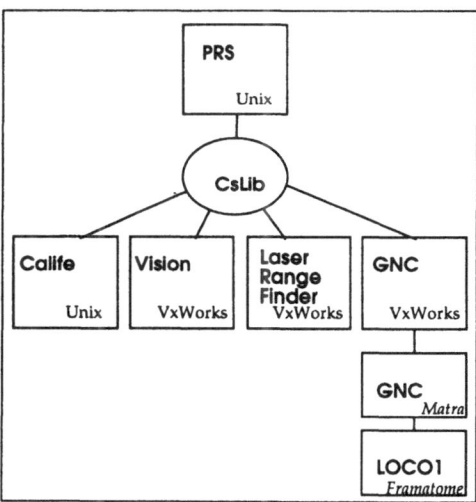

Figure 14. software sub-systems

The Eden experimentation is being developed using three software systems (figure 14):

Calife: A multi-sensor perception software integrating all the modeling and perception algorithms [17] developed at LAAS.

CsLib: A real-time client/server communication software also developed at LAAS, used to link the different modules to the execution control system [18].

C-PRS: An implementation in C of PRS (Procedural Reasoning System [16], developed at SRI)[10] and costumized at LAAS to implement the execution control system.

6. Conclusion

The EDEN experiment described in this paper implements all the capacities for autonomous navigation in natural environments. The main issues that we are currently developing further are related to the choice of observation points for a better data acquisition (currently path search to the goal and stop before the unknown regions); perception cost and gain should be integrated in the path search itself. Integration of color stereo, and the fusion of vision with depth imaging is another important issue being investigated, as well as localization using natural features. A full integration of the task planning and programming environment is currently carried out.

The results of this experiment benefit to the developement of intervention robots that have to perform their navigation in unknown environments, in particular autonomous planetary rovers.

Acknowledgments. The authors wish to thank all the members of the EDEN team.

References

[1] T. Kanade. Panel discussion: Possibilities in alv research. In *IEEE International Conference on Robotics and Automation, San Francisco (USA)*, April 1986.

[2] Charles Thorpe, Martial H. Hebert, Takeo Kanade, and Steven A. Shafer. Vision and navigation for the Carnegie-Mellon navlab. *IEEE Transaction on Pattern Analysis and Machine Intelligence*, 10(3), May 1988.

[3] C.R. Weisbin, M. Montenerlo, and W. Whittaker. *Evolving Directions in NASA's Planetary Rover Requirements end Technology*, chapter 4. Centre National d'Etudes Spatiales, France, Sept 1992.

[4] B. Wilcox and D. Gennery. A mars rover for the 1990's. *Journal of the British Interplanetary Society*, 40:484–488, 1987.

[5] E. Krotkov, J. Bares, T. Kanade, T. Mitchell, R. Simmons, and R. Whittaker. Ambler: A six-legged planetary rover. In *'91 International Conference on Advanced Robotics (ICAR),Pisa (Italy)*, pages 717–722, June 1991.

[6] E. Schalit. Arcane : Towards autonomous navigation on rough terrains. In *IEEE International Conference on Robotics and Automation, Nice, (France)*, page 2568 2575, 1992.

[7] R. Chatila, B. Dacre-Wright, M. Devy, P. Fillatreau., S. Lacroix, F. Nashashibi, P. Pignin, T. Siméon. Genération des déplacements autonomes. In *Projet I.Ares, Rapport de Recherche LAAS no. 93-272*, July 1993.

---

[10]C-PRS is a product of ACS Technology.

[8] T. Siméon and B. Dacre Wright. A Practical Motion Planner for All-terrain Mobile Robots. In *IEEE International Workshop on Intelligent Robots and Systems (IROS '93) Japan*, July 1993.

[9] Z. Zhang. Recalage 3D. Programme VAP : Rapport final de phase 5, Institut National de Recherche en Informatique et en Automatique - Sophia Antipolis, 1992.

[10] P. Fillatreau and M. Devy. Localization of an autonomous mobile robot from 3d depth images using heterogeneous features. In *IEEE International Workshop on Intelligent Robots and Systems (IROS '93), Yokohama, , Japan)*, July 1993.

[11] B. Dacre Wright. Planification de trajectoires pour un robot mobile sur terrain accidenté. These de Doctorat de l'Ecole Nationale Supérieure des Télécommunications (Paris), L.A.A.S., Oct. 1993.

[12] J. Barraquand and J.-C. Latombe. Robot motion planning: A distributed representation approach. In *International Journal of Robotics Research*, 1991.

[13] R. Chatila, R. Alami, B. Degallaix, and H. Laruelle. Integrated planning and execution control of autonomous robot actions. In *IEEE International Conference on Robotics and Automation, Nice, (France)*, 1992.

[14] R. Alami, R. Chatila, and B. Espiau. Designing an intelligent control architecture for autonomous robots. In *'93 ICAR, Tokyo, Japan*, Nov. 1993.

[15] G. Giralt, R. Chatila, and R. Alami. Remote intervention, robot autonomy, and teleprogramming: Generic concepts and real-world application cases. In *'93 IROS, Yokohama, Japan*, July 1993.

[16] M.P.Georgeff and F. F. Ingrand. Decision-Making in an Embedded Reasoning System. In *11th International Joint Conference on Artificial Intelligence (IJCAI), Detroit, Michigan (USA)*, 1989.

[17] F. Nashashibi, M. Devy, and P. Fillatreau. Indoor Scene Terrain Modeling using Multiple Range Images for Autonomous Mobile Robots. In *IEEE International Conference on Robotics and Automation, Nice, (France)*, pages 40–46, May 1992.

[18] R. Chatila and R. Ferraz De Camargo. Open architecture design and inter-task/intermodule communication for an autonomous mobile robot. In *IEEE International Workshop On Intelligent Robots and Systems, Tsuchiura, Japan*, July 1990.

A Unified Approach to Planning, Sensing and Navigation for Mobile Robots

Alexander Zelinsky[1] and Shin'ichi Yuta[2]

Intelligent Machine Behaviour Section[1]
Electrotechnical Laboratory
Tsukuba 305, Japan

Intelligent Robotics Laboratory[2]
University of Tsukuba
Tsukuba 305, Japan

Abstract - Much of the focus of the research effort in path planning for mobile robots has centred on the problem of finding a path from a start location to a goal location, while minimising one or more parameters such as length of path, energy consumption or journey time. Most of the experimental results reported in the literature have centred on simulation results. Only a small subset of the reported results have been implemented on real autonomous mobile robots. It is the goal of our research program to develop path planning algorithms whose correctness and robustness can be tested and verified by implementation on our experimental, self-contained and autonomous mobile robot - the Yamabico.

1. Introduction

Much of the focus of the research effort in path planning for mobile robots has centred on the problem of finding a path from a start location to a goal location, while minimising one or more parameters such as length of path, energy consumption or journey time. Most of the experimental results reported in the literature have centred on simulation results. Only a small subset of the reported results have been implemented on real autonomous mobile robots. It is the goal of our research program to develop path planning algorithms whose correctness and robustness can be tested and verified by implementation on our experimental, self-contained and autonomous mobile robot - the Yamabico which was reported by Yuta *et.al.* in [1].

The primary drawbacks of most reported path planning methods is that they operate in known environments and only find the shortest path between the start and goal positions. However completely known environments rarely exist in practise, for example a map of a building or factory can be readily obtained and used to perform path planning for a mobile robot. However it is impossible to guarantee that the map is accurate, since it does not record the presence of people or obstacles such as boxes left in passage ways. Environments are in reality not static, they are dynamic i.e. constantly changing. This introduces uncertainty into the problem of path planning. To be useful an autonomous mobile must be able to handle the unknown. Path planning in unknown environments as studied by Chatila [2], Crowley [3], Moravec [4] and Thorpe [5], has taken the approach that the environment is completely unknown and the mobile robot must incrementally map the environment using the robot's onboard sensors. It can be argued that this is the correct approach to the problem. However, it has not produced significant results. This is due to the limitations of the current generation of sensing technology. Current sensors either have serious limitations in their accuracy e.g. ultrasonic sensors or they require enormous computation and memory needs e.g. vision sensors. Building accurate maps with current sensing technology is a tall order. At present there are no methods available which allow a mobile robot to construct an accurate representation of the environment quickly using onboard processing.

Planning paths that only minimise the distance to a goal are usually unsuitable for implementation on an actual mobile robot. The safety of the robot is important particularly since there are uncertainties in the sensor data, such as the exact shape and position of obstacles. This problem is compounded by the uncertainty in the dynamic control of a robot i.e. the precise position of the robot is not always known by the robot's control system. Thus both minimum distance to a goal and safety of the robot need to be considered simultaneously during path planning. This problem could be solved by adding an external localisation system which would maintain an accurate estimate of robot's position. This involves adding beacons or artificial landmarks to the environment. This adds unnecessary artificial structure to the environment. One goal of our research is to construct mobile robot navigation schemes that operate in unstructured indoor environments.

In this paper we present an approach to the implementation of a navigation system on a mobile robot using a unified approach to planning. A unified approach uses a combination of known and unknown environment path planning techniques. The environment in which a mobile robot must operate in is usually well known e.g. a factory layout. The plan of the environment e.g the design drawings of a building, can be supplied to the robot's path planner. The robot will plan a global path to the goal through the known environment using a vertex graph model and the A* algorithm. However, this plan is insufficient to allow a robot to execute the path relying solely on dead reckoning. Therefore, the global path planner must also consider which landmarks in the environment should be tracked for position estimation and the sensing positions to optimally observe the selected landmarks. Hence, the model of the environment must not only include geometric data, but it must also include the types of surfaces and the characteristic features of the landmarks which can be measured by sensors. Once a plan for the robot's mission has been constructed, the robot executes the planned path while observing specific landmarks.

If the robot unexpectedly encounters an obstacle which interferes with the execution of the planned mission, the robot must avoid the obstacle. In the literature there have been various local obstacle avoidance procedures proposed. A popular approach has been to use potential fields that were pioneered by Khatib [6]. The problems of using potential fields is well documented in the literature by Koren *et. al.* [7]. The worst problem affecting potential fields is the local minima problem. The other approach to local obstacle avoidance is heuristic navigation. Heuristic navigation as studied by Borenstein *et. al.* [8] and Chattergy [9] guide the robot to the goal by using rules to decide, based on local sensor information, which path of those available is "best". Heuristic path planners can solve a wide variety of path planning problems, but a problem can always be found where a particular heuristic strategy fails. The appealing feature of heuristic obstacle avoidance procedures is that they require little processing and can produce rapid "reactive" behaviours. However, in these approaches since no planning is done the resulting execution paths may not be optimal and may lead to a failure of the mission.

In this paper we present a local obstacle avoidance scheme called "reactive planning". In this scheme we use the "path transform" which was developed by the first author [10]. The path transform is a grid based planning method. This method considers both minimum distance to a goal and the safety of the robot. The path transform can be regarded as a "numeric potential field" which has the desirable properties of potential field path planners without suffering the local minima problem. Once the robot detects an obstacle with its sensors, it builds a local map of the environment which includes an accurate representation of the freshly sensed obstacle. Making a grid based map using sensor data is a straight forward procedure. Using a grid based planner over a small area is computationally inexpensive. The robot then plans a path which avoids the obstacle and allows the robot to rejoin the originally planned global path. The reactive planning approach yields a fast, efficient and robust local obstacle avoidance method.

This paper will present the research results of implementing the unified planning approach to mobile robot navigation on the Yamabico. Video results showing experimental results of planning global paths and sensing points, execution of the global path, and reactive planning will be shown.

The problem of planning a path of complete coverage of an environment by a mobile robot has received little research attention. A path of complete coverage is a planned path in which a robot sweeps all areas of free space in an environment in a systematic and efficient manner. Possible applications for paths of complete coverage include autonomous vacuum cleaners, lawn mowers, security robots, land mine detectors etc. This paper will also present a solution to this problem based upon an extension to the "path transform". This solution is in keeping with unified approach to path planning. This paper will present the experimental results showing planning and execution of complete coverage paths.

The Yamabico robot is equipped with two driving wheels mounted on a central axis and has optical and ultrasonic range sensors (see Figure 1). The ultrasonic sensors are used to track landmarks and detect the presence of unforeseen obstacles, thus preventing collisions. The optical range sensor is also used to detect unexpected obstacles. However, it's primary purpose is to accurately sense the obstacles, and feed this data to the reactive planner. Each hardware function of the robot is modularised to run on a single board computer. A master module implements decision making and coordinates the actions of each hardware module via a shared communications bus. The master module is programmed to control the operations of the robot with the sensor based ROBOL/0 language developed in your laboratory by Suzuki *et. al.* [11]. Software on the master module runs under our specially developed multi processing MOSRA operating system.

Figure 1. The Yamabico Robot

2. Path and Sensing Planning

The Yamabico mobile robot navigates by using a a map of the operational environment. The robot's map is a 2 dimensional linked list of vectors. The vectors do not only represent the geometry of the environment, they also record which obstacle surfaces can be sensed by the robot's sensors. For example, the 2 dimensional projection onto the floor plane of a table will result in a rectangular shape. The robot should plan a path that avoids the table, however it should not use the table to perform landmark tracking with its ultrasonic sensors. For a real autonomous robot to execute a planned path it must able to correct it's position during path execution time. However, the robot must plan where it should correct its estimated position. Path planning is done by building a visibility graph and then performing a graph search. Refer to Figure 2 for an example of a vector map of an environment and the corresponding visibility graph which is used for path planning (the heavy line shows the solution path). Details of path planning with the consideration of the cost of is presented by Nagatani *et. al.* in [12].

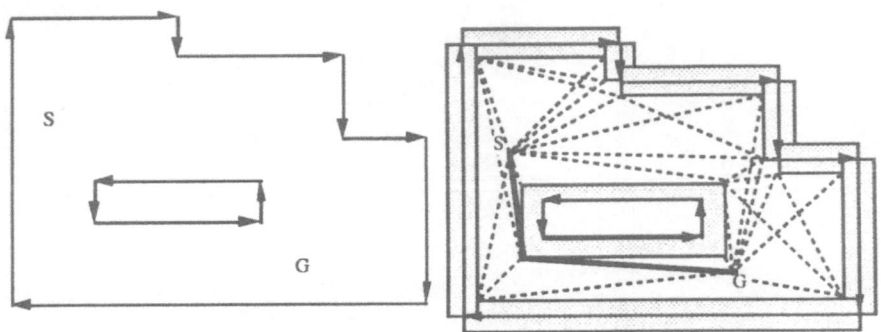

Figure 2.Vector Map and Path Planning Visibility Graph

Once a path has been planned the Yamabico mobile robot executes the path. The robot monitors the execution of the path by using dead reckoning and observing landmarks with its ultrasonic sensors. The robot uses the ultrasonic sensors to measure distances to landmarks, such as walls. The robot's vector map is used to look up the expected range measurement to the landmark. This measurement is compared against the actual range measurement to the landmark. After the sighting of the landmark the estimate of the robot's position in the environment is adjusted for errors. Refer to Figure 3 for an example of the robot monitoring the execution of a path from S to G. In this example the robot attempts to execute the path shown with the broken lines. The actual path executed by the robot is shown with the unbroken line. The robot senses for landmarks and corrects the robot's position at the tracking points T.

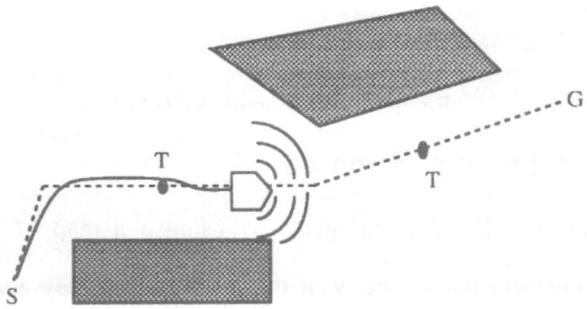

Figure 3. Landmark Tracking during Path Execution

3. Reactive Planning

A novel approach to the problem of findpath using distance transforms (Refer Figure 4) was presented by Jarvis *et. al.* [13]. This approach considers path planning to be finding all paths from the goal location back to the start location. This planning approach propagates a distance wave front through all free space from the goal. From any starting point in the environment the shortest path to the goal is traced by following path of steepest descent. If no downhill path exists then we can conclude the goal is unreachable. Despite the disadvantage of computing all paths from the goal location the distance transforms have a number of advantageous properties. These include; supporting multiple robots and multiple goals. For example paths which favour or avoid certain areas in the environment. The Jarvis *et. al.* approach suffers the problem of moving too close to obstacles while it minimises the length of the path to the goal.

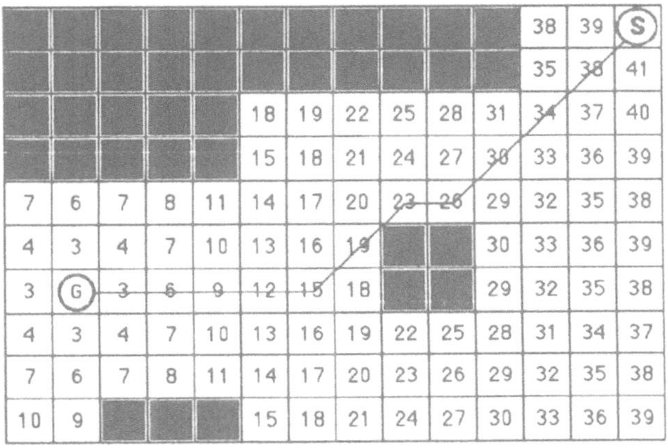

										38	39	S
										35	38	41
				18	19	22	25	28	31	34	37	40
				15	18	21	24	27	30	33	36	39
7	6	7	8	11	14	17	20	23	26	29	32	35	38
4	3	4	7	10	13	16	19			30	33	36	39
3	G	3	6	9	12	15	18			29	32	35	38
4	3	4	7	10	13	16	19	22	25	28	31	34	37
7	6	7	8	11	14	17	20	23	26	29	32	35	38
10	9				15	18	21	24	27	30	33	36	39

Figure 4. Distance Transform

Using the path transform approach presented. by Zelinsky in [10] instead of propagating a distance wave front from the goal, another wave front is propagated which is a combination of the distance from the goal together with a measure of the discomfort of moving near obstacles. This has the effect of producing a distance transform which has the properties of potential fields *without* local minima.

The distance transform is extended to include safety from obstacles information in the following way. Firstly, the distance transform is inverted into an "obstacle transform" where the obstacle cells become the goals. The resulting transformation yields the minimum distance for each free cell from the boundary of the closest obstacle cell. Lastly, a second distance transform is generated through free space from the goal location using a new cost function. This cost function is refered to as the "path transform" (*PT*). The path transform for a cell *c* is defined as:

$$PT(c) = \min_{p \in P} \left(length(p) + \sum_{c_i \in p} \alpha obstacle(c_i) \right)$$

where P is the set of all possible paths from the cell c to the goal, and $p \in P$ i.e. a single path to the goal. The function *length(p)* is the length of path p to the goal. The function *obstacle(c)* is a cost function generated using the values of the obstacle transform. It represents the degree of discomfort the nearest obstacle exerts on a cell c. The weight α is a constant ≥ 0 which determines by how strongly the path transform will avoid obstacles.

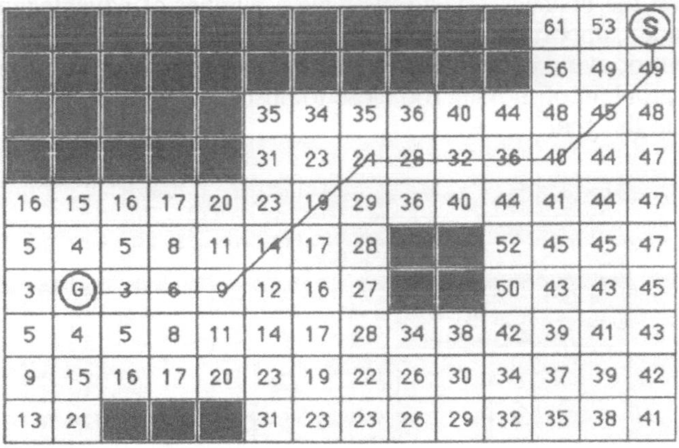

											61	53	(S)
											56	49	49
					35	34	35	36	40	44	48	45	48
					31	23	24	28	32	36	48	44	47
16	15	16	17	20	23	19	29	36	40	44	41	44	47
5	4	5	8	11	14	17	28			52	45	45	47
3	(G)	3	6	9	12	16	27			50	43	43	45
5	4	5	8	11	14	17	28	34	38	42	39	41	43
9	15	16	17	20	23	19	22	26	30	34	37	39	42
13	21				31	23	23	26	29	32	35	38	41

Figure 5. Path Transform with $\alpha = 0.5$

The value stored in the path transform of each cell represents the minimum propogated cost to the goal. The propogated cost is a weighted sum of the distance to the goal and the cost of approaching obstacles. Since the path transform determines the costs of all paths to the goal from each cell it therefore does not contain local minima. The path to the goal is found by tracing the path of steepest descent. Examples of the path transform with different values of α are shown in Figures 5 and 6. Figure 5 shows the path transform with $\alpha = 0.5$, resulting in a solution path that takes a safer path to the goal. Figure 6 shows the path transform with $\alpha = 1.0$, resulting in a solution path which takes the safest path to the goal.

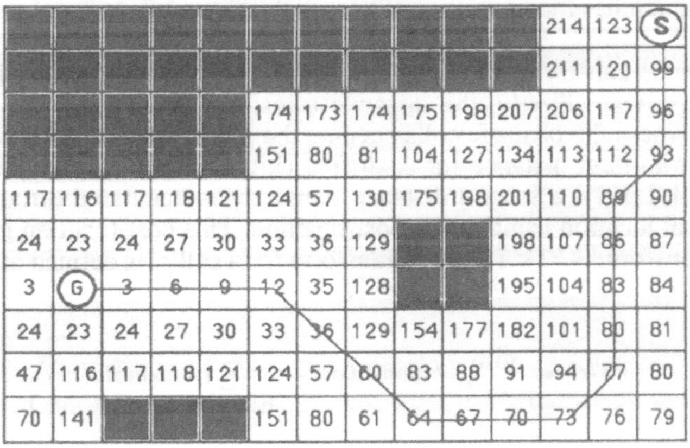

											214	123	(S)
											211	120	99
					174	173	174	175	198	207	206	117	96
					151	80	81	104	127	134	113	112	93
117	116	117	118	121	124	57	130	175	198	201	110	88	90
24	23	24	27	30	33	36	129			198	107	86	87
3	(G)	3	6	9	12	35	128			195	104	83	84
24	23	24	27	30	33	36	129	154	177	182	101	80	81
47	116	117	118	121	124	57	60	83	88	91	94	77	80
70	141				151	80	61	64	67	70	73	76	79

Figure 6. Path Transform with $\alpha = 1.0$

The cost function *obstacle(c)* that was found to be most effective was a cubic function that ranged from zero at a fixed distance, set by the user, to a maximum cost at zero distance. Such a cost function has the advantages of: creating a saddle between the repulsion peaks of neighbouring obstacles, ease of computation, and has its effects bounded in a local area.

There has been other work which is similar to the path transform reported by Barraquand and Latombe [14]. Barraquand and Latombe construct a numeric potential field using a grid representation. This is done in three (3) steps. Firstly, they compute an "obstacle transform" of the free space cells, from which they extract a "distance skeleton" which represents a digitised Voronoi diagram of the free space in the environment. Joining the highest values in the obstacle transform with line segments will yield a distance skeleton. They then connect the goal cell to the distance skeleton. Secondly, they compute a distance transform from the goal cell to all cells that are members of the distance skeleton. Thirdly, they compute another distance transform from the member cells of the distance skeleton to all the remaining free space cells in the environment. The end result is a "numeric" Voronoi diagram of the environment. This method is more complicated to compute than the path transform. Also, since it maximises clearance from obstacles it can divert the solution too far from the shortest path. Another drawback is because Barraquand and Latombe do not specifically consider clearance information to generate their numeric potential, their method can guide the robot through narrow free space channels that are close to the goal there by endangering the robot. They counter this problem by removing channels or branches of the Voronoi diagram that are too narrow. In effect they grow selected obstacles, and thereby lose the completeness of their solution. The path transform does not suffer this drawback.

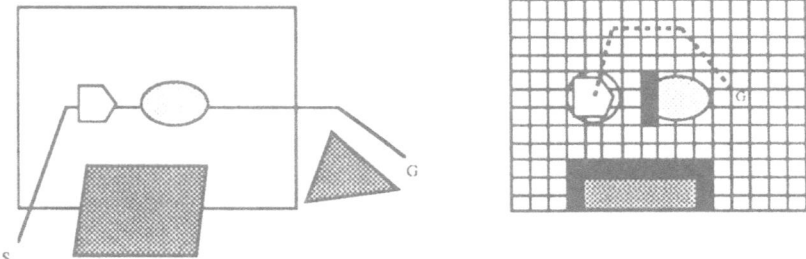

Figure 7. Reactive Planning

The Yamabico mobile robot navigates by following the paths planned by the global planner, and uses the path transform to deal with unexpected obstacles. Once a path has been planned the Yamabico mobile robot executes the path. The robot monitors the execution of the path by using dead reckoning and observing landmarks, such as walls with its ultrasonic sensors. If an unexpected obstacle is encountered along the planned path trajectory by the ultrasonic or laser sensing systems, the Yamabico stops. The robot then uses its laser range finding system to determine the shape and location of the obstacle. The robot scans the environment to the left and right of the obstacle. A grid map covering an area of 4.0m x 3.0m with a grid cell size of 0.20m is projected onto the environment, the map is centred on the mobile robot. The relevant portions of the vector map together with the laser sensor data are projected onto the grip map. A goal position is placed 1.5m behind the sensed obstacle along the original planned trajectory. The Yamabico robot is almost cylindrical in shape and has a diameter of approximately 0.36m. Collision testing can therefore be efficiently implemented by growing the obstacles in the environment by 0.20m. Refer to Figure 7 for an example of reactive planning using the path transform. The first diagram shows the environment, the planned path and the unexpected obstacle. The second diagram shows the local grid map, with the sensor data and environment map data, and the obstacle avoidance path.

The proposed collision avoidance scheme described above was initially implemented on AMROS simulator developed in our laboratory by Kimoto *et. al.* [14]. This simulator models the kimematics of the the locomotion system, and has realistic models of the robot's ultrasonic and laser sensors. The simulator is programmed in the exactly same manner as the Yamabico robot. The simulator proved to be a valuable debugging tool. Collision avoidance software could be tested on the simulator before it was used on the robot. Figure 5 shows an example of output from the AMROS Simulator of the Yamabico navigating down an obstacle strewn corridor. The corridor is 13m long and 2.5m wide. The robot's navigation task is to move down the corridor, as each unexpected obstacle is encountered the robot makes a local map, and places a goal behind the obstacle further down the corridor. Using this strategy the robot is able to avoid three obstacles, back its way out a dead-end, and deduce that the corridor is blocked off, and abandons the mission. Figure 8 shows output from the AMROS simulator.

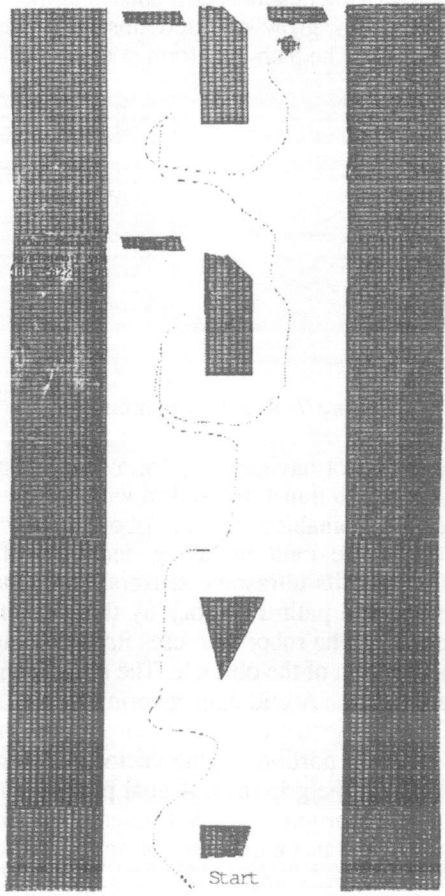

Figure 8. Simulation Results

Finally the reactive planner was implemented and tested on the Yamabico. The Yamabico was successfully able to navigate down a corridor 20m long containing numerous unknown obstacles. Experimental results showed that once the robot detected an unexpected obstacle with its perception sensors, a 100 degree laser scan of the environment in front of the robot (50 degrees left and right of the robot),making a local map of the environment and computing the path transform can be done in less than 3 seconds.

4. Planning Paths Of Complete Coverage

Much of the focus of the research effort to date has centred on the problem of finding a path from a start location to a goal location, while minimising one or more parameters such as length of path, energy consumption or journey time. A mobile robot should be capable of finding paths which ensure the complete coverage of an environment. The path transform can be adapted to find such a path. To find the path of complete coverage, instead of descending along the path of steepest descent to the goal, the robot follows the path of steepest ascent. In other words the robot moves away from the goal keeping track of the cells it has visited. The robot only moves into a grid cell which is closer to the goal if it has visited all the neighbouring cells which lie further away from the goal. A full discussion of paths of complete coverage is provided in Zelinsky et. al. [16]. Refer to Figure 9 for an example of the path of complete coverage.

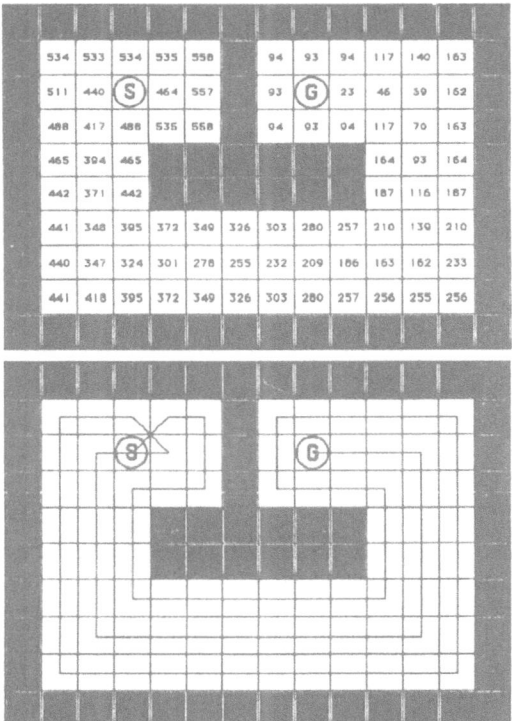

Figure 9. Path of Complete Coverage

To correctly execute the planned path of complete coverage the robot must navigate with great accuracy. To rely on dead reckoning alone to execute a long and complicated path is unsatisfactory. The Yamabico periodically re-adjusts its position estimate. This is done using the techniques described earlier in this paper. The precomputation of the landmark sensing points can be elegantly done by using data needed to compute the path transform. The path transform is computed by using a combination of the obstacle and distance transforms. The obstacle transform knows the exact distance that each free space grid cell is from the closest obstacle filled grid cell. We modify the obstacle transform to include directional information i.e. in which direction the closest obstacle lies. The Yamabico is then able to correct its position estimate dynamically. Refer to Figure 10. for output from the AMROS simulator of a path of complete coverage.

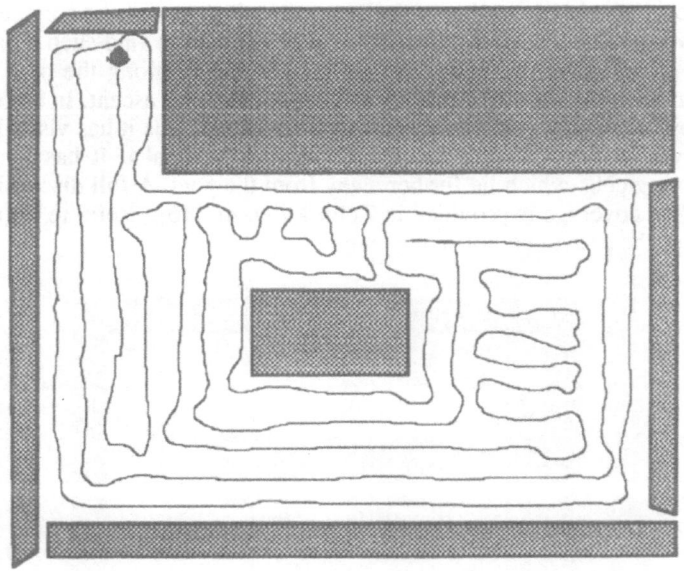

Figure. 10 Simulation Result

The path of complete coverage was successfully implemented in an environment which was similar to the one shown in the simulation.

5. Conclusions

This paper presented a technique for unifying navigation schemes for a mobile robot in known and unknown environments. A method was presented to join global and local path planners that takes into account maintaining the estimated position of the robot. An efficient method for local path planning based on numeric potential fields called the path transform was also presented. It was also shown how the path transform could be easily adapted to create a new type of path which ensures complete coverage of the environment by a mobile robot. The validity of the reported approaches was demonstrated by implementation of the navigation schemes on our self-contained autonomous robot - the Yamabico.

References

[1] S. Yuta, S. Suzuki and S. IIda, "Implementation of a Small Size Experimental Self-Contained Autonomous Robot - Sensors, Vehicle Control, and Description of Sensor based Behaviour", Proceedings of 2nd International Symposium on Experimental Robotics, June 25-27, 1991, Toulouse, France.

[2] R. Chatila, "Path Planning and Environment Learning", European Conference on Artificial Intelligence, pp211-215, July 1982.

[3] J.L. Crowley, "Navigation for an Intelligent Mobile Robot", IEEE Journal of Robotics and Automation, RA-1 No. 1, pp31-41, March 1985.

[4] H.P. Moravec, "Obstacle Avoidance and Navigation in the Real World by a Seeing Rover", Phd dissertation, Stanford University, September 1980.

[5] C.E. Thorpe, "FIDO:Vision and Navigation for a Robot Rover", Phd dissertation, Carnegie Mellon University, Department of Computer Science, December 1984.

[6] O. Khatib, "Real-Time Obstacle Avoidance for Manipulators and Mobile Robots", International Journal of Robotics Research, Vol. 5 No.1, pp90-98, 1986.

[7] Y. Koren and J. Borenstein, "Potential Field Methods and their Inherent Limitations for Mobile Robot Navigation", Proceedings of IEEE International Conference on Robotics and Automation, pp1398-1404, April 1991.

[8] J. Boreinstein and Y. Koren, "Real-Time Obstacle Avoidance for Fast Mobile Robots in Cluttered Environments", Proceedings of IEEE International Conference on Robotics and Automation, May 1990, Cincinatti, pp 572-577.

[9] R. Chattergy, "Some Heuristics for the Navigation of a Robot", International Journal of Robotics Research, Vol. 4 No.1, pp59-66, 1985.

[10] A. Zelinsky, "Environment Exploration and Path Planning Algorithms for a Mobile Robot using Sonar", Phd dissertation, University of Wollongong, Department of Computer Science, October 1991.

[11] S. Suzuki, M.K. Habib, J. Iijima and S. Yuta, "How to Describe the Mobile Robot's - Sensor based Behaviour?", Journal of Robotics and Autonomous Systems, No.7, 1991, North-Holland, pp 227-237.

[12] K, Nagatani and S. Yuta, "Path and Sensing Point Planning for Mobile Robot Navigation to Minimize the Risk of Collision", Proceedings of International Conference on Intelligent Robots and Systems (IROS), July 26-30, 1993, Yokohama, pp2198-2206.

[13] R.A. Jarvis and J.C. Byrne, "Robot Navigation: Touching, Seeing and Knowing", Proceedings of 1st Australian Conference on Artificial Intelligence, November 1986.

[14] J. Barraquand and J.-C. Latombe, "Robot Motion Planning: A Distributed Representation Approach", International Journal of Robotics Research Vol. 10 No.6, pp628-649, 1991.

[15] K. Kimoto and S. Yuta, "A Simulator for Programming the Behavior of an Autonomous Sensor-Based Mobile Robot", Proceedings of International Conference on Intelligent Robots and Systems (IROS), July 7-10, 1992, Raleigh, pp 1431-1438.

[16] A. Zelinsky, R.A. Jarvis, J.C. Byrne and S. Yuta, "Planning Paths of Complete Coverage of an Unstructured Environment by a Mobile Robot", to be presented , International Conference of Advanced Robotics - ICAR '93, November 2-5, 1993, Tokyo.

Combining Physical and Geometric Models to Plan Safe and Executable Motions for a Rover Moving on a Terrain

Moëz CHERIF, Christian LAUGIER
Christine MILESI-BELLIER

LIFIA-INRIA Rhône-Alpes
46, av. Félix Viallet, 38031 Grenoble Cedex 1, FRANCE
E-mail: Moez.Cherif@imag.fr

Bernard FAVERJON

ALEPH Technologies
16, rue du Tour de l'Eau, 38400 St Martin d'Hères, FRANCE

Abstract—
  This paper deals with the problem of motion planning for a mobile robot moving on a three dimensional terrain. The main contribution of this paper is a planning method which takes into account non-trivial features such as: robot dynamics, physical interaction between the robot and the terrain, no-collision and kinematic constraints. The basic idea of our method is to integrate geometric and physical models of the robot and the terrain in a two-stage trajectory planning process. This process combines a discrete search strategy and a continuous motion generation method which is based upon the control of the model of the robot. We will describe how each level operates and how they interact in order to generate a sequence of sub-trajectories. We will also show how both of the robot and the terrain are modelled, and how the interactions between them are dealt with and used during the motion generation.

1. Introduction

1.1. Overview of the Problem

This paper addresses the problem of planning motions of a vehicle A moving on a hilly three dimensional terrain T. In such a context, the mechanical characteristics of the robot and the physical interactions between the wheels and the ground play a major role. Indeed friction, sliding and skidding phenomena may modify the behaviour of the vehicle dramatically. Therefore, such parameters have to be accurately modelled and integrated within the motion planning schema.

  The problem to solve is to find a *"safe"* and *"executable"* motion of A between an initial configuration $q_{initial}$ and a goal configuration q_{final}. A motion is *"safe"* if it verifies the geometric/kinematic constraints of the robot. That means no-collision with the terrain and the obstacles, no tip-over, maintaining contacts between the ground and the wheels, and satisfaction of non-holonomic constraints. A motion is considered *"executable"* if the dynamic of A and the physical interactions between its wheels and the ground are taken into account during the motion generation. Since the formulation of these last features cannot be done using purely geometric and kinematic models, appropriate *physical models* have to be used in order to capture all the features of the movement of the robot.

1.2. Related Works

During the last decade, most of the work in robot motion planning has focused on solving the problem of generating collision-free trajectories for mobile robots moving on a planar terrain. This problem is formulated in terms of geometric (e.g. no-collision) and kinematic (e.g. non holonomic) constraints to be satisfied during the robot motion. Very few results have been obtained yet, when additional dynamic constraints have to be processed. Generally, all the implemented motion planners apply some restrictive hypotheses to reduce the algorithmic complexity and to make the problem tractable (see [13]).

More recently, some interesting contributions addressing motion planning for a mobile robot moving on a rough terrain have been reported ([16][17][7][14] and [18]). A first approach introduces elementary dynamic constraints in the time optimal motion planning, and uses a continuous three dimensional B-patch model of the terrain and a two dimensional B-spline curve to describe a path [16][17]. It is based upon a two-step method to search a time optimal trajectory on this terrain (determination of an optimal path and optimization of the trajectory on each involved patch). It combines the kinematics and a simple dynamic model of the vehicle with the terrain description in order to compute a velocity curve which represents the upper bounds of the speed of the vehicle during its motion along a path. The robot dynamics is reduced to some basic constraints which are associated to the mass center. The considered robot/terrain interactions are mainly restricted to the evaluation of some simple stability criteria based on the formulation of no-sliding, no tip-over and no-collision constraints. In order to simplify the analysis of the reaction and friction forces, the contact points are assumed to be planar and the interaction forces are transfered to the vehicle mass center. As mentionned in [17], such assumptions restrict the terrain to be smooth and the obstacles to be large compared to the robot size.

Another approach consists in applying graph search techniques operating on a discretized model of the surface of the terrain [7][14][18]. In this case, the contact between a wheel and the ground is reduced to a single point and the stability criteria are evaluated with respect to the planar faces which approximate the surface of the terrain. For instance, [18] assumes that the robot does not tip-over if the projection of its gravity center belongs to the convex hull of the projection of the contact points, and that the stability analysis is satisfied by only considering the static equilibrium of the robot on the ground. Such assumptions are insufficient to generate an executable motion since the behaviour of the robot strongly depends on its dynamics and on the effects of the physical interactions with the terrain (e.g. sliding or skidding of the wheels, or local deformation of the ground).

1.3. Contribution of the Paper

Despite the ability of the above-mentioned approaches to solve some instance of the path planning problem (verifying geometric/kinematic constraints), the generation of "*executable motions*" for a vehicle moving in a natural environment is far to be fully accomplished. This can be explained by the fact that the behaviour of the robot results from the combination of various geometric and physical criteria: the mechanical architecture of the vehicle, the characteristics of the motion control law which is applied, the vehicle/terrain interactions, and the strategic orders given to the robot.

The main contribution of this paper is to propose a planning method which takes into account such non-trivial features. It is based upon a two-stage approach combining a discrete search strategy and a continuous motion generation technique. The first step (the "*Explore*" function) determines a set of intermediate safe and stable configurations of the robot using the geometric/kinematic models of the robot and of the terrain. The second step (the "*Move*" function) computes an executable motion using the dynamic model of the robot and appropriate physical models of the terrain and of the wheels/ground interactions.

This paper is organized as follow: section 2 describes the used geometric/kinematic, dynamic, and physical models. Then, sections 3, 4, and 5 present the main features of the proposed method and the two basis functions "*Explore*" and "*Move*". Finally, some experiments obtained with a six wheeled vehicle moving on a three dimensional terrain are presented in section 6.

2. Task Modeling Using Geometric and Physical Models

2.1. The Geometric/Kinematic Model $G(\mathcal{A})$ of the Robot

The considered mobile robot $\mathcal{A}$ is represented as an articulated object described by a set of rigid bodies Ω_i (as shown in figure 1). Each element Ω_i corresponds to a given component of $\mathcal{A}$ such as the main body, a link of the chassis, an axle, or a wheel. The geometric model $G(\mathcal{A})$ is represented as a binary tree where the leaves correspond to the components Ω_i and each subtree is associated to a subpart of the robot. An arc connecting two components (or subparts) of the robot represents either a joint (e.g. rotoïd joint used for the articulation between a wheel and an axle), or a simple union (e.g. cylindrical and conical elements used to build a wheel as shown in figure 1) depending on whether one element can move relatively to the other one. $G(\mathcal{A})$ is used by the "*Explore*" function as presented in sections 3 and 4.

In this paper, we have considered a non-holonomic mobile robot a locomotion system composed of six motorized wheels — a cantral pair of wheels connected to the main body of $\mathcal{A}$ and two others pairs connected respectively to the front and the rear axles. Each of these axles is articulated so that it can have a roll and a yaw movement. Finally, a joint is associated respectively to the front and the rear links of the chassis in order to enable a "peristaltic" effect. This allows to adapt the configuration of the robot to the local irregularities of the terrain.

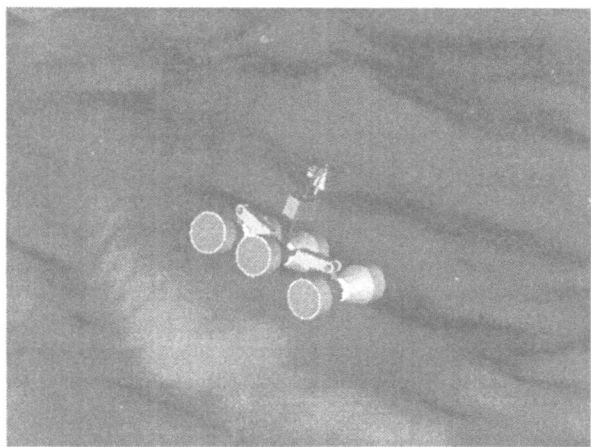

Figure 1. The six-wheeled mobile robot $\mathcal{A}$ on the terrain

A full configuration q of $\mathcal{A}$ is given by $n_\delta + 6$ parameters: *six* parameters $(x, y, z, \theta, \phi, \psi)$ specifying respectively the position and the orientation of the main body in a fixed reference frame, and n_δ parameters specifying the value of the joints δ_k of the robot.

2.2. The Geometric Model $G(\mathcal{T})$ of the Terrain

The terrain $\mathcal{T}$ is initially represented by an elevation map in z associated with a two dimensional regular grid in (x, y) with respect to a fixed reference frame. In the current implementation, the geometric model $G(\mathcal{T})$ of $\mathcal{T}$ is represented by a set of patches obtained from the set of (x, y) (see figure 1). $G(\mathcal{T})$ will be used in the sequel by the *"Explore"* function during the contact points computation and the collision avoidance.

2.3. The Dynamic Model $\Phi(\mathcal{A})$ of the Robot

Modeling the dynamics of the vehicle $\mathcal{A}$ requires to consider both its geometry and some additional information concerning its physical properties (e.g. the mass and the inertia properties of its elements, the physical characteristics of its articulated mechanisms). The dynamic model $\Phi(\mathcal{A})$ of $\mathcal{A}$ will be used in the sequel by the *"Move"* function. We describe in this section the model we have implemented (see [4]). Two main steps have to be applied to construct $\Phi(\mathcal{A})$:

2.3.1. Representation of the Mechanical Structure of $\mathcal{A}$

Two types of constructions are required to model the mechanical structure of the vehicle: the rigid components Ω_i (like in $G(\mathcal{A})$) and the articulated mechanisms associated to the joints δ_k of $\mathcal{A}$.

Each articulated mechanism is represented by a network $\Phi(\delta_k)$ combining a set of connectors c_r and a set of specific points selected on the rigid bodies Ω_i corresponding to the joint δ_k. The connectors are defined in terms of visco-

460

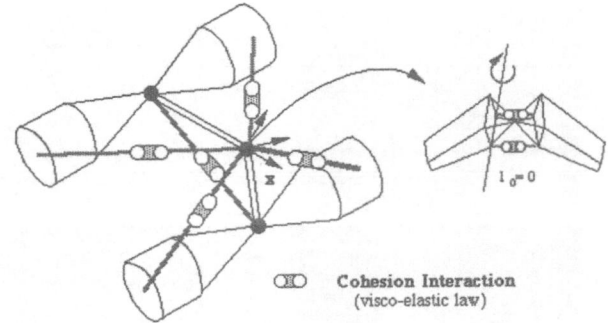

Figure 2. Modeling a 3D rotoïd mechanism

elastic laws (i.e. combination of springs and dampers). For instance, a 3D rotoïd mechanism is represented by two rigid bodies connected through two pairs of points respectively selected on them and belonging to the rotation axis. This construction is illustrated by figure 2 in the case of the articulation between the front and the middle axles.

We have also used *"conditional connectors"* to model the mechanical stops constraining the motions of the joints. A conditional connector basically acts exactly like a spring and/or damper connector but in a specified domain of the DOFs of the considered joint. This is illustrated by figure 3 for a 2D rotoïd joint. The effect of c_{stop} is taken into account only when the distance $l_{i,j}(t)$ between $P_{i,k}$ and $P_{j,k'}$ is within a given $D_{c_{stop}}$ depending on the limits $l_{0,c_{stop}}$ and $L_{0,c_{stop}}$ (corresponding respectively to the minimum and maximum values of the joint).

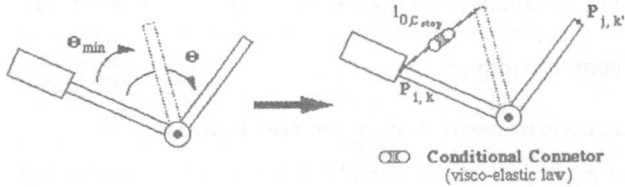

Figure 3. Modeling a mechanical stop of a 2D rotoïd joint

2.3.2. Formulation of the Dynamics of $\mathcal{A}$

We describe now the motion equations of $\mathcal{A}$. These are given by considering the dynamic motion of all the components of $\mathcal{A}$. Let's consider the motion in a $3D$-space of a given rigid body Ω_i having a mass m_i. The configuration of Ω_i at time t with respect to a fixed coordinate frame is given by the two column position vector $r_i(t) = (x_i(t), y_i(t), z_i(t))^T$ and rotation vector $\alpha_i(t) = (\alpha_{i,x}(t), \alpha_{i,y}(t), \alpha_{i,z}(t))^T$. The full state of $\mathcal{A}$ is then given by the set of all the states $(r_i(t), \alpha_i(t), \dot{r}_i(t), \dot{\alpha}_i(t))$ of Ω_i.

Using this model, it is obvious that the dimension of the state space of $\mathcal{A}$

becomes equal to $12\, n_\Omega$ where n_Ω is the number of the Ω_i components used to build the physical model of $\mathcal{A}$. Even the high dimension of such state space, this formulation enables to couple directly each element of the robot to those others that are in contact with it. It is different from the dynamic model based on a joint space formulation where each link is related to the one immediately before it, and each motion is sensed in relative and requires successive transformations processing which can be time consuming.

The motion equations of a solid Ω_i are specified by the classic Euler/Newton equations:

$$F_i \;=\; m_i\, \ddot{r}_i(t) \tag{1}$$

$$T_i \;=\; \dot{L}_i(t) \tag{2}$$

where F_i and T_i are respectively the sums of forces and torques applied on Ω_i, and $L_i(t)$ is the angular moment about the center of mass of Ω_i, G_{Ω_i}. $\dot{L}_i(t)$ can be written in a linear form as follows:

$$\dot{L}_i(t) \;=\; I_{\Omega_i}\, \ddot{\alpha}_i(t) \tag{3}$$

where I_{Ω_i} is the inertia tensor of Ω_i about the fixed frame axes x, y, z. Even $L_i(t)$ can be related in another form to the angular acceleration (see [8]), we assume that the type of rotation of Ω_i makes the Coriolis and centrifugal effects negligible (because the robot is supposed to move with a small velocity). F_i and T_i are computed using the *Euler's* principle of superposition, by combining the applied forces to an equivalent force acting on G_{Ω_i} and generating the associated torques relative to the points of application. Then, F_i and T_i can be written as follows:

$$F_i \;=\; F_d + \sum_{force\, j} F_{i,j} \tag{4}$$

$$T_i \;=\; \sum_{force\, j} (G_{\Omega_i} P_{i,j} \times F_{i,j}) + \sum_{torque\, k} T_{i,k} \tag{5}$$

where $T_{i,k}$ are the torques acting on Ω_i, $P_{i,j}$ are the points where the forces $F_{i,j}$ are applied, $G_{\Omega_i} P_{i,j}$ is the vector from G_{Ω_i} to $P_{i,j}$, and $\times$ is the cross product. The term F_d includes the forces generated by the gravity and additional forces such as the viscosity of the environment. The set of forces $F_{i,j}$ is obtained from contact effects on Ω_i: action of the c_r connectors, connecting Ω_i to some others components of $\mathcal{A}$, and/or interactions with the ground for the wheel components. The torques $T_{i,k}$ are derived from the control parameters of the wheels motorization and are associated only to the components Ω_i used to modelize the wheels of $\mathcal{A}$.

2.4. The Physical Model $\Phi(\mathcal{T})$ of the Terrain

Since we have to integrate the robot/terrain interactions within the motion planning process, it is necessary to build a representation $\Phi(\mathcal{T})$ which is able to capture both the geometric and the physical properties of $\mathcal{T}$ and which allows

the formulation of such interactions. For that purpose, we have implemented
a model based upon the concept of *"physical models"*, initially proposed for
computer graphics applications (see [11]).

The basic idea is to consider that motions and/or deformations of physical objects result from the application of physical laws involving a set of forces
whose application points depend on the intrinsic structure of the interacting objects. A practical way to modelize this phenomenon is to discretize the objects
of the scene appropriately, and to couple each component of the model with a
set of differential equations combining two dual variables: the force F and the
position P (or the velocity V). It roughly consists in selecting an appropriate
set of interconnected punctual masses to represent the objects. It has been
shown [11] that two types of generic components are sufficient to model a large
variety of physical situations using this approach : (1) the "matter component"
represented by a punctual mass obeying the Newtonian dynamics, and (2) the
"interaction component" which corresponds to appropriate physical laws associated to two interacting matter components (e.g. linear viscous/elastic interactions, cohesion properties, and elastic collisions, and complex interaction
laws such as dry friction applied to characterize dynamic phenomena associated to the vehicle motions (see section 5.1). Using this approach, the physical
model of a given object can be seen as a network of punctual masses where the
arcs define the interactions between the components.

Two main steps are then used to build such physical model: (1) the determination of the mass distribution representing the discretization of the object
in terms of the matter components, and (2) the characterization (determination
of the network structure) and the formulation of the interaction components
between the matter components of the object themselves (e.g. cohesion of the
components representing a terrain), and of the interactions with their environment (e.g. interactions between a rock and a wheel of the robot, or a region
with moving rocks).

Performing the first step when we build the physical model $\Phi(T)$ of the
terrain requires to take into consideration several criteria such as the terrain
surface shape, the average distribution of the contact points between the wheels
and the ground, and the complexity of $\Phi(T)$ (i.e. the number of "matter components"). In the current implementation of the system, the mass distribution
is processed by converting the geometric model $G(T)$ into a set ST of spheres
S_i whose profile approximates the surface of T (see figure 4). This is done in
such a way that each point of the terrain surface (i.e. list of the points given by
$G(T)$) should be located on the shape of at least a sphere of ST. Each punctual mass $\Phi(P_i)$ corresponds then to a sphere S_i and is located at its center.
Each element S_i of ST defines a spherical "occupancy area" associated to the
component $\Phi(P_i)$ of the model (this area is generally characterized by an elastic
non-penetration law). Consequently, $\Phi(P_i)$ can be seen as a "ball" which in
turn can be represented by the (m_i, R_i, P_i), where m_i is the weight of $\Phi(P_i)$, R_i
the radius of S_i and P_i the position of its center. Figure 5 shows the physical
model of the terrain T and the associated set of spheres ST. $\Phi(T)$ is represented by a network of elementary components $\Phi(P_i)$ characterized by a set

of arcs a_{ij} representing the internal physical interactions (given by elementary interaction or composite complex laws as mentionned earlier).

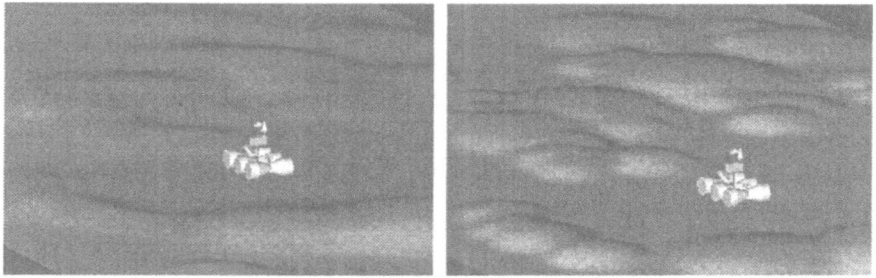

Figure 4. Decomposition of the terrain into spheres

The efficiency of such a representation is related to the fact that it allows to maintain the geometric features of the motion planning problem (i.e. checking the geometric constraints as the no-collision), and easily build the physical model $\Phi(\mathcal{T})$, and to formulate the interactions with the robot. Indeed, describing $\mathcal{T}$ in terms of a set of spheres leads to handle a little amount of information to represent its geometric shape (as shown in figure 5). Furthermore, the combination of such simple primitives with an appropriate hierarchical description allows to obtain efficient models which can be used in the collision detection and the obstacle avoidance (as in [19]). This can be illustrated for instance during the distance computation between different objects, since we have just to handle the set of the centers of the spheres and their radii.

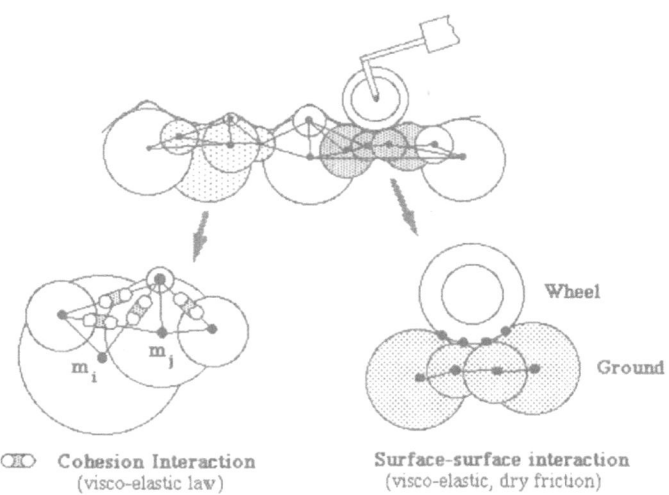

Figure 5. Physically-based model of the terrain

3. The Motion Planning Approach

3.1. The General Algorithm

Using purely physical models $\Phi(\mathcal{A})$ of the robot and $\Phi(\mathcal{T})$ of the terrain to solve the global motion planning problem (between $q_{initial}$ and q_{final}) may lead to important computational burden. Indeed, finding an executable and safe trajectory requires to operate in the "state-space" of the robot and to handle complex differential equations depending on the constraints of the problem.

This is why, we propose in the sequel an iterative algorithm based upon two complementary steps. The first step ("*Explore*") consists in exploring a reduced search space of the robot in order to determine a sequence $\{q^1, q^2, \ldots, q^{N_p}\}$ of appropriate safe intermediate configurations of $\mathcal{A}$ representing a first approximation of the searched solution. We recall here that a configuration q of the robot is given by the set of $(x, y, z, \theta, \phi, \psi)$ and the joints parameters. The second step ("*Move*") introduces the physical constraints (i.e. dynamic of $\mathcal{A}$ and the physical vehicle/terrain interactions) to solve a motion generation problem $(\mathcal{P}_i)$ for each pair (q^i, q^{i+1}). We should note here that the general scheme of our approach is different from methods operating in two stages: (1) processing a path, and after (2) simulating the robot motion in order to generate a full trajectory ([18][1]). As shown in figure 6, the physical models of the task are taken into consideration at each iteration of the algorithm by processing the "*Move*" function.

For instance, let q^i_{next} be a stable configuration of the robot generated by "*Explore*" during the searching process starting from a current configuration q^i. If the motion generation fails when we solve $(\mathcal{P}_i)$ between q^i and q^i_{next} — because of obstacles and/or possible sliding or skidding of the wheels of $\mathcal{A}$ —,

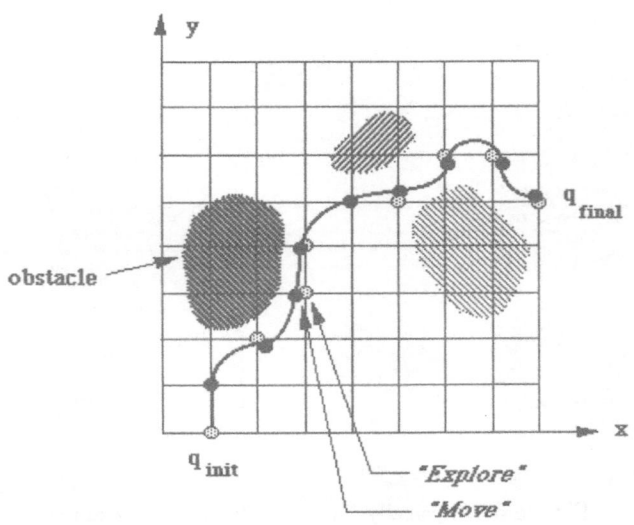

Figure 6. General scheme of the approach

q_{next}^i is considered unreachable. In that case, another stable successor of q^i is chosen and has to be validated by the *"Move"* function in the next iteration. A motion generation failure is detected if we cannot find a trajectory so that the robot can move until a configuration q_{next}' located in the neighbourhood[1] of q_{next}^i. Let consider q_{next}^i a reachable configuration. The next iteration of the motion planning algorithm proceeds by exploring the search space starting from q_{next}^i (using *"Explore"*) and searching a continuous motion using *"Move"* starting from q_{next}' (see figure 6). As mentionned earlier, this means that the exploration of the search space gives us only an approximation of the motion planning solution and the real trajectory of the robot is provided by the motion generation step when the physical models are processed. The main advantage of such algorithm is to make the global motion planning under dynamic constraints more tractable by using locally the proposed physical models during the motion planning time.

3.2. The *"Explore"* Function

As mentioned earlier, only safe configurations q for which the robot is put on the terrain and verifies no-collision and stability criteria (relative to tip-over and static equilibrium constraints) have to be considered. Such a full configuration is called in the sequel a *"placement"*. In that case, the parameters of q are not independent and the inter-relations between them depend on the vehicle/terrain relation (i.e. distribution of the contact points according to the geometry of the terrain). These interdependancies between some of the configuration parameters allow us to reduce the dimension of the search space $\mathcal{C}_{search}$ when searching for stable placements of the robot (see [1]): knowing a set $\tilde{q} = (x, y, \theta)$ of the configuration parameters (horizontal position and orientation of the main body), it is possible to compute the other parameters by evaluating the constraints imposed by the wheels/ground contacts, the joints characteristics and the stability. This method is described in section 4.

The *"Explore"* function basically operates like the path planning method developed in [18] (and also in [10]). It proceeds by using an A^* algorithm to find the near-optimal cost path in a graph $\mathcal{G}$ incrementally generated after a rough construction of the search space $\mathcal{C}_{search}$ and starting from $\tilde{q}_{initial}$. Let $\tilde{q}^i$ be the current sub-configuration of the robot at a given iteration of the algorithm. The expansion of the node N_i of $\mathcal{G}$ associated to $\tilde{q}^i$ is realized by applying a local criteria on $\tilde{q}^i$ and a simple control u_i (as used in [18]) during a fixed period of time ΔT_i, in order to determine the set of its *"possible successors"*. Each successor is considered as the basis for a possible stable placement of the robot and its admissibility is evaluated using the method described in section 4. Only admissible successors will be then considered for the next step and are stored into a list. At the next iteration, the *"best"* successor $\tilde{q}_{next}^i$ (according to a cost function) is selected before processing the motion generation step (the *"Move"* function) in order to verify if it can be reached by the robot,

---

[1] We assume that the neighbourhood of a configuration q is given by the domain defined on the sub-set of parameters $\tilde{q} = (x, y, \theta) : [x - \delta x, x + \delta x] \times [y - \delta y, y + \delta y] \times [\theta - \delta \theta, \theta + \delta \theta]$, where δx, δy and $\delta \theta$ are given constants.

and becoming in turn the current sub-configuration. A cost function $f(N_i)$ is assigned to each node N_i of $\mathcal{G}$ in order to guide the search process. In the current implementation, we have used a cost function equal to the trajectory length executed by the robot.

3.3. The "*Move*" Function

Since only discrete static placements of $\mathcal{A}$ have been generated during the first step, it remains to generate continuous motion between the selected configurations. This requires to take into account the dynamic characteristics of the vehicle and of the wheels/ground interactions. The purpose of the "*Move*" function consists in verifying locally the existence of an *executable* motion of $\mathcal{A}$ between the current configuration q^i and its successor q^i_{next} when the physical models $\Phi(\mathcal{A})$ and $\Phi(\mathcal{T})$, the wheels/ground interactions and the motion control law are taken into account.

As the dynamics of $\mathcal{A}$ and the physical interactions with $\mathcal{T}$ are processed in term of the differential equations of motions of the components Ω_j of $\Phi(\mathcal{A})$, the problem $(\mathcal{P}_i)$ of finding an *executable* motion between q^i and q^i_{next} has to be formulated in the *state-space* of $\mathcal{A}$ (described in section 2.3). Furthermore, the main difficulty to solve $(\mathcal{P}_i)$ comes from the fact that the geometric/kinematic constraints and the dynamic vehicle/terrain interactions have to be simultaneously considered during the trajectory generation process. In order to take into consideration such features and to achieve such a process, we propose two methods which will be detailed in section 5:

- The first method is based on the combination of a classical geometric based trajectory generation and a motion generation process having the ability of taking into account the physical models of the task and the robot/terrain interactions. The proposed algorithm consists in generating a sequence of nominal sub-trajectories Γ_j verifying the geometric/kinematic constraints and guiding the robot from the current sub-configuration $\tilde{q}^i$ to its successor $\tilde{q}^i_{next}$. The full motion of $\mathcal{A}$ is then obtained by processing its dynamic and its interactions with $\mathcal{T}$ along the set of the Γ_j.

- The second method consists in using optimal control theory to solve the robot motion generation problem between q^i and q^i_{next} for a motion time equal to ΔT_i. The purpose of such an approach is to find an optimal trajectory of the robot and the associated controls (according to a cost function such as control energy or motion time) using a *state-space* formulation and without precomputed nominal trajectories.

4. Computing a Stable Placement

4.1. The Basic Idea

The aim of this step of the "*Explore*" function is to find a full stable configuration q^i of $\mathcal{A}$ corresponding to the generated sub-configuration $\tilde{q}^i$, or to conclude that $\tilde{q}^i$ is not admissible.

Initially, the robot is characterized by its location (x^i, y^i, θ^i) and a set of default values associated to the remaining parameters of q^i so that the robot is above the ground. The principle of the algorithm is then to simulate the falling of the robot on the ground. This is done by minimizing the potential energy of the robot under a set of constraints expressing the fact that contact points must stay above the ground (constraints along the z direction). It is important to note that we are not writing here the true static equilibrium of the system. No friction forces are modelled, and we only constrain the wheels to lie on the surface of the ground. The only thing that prevents the vehicle slipping along a slope is the fact that (x^i, y^i, θ^i) is maintained constant during the minimization process. The solution is obtained using an iterative process in which a quadratic criterion is minimized under linear constraints that locally approximate the potential energy and the contact constraints as described in [1]. An implementation based on the Bunch and Kaufman algorithm [3] has been used.

4.2. Writing the Potential Energy Criterion

The potential energy consists of two terms: the gravity term and the spring energy term. The gravity term is computed as the sum of the gravity term for each rigid body of the robot. For a given body Ω_k, it is simply given by $m_k\, g\, z_{G_k}$, where z_{G_k} is the altitude of its gravity center. z_{G_k} is a function of the configuration variables and its derivative can be easily computed from the Jacobian matrix J_C of the body Ω_k. Thus a linear approximation of the gravity term is obtained around the current configuration. The spring energy term is the sum of the energy for each spring joint (i.e. a joint for which the applied effort is proportional to the articular position). For such a joint j, it is given by a quadratic function of the configuration variables as: $K_j\,(q_j^i - q_{j_0}^i)^2$.

4.3. Writing the Contact Point Constraints

In order to determine efficiently the contact points between the wheels and the ground, we use hierarchical models of the environment and of the robot bodies. For a given configuration of the robot, our distance computation algorithm determines all pairs of convex primitives for which the distance is smaller than a given threshold. This threshold is defined as the maximum distance variation that may occur during an iteration of the minimization. For such a pair, we write that the distance must remain greater than zero. It can be shown that the distance variation can be computed as a linear function of the configuration variables by means of Jacobian matrices [6]. Let C be the point of the robot primitive which is the nearest of the ground primitive, J_C the Jacobian matrix relating configuration variable derivatives to the derivatives of coordinates of point C, and let n be the normal to the ground at the contact point. Then, we have $\delta d = n^t J_C \delta q$. Thus, we can express the distance constraint as a linear constraint in the configuration variable variations.

4.4. Stability and Slipping Analysis

Once the contact points are known, we have to evaluate the physical feasibility of the configuration. It includes the stability of the robot, but also the non-slipping conditions of the wheels on the ground. This computation is based on the application of the Coulomb's friction law for each contact point. The equilibrium state of $\mathcal{A}$ is given by the equations expressing that the sum of forces and torques at the gravity center of the vehicle is zero. These forces and torques are obtained by the reaction of the terrain and the control torques applied to the wheels. The configuration is then considered to be admissible if the convex hull defined by the intersection of all the equations and inequations constraints defined by the formulation of the Coulomb's law and contact conditions is non-empty. It means that there exists a set of reaction forces that satisfies the equilibrium equations together with the non-sliding conditions. These last conditions are given by a set of inequations formulating the inclusion of the reaction forces in the friction cone associated to the Coulomb's law. In other words, it means that the vehicle can remain safely stopped in this configuration. In the other case, the configuration is considered as not admissible.

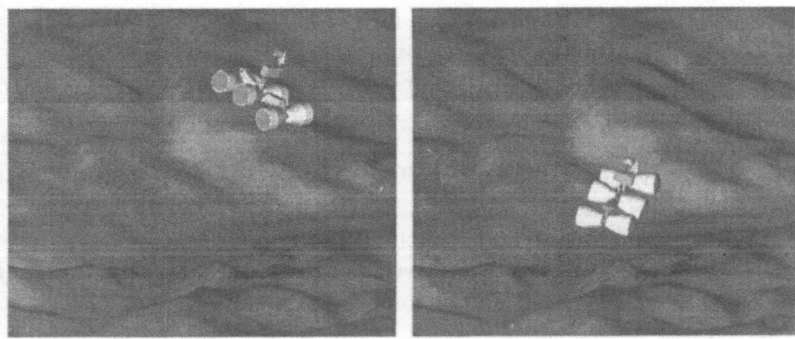

Figure 7. Different placements of the robot $\mathcal{A}$ on the terrain

5. Solving the Motion Generation Problem

5.1. Processing the Wheels/Terrain Interactions

The most difficult problem to solve when dealing with a vehicle moving on a terrain, is to define realistic vehicle/terrain interaction laws. Such interaction laws depend on both, the distribution of the contact points, and the type of the surface-surface interactions to be considered.

The first problem is automatically solved by the distance computation defined on the hierarchical description of the components Ω_j of $\mathcal{A}$ and the decomposition ST of the terrain into a set of spheres. The existence of contact points between the wheels of $\mathcal{A}$ and the spheres S_i of ST is processed considering given distance criteria, whose satisfaction leads to activate the associated interaction law. As mentioned in section 2.4, theses distance criteria depend on the radii of the different spheres S_i.

The surface-surface interactions are processed using two types of constructions: a visco-elastic law associated to a non-penetration threshold, and a surface adherence law that produces sliding and gripping effects. The first point consists in modeling the non-penetrability between the wheels of $\mathcal{A}$ and the ground by associating to the set of S_i involved in the contact, a potential acting as a conditional spring. In order to solve the second point, we make use of a *finite state automaton* (also used in [9] and [12]) since complex phenomena like dry friction involves three different states basically : no contact, gripping, and sliding under friction constraints. The commutation from one particular state to another is determined by conditions involving gripping forces, sliding speed, and relative distances (see figure 8). Each state is characterized by a specific interaction law. For instance, a viscous/elastic law is associated to the gripping state, and a Coulomb equation is associated to the sliding state (see [12]).

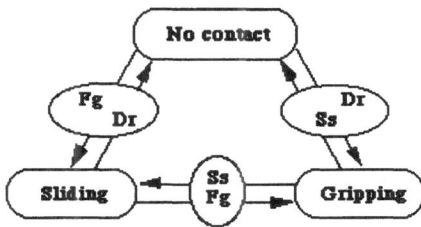

Figure 8. Finite state automaton to model dry friction phenomena

5.2. Motion Generation by Tracking a Sequence of Nominal Subtrajectories

5.2.1. The Basic Idea

The purpose of the motion generation step is to determine if it exists a feasible motion which allows to move from q^i to q^i_{next}, and to determine the characteristics of the control to apply on $\mathcal{A}$ in order to perform this motion. The main difficulty comes from the fact that the related method has to simultaneously take into account the geometric/kinematic characteristics of the problem and the vehicle/terrain interactions.

A first way to solve this problem is to apply an iterative algorithm combining a classical geometric based trajectory generation and a motion generation process having the ability of computing the physical interactions. Let δt be the time increment considered ($\delta t = \Delta T_i / k$), and let $s_i(n\delta t)$ be the current state of $\mathcal{A}$ at the step n of the algorithm. We note in the sequel $P_i(n\delta t)$ and $\theta_i(n\delta t)$ respectively the position and the orientation of the robot when it is in state $s_i(n\delta t)$.

5.2.2. Geometric Based Trajectory Generation

It consists in hypothesizing a nominal trajectory Γ_n allowing $\mathcal{A}$ to move from $s_i(n\delta t)$ to $s_{i,next}$ while verifying the kinematic constraints (i.e. non holonomic constraint), but without taking into account the dynamic constraints coming from the vehicle/terrain interactions. Such a trajectory is made of straight line segments and circular arcs. The approach can easily be implemented using a technique inspired from the Dubins approach [5]. In the current implementation, the gyration radius R is chosen in such a way that the vehicle tracks an optimal trajectory while minimizing the sliding effects of the wheels. When the distance between $P_i(n\delta t)$ and $P_{i,next}$ is large enough, the velocities of the controlled wheels are all positive. Otherwise, R is set to half of this distance and the velocities of the opposite controlled wheels have opposite signs, and consequently the vehicle will skid and use more energy to execute the motion.

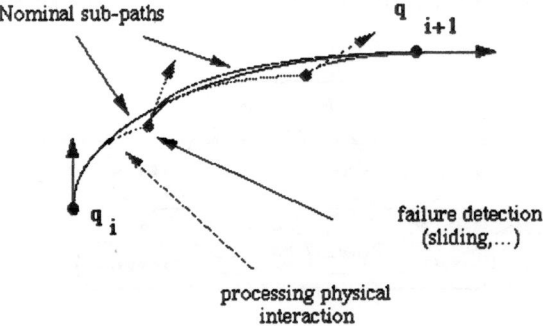

Figure 9. The motion generation scheme

5.2.3. Motion Generation Using Physical Models

The motion generator operates upon the models $\Phi(\mathcal{A})$ and $\Phi(\mathcal{T})$. It takes as input the velocity controls applied on each controlled wheel during the time increment δt. This velocity controls are computed from the linear and steering speeds which are associated to the reference point of $\mathcal{A}$ when moving on Γ_n (such controls are supposed to be constant during δt). For the purpose of the physical simulation, these controls are converted into a set of torques $u(t)$ to apply on $\Phi(\mathcal{A})$. Then the prediction of the motion of $\mathcal{A}$ during the time increment δt is performed by simulating the evolution of both $\Phi(\mathcal{A})$ and $\Phi(\mathcal{T})$ under the effects of $\mathcal{T}$ and the gravity (as used in [12]). Since the motion generation integrates physical phenomena like sliding or skidding, the state $s_i^*(n\delta t)$ of $\mathcal{A}$ obtained after n successive motions may be different from the nominal state $s_i(n\delta t)$. If $P_i^*(n\delta t)$ and $\theta_i^*(n\delta t)$ are too far from their nominal values, then the motion generation fails and the processed sub-goal represented by $s_{i,next}$ is rejected. Otherwise, the algorithm is iterated until the neighbourhood of $s_{i,next}$ is reached (see figure 9).

This approach works fairly well when the part of the terrain crossed by the vehicle during the time ΔT_i is not too chaotic (since the geometric step of the algorithm operates under some simplifying assumptions). This is why we are currently implementing a new motion generation technique based upon optimal control theory. Such a technique is described in the next section.

5.3. Motion Generation by Optimal Control

We present here a method based upon optimal control theory to solve locally the robot motion generation problem between the states s_i and s'_{next} for a motion time equal to ΔT_i. ΔT_i is provided by "*Explore*" during the expansion of the nodes of $\mathcal{G}$. We can then write $s(\Delta T_i) = s'_{next}$

Without loosing in generality, we describe in the following the optimal control technique for a single rigid body Ω_j associated to a wheel of $\mathcal{A}$. We will consider the case where the trajectory minimizes the control energy of the robot to reach a state $s(\Delta T_i)$ so that the corresponding sub-configuration $\tilde{q}(\Delta T_i)$ belongs to the neighbourhood of $\tilde{q}^i_{next}$. Let $u^j(t)$ be the controlled torques associated to the motorization of Ω_j. The cost function J^j_i corresponding to the motion of Ω_j can be formulated in a quadratic form as:

$$J^j_i = \frac{1}{2}\left[e(\Delta T_i)^T D\, e(\Delta T_i) + \int_0^{\Delta T_i} u^j(t)^T M\, u^j(t)\, dt\right] \tag{6}$$

where the matrices M and D are supposed to be positive definite and constant in time, and $e(\Delta T_i)$ is equal to the vector $(r_j(\Delta T_i) - r_{j,i,next}, \alpha_j(\Delta T_i) - \alpha_{j,i,next})$ [2]. Let $s^j(t) = (r_j(t), \alpha_j(t), \dot{r}_j(t), \dot{\alpha}_j(t))^T$ be the state at time t of Ω_j, the motion equations (1) and (2) can be reformulated in a matrix form as:

$$\frac{ds^j(t)}{dt} = \begin{bmatrix} 0 & 0 & I & 0 \\ 0 & 0 & 0 & I \\ 0 & 0 & 0 & 0 \\ 0 & 0 & 0 & 0 \end{bmatrix} s^j(t) + \begin{bmatrix} 0 \\ 0 \\ \frac{1}{m_j}F \\ I_{\Omega_j}^{-1}T \end{bmatrix} \tag{7}$$

where the terms *Null* and I represent the 3×3 zero and identity matrices respectively. The equation (7) can been written in an explicit linear form as:

$$\frac{ds^j(t)}{dt} = A\, s^j(t) + B\, u^j(t) + C(t) \tag{8}$$

where the matrix B is constant in time and determines the form $u^j(t)$ is related to the motion equation (i.e. how it is applied), and $C(t)$ is the column vector which describes the remaining non controlled forces and torques. These forces and torques are provided by the contact interactions with the ground (described in section 5.1), and by the internal connectors used to link Ω_j to the other components of $\mathcal{A}$ such as the chassis or the axles (as presented in section 2.3). The problem of finding a trajectory of Ω_j and the associated controls can then be formulated as $(\mathcal{P}^j_i)$: "*Given the motion equation (8) of Ω_j, and the boundary condition $s^j(0) = s^j_i$, find the control function $\bar{u}^j(t)$ such that J^j_i is minimized*".

---

[2] According to the description of the state-space of $\mathcal{A}$ presented in section 2.3, $r_{j,i,next}$ and $\alpha_{j,i,next}$ can be determined using the geometric model of $\mathcal{A}$ and the placement q^i_{next}.

The optimal control $\bar{u}^j(t)$ is explicitly given by the set of the following equations [2]:

$$\bar{u}^j(t) = -M^{-1}B^T(K(t)s^j(t) - R(t)) \tag{9}$$

$$\frac{dK(t)}{dt} + K(t)A + A^TK(t) - K(t)S(t)K(t) = 0 \tag{10}$$

$$\frac{dR(t)}{dt} + (A^T - K(t)S(t))R(t) - K(t)C(t) = 0 \tag{11}$$

$$S(t) = B(t)M^{-1}B^T(t) \tag{12}$$

where $K(t)$ is the positive definite solution of the Riccati equation (10) and the column vector $R(t)$ is obtained by solving (11), considering the boundary conditions $K(\Delta T_i) = D$ and $R(\Delta T_i) = Ds^j_{i,next}$. The optimal trajectory is the solution of the linear differential system:

$$\frac{ds^j(t)}{dt} - (A - S(t)K(t))s^j(t) - S(t)R(t) + C(t) = 0 \tag{13}$$

which can be solved forwards since the initial state s^j_i is known. The differential systems (10) and (11) should be integrated backwards in time since the boundary conditions are defined on the final time ΔT_i.

6. Implementation and Experiments

The approach presented in this paper has been implemented on a Silicon Graphics workstation within the robot programming system ACT³ [15]. We illustrate in the sequel two cases of the motion planning of $\mathcal{A}$ with maneuvring (fig. 11) and without maneuvring (fig. 12). The continuous motion of $\mathcal{A}$ between the intermediate placements (computed by *"Explore"*) is generated by the *"Move"* function using the dynamic model of the vehicle and the physical models of the terrain described in section 2. Figure 10 shows the initial and the final placements of the robot on a simplified numerical model of the terrain. For both cases, the generated trajectories are composed of a sequence of straight line segments and circular arcs (computed according to the non-holonomic constraint of $\mathcal{A}$). In figure 11, we illustrate the case of maneuvring when controls of opposite signs are applyed on the opposite wheels. Figure 12 shows the robot motion without maneuvers; the different gyrations are executed when positive controls of different magnitude are applyed on the opposite wheels.

---

³The ACT system is developed and currently marketed by the Aleph Technologies Company.

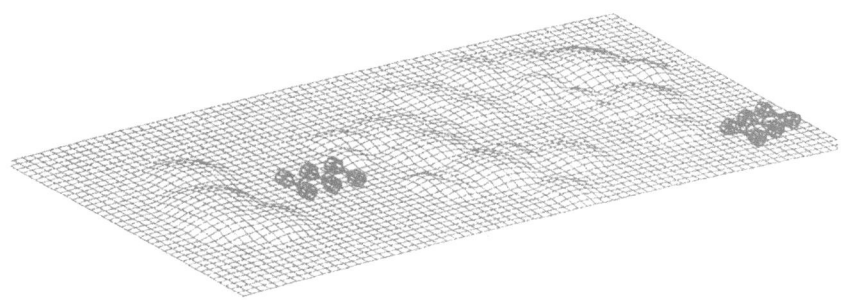

Figure 10. The initial and final configurations of the robot $\mathcal{A}$ on the terrain (respectively on the left and on the right)

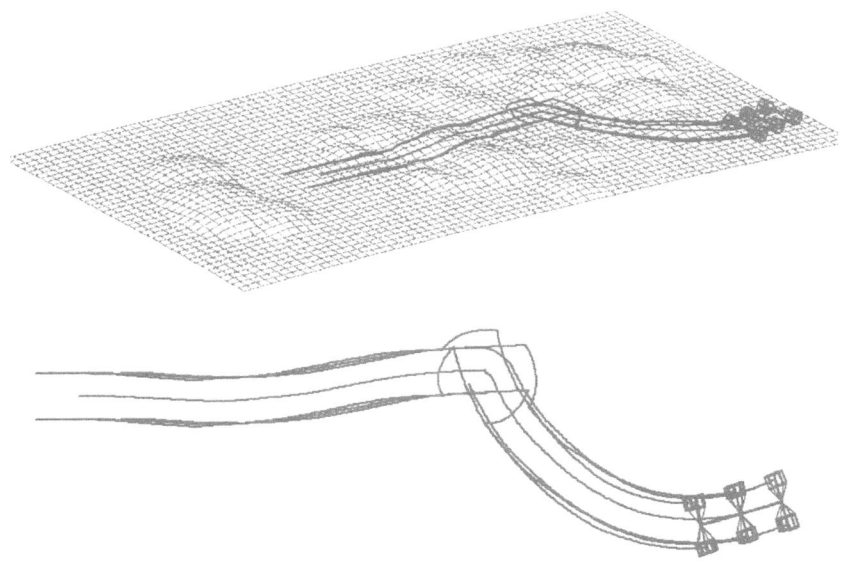

Figure 11. The trajectories executed by the main body and the six wheels of $\mathcal{A}$ in the case of maneuvers

Acknowledgment

The work presented in this paper has been partly supported by the CNES (Centre National des Etudes Spatiales) through the RISP national project and the MRE (Ministère de la Recherche et de l'Espace) . It has been also partly supported by the Rhône-Alpes Region through the IMAG/INRIA Robotics project SHARP.

474

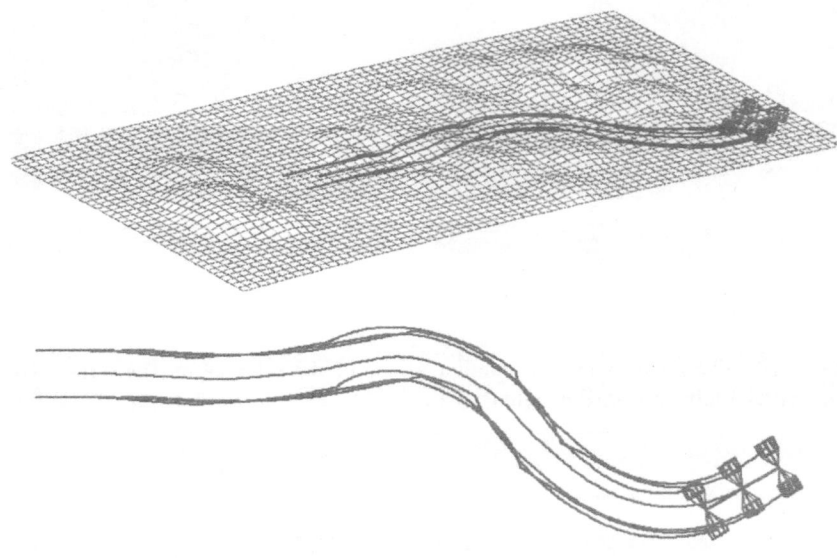

Figure 12. The trajectories executed by the main body and the six wheels of $\mathcal{A}$ in the case of no-maneuvers

References

[1] Ch. Bellier, Ch. Laugier, B. Faverjon, *"A kinematic Simulator for Motion Planning of a Mobile Robot on a Terrain"*, IEEE/RSJ Int. Workshop on Intelligent Robot and Systems, Yokohama, Japan, July 1993.

[2] A.E. Bryson, Y. Ho, *"Applied Optimal Control: Optimization, Estimation and Control"*, Hemisphere Publishing Corp., 1975.

[3] J.R. Bunch, L. Kaufman, *"A Computational Method for the Indefinite Quadratic Programming Problem"*, in Linear Algebra and its Applications, 34:341-370, 1980.

[4] M. Cherif, Ch. Laugier, *"Using Physical Models to Plan Safe and Executable Motions for a Rover Moving on a Terrain"*, First Workshop on Intelligent Robot Systems, J.L.Crowley and A. Dubrowsky (Eds), Zakopane, Poland, July 1993.

[5] L.E. Dubins, *"On Curves of Minimal Length with a Constraint on Average Curvature, and with Prescribed Initial and Terminal Positions and Tangents"*, in American Journal of Mathematics, 79:497-516, 1957.

[6] B. Faverjon, P. Tournassoud, *"A Local Based Approach for Path Planning Manipulators with a High Number of Degrees of Freedom"*, IEEE Int. Conf. on Robotics and Automation, Marsh 1987, Raleigh, USA.

[7] D. Gaw, A. Meystel, *"Minimum-time Navigation of an Unmanned Mobile Robot in a 2-1/2D World with Obstacles"*, IEEE Int. Conf. on Robotics and Automation, April 1986.

[8] H. Goldstein, *"Classical Mechanics"*, 2nd edition, Addison-Wesley, Massachusetts, 1983.

[9] J.K. Hahn, *"Realistic Animation of Rigid Bodies"*, Computer Graphics (Siggraph'88), 22(4):299-308, August 1988.

[10] M. Iagolnitzer, F. Richard, J.F. Samson, P. Tournassoud, *"Locomotion of an All-terrain Mobile Robot*, IEEE Int. Conf. on Robotics and Automation, Nice, France, May 1992.

[11] S. Jimenez, A. Luciani, C. Laugier, *"Physical Modeling as an Help for Planning the Motions of a Land Vehicle"*, IEEE/RSJ Int. Workshop on Intelligent Robot and Systems, Osaka, Japan, November 1991.

[12] S. Jimenez, A. Luciani, C. Laugier, *"Predicting the Dynamic Behaviour of a Planetary Vehicle Using Physical Models"*, IEEE/RSJ Int. Workshop on Intelligent Robot and Systems, Yokohama, Japan, July 1993.

[13] J.C. Latombe, *"Robot Motion Planning"*, Kluwer Academic Publishers, Boston, USA, 1991.

[14] A. Liegeois, C. Moignard, *"Minimum-time Motion Planning for Mobile Robot on Uneven Terrains"*, in Robotic Systems, Kluwer 1992, pp. 271-278.

[15] E. Mazer, J. Troccaz and al., *"ACT: a Robot Programming Environment"*, IEEE Int. Conf. on Robotics and Automation, Sacramento, April 1991.

[16] Z. Shiller, J.C. Chen, *"Optimal Motion Planning of Autonomous Vehicles in Three Dimensional Terrains"*, IEEE Int. Conf. on Robotics and Automation, May 1990.

[17] Z. Shiller, Y.R. Gwo, *"Dynamic Motion Planning of Autonomous Vehicles"*, IEEE Trans. on Robotics and Automation, vol 7, No 2, April 1991.

[18] Th. Simeon, B. Dacre Wright, *"A Practical Motion Planner for All-terrain Mobile Robots"*, IEEE/RSJ Int. Workshop on Intelligent Robot and Systems, Yokohama, Japan, July 1993.

[19] J. Tornero, G.J. Hamlin, R.B. Kelley, *"Spherical-object Representation and Fast Distance Computation for Robotic Applications"*, IEEE International Conference on Robotics and Automation, Sacramento, USA, May 1991.

Autonomous Walking in Natural Terrain:
A Retrospective on the Performance of the Ambler

Eric Krotkov   Reid Simmons

Robotics Institute
Carnegie Mellon University
Pittsburgh, PA 15213

Abstract— The objective of our research is to enable robotic exploration of remote, natural areas. Toward this end, we have developed an integrated walking system for the Ambler—a six-legged walking robot designed for planetary exploration—and tested it in indoor and outdoor autonomous walking trials. In this paper, we summarize our approach to autonomous walking, compare it to other approaches from the literature, and present results of the walking trials. Then we assess the performance relative to our objectives for self-reliant control, power-efficient locomotion, and traversal of rugged terrain.

1. Introduction

In our extensive reporting on the Ambler walking robot, we have concentrated on technical approaches (mechanisms, perception, planning, real-time control and task-level control) and experimental results. In this paper, we aim instead for an "intellectual audit" that compares the experimental results with the initial objectives, and reckons the difference. We find this topic especially appropriate because the Ambler has recently retired from active duty.

In the next section, we articulate the broad goals, the principal challenges, and the distinguishing features of our approach. In Section 3, we review related research, and identify the respects in which the Ambler walking system is unique. In Section 4, we briefly introduces the integrated walking system and documents the main experimental walking results. We conclude in Section 5 with a summary, a critical evaluation (the audit), and a look to the future.

2. Objective, Challenges, and Approach

2.1. Objective

The chief objective of this work is to enable robotic exploration of remote, natural areas. We consider exploration missions with two defining characteristics:

1. Remote area. By "remote" we mean conditions under which operator intervention is either not feasible (typically because of delay) or not cost-effective (typically because of the high cost of ground support).

2. Natural environment. The forces of nature produce irregular shapes and materials, causing natural terrain to exhibit roughness over a wide range of scales [5]. There are no prepared surfaces in nature.

To motivate the objective and make it more concrete, we consider the planet Mars as an instance of an area that is remote, natural, and unexplored. Others have detailed the expected benefits of observing geophysical, meteorological, and biological conditions on Mars, and have assessed the risks and costs of sending human crews [19]. Robotic exploration of Mars provides an alternative that eliminates risk to human lives and promises to substantially reduce mission cost.

The motivation of Mars exploration exists, but it is not unique. Other natural, remote areas such as the Moon, seabeds, and subsurface mines would suit our purposes equally well. Further, Mars exploration is a motivation, not an objective. As a consequence, the research does not require flight-qualified components for computing, data storage, or communications, nor does it require formulation for extremes of gravity, vacuum, or other conditions.

2.2. Challenges

From the stated characteristics of the mission follow specific challenges to be addressed:

1. Self-reliant control. The remoteness of the area to be explored impacts the feasibility and cost of different control regimes, ranging from pure teleoperation (reliance on operator) to pure rover autonomy (reliance on self). Regimes from the teleoperation end of the spectrum are less feasible, because round-trip signal delays preclude the rover from responding quickly to hazardous conditions (e.g., sinking into quicksand), and less cost-effective, because it requires costly ground support. The challenge is self-reliant control that enables timely rover responses, requires little ground processing, and enables operation for extended periods and distances. The longer the duration and wider the coverage, the greater the likelihood of meaningful return from exploration missions.

2. Limited power. The remoteness of the area requires the rover to supply its own power, which in turn dictates power budgets that are meager by laboratory standards. Limited power places a premium on efficiency of operation and penalizes unproductive behavior such as bumping into obstacles (which transfers energy to the terrain) and wandering repeatedly in the same small subregion (which serves no exploration purpose).

3. Rugged terrain traversal. The surface of Mars viewed by the Viking landers is rugged, uneven, and rocky (Figure 1). This terrain challenges most existing mobile robot mechanisms [2]. Even with a suitable form of locomotion, the terrain poses significant stability control challenges, since the natural, irregular materials may shift, fracture, and deform. The irregular terrain also challenges established machine perception techniques, because it does not satisfy standard constraints on shape (e.g., symmetry) or surface properties (e.g., smoothness), nor does it admit controlled lighting or fixtured objects.

Just as challenges follow from the characteristics of the mission, so do challenges follow from the mechanical configuration of the Ambler as a hexapod. Designed for mobility and efficiency, the Ambler features orthogonal legs, a circulating gait, and

478

Figure 1. Martian surface

level body motion [2]. Unlike rolling and crawling mechanisms, walkers are able to select favorable footholds, and to avoid transferring energy to terrain by unsupportive contact. In order to capitalize on these strengths, a key challenge is to accurately model the terrain so that footholds and obstacles can be identified. Other challenges include maintaining stability while advancing, planning the sequence of leg and body motions required to advance or to turn, and coordinating degrees of freedom absent in rollers and crawlers.

2.3. Approach

Our research has addressed each of these challenges, enabling high-performance autonomous walking. We identify two key distinguishing features of the approach:

1. Autonomy. Our strategy is to meet the challenge of self-reliant control by seeking to maximize robot autonomy. The approach is to develop a system that both plans (deliberates, reasons) and reacts (responds reflexively to a stimulus), enabling the Ambler to operate for extended periods. In this approach, planning algorithms evaluate conservatively each Ambler motion, checking them several times for feasibility and stability. While executing the plan, if robot sensors indicate signs of trouble, then reactive behaviors stabilize the robot.

2. Explicit models. We model explicitly terrain geometry, mechanism kinematics, vehicle stability margins, and sensor uncertainty. These models play critical

roles in meeting the challenges derived from the mission characteristics and the robot form:

- Models of terrain geometry enable planners to select favorable footholds, thus meeting the challenge of rugged terrain traversal and capitalizing on the high-mobility characteristics of walking robots.

- Models of terrain geometry also permit planners to select leg and body motions minimizing terrain contact, thus meeting the challenge of limited power.

- Models of mechanism kinematics allow controllers to accurately execute the plans, thus addressing the challenges of rugged terrain traversal and limited power.

- Models of stability margins allow planners and controllers to ensure the safety of different motions, thus meeting the challenge of extended operation under self-reliant control.

The complexity of the models is high compared to purely reactive approaches. But the complexity produces significant benefits: using models to plan and analyze moves contributes to the rover safety and power efficiency that are overriding concerns for planetary missions.

3. Related Research

Researchers have advanced a spectrum of mobile robot concepts suited for planetary exploration. In this section we review relevant mobile robots that have been built and operated.

Sustained effort at NASA's Jet Propulsion Laboratory has developed a family of rovers including the Surveyor Lunar Rover Vehicle, Robby, Rocky, and Go-For. Using these wheeled mechanisms as testbeds, JPL researchers have developed control architectures — including Computer-Aided Remote Driving [25], behavior control [10, 11], and Semi-Autonomous Navigation (SAN) [26] — for sharing mission control between human operators and robots. Using these control architectures, various vehicles have successfully demonstrated navigation in desert settings. Still, the degree of autonomy is limited even with SAN, which requires more computation and look-ahead planning than the other control architectures.

Researchers at MIT have argued that small rovers on the order of 1 to 2 kg are suitable for planetary exploration [3] and lunar base construction [4]. They have developed the Attila family of small six-legged walkers, and are striving to make the robots totally autonomous units using the subsumption architecture and "behavior languages" [1]. Although the degree of autonomy inherent in the approach is high, the potential for efficiency appears to be low: local behaviors may cause the rover to take suboptimal or even dangerous actions, and may cause the rover to transfer substantial energy to the terrain by repeatedly bumping into obstacles. Further, the capability of performing meaningfully complex missions in realistic settings is yet to be demonstrated.

Other researchers have built and operated prototype planetary rovers, including the Marsokhod [9, 16], and the Walking Beam [7]. To date, these research efforts have concentrated primarily on mechanical design and real-time control, and have not yet addressed or achieved autonomous performance.

Mobile roboticists have developed a number of integrated systems that exhibit (or potentially exhibit) relatively high degrees of autonomy. Examples of indoor robots with autonomous capabilities include Hilare [6], Carmel, Xavier, and many others. Examples of outdoor robots with autonomous capabilities include the Autonomous Land Vehicle [24], the Navlab [23], and VaMoRs [8]. These robots typically operate in environments relatively rich in structure (e.g., well-defined hallway corridors, or roadways with lane markings) compared to the surface of the Moon or Mars.

Researchers have developed a variety of walking mechanisms with the high mobility required by planetary rovers. These walkers include the Adaptive Suspension Vehicle [22], Titan I through Titan VI [12, 13], hopping machines [14, 17], the Odex [18], the Recus [15], and the Aquabot [15]. Many of these efforts emphasize electromechanical design and performance. Thus, these robots tend to be either teleoperated or operate under supervisory control, so the degree of autonomy achieved is low.

In summary, related research (i) produces mechanisms with high mobility but without high autonomy, (ii) researches autonomous units but without mission capabilities, and (iii) develops autonomous systems with mission capabilities but only in structured environments. To our knowledge, there are no robot systems other than the Ambler with high mobility, high autonomy, and mission capabilities for unstructured, remote, natural environments.

4. Autonomous Walking

To date, the Ambler has walked autonomously a total of over 4 km, much of it over rugged, difficult terrain. Table 1 reports statistics of several walking trials, which were selected to indicate the progressive escalation of challenges and capability.

The Ambler integrated walking system [21] consists of a number of distributed

Date	Terrain	Body Moves	Planar Travel (m)	Body Rotation (rad)	Leg Travel (m)
1990	Indoor obstacle course	27	11	2	—
1991	Outdoor obstacle course	100	25	2	—
1992	Indoor obstacle course	397	107	55	901
1992	Outdoor field	151	46	—	—
1992	Outdoor field	1219	527	20	3330

Table 1. Statistics from selected walking trials

The planar travel term represents the planar distance traveled by the Ambler body. The leg travel term represents the sum of the planar distances traveled by the legs, measured from pick-up to set-down. A null entry indicates insufficient data to calculate statistics. Most of the leg travel occurs during circulation.

Figure 2. Ambler traversing indoor obstacle course
Ambler on sandy terrain with meter-tall boulders.

modules (processes), each with a specific functionality: the Task Controller coordinates
the distributed robot system; the Real-Time Controller implements motion control; the
perception modules acquire and store images, and construct terrain elevation maps; the
planning modules plan footfalls, leg trajectories, and gaits; and the graphical user
interface enables simple interaction.

A typical trial begins by executing perception, planning, and real-time processes
on the on-board computers. For convenience, the standard outputs of these processes
are displayed on windows on off-board workstations. A human operator enters a path
as a sequence of arcs of circles, and issues the command to start walking. The Ambler
then operates autonomously, planning and executing every footfall, leg move, body
move, leveling maneuver, and other action or reaction.

4.1. Indoor

For indoor trials, the Ambler operated on obstacle courses fashioned from 40 tons of
sand, 20 tons of boulders, a 30° wooden ramp, and various other objects (Figure 2).
The courses typically include rolling, sandy terrain with several boulders 1 m tall, ten
or so boulders 0.5 m tall, a ditch, and a ramp. The largest of the boulders is 1.5 m tall,
4 m long, and 2 m wide. With such materials, the obstacle courses pose significant
barriers to locomotion. We know of no other robot that could surmount all of the
obstacles in the course.

Traversing a variety of these obstacle courses, the Ambler has demonstrated long-
term autonomous walking. In one trial, the Ambler took 397 steps and traveled about
107 meters following a figure-eight pattern, each circuit of which covers about 35 m
and 550 deg of turn. Figure 3 illustrates the elevation map constructed and used by the

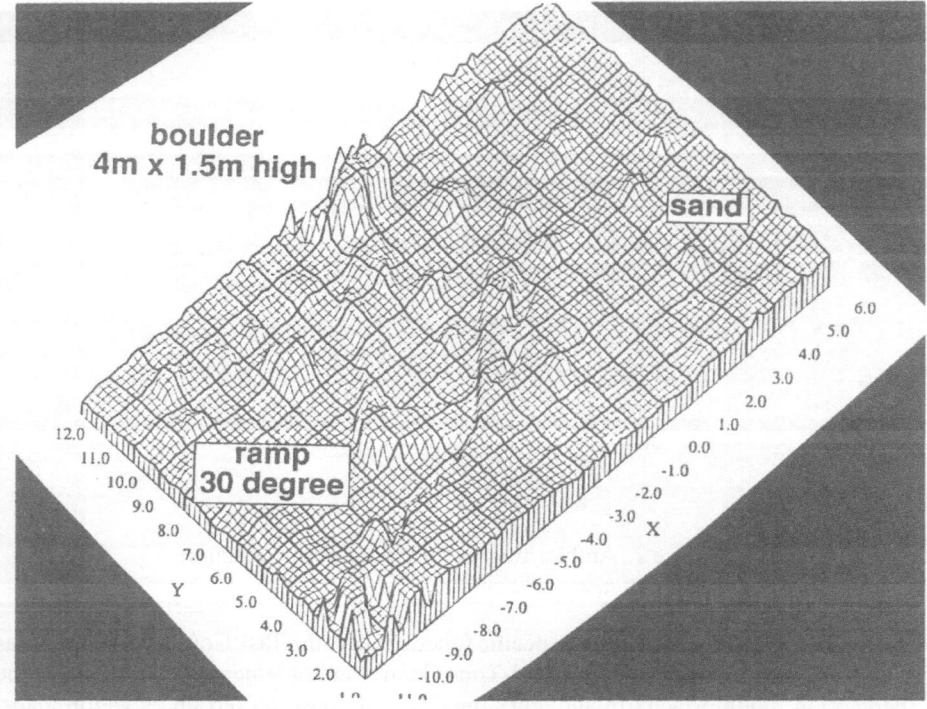

Figure 3. Elevation map of indoor obstacle course

Ambler. Completing more than two circuits during the trial, the Ambler turned nearly 9 complete revolutions, thus devoting a significant fraction of the trial to turning rather than to advancing. The two circuits were different in the sense that the robot did not take the same step twice.

4.2. Outdoor

For one set of outdoor trials, the Ambler operated on rolling, grassy terrain (Figure 4). The site is a field cleared in a hilly, wooded area at an industrial park. Although it does not contain obstacles like those encountered in the indoor trials, the site poses its own challenges: steeper slopes and side-slopes, and soft ground.

In the longest of these trials, the Ambler took 1219 steps, traveling 527 meters horizontally and 25 meters vertically. It followed a meandering course that included first climbing a hill and descending from it, then roughly following an iso-elevation contour for 250 meters, executing a point turn of roughly π radians, and following an iso-elevation contour back to the starting region. This course is significantly simpler to follow than the indoor figure-eight pattern because it involves less turning and less acute turning. In the process, the robot ascended a 30 percent grade, and traversed side-slopes on 15 percent grades.

Figure 4. Ambler walking 527 meters in field

5. Discussion

The results of the indoor and outdoor walking trials show that the approach meets the challenges identified earlier:

1. Self-reliant control. The approach meets this challenge, since the integrated walking system exhibits a high level of autonomy in planning and executing motions (e.g., traversing a specified path), reacting to unsafe conditions (e.g., leveling), learning good footfalls on the basis of experience, and operating for extended periods.

2. Limited power. The approach meets this challenge by reducing unwanted terrain contact to a negligible (but non-zero) level. Thus, the system transfers far less energy to the terrain than would a purely reactive or purely randomized robot.

3. Rugged terrain traversal. The approach enables locomotion over challenging obstacle courses, soft ground, and significant slopes and side-slopes.

484

Based on this analysis, we find the Ambler system to be well-suited for the class of exploration missions originally intended. Thus, as "intellectual auditors" we are satisfied that the account balances. However, we *were* surprised by the level of effort required to meet all of the challenges. We knew that we could use a simple system to accomplish simple tasks, but we did not appreciate in advance how much more complex the system would be to accomplish the moderately difficult tasks attempted.

We continue to pursue this line of research (see [20] for details). In our ongoing work, we are developing a legged robot system that will perform multi-day, multi-kilometer missions, with the robot acting as a capable, remote science assistant. We seek to achieve levels of autonomy comparable to the Ambler, but with a system characterized by simplicity and extreme reliability.

Acknowledgements

Many members of the Planetary Rover project have contributed to the results reported here. In particular, we acknowledge the efforts of Brian Albrecht, Chris Fedor, Regis Hoffman, Henning Pangels, Gerry Roston, David Wettergreen, and Red Whittaker. This research is sponsored by NASA under grant NAGW-1175.

References

[1] C. Angle and R. Brooks. Small Planetary Rovers. In *Proc. IEEE Intl. Workshop on Intelligent Robots and Systems*, pages 383–388, Tsuchiura, Japan, July 1990.

[2] J. Bares and W. Whittaker. Configuration of Autonomous Walkers for Extreme Terrain. *Intl. J. Robotics Research*, To appear, 1993.

[3] R. Brooks and A. Flynn. Fast, Cheap, and Out of Control: A Robot Invasion of the Solar System. *J. British Interplanetary Society*, 42(10):478–485, 1989.

[4] R. Brooks, P. Maes, M. Mataric, and G. More. Lunar Base Construction Robots. In *Proc. IEEE Intl. Workshop on Intelligent Robots and Systems*, pages 389–392, Tsuchiura, Japan, July 1990.

[5] P. Burrough. Fractal Dimensions of Landscapes and Other Environmental Data. *Nature*, 294:240–242, 1981.

[6] R. Chatila and J.-P. Laumond. Position Referencing and Consistent World Modeling for Mobile Robots. In *Proc. IEEE Intl. Conf. on Robotics and Automation*, pages 138–145, St. Louis, Missouri, 1985.

[7] W. Chun, S. Price, and A. Spiessbach. Design and Construction of a Quarter Scale Model of the Walking Beam. Martin Marietta Space Systems, 1989.

[8] E. Dickmanns and B. Mysliwetz. Recursive 3-D Road and Relative Ego-State Recognition. *IEEE Trans. Pattern Analysis and Machine Intelligence*, 14(2), February 1992.

[9] L. D. Friedman. What Now With the Soviets? *The Planetary Report*, 11(4):4–7, July/August 1991.

[10] E. Gat. ALFA: A Language for Programming Reactive Robotic Control Systems. In *Proc. IEEE Intl. Conf. Robotics and Automation*, pages 1116–1121, Sacramento, California, May 1991.

[11] E. Gat. Robust Low-Computation Sensor-Driven Control for Task-Directed Navigation. In *Proc. IEEE Intl. Conf. Robotics and Automation*, pages 2484–2489, Sacramento, California, May 1991.

[12] S. Hirose. A Study of Design and Control of a Quadruped Walking Vehicle. *Intl. Journal of Robotics Research*, 3(2):113–133, Summer 1984.

[13] S. Hirose, K. Yoneda, and K. Arai. Design of Prismatic Quadruped Walking Vehicle TITAN VI. In *Proc. Intl. Conf. Advanced Robotics*, pages 723–728, 1991.

[14] J. Hodgins and M. Raibert. Adjusting Step Length for Rough Terrain Locomotion. *IEEE Trans. Robotics and Automation*, 7(3):289–298, June 1991.

[15] Y. Ishino, T. Naruse, T. Sawano, and N. Honma. Walking Robot for Underwater Construction. In *Proc. Intl. Conf. Advanced Robotics*, pages 107–114, 1983.

[16] A. Kemurdjian, V. Gromov, V. Mishkinyuk, V. Kucherenko, and P. Sologub. Small Marsokhod Configuration. In *Proc. IEEE Intl. Conf. Robotics and Automation*, pages 165–168, Nice, France, May 1992.

[17] M. Raibert. *Legged Robots That Balance*. MIT Press, Cambridge, Massachusetts, 1986.

[18] Russell. Odex I: The First Functionoid. *Robotics Age*, 5(5):12–18, September/October 1983.

[19] U. S. Congress Office of Technology Assessment. *Exploring the Moon and Mars: Choices for the Nation*. U.S. Government Printing Office, Washington, DC, July 1991. OTA-ISC-502.

[20] R. Simmons and E. Krotkov. Autonomous Planetary Exploration. In *Proc. Intl. Conf. Advanced Robotics*, Yokohama, Japan, November 1993.

[21] R. Simmons, E. Krotkov, W. Whittaker, B. Albrecht, J. Bares, C. Fedor, R. Hoffman, H. Pangels, and D. Wettergreen. Progress Towards Robotic Exploration of Extreme Terrain. *Journal of Applied Intelligence*, 2:163–180, Spring 1992.

[22] S. Song and K. Waldron. *Machines that Walk: The Adaptive Suspension Vehicle*. MIT Press, Cambridge, Massachusetts, 1989.

[23] C. Thorpe, editor. *Vision and Navigation: The Carnegie Mellon Navlab*. Kluwer, 1990.

[24] M. Turk, D. Morgenthaler, K. Gremban, and M. Marra. VITS- A Vision System for Autonomous Land Vehicle Navigation. *PAMI*, 10(3), May 1988.

[25] B. Wilcox and D. Gennery. A Mars Rover for the 1990's. *Journal of the British Interplanetary Society*, 40:484–488, 1987.

[26] B. Wilcox, L. Matthies, D. Gennery, B. Cooper, T. Nguyen, A. Litwin, A. Mishkin, and H. Stone. Robotic Vehicles for Planetary Exploration. In *Proc. IEEE Intl. Conf. Robotics and Automation*, pages 175–180, Nice, France, May 1992.

Development of a Reactive Mobile Robot Using Real Time Vision

Patrick Rives* , Roger Pissard-Gibollet**, Konstantinos Kapellos**

* INRIA Sophia Antipolis	** ISIA-Ecole des Mines de Paris
2004 Route des Lucioles	Rue Claude Daunesse
06565 Valbonne Cedex - France	06565 Valbonne Cedex - France
rives@sophia.inria.fr	pissard(kapellos)@isia.cma.fr

Abstract— In this paper, we discuss both theoretical and implementation issues of a vision based control approach applied to a mobile robot. After having briefly presented a visual servoing framework based on the task function approach, we point out some problems of its application to the case of mobile nonholonomic robots. We will show how by using additional degrees of freedom provided by an manipulator arm, we can overpass these difficulties by introducing redundancy with respect to the task. The second part of the paper deals with the development of an experimental testbed specially designed to rapidly implement and validate reactive vision based tasks. It is constituted by a mobile robot carrying an hand-eye system using a dedicated vision hardware based on VLSI's and DSP 96002's processors which allows to perform image processing at video rate. Programming aspects are fully taken into account from tasks level specification up to real time implementation by means of powerfull software tools handling explicitely reactivity and real time control issues.

1. Introduction

Recent improvements in the field of vision sensors and image processing allow to hope that the use of vision data directly into the control loop of a robot is no more an utopic solution. Commonly, the general approach in robot vision is the following: processing vision data into the frame linked to the sensor, converting data into the frame linked to the scene by means of inverse calibration matrix and computing the control vector of the robot into the frame linked to the scene. This scheme works in open loop with respect to vision data and cannot take into account errors in the measurement or in the estimation of the robot position. Such an approach needs to perfectly overcome the constraints of the problem: the geometry of the sensor (for example, in a stereovision method), the model of the environment and the model of the robot. In some cases, this approach is the only one possible but, in many cases, an alternative way consists in specifying the problem in terms of control directly into the sensor frame. This approach is often referred as visual servoing [WEI 84] or sensor based control [ESP 87]; in this case, a closed loop can be really performed with

---

[0]The work on mobile robot is supported by firm ECA (Toulon, France) under a CIFRE contract. The work on Orccad system is made in cooperation with Aleph Technologies (Grenoble, France).

respect to the environment and allows to compensate the perturbations by a robust control scheme. Moreover, this approach seems to be a fruitful way to implement robotics tasks based on *reactivity concept* [BRO 86]. The work described in this paper deals with such an approach applied to autonomous mobile robots using vision sensor. More particularly, we will focus on the development of our experimental testbed including a mobile robot carrying a hand-eye system and the dedicated vision hardware which has been developped for testing reactive approaches based on visual servoing.

Part I
Theoretical Aspects

2. Sensor Based Tasks Framework

As a preliminary, it must be noted that the framework which will be presented here, only addresses local level task specification aspects using sensor based control approaches. Therefore, it excludes global level task specification aspects like mission planning or off-line programming. At this level of specification, a robotics application will be stated as a concatenation of elementary sensor based tasks. At each elementary task is associated a servoing law, continuous on a time interval $[T_{begin}, T_{end}]$, and built from the sensor's outputs. The transitions between the elementary tasks are scheduled by the events occuring during the task (reactive approach).

2.1. Elementary Sensor based Task Specification:

At this level, we want to provide a formalism to easily specify servoing tasks controlling the local interactions between the robot and its environment. A natural way is to specify these interactions in terms of kinematics constraints between a frame linked to the robot and a certain frame associated to a part of the local environment. By analogy with the *Theory of Mechanism*, we call this set of constraints : a **linkage**. The class of the linkage is so defined by the dimension of the subspace of compatible velocities which are let unconstrained by the linkage. This subspace of compatible velocities can be also viewed as a screw space and it fully characterizes the interactions between the robot and its environment. This concept of linkage plays an essential role in the synthesis of control laws using contact sensors for assembly applications. It can be easily extended to the case of no contact sensors like proximity or vision sensors. So, we can define the notion of *virtual linkage* by replacing the mechanical constraints by, for example, optical constraints like bringing the optical axis of a camera into alignment with a certain axis of a frame attached to the scene. After specifying an elementary task in terms of a peculiar virtual linkage to be reached between the frame attached to the robot and the frame attached to the environment, we need to realize this virtual linkage by using sensors data. To do that, it is necessary to identify the mapping between the configuration space, characterizing the relative position of the robot with respect to its local environment, and the sensor response space.

2.2. Modeling the Interactions between the Sensor and its local environment

We only give the main characteristics of this approach, more inquires can be found in [SAM 90], [CHA 90] and [ESP 92]. A sensor (S), providing a multidimensional signal $\underline{s}$ is linked to a robot (R) with related frame F_S. (S) interacts with an environment at which is associated a frame F_T. The position (location and attitude) of (R) is an element $\bar{r}$ of SE_3 (Special Euclidean Group) which is a six dimensional differential manifold. Let us assume that the output of (S) is fully characterized by the relative position of (S) with respect to (T) (for example, if we use vision sensor, we don't consider image changes dues to lightning variations). Using this assumption, the sensor's output is modeled as a function $s(\bar{r}, t)$ which is fully defined by the nature of the sensor, the position of the robot $\bar{r}$ and the time t representing the own motion of the part of environment associated to the frame F_T. This function represents a mapping between the configuration space of the robot and the sensor response space. For many types of sensors (proximity, vision, force...), an analytical form of this function can be found. Due to the properties of the screw spaces (see [ESP 92] for more details) the differential form of $s(\bar{r}, t)$, can be expressed as a screw product :

$$\dot{\underline{s}} = H \bullet T_{CT}, \tag{1}$$

where

H is called the *interaction screw*, T_{CT} is the velocity screw characterizing the relative motion between F_S and F_T and $\bullet$ represents the screw product.

Considering the bilinear form L^T of the interaction screw H expressed in the sensor frame F_C, and with an obvious breach of notation, we obtain :

$$\dot{\underline{s}} = L^T . T_{CT}, \tag{2}$$

Interaction matrix may be derived for many exteroceptive sensors. In the case of vision sensor, the vector $\underline{s}$ may be constituted from the projection of geometrical primitives in the environment like points, lines, circle... An analytical expression for the interaction matrix when the image features are general algebraic curves can be also obtained (for more details, see [CHA 90]). As a main result, we can note that the interaction screw formalism is a powerful means to model the sensor behavior. It relates both the value of the signal to the relative position between the sensor and the target and its variation to their relative motions. Moreover, due to the properties of the screw space, it can be easily expressed in terms of virtual linkage.

2.3. Relating Interaction Matrix and Virtual Linkage

Let us come back to the specification of a robotics application as a concatenation of elementary subtasks expressed in terms of a succession of virtual linkages to be realized between the robot and its environment. With no lack of generality, we now focus on the use of vision sensors. Let us consider the case where visual signal output is invariant ($\underline{\dot{s}} = 0$) with respect to a set of camera motions defined by T^*_{CT} :

$$\underline{\dot{s}} = L^T . T^*_{CT} = 0 \tag{3}$$

It can be easily seen that :

$$T^*_{CT} = Ker(L^T) \tag{4}$$

where $Ker(L^T)$ is the kernel of interaction matrix.

Now, recalling the definition of the virtual linkage, we find out that the subspace spanned by T^*_{CT} is precisely the subspace of compatible velocities which leads no change in the image during the motion of the camera. According to this fact, it becomes possible to define a virtual linkage by a set of elementary visual signals in the image invariant with respect to a motion belonging to the subspace of compatible velocities spanned by the virtual linkage. To perform a specific virtual linkage, many different features in the image can be used like points, lines, area, inertial moments and so on... In practice, their choice is guided by the application : a priori knowledge about the scene, low cost image processing, robustness with respect to the measurement noise...

By this formalism, this approach gives a fruitful way for specifying an elementary vision based task in terms of an *image target* to be reached when the task is well performed. We think it is a straightforward manner to implement reactivity approaches using vision sensors. In the next section, we will show how such an approach can be embedded in a robust closed loop control scheme using the *task function formalism* [SAM 90].

2.4. Sensor Based Task Function

A now classic approach in robotics it to consider the process of achieving a task such as tracking or positioning as a problem of regulation to zero of a certain output function: *the task function*. For example, a trajectory following task in cartesian space can be expressed as a regulation to zero of the task function: $e(t) = \bar{r}(t) - R^*(t)$, where R^* is the desired trajectory expressed in SE_3. We consider that the task is perfectly achieved if $\|e(t)\| = 0$ during the time interval $[T_{begin}, T_{end}]$ (see [SAM 90] for full details about the task function approach in robot control). The application of the task function approach to vision based tasks is straightforward: the task is now described as a desired values for a certain set of elementary visual signals to obtain. In mathematical terms the general definition of a visual task function vector e is :

$$e(t) = C(s(\bar{r}, t)) - s^*(t)) \tag{5}$$

where

- s^* can be considered as a desired value of the vector of the elementary visual signals (reference image target) to be reached in the image.

- $s(\bar{r}, t)$ is the value of the visual features currently observed by the camera, which only depends on the position between the camera and the scene.

- C is a matrix which allows, for robustness issues, to take into account more visual features than necessary. This matrix must have some properties as we will see later, for ensuring convergency at control level.

The quantity which arises naturally in our output regulation problem is the *task error* $\dot{e}$. For simplicity, we deal with a very simple example of *velocity control scheme* using visual feedback (for more advanced control schemes, see [SAM 90]). Assuming the presence of an appropriate robot's velocity servoing loop in Cartesian space, we can use as control input, the velocity T_c of the camera. We thus may choose the following control law :

$$T_c = -\lambda e \tag{6}$$

with $\lambda > 0$.

Let us consider the assumption that e is only depending of $(\bar{r}, t)$, we get :

$$\dot{e} = \quad \frac{\partial e}{\partial s} \cdot \frac{\partial s}{\partial \bar{r}} \cdot T_c + \frac{\partial e}{\partial t} \tag{7}$$

$$\dot{e} = \quad C.L^T.T_c + \frac{\partial e}{\partial t} \tag{8}$$

where $\frac{\partial e}{\partial t}$ parametrizes an eventual motion of the target. In the case of a motionless target $\frac{\partial e}{\partial t} = 0$, and we obtain :

$$\dot{e} = \quad C.L^T.T_c = -\lambda\, C.L^T.e \tag{9}$$

An exponential convergence will thus be ensured under the sufficient condition:

$$CL^T > 0 \tag{10}$$

in the sense that a $n \times n$ matrix A is positive if $x^T A x > 0$ for any nonzero $x \in \mathbf{R}^n$.

A good and simple way to satisfy this convergence condition in the neighbourhood of the desired position is to choose for the matrix C the generalized inverse of the interaction matrix associated to s^* :

$$C = L^{T+}_{|s=s^*}. \tag{11}$$

2.5. Vision Based Control Applied to a Mobile Robot

Applying vision based control theory to mobile robots provides an interesting framework for implementing reactive applications. However it is necessary to take some care due to the presence of nonholonomic constraints. Because of this, it is now wellknown [SAM 91] that pure state feedback stabilization of a two-wheels driven nonholonomic cart's configuration around a given terminal configuration is not possible, despite the controllability of the system.

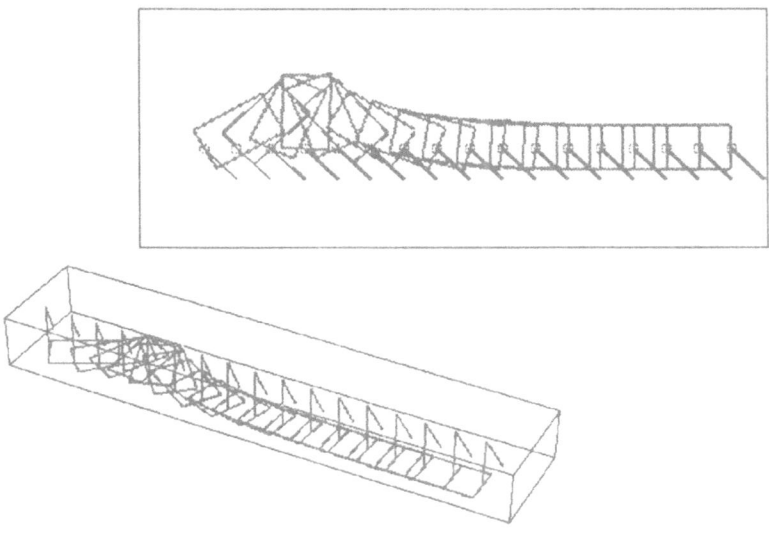

Figure 1. Wall following

As a consequence, we cannot ensure a convergence of the task error by using simple gradient-like methods in the image frame, for a camera rigidly attached to the mobile base. One way to avoid this problem consists in adding some degrees of freedom to the camera by mounting it on the wrist of a manipulator. Then, if we consider the whole mechanical system constituted by the cart and its manipulator like a single kinematic chain, we can show that the mapping between the configuration space (d.o.f of the cart + d.o.f of the manipulator) and the position (attitude and location) of the camera is almost everywhere regular (excepted in some isolated configuration) [RIV 92] and it becomes possible to fully control the motion of the camera without being limited by the cart nonholonomic constraints. By this way, we can execute visual servoing tasks directly defined in the camera frame in the same manner that in the case of holonomic systems. Obviously, in such an approach, we will not be able to control explicitly the trajectory of the cart but, by using some trivial geometric considerations, we can bound the volume which will be occupied by the cart during its displacement. We apply this approach to our system constituted by a two wheels driven cart with a camera mounted on a two d.o.f manipulator. The figure 1 shows simulation results of such an approach where the vision based task consists of wall following by servoing on the two parallel lines corresponding to the skirting boards at the bottom of the walls.

3. ORCCAD Principles

As previously seen, elementary servoing tasks, such as visual servoing, can be efficiently solved in real-time using closed loops.

On the other hand, to execute an application, considered like a temporal arrangement of elementary servo-control tasks, the robotic system must be able to react to occurences of various events. Events from the environment may trigger or kill elementary servo-control tasks or activate a particular recovering proccess. Also, events generated by the robotic system itself, informing for the normal or failed termination of elementary servoing tasks, must be as well considered.

Therefore a rigourous methodology is required to design such a robotic process. It must handle continuous and discret time aspects. A reliable robot programming environment must be provided. These requirements are at the basis of the development of the ORCCAD system [SIM 93], a generic robot controller handling reactivity and real-time issues.

We give a short overview of the ORCCAD's general concepts and refer the reader to [SIM 93] and [COS 92] for a detailed presentation. We will mainly focus our description on its key entity that models servo-control tasks and their associated event-based behaviour, the so called Robot-task.

3.1. Robot-task

A Robot-task is formally defined as the parametrized entire specification of:

- an elementary servo-control task, i.e the activation of a control scheme structurally invariant along the task duration

- a logical behavior associated with a set of signals liable to occur previously to and during the task execution.

The Event-based Behavior In a way, a Robot-task is atomic for the application level. However it follows an internal sequencing which has not to be seen in normal (failure-free) circumstances. Nevertheless the Robot-task has also to exchange information with other Robot-tasks, in order to synchronize and/or condition their activation. In the ORCCAD framework these two aspects are considered in a single way. The used approach is the one of reactive systems with the synchrony assumption [BEN 91]: signals are emitted without any absolute reference time from and to an automaton which specifies the Robot-task behavior.

In ORCCAD, all signals and associated processings belong to the following well defined categories :

- the pre-conditions: Their occurence is required for starting the servoing task. They may be pure synchronization signals or signals related to the environment usually obtained by a sensor.

- the exceptions: are generated during the execution of the servoing task and indicates a failure detection. Three different types are defined :

  - type 1 processing: corresponding to the modification of the parameters value of the control law (gain tuning, for example).

 — type 2 processing: starting the activation of a new Robot-task. The current one is therefore killed. When the ending is correct, the nominal post-conditions are fulfilled. Otherwise, a specific signal is emitted to the application level, which knows the recovering process to activate.

 — type 3 processing: when a fatal error is occuring. Then everything is stopped.

• the post-conditions: They are either logical synchronization signals emitted by the Robot-task automaton itself in case of correct termination, or signals related to the environment handled as conditions for a normal termination of the Robot-task.

Part II
Implementation Aspects

4. Description of the Robotic System

We have developped a highly versatile testbed in order to validate reactive vision based approaches in real experiments. This robotic system uses a cart-like mobile robot carrying a two d.o.f head with a CCD camera. The typical application domain concerns automatic exploration of an indoor environment. So we have designed a mobile robot light enough (75 Kg), easy to handle (cart-like), and with reduced size (lenght: 700 mm, width: 500 mm, height: 600 mm).

The on board computer architecture is built around a VME backplane which can accept until 12 VME boards. For these experiments we use a Motorola MVME 167 and a 6 axes custom-made servocontrol board based on dedicated Hewlett Packard components (HCTL 1100) and allowing to control different type of motors (DC, DC brushless, step motor).

The mobile robot disposes of two switchable power supplies made of an extern supply and two on board batteries. During the development step, an umbilical link is used both for power and communication (ethernet thin link, three video channels and two serial links). The Robot-task is designed and checked on a Sun Sparc workstation; then it is downloaded to the robot through ethernet. During the execution, the monitoring of the robot can also be done through ethernet. Figure 2 presents an overview of the whole experimental testbed.

Implementing efficiently a vision based control approach needs to solve several problems at the image processing level. The major difficulties are induced by strong real time constraints and the processing of large dataflow from the sequence of images. To overpass this difficulties, we have developped an original machine vision ([MAR 91], [RIV 93]) characterized by its modularity and its real time capabilities (figure 3). Its architecture is based both on VLSI chips

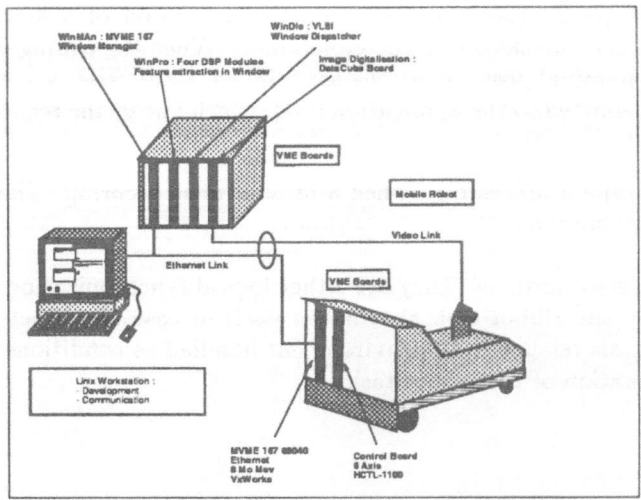

Figure 2. robotic system

for low level processing and on multi DSP's processors for more elaborated processings. It implements the concept of active window. An active window is attached to a particular region of interest in the image and has in charge to extract a desired feature in this region of interest. At each active window is associated a temporal filter able to perform the tracking of the feature along the sequence. Several active windows can be defined, at the same time, in the image and different processings can be done in each window. For facilities of development, this machine vision has been implemented in an independant VME rack outside of the robot. However, for experiments which require high autonomy, these boards can be plugged in the on board rack.

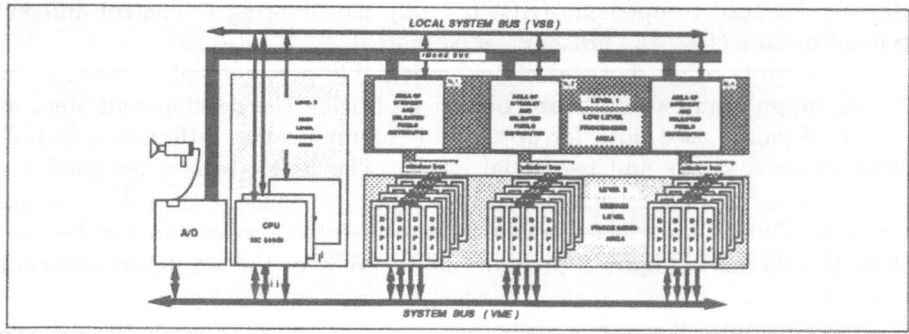

Figure 3. Overview of the machine vision

5. Visual Servoing Robot-task Design

The ORCCAD system provides a set of CAD tools allowing to design, check and implement Robot-tasks. An object-oriented approach has been choosen for modeling the Robot-task. A given Robot-task is then fully specified by instantiation of required objects leaves of a set of class hierarchies structuring a robotic system. These objects are defined by the functional and the temporal set of attributes and methods concerning coherence testing, automatic code generation for simulation and execution, ...

This modelling is at the basis of an HMI allowing the control expert to specify a Robot-task. Let us now describe the different stages of a visual servoing Robot-task design.

Continuous time specification At this stage of design the end-user's specification, done in natural language, provides information about actions to be performed, information about the robotic system to be used, events to be considered. Here, the end's user task specification could be *the robot goes in front of the door* or *the robot follows wall*.

For a control expert, a first stage, is to see if the problem is well conditionned and to choose an adequate control law. In our case, we will use a visual servoing technique to complete the task in a robust way with respect to robot's environment changes.

The continuous-time control law achieving the specified action is specified in term of block-diagrams and/or equations. For a two-wheeled mobile robot driven by visual servoing Robot-task, the control law computes the motor velocities (wheels and head axis) to be applied from an error between the desired and the current value of the visual feature to be reached. Usually, such a scheme may be splitted in several functionnal modules which exchange data. In figure (4) we could see the functionnal decomposition where :

- *GT_VS* generates a visual features trajectory in the image space to be followed : $s^*(t)$.

- *WINFEAT* interfaces the Robot-task with the real time vision system. It furnishes the current visual features parameters $s(\bar{r}, t)$.

- *ERROR_VS* computes the error signal in the image space: $\epsilon = s(\bar{r}, t) - s^*(t)$.

- *TF_VS* computes the control law using the task function approach : $e = C\epsilon$ and $T_c = -\lambda e$.

- *INVJAC* computes the jacobian robot matrix J_R^{-1} ([RIV 92]).

- *CIN* sends control-inputs to the robot's interface : $\dot{q} = J_R^{-1} T_c$.

- *ANIS* interfaces or simulates the whole mechanical system (mobile robot and its head) to be controlled.

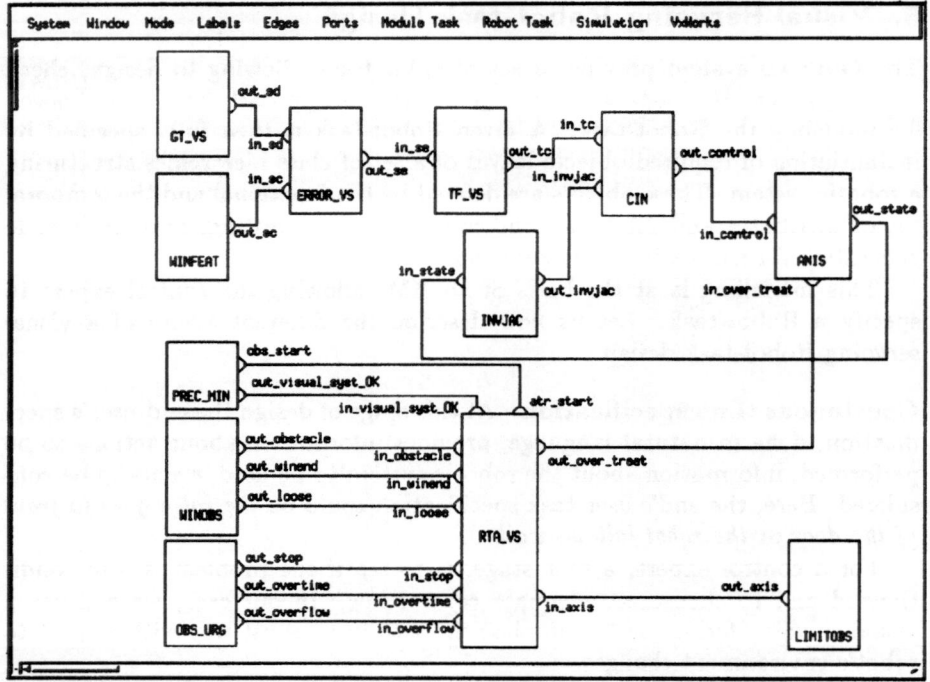

Figure 4. visual servoing Robot-task

At this stage of the control law conception the Robot-task designer chooses the classes to be instantiated and defines the objects (so called F-objects) considering the functional set of attributes (minimum values for closed loop gains, initial and final positions, location of sensors ...). The ORCCAD system guides, as far as possible, the user's choices and checks the specifications coherence.

The result of this stage of design is the specification of the functional graph of the Robot-task which is an oriented graph, the nodes of which are the F-objects instantiated for the Robot-task, and the edges of which correspond to the existence of formal data transfers between F-objects.

Time constrained specification The passage from the above continuous-time specification to a description taking into account implementation aspects is done by specifying at this level the *temporal properties*, i.e sampling periods, durations of computations, communication and synchronization between the processes, data processing code, ...

At this stage of specification the basic entity manipulated is the Module-task. It is a real-time task used to implement an elementary functional module of the control law. It owns temporal properties, like its computation duration and its activation period. Since the Module-tasks may, possibly, be distributed over a multiprocessor target architecture they communicate using typed ports (8 communication and synchronization mechanisms are provided). Each Module-task owns one input port for each data it handles and one output port for each data it produces. The structure of the **Module-tasks** separates calculations, related to control algorithms issues, and communications, related to implementation issues and operating system calls (see fig 5 illustrating the structure of a periodic Module-task).

```
Initialization code
Every (T-period){
Reading all input ports
Data processing code
Writing all output ports}
```

Figure 5. Module-task structure

Since I/O ports are the only visible parts of the **Module-tasks** and their connections are given by the designer of the Robot-task, the possibility to reuse the Module-tasks as objects in different Robot-task schemes exists with no need of internal data structure modification.

At this step the Robot-task designer adds to the already created F-objects properties characterizing ports (name, synchronization type, name and type of exchanged variables), localization of associated codes (initialization, computation), period and duration of the Module-taskse and name of the dedicated processor creating the MT-objects.

In the visual servoing task, all communication types are *asynchronous/asynchronous* excepted between the *CIN* and *ANIS* Module-task where the synchronisation is *asynchronous/synchronous*. Sampling period for visual Module-tasks (*GT_VS, WINFEAT, ERROR_VS, TF_VS*) is lower (40 ms) than for robot Module-tasks (*INJ_JAC, CIN*) which is 10 ms. The robot interface Module-task *ANIS* is not periodic but it is synchronised by the *CIN* Module-task.

The result of this stage of design is the time-constrained specification graph of the Robot-task which is an oriented graph, the nodes of which are the MT-objects instantiated for the Robot-task, each arrow corresponds to the transfer of a variable from a Module-task to another one.

The Event-based Behavior of the Robot-task The servoing visual Robot-task behavior is driven by nine events which allow to monitor the task :

- two pre-condition events (monitoring by the observer *PREC_MIN*) to activate the servoing control.

- one exception servoing control of "type 1" indicates that odometric registers are overflowing and implies the *ANIS* Module-task to reconfigure the registers. The servoing control remains active.

- two events (monitored by *OBS_URG*) stop the robot task when
    - execution time is elapsed,
    - the user decides to stop on keyboard.

- head axes limits generated by *LIMITOBS*,

- events on windows image managed by *WINOBS* when :

    visual features are lost,

    visual features will soon disapear,

    a visual obstacle occurs in a window.

In the future, we will be further defined the Robot-task behavior by adding more observers.

These specifications are encoded using the synchronous language ESTEREL ([BER 92]). The compiler transforms an ESTEREL program into a finite determinist automaton. On the resulting automaton automatic proofs can be performed using verification systems (AUTO described in [ROY 90]) and therefore one can be sure of the logical correctness of the program.The Robot-task automaton is illustrated on figure 6.

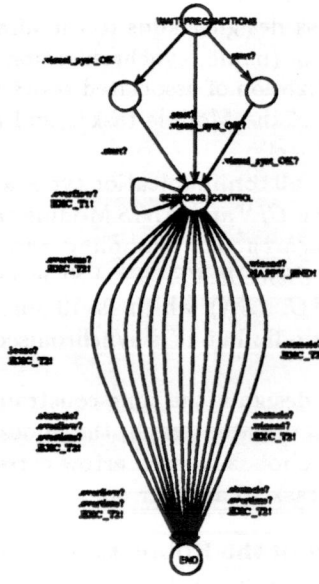

Figure 6. Robot-task automaton

The automaton code and its interfacing with the environment code is automatically generated in the ORCCAD system using information about the events considered by the user.

Robot-task simulation The robotic system, belongs to the physical world, and can generally be described by a set of differential derivate equation in continuous time ; On the other side, the controller works basically in the frame of discrete time. SIMPARC (SImulation for MultiProcesseurs Advanced Robot Controller), see ([AST 92]), is an hybrid simulator from time point of view which mixes both continuous and discrete simulation.

Using the time constrained specification graph of the Robot-task specified by the control expert, it is possible to automaticaly generate code for SIMPARC simulation. Simulation results help the user to verify and tune the control law and to choose temporal properties to be assigned to the Module-tasks : activation period, synchronization type.

Robot-task implementation When the Robot-task simulation results are satisfying, an other feature of the designing tool can also generate a downloading code to be executed by the robot system. Module-tasks are VxWorks real time tasks running on the robot CPU board. They communicate between each other using specified synchronization through shared memory.

All the design process has been successfully validated in a real robot's positioning experiment with respect to a target by using vision based control.

6. Conclusion

In this paper, we both presented the theoretical framework and the implementation tools allowing to validate reactive sensor based approaches applied to mobile robots. We state a robotic action as a set of elementary servoing tasks (continuous time aspect) associated with a logical behavior events driven (reactive aspect). On the first hand, we propose a powerful formalism to build visual servoing tasks allowing to explicitely control the robot in closed loop with regard to the environment. On the other hand, we use synchronous programming approach to solve the reactive aspect of the task. We have also developped an experimental testbed constituted by a mobile robot carrying a two d.o.f head with a CCD camera associated to a dedicated real time image processing system. All the programming aspects are taken into account by the ORCCAD system providing powerfull tools for designing, implementing and checking robotics applications. Presently, the integration of our testbed is almost finished and we expect a complete visual servoing experiment for the end of the year.

References

[AST 92]   C. Astraudo, J.J. Borrelly :*Simulation of Multiprocessor Robot Controllers*, Proceedings IEEE Int. Conf. on robotics and automation, Nice, May 92.

[BEN 91]   A. Benveniste, G. Berry :*The synchronous approach to Reactive and Real-Time Systems*, Proc. of the IEEE, vol 79, n 9, September 1991.

[BER 92]   G. Berry, G. Gonthier :*The synchronous Esterel Programming Language : Design, Semantics, Implementation*, Science of Computer Programming, Vol 19, no 2, pp 87-152, Nov. 1992.

[BRO 86]   R. Brooks : *A robust layered control system for a mobile robot*, IEEE Transaction on Robotics and Automation, vol 2, n1, pp 14-33, 1986.

[CHA 90] F. Chaumette : *La relation vision-commande: théorie et applications à des tâches robotiques*, Ph-D Thesis, University of Rennes I, France, July 1990.

[COS 92] E. Coste-Manière, B. Espiau, D. Simon :*Reactive objects in a task level open controller*, Int. Conf. IEEE on Robotics and Automation, Vol 3, pp 2732-2737, Nice, May 1992.

[ESP 87] B. Espiau :*Sensory-based Control: Robustness issues and modelling techniques. Application to proximity sensing*, NATO Workshop on Kinematic and Dynamic Issues in Sensor Based Control, Italy, October 1987.

[ESP 92] Espiau B., Chaumette F., Rives P. : *A New Approach to Visual Servoing in Robotics*, IEEE Transaction on Robotics and Automation, Vol. 8, No. 3, June 1992

[MAR 91] Martinet P.,Rives P.,Fickinger P.,Borrelly J.J : *Parallel Architecture for Visual Servoing Applications*, Workshop on Computer Architecture for Machine Perception, Paris, Dec. 1991

[RIV 89] P. Rives, F. Chaumette, B. Espiau : *Visual Servoing Based on a Task Function Approach*, 1stISER International Symposium on Experimental Robotics, Montreal, 1989

[RIV 92] P. Rives, R. Pissard-Gibollet : *Reactive Mobile Robots Based on a Visual Servoing Approach*, AIS'92, Perth, July 1992

[RIV 93] P. Rives, J.J. Borrelly, J. Gallice, P. Martinet : *A Versatile Parallel Architecture for Vision Based Applications*, Workshop on Computer Architecture for Machine Perception, New Orleans, Dec 15-17, 1993

[ROY 90] V. Roy, R. de Simone :*Auto and autograph*, In R. Kurshan, editor, *proceedings of Workshop on Computer Aided Verification*, New Brunswick, June 1990.

[SAM 91] C. Samson, K. Ait-Abderrahim : *Feedback control of a nonholonomic wheeled cart in cartesian space*, Proceedings IEEE Int. Conf. on robotics and automation, Sacramento, pp 1136-1141, April 1991.

[SAM91b] C. Samson :*Time-varying feedback stabilization of nonholonomic car-like mobile robots*, INRIA research report, n 1515, Septembre 91.

[SAM 90] C. Samson, B. Espiau, M. Le Borgne : *Robot control: the task function approach*, Oxford University Press, 1990.

[SIM 93] D. Simon, B. Espiau, E. Castillo, K. Kapellos :*Computer-aided design of generic robot controller handling reactivity and real-time control issues*, IEEE Trans. on Control Systems and Technology, to appear, fall 1993.

[WEI 84] L. E. Weiss : *Dynamic visual servo control of robots. An adaptive image based approach*. Carnegie Mellon Technical Report, CMU-RI-TR-84-16, 1984.

Mobile robot miniaturisation:
A tool for investigation in control algorithms.

Francesco Mondada, Edoardo Franzi, and Paolo Ienne

Swiss Federal Institute of Technology
Microcomputing Laboratory
IN-F Ecublens, CH-1015 Lausanne
E-mail: Francesco.Mondada@di.epfl.ch
Edoardo.Franzi@di.epfl.ch
Paolo.Ienne@di.epfl.ch

Abstract— The interaction of an autonomous mobile robot with the real world critically depends on the robots morphology and on its environment. Building a model of these aspects is extremely complex, making simulation insufficient for accurate validation of control algorithms.

If simulation environments are often very efficient, the tools for experimenting with real robots are often inadequate. The traditional programming languages and tools seldom provide enought support for real-time experiments, thus hindering the understanding of the control algorithms and making the experimentation complex and time-consuming.

A miniature robot is presented: it has a cylindrical shape measuring 55 mm in diameter and 30 mm in height. Due to its small size, experiments can be performed quickly and cost-effectively in a small working area. Small peripherals can be designed and connected to the basic module and can take advantage of a versatile communication scheme. A serial-link is provided to run control algorithms on a workstation during debugging, thereby giving the user the opportunity of employing all available graphical tools. Once debugged, the algorithm can be downloaded to the robot and run on its own processor.

Experimentation with groups of robots is hardly possible with commercially available hardware. The size and the price of the described robot open the way to cost-effective investigations into collective behaviour. This aspect of research drives the design of the robot described in this paper. Experiments with some twenty units are planned for the near future.

1. Introduction

Today the mobile robotics field receives great attention. There is a wide range of industrial applications of autonomous mobile robots, including robots for automatic floor cleaning in buildings and factories, for mobile surveillance systems, for transporting parts in factories without the need for fixed installations, and for fruit collection and harvesting. These mobile robot applications are beyond the reach of current technology and show the inadequacy of traditional design methodologies. Several new control approaches have been attempted to improve robot interaction with the real world aimed at the autonomous achievement of tasks. An example is the *subsumption architecture* proposed by Brooks [1] which supports parallel processing and is modular as well as robust. This approach is one of the first solutions systematically implemented on real robots with success. Other researchers propose new computational approaches like *fuzzy logic* [2] or *artificial neural networks* [3].

The interest in mobile robots is not only directed toward industrial applications. Several biologists, psychologist and ethologists are interested in using mobile robots to validate control structures observed in the biological world. Franceschini [4] uses a robot to validate the structure of the retina observed on a fly, Beer [5] to replicate the mechanism that coordinates leg movements in walking insects, Deneubourg [6] to get a better understanding of collective behaviour in ant colonies.

All these research activities are based on mobile robot experimentation. A simpler way to validate control algorithms is to use simulations, but the simplifications involved are too important for the results to be conclusive. The control algorithm embedded in the robot must consider its morphology and the properties of the environment in which it operates [7]. Real world features and anomalies are complex and difficult to modelise, implying that the experimentation of control algorithms through simulation can only be used in preliminary study but cannot prove the success of the control algorithm in the real world. The sole way to validate an algorithm to deal with these problems is to test it on a real robot [8].

Many robots have been designed to perform experiments on control algorithms but only a few make cost-efficient experiments possible. Brooks has designed several robots with effective electronics and mechanics [9]. The control algorithms are programmed in the subsumption behavioural language, taking into account software modularity, and real-time and parallel processing. Unfortunately, during experiments, only a few tools are available to improve the understanding of the control process. Moreover, the custom programming language makes code portability and algorithm diffusion difficult. Steels [10] uses a video-camera to record robot actions during experiments but all the data concerning the robot control process are available only at the end of the experiment. Other platforms, such as the Nomad robot [11], make real-time interaction possible via a radio link, and have standard programming languages, but the size of the robot and the environment it requires make experimentation uncomfortable.

The lack of a good experimentation mobile robot for single-robot experiments, means that it is impossible today to perform collective-behaviour experiments. The programming environment and the real-time visualisation tools are totally insufficient for this purpose.

The development of the miniature mobile robot *Khepera* addresses the problems mentioned above. Its hardware is designed so that it is small enough for the operation of several at the same time and in small areas, for example on a desk-top. Modularity allows new sensors and actuators to be easily designed and added to the basic structure. A versatile software structure is provided to help the user to debug the algorithms and to visualise the results.

2. Hardware

Miniaturisation is an important challenge for industry: CD players, computers, video cameras, watches and other consumer products need to implement many functionalities in a small volume. In the robotics field many applications need small actuators, small teleoperated machines or tiny autonomous robots. Dario [12] gives a comprehensive description of the research field and of the available technology. In the Khepera design, miniaturisation is the key factor in making cost-effective experimentations possible both for single or multiple robot configurations.

2.1. Generalities

The robot presented in this paper is only a first step in the direction of miniaturisation. Dario define this category of robots as *miniature robots*. They measure no more than a few cubic centimetres, generate forces comparable to those applied by human operators and incorporate conventional miniature components. The next miniaturisation step needs special fabrication technologies, today in development. Khepera uses electronic technology available today: the new family of 683xx microcontrollers from Motorola makes the design of complete 32 bit machines extremely compact. *Surface mounted devices* (SMD) allow an important increase in component density on printed circuit boards. New compact sensors, including some signal preprocessing on the sensing chip, reduce the need of additional circuitry. Only the mechanical parts (wheels, gears, manipulator) are built expressly for Khepera, as well as the magnetic sensors for counting the wheel revolutions.

The design of such miniaturised robots demands a great effort spanning several fields. The result is a complex compromise between functionalities to be implemented, available volume, current technology, power requirements, etc.

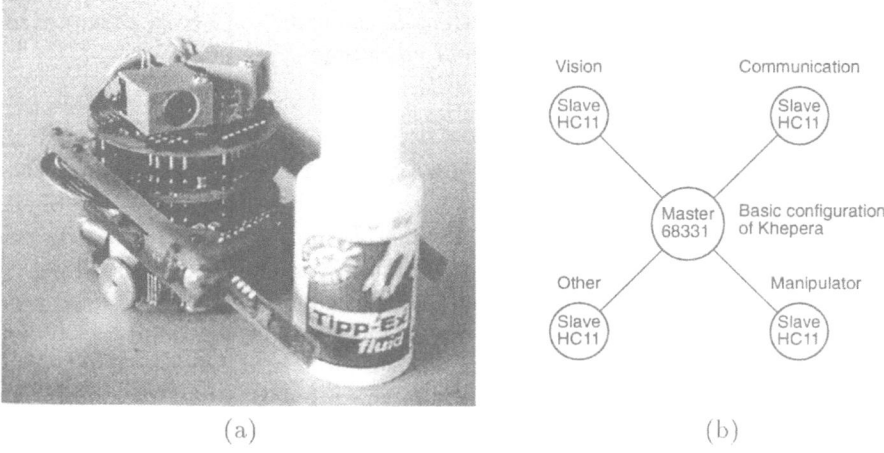

(a)   (b)

Figure 1. (a) The Khepera robot. (b) Communication network topology.

504

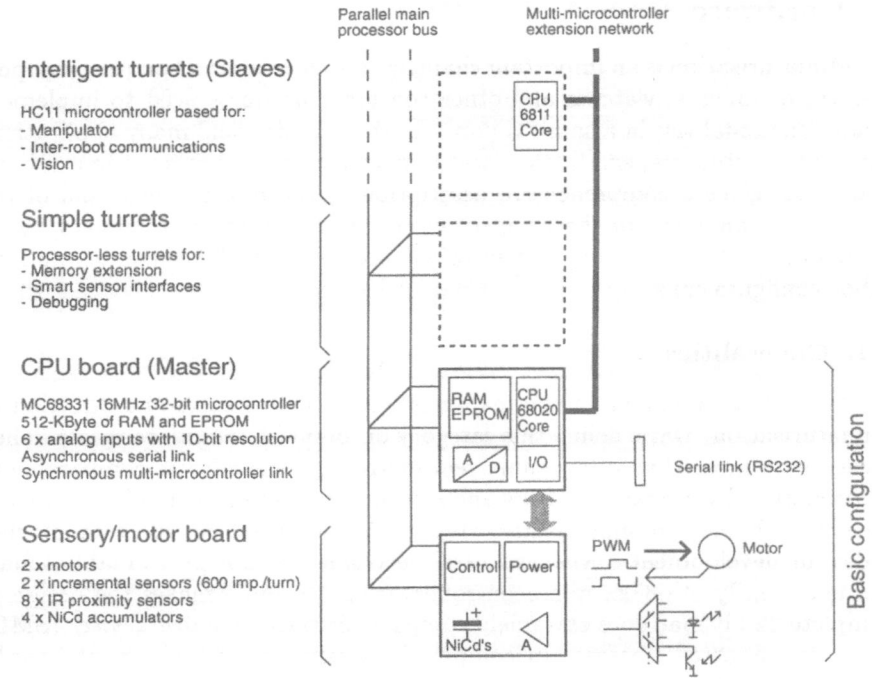

Parallel main
processor bus

Multi-microcontroller
extension network

Intelligent turrets (Slaves)

HC11 microcontroller based for:
- Manipulator
- Inter-robot communications
- Vision

CPU
6811
Core

Simple turrets

Processor-less turrets for:
- Memory extension
- Smart sensor interfaces
- Debugging

CPU board (Master)

MC68331 16MHz 32-bit microcontroller
512-KByte of RAM and EPROM
6 x analog inputs with 10-bit resolution
Asynchronous serial link
Synchronous multi-microcontroller link

RAM
EPROM

CPU
68020
Core

A/D   I/O

Serial link (RS232)

Sensory/motor board

2 x motors
2 x incremental sensors (600 imp./turn)
8 x IR proximity sensors
4 x NiCd accumulators

Control Power

PWM → Motor

NiCd's A

Basic configuration

Figure 2. Khepera hardware architecture.

Khepera is composed of two main boards (figure 2). Application-specific extension turrets for vision, for inter-robot communications, or which are equipped with manipulators can be directly controlled via the Khepera extension busses. Khepera can be powered by an external supply when connected for a long time to a visualisation software tool; however, on-board accumulators provide Khepera with thirty minutes of autonomous power supply.

2.2. Distributed processing

One of the most interesting features of Khepera is the possibility of connecting extensions on two different busses. One parallel bus is available to connect simple experimentation turrets. An alternative and more sophisticated interface scheme implements a small local communication network; this allows the connection of intelligent turrets (equipped with a local microcontroller) and the migration of conventional or neural pre-processing software layers closer to the sensors and actuators. This communication network (figure 1(b)) uses a star topology; the main microcontroller of the robot acts as a master (at the centre of the star). All the intelligent turrets are considered as slaves (on the periphery of the star) and use the communication network only when requested by the master.

This topology makes it possible to implement distributed biological controls,

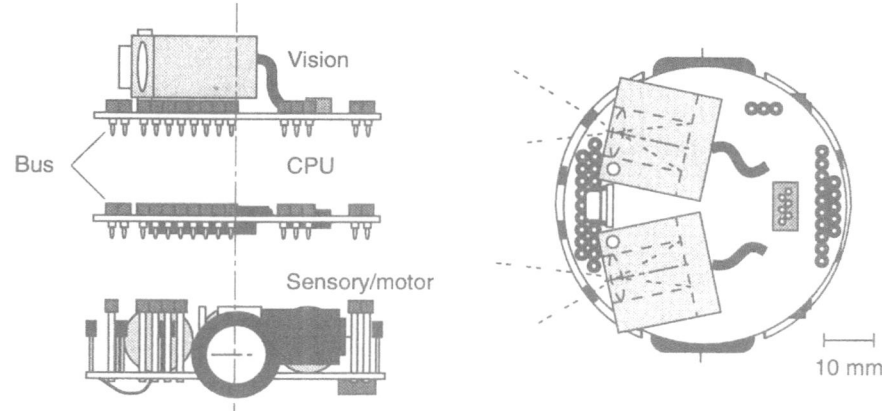

Figure 3. Khepera physical structure: Basic sensory/motor, CPU and vision boards.

such as arm movement coordination or feature extraction and pre-processing in the vision, as observed in a large number of insects. The multi-microcontroller approach allows the main microcontroller of Khepera to execute only high level algorithms; therefore attaining a simpler programming paradigm.

2.3. Basic configuration

The new generation of Motorola microcontrollers and in particular the MC68331 makes it possible to build very powerful systems suitable for miniature neural control. Khepera takes advantage of all the microcontroller features to manage its vital functionality. The basic configuration of Khepera is composed of the CPU and of the sensory/motor boards.

The CPU board is a complete 32 bit machine including a 16 MHz microcontroller, system and user memory, analogue inputs, extension busses and a serial link allowing a connection to different host machines (terminals, visualisation software tools, etc.). The microcontroller includes all the features needed for easy interfacing with memories, with I/O ports and with external interrupts. Moreover, the large number of timers and their ability to work in association with the I/O ports indicate that this device is the most important component in the design.

The sensory/motor board includes two DC motors coupled with incremental sensors, eight analogue *infra-red* (IR) proximity sensors and on-board power supply. Each motor is powered by a 20 kHz *pulse width modulation* (PWM) signal coming from a dedicated unit of the microcontroller. These signals are boosted by complete four-quadrant NMOS H bridges. Incremental sensors are realised with magnetic sensors and provide quadrature signals with a resolution of 600 impulsions per wheel revolution. IR sensors are composed of an emitter and of an independent receiver. The dedicated electronic interface is built with multiplexers, sample/hold's and operational amplifiers. This allows the

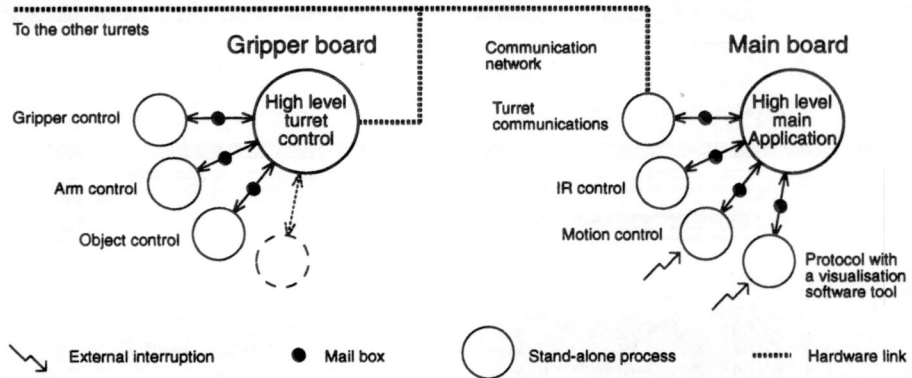

Figure 4. Hierarchical software structure.

measurement of the absolute ambient light and the estimation, by reflection, of the relative position of an object from the robot.

2.4. Additional turrets

To make experiments involving environment recognition, object detection, object capture and object recognition possible, two intelligent turrets have been designed and built: one for stereoscopic vision, the other containing a manipulator.

The stereoscopic vision turret employs two 64 pixel linear photoelement arrays and a dedicated optical element. The analogue value of each pixel is coded on 16 grey levels. To obtain useable data under a wide spectrum of enlightenment conditions, an additional sensor is used to perform as an automatic iris: the integration time necessary for the photoelement arrays is controlled by intensity of the ambient light. Mondada *et al.* [8] proved the validity of this stereoscopic vision in robot navigation (spatial frequency filtering was used in obstacle detection and avoidance).

The manipulator turret makes Khepera capable of an interaction with objects of its environment. Two DC motors control the movements of the manipulator (elevation and gripping). Different classes of objects can be detected by the gripper sensors which measure sizes and resistivities.

Robots displaying collective behaviour need means to perform inter-robot communications and localisation. Turrets providing Khepera with these functionalities are under study at the time of writing.

3. Software

Managing all the Khepera resources is a complex task. The large number of asynchronous events to control, and the necessity to share some critical interfaces led to the development of a complete low-level software organised as a collection of basic I/O primitives [13].

3.1. Hierarchical software structure

The multi-microcontroller approach and the complex tasks to manage required a hierarchical approach to the software structure. The concept chosen applies when intelligent turrets (equipped with a microcontroller) are used. Two software structures are implemented: a single high-level application program and a number of stand-alone local processes (figure 4). Stand-alone local processes (e.g., for IR sensor sequencing, motion control, wheel incremental-sensor counting, etc.) are executed cyclically according to their own event timer and possibly in association with external interrupts. The high-level application software run the control algorithm and communicate with the stand-alone local processes via a mail-box mechanism. This decoupling of low- and high-level tasks makes the development of complex applications quick and easy.

3.2. Control of Khepera

Experiments with Khepera are performed in two different ways: by running algorithms on autonomous robots or in connection with visualisation software tools.

As already mentioned, the details of the basic input/output activities are managed through a library of stand-alone processes. During the development, the standard RS232 link is used, through a generic high level protocol, to communicate with these processes from a workstation. The application software is therefore run on the workstation and calls to the basic primitives make it possible to monitor the robot activity possible. All standard and specialised visualisation tools can be employed to simplify the control algorithm debugging.

Because the application software is written in standard C language, debugged algorithms can easily be converted to run on the Khepera CPU. Applications can be downloaded to Khepera and the robot becomes autonomous from the development environment.

4. Experimentation environment

The quality of the measurements obtained in robot experiments and the efficiency of the whole experimentation process critically depends on the structure of the working environment. Tools currently available for simulation are far better developed than those used for experimenting with real robots. The real time interaction with the control process and the continuous visualisation of the parameters make possible a faster and better understanding of the mechanisms involved. For these reasons, it is necessary to develop better visualisation and interactive software tools adapted to the experimentation tasks.

The simplest way to implement a comprehensive graphical interface is to use a scientific workstation. This must be connected to the robot to collect the data for display and to communicate the orders coming from the user. The physical arrangement of all elements involved in the experiment must be compact, allowing a complete and comfortable control. Thanks to miniaturisation, this can be obtained as illustrated in figure 5: the entire configuration, including robot, environment and workstation, is conveniently arranged on a

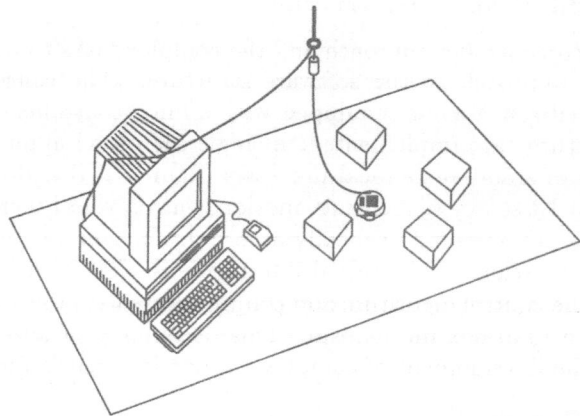

Figure 5. Khepera experimentation environment.

normal table. In the arrangement shown in figure 5 the serial link cable does not disturb the movement of the robot. A device to prevent the cable from rolling up is placed at mid-length on the serial cable. For experiments involving more than one robot, the wired serial link can no longer be used. Special radio communication modules are being developed for this purpose. This additional module will provide means to control several Khepera at the same time.

With a wired or a radio serial link, the data flow between workstation and robot must be as little as possible. To minimise this flow without restricting user ability, the control algorithm runs on the workstation and communicates to the robot only the data concerning sensors and motors. This configuration is optimal when an important number of parameters must be displayed and controlled.

Several programming and visualisation tools are used to perform experiments with Khepera. Here, three programming styles will be presented: the first uses a classical programming language to build stand-alone applications, the second a complete graphical programming interface and the third is a compromise between the previous two, making the best of both.

4.1. Complete applications

A first programming possibility is to code the control algorithm and the user interface in a traditional way. This is a good choice for software engineers or researchers who have already developed a visualisation and control interface, for instance in a simulation environment. It is often the case that, when a researcher starts to perform real robot experiments, a simulator has already been developed and used for preliminary studies. In this situations, the simulator can easily be adapted by replacing the simulated actions with calls to the interface with the real robot. This can usually be made without modifying the user interface.

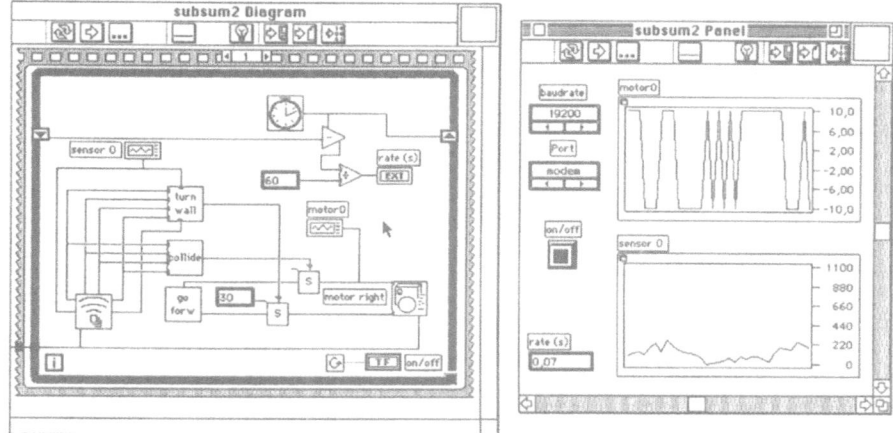

Figure 6. LabVIEW display.

Some very interesting results have been achieved with this approach on the neural networks simulator developed by Ph. Gaussier [14]. The simulator is used as a tool to design neural networks for robot control. A real time visualisation interface permits a verifiable step-by-step learning process on the robot.

A similar experience in interfacing Khepera with a simulator is in progress using the simulator BugWorld, developed at the Institute für Informatik of the University of Zürich. In this case, the control interface will be complemented with a measurement system which enables the user to plot the trajectory of the robot in real time on the host screen.

4.2. LabVIEW

The software package *LabVIEW* is a commercial product from National Instruments [15] and runs on several host machines (PC, Macintosh or Sun workstations). LabVIEW has been developed as an environment for the design of *virtual instruments* (VI). Every VI comprises a control panel and an interconnection diagram. On the panel, the user can interactively specify graphical devices for input (e.g., sliding cursors, buttons, text controls) and for output (e.g., gauges, images, graphs). In the diagram, the user graphically enters the functionality of the VI. A library of standard functionalities is available to perform this task: icons performing numerical functions, string treatment, matrix computations, etc. can be linked together to design an algorithm. An icon can be associated with a complete VI and used hierarchically in the diagram of another instrument, thus allowing a modular approach. Moreover, modules can be written in standard programming languages, such as C or Pascal.

An sample experiment is shown in figure 6. The diagram represents the computation of a subsumption-based algorithm [1]. Two modules, collide and

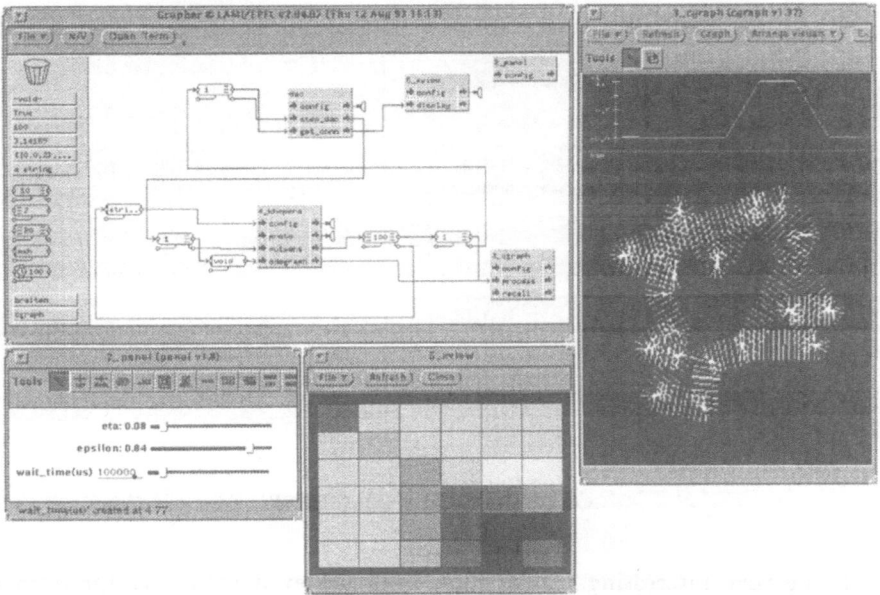

Figure 7. Grapher display.

turn wall, take inputs from the sensors (bottom left icon) and are connected to feed appropriate commands to the motors (right icon). The sensors and motor icons communicate with the robot through the wired serial link. The two graphs on the panel visualise the state of one motor and of one sensor. The modules in the top right part of the diagram evaluate the period required to recompute the algorithm; this is displayed at the bottom left of the panel.

LabVIEW is an optimal tool for the design of experimentation environments without the use of programming languages. The complete graphical interface helps specifying the interaction items of the panel but becomes inefficient when designing complex control algorithms. In this case, it is more efficient to design modules using standard programming languages. The only disadvantage of LabVIEW version 2 is that the display possibilities are somehow limited. The version 3 will provide more interactive capabilities and will be a better design tool for mobile robot experimentation.

4.3. Grapher

Grapher [16] is an experimentation tool developed at LAMI by Y. Cheneval and L. Tettoni for the Esprit Elena Project. Due to code optimisation and to improve performance, the software package is available only on SUN SparcStations. In the Grapher environment, an experiment is defined interconnecting modules that perform sub-tasks such as pure computation, visualisation or control. The programming of these modules is done in C language. The inter-

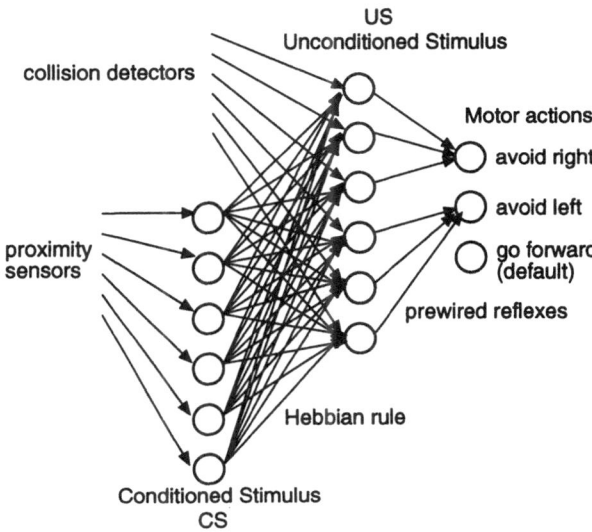

Figure 8. Distributed Adaptive Control experiment architecture.

connections are graphically specified by the user. The wiring diagram is on a single level, therefore preventing a hierarchical approach. For this reason, and to avoid over-complicated wiring schemes, the modules perform quite complex tasks. To facilitate the development, a large number of standard modules are available, making the best of the available hardware possibilities; the performance and flexibility of the visualisation is particularly impressive. Comparing Grapher to LabVIEW, the former is less intuitive, needs some programming knowledge but make complex experimentation efficient. The experiment described in the next section illustrates this aspect.

5. Experimentation in Distributed Adaptive Control

As an example of the development techniques outlined above, the environment to evaluate a control architecture will be presented in this section. The control mechanism is developed according to the design methodology of *distributed adaptive control* [17]. This approach is in turn derived from a distributed self-organising model of the behavioural phenomenon of classical conditioning [18] [19]. The example involves an autonomous agent that can learn to avoid obstacles using collision and proximity sensors.

The control structure consists of a neural net with three groups of neurons (figure 8) named *Unconditioned Stimulus* (US), *Conditioned Stimulus* (CS) and *Motor actions* (M). Neurons of the US group are directly connected with collision sensors, simulated here by the saturation of the proximity sensors. A prewired connection between the US and the M groups implements the robot basic reflex of avoiding obstacles at the time of a collision. Neurons of the CS group obtain their inputs from the proximity sensors. The learning is performed

512

with an Hebbian rule on the connections between CS and US. The weights $K_{i,j}$ of these connections are updated according to:

$$\Delta K_{i,j} = \frac{1}{N}\left(\eta \cdot s_i s_j - \epsilon \cdot \bar{s} \cdot K_{i,j}\right) \qquad (1)$$

where N defines the number of units in the CS, η is the learning rate, ϵ the decay rate, and $\bar{s}$ the average activation in the group US.

This way, the robot can develop a conditional response, learning to avoid obstacles using the proximity sensors without producing collisions. During the experimentation it is interesting to observe the evolution of the learning process on the matrix K, which depends on η and ϵ.

The software environment used for this experiment is Grapher (see section 4.3). Figure 7 shows the experiment display on a SUN SparcStation. The principal window, on the top left, illustrates the functional diagram of the experiment: The dac module (centre top) performs the computation of the algorithm and interacts with the module khepera (centre bottom) to control the robot. Three other modules permit user interface. The panel module (top right) allows the user to control η, ϵ and the algorithm computation period by means of sliding cursors. The xview module displays the K matrix in the centre bottom window. Finally, the cgraph module (bottom right) displays the sensor state and the trajectory of the robot, as visible in the rightmost window.

If the control algorithm C source with no display capabilities is available, the experimental set-up can be designed in less than one day. The user gains complete control of the physical robot, the development environment and all the parameters of the algorithm in real time, thus obtaining an optimal visualisation of the process.

6. Conclusions

The miniaturisation of Khepera makes a compact and efficient experimentation environment possible. Associated with effective software tools, this robot is an optimal platform to test control algorithms. The modularity at the hardware, software and control tools levels gives to the user the necessary flexibility to perform accurate experiments quickly. An example of experimentation environment has been presented. The reduced size and cost of the miniature robots described make possible experimentation on collective behaviour among groups of robots. This will be the main research activity in the near future.

Acknowledgements

The authors would like to thank André Guignard for the important work in the design of Khepera and Jelena Godjevac, Paul Verschure, Claude Touzet, Philippe Gaussier, Stéphane Zrehen, Yves Cheneval and Laurent Tettoni for help in testing the algorithms and in the development of the experimentation tools. This work has been supported by the Swiss National Research Foundation (project PNR23).

References

[1] R. A. Brooks. A robust layered control system for a mobile robot. *IEEE Robotics and Automation*, RA-2:14–23, March 1986.

[2] J. Heller. Kollisionsvermeidung mit fuzzy-logic. *Elektronik*, 3:89–91, 1992.

[3] U. Nehmzov and T. Smithers. Using motor actions for location recognition. In F. J. Varela and P. Bourgine, editors, *Proceedings of the First European Conference on Artificial Life*, pages 96–104, Paris, 1991. MIT Press.

[4] N. Franceschini, J.-M. Pichon, and C. Blanes. Real time visuomotor control: From flies to robots. In *Proceedings of the Fifth International Conference on Advanced Robotics*, pages 91–95, Pisa, June 1991.

[5] R. D. Beer, H. J. Chiel, R. D. Quinn, K. S. Espenschied, and P. Larsson. A distributed neural network architecture for hexapod robot locomotion. *Neural Computation*, 4:356–65, 1992.

[6] J. C. Deneubourg, S. Goss, N. Franks, A. Sendova, A. Franks, C. Detrin, and L. Chatier. The dynamics of collective sorting: Robot-like ant and ant-like robot. In J. A. Mayer and S. W. Wilson, editors, *Simulation of Adaptive Behavior: From Animals to Animats*, pages 356–365. MIT Press, 1991.

[7] R. A. Brooks. Intelligence without representation. *Artificial Intelligence*, 47:139–59, 1991.

[8] F. Mondada and P. F. M. J. Verschure. Modeling system-environment interaction: The complementary roles of simulations and real world artifacts. In *Proceedings of the Second European Conference on Artificial Life*, Brussels, 1993.

[9] R. A. Brooks. Elephants don't play chess. *Robotics and Autonomous Systems*, 6:3–15, 1990. Special issue.

[10] L. Steels. Building agents out of autonomous behavior systems. In *The Biology and Technology of Intelligent Autonomous Agents*. NATO Advanced Study Institute, Trento, 1993. Lecture Notes.

[11] Nomadic Technologies, Inc., Palo Alto, Calif. *The NOMAD Robot*. Data-sheet.

[12] P. Dario, R. Valleggi, M. C. Carrozza, M. C. Montesi, and M. Cocco. Microactuators for microrobots: A critical survey. *Journal of Micromechanics and Microengineering*, 2:141–57, 1992.

[13] E. Franzi. Low level BIOS of minirobot Khepera. Internal report R93.28, LAMI - EPFL, Lausanne, 1993.

[14] P. Gaussier. *Simulation d'un système visuel comprenant plusieurs aires corticales: Application à l'analyse de scènes*. PhD thesis, Paris XI - Orsay, Paris, November 1992.

[15] National Instruments Corporation. *LabVIEW 2*, January 1990. User Manual.

[16] Y. Cheneval, P. Bovey, and P. Demartines. Task B2: Unified Graphic Environment. Delivrable R1-B2-P, ESPRIT Elena Basic Research Project no. 6891, June 1993.

[17] P. F. M. J. Verschure, B. J. A. Koese, and R. Pfeifer. Distributed adaptive control: The self-organization of structured behavior. *Robotics and Autonomous Agents*, 9:181–96, 1992.

[18] P. F. M. J. Verschure and A. C. C. Coolen. Adaptive fields: Distributed representations of classically conditioned associations. *Network*, 2:189–206, 1991.

[19] I. P. Pavlov. *Conditioned Reflexes*. Oxford University Press, London, 1927.

Experimental Investigations of Sensor-Based Surface Following Tasks by a Mobile Manipulator

David B. Reister, Michael A. Unseren, James E. Baker,
and François G. Pin

Autonomous Robotic Systems Group
Oak Ridge National Laboratory
P.O. Box 2008
Oak Ridge, TN 37831-6364

Abstract — This paper discusses experimental investigations of the
feasibility and requirements of simultaneous external-sensor-based-control
of the wheeled platform and the manipulator of a mobile robot. The
experiments involve 3-D arbitrary surface following by the manipulator while
the platform moves along a predefined trajectory.

A variety of concave and convex surfaces were used in the experiments,
during which target and measured values of the platform and arm positions
and orientations, together with the surface absolute location and normal
estimates, were logged at 10 Hz. For all experiments, the data logs showed
significant noise, at high frequency, in the calculated surface normal values
despite smooth tracking of their target values by the arm and the platform,
with typical closed loop delays between target and achieved values of the order
of 100 msec. This high-frequency noise in the calculated values is conjectured
to result mainly from the arm's transmission cables compliance and backlash
in the spherical wrist gears. On the other hand, the end-effector distance to
the surface showed some low frequency errors of the order of $\pm20\%$. The two
major sources of these low frequency errors appeared to reside respectively
in the low values of the velocity bound and gain parameters utilized to filter
the high frequency noise in the calculated normal values prior to using them
as input to the arm control, and in the rolling contact of the platform's
rubber-coated wheels on the ground where significant errors in the platform's
positions and orientations can accumulate.

1. Introduction

Mobile manipulators, i.e., manipulators mounted on mobile platforms,
are attracting significant interest in the industrial, military, and public
service communities because of the significant increase in task capabilities
and efficiency which results from their large-scale mobility combined
with manipulation abilities. When the platform and manipulator move
simultaneously, the motion planning and control of the two subsystems
cannot be fully decoupled; in particular the position estimations for trajectory
tracking derived at loop rate from internal joint sensors need to be carefully
integrated: while instantaneous position estimates are direct functions of
the joint sensor data for the manipulator subsystem, they are functionals
of the wheel rotation and steering sensor data for the platform, typically
requiring time integration of these values over the entire trajectory history
due to the non-holonomic nature of the wheeled system. As a result, motion
accuracies of human-size platforms are generally two or three orders of
magnitude worse than the accuracies of common manipulators. This paper

discusses experiments performed with the HERMIES-III autonomous mobile manipulator, investigating the feasibility and requirements of simultaneous external-sensor-driven platform and redundant manipulator motions, and the interaction of the two subsystems' internal sensor-based position estimation. The experiments involved precise following of arbitrary 3-D surfaces by the end-effector of the redundant manipulator while the platform performs trajectory tracking along a predefined path.

2. Experimental Configuration

HERMIES-III [1],[2] is a human-size mobile manipulator test-bed (see Fig. 1) incorporating the seven degree-of-freedom (d.o.f.) CESARm research manipulator on a four d.o.f. platform (with two independently-driven and steerable wheels). The software and hardware configuration of HERMIES-III utilizes the HELIX communication protocol [3], allowing for a fully distributed and message passing modular control system on several connected VMEbus racks. The platform's control drivers and sensor-feedback odometry modules operate at 20 Hz. The redundancy resolution module for the CESARm includes 3-D position and 3-D orientation control and utilizes a minimum Euclidean norm-based algorithm running at 50 Hz on a 68020 processor [4]. The forward kinematics calculations run at 150 Hz on another 68020 processor.

For the experiments, a very accurate (.05 mm precision), single point LED triangulation-type range finder [5],[6] was held in the gripper of the CESARm (see Fig. 1). The platform was assigned to perform a specified trajectory on the floor, while the CESARm end-effector's task was to follow an *a priori* unknown surface, maintaining both constant distance and constant orientation from the surface using the range finder data. Because the LED range sensor is unidirectional and provides data from a single beam, estimation of the surface normal (necessary to maintain constant orientation) required an estimate of the absolute displacement of the measurement point on the surface expressed in the reference (or world) coordinate system. This estimation therefore required propagation of position and orientation estimates through the entire platform-arm-gripper-sensor-beam chain.

2.1. Surface Calculation

The distance sensor is located on the end-effector of the CESARm that is mounted on the mobile platform of HERMIES-III. The goal is to measure the curvature of an arbitrary surface and keep the sensor at a fixed distance from the surface and normal to the surface. To measure the curvature of the surface, we must measure points on the surface in the world coordinate system. The system has three coordinate systems: world, platform, and arm (see Fig. 2). In this subsection we show how the measured data is used to calculate a point on the surface in each of the coordinate systems.

Our objective is to calculate a point on the surface of an object in the world coordinate system (x_s, y_s). In the experiments dealt with here, we assume that the shape of the object does not depend on the z coordinate (e.g., that the unknown object is a cylinder). Thus, all of our calculations will be in 2D geometry. The measured data are the distance to the surface (D), the configuration of the last link of the manipulator measured in the arm coordinate system (x_b, y_b, θ_b), and the location of the platform in the world coordinate system (x_p, y_p, ϕ_p). The arm is controlled in 3D space $(x, y, z,$ roll, pitch, yaw) and θ_b is the measured yaw angle.

516

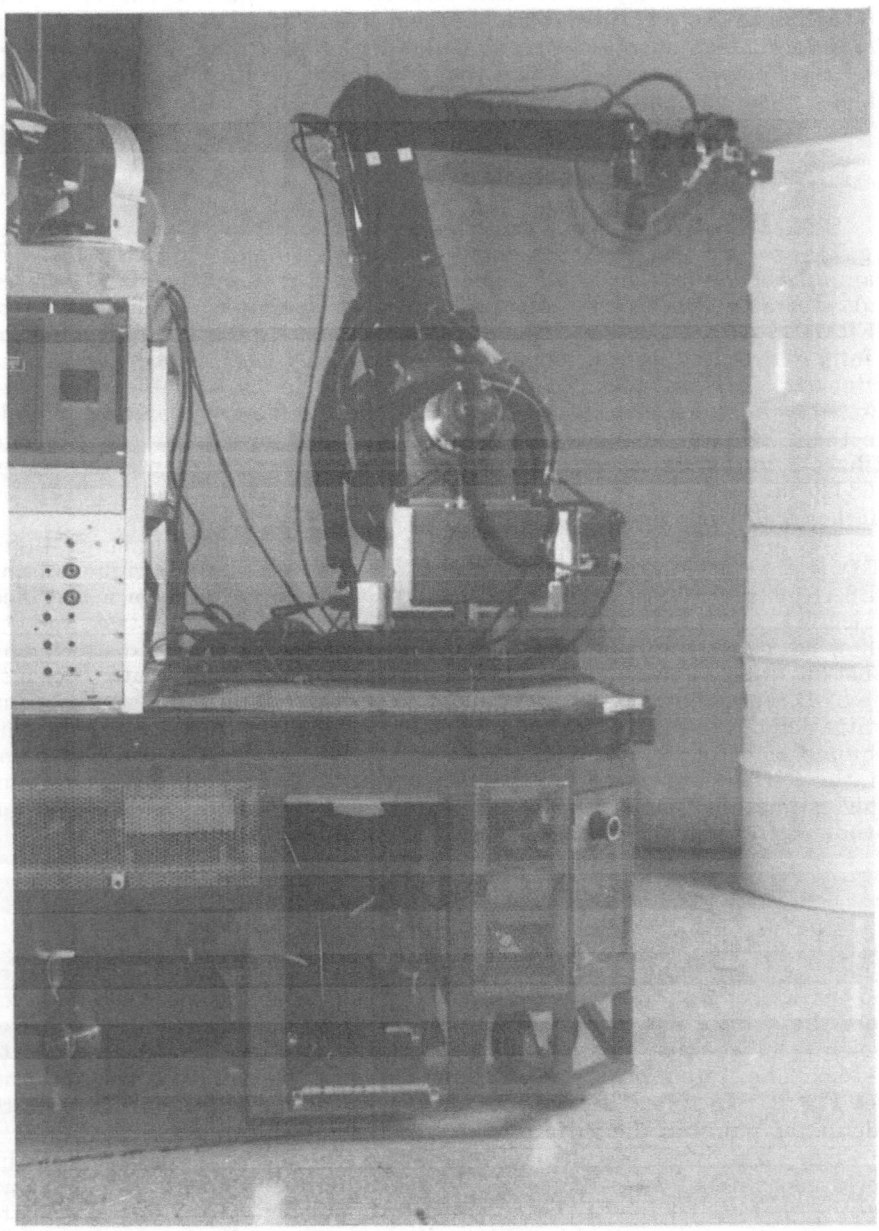

Fig. 1. The HERMIES-III robot follows an arbitrary unknown surface.

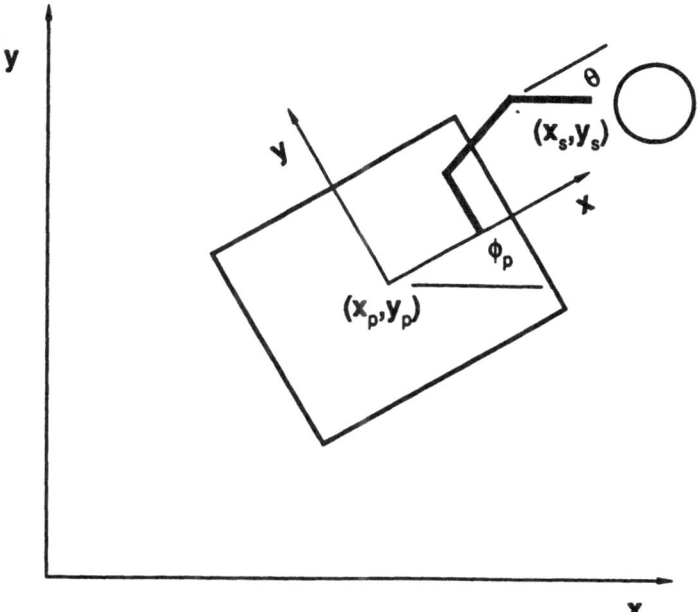

Fig.2. The experimental configuration.

The base of the arm is attached to the platform. Thus, the transformation between the arm coordinates and the platform coordinates does not vary. Unfortunately, the transformation requires both a translation and a rotation of 180 degrees. If (x_h, y_h, θ_h) is the configuration of the last link of the manipulator measured in the platform coordinate system:

$$x_h = -x_b + B \tag{1}$$

$$y_h = -y_b \tag{2}$$

$$\theta_h = \theta_b - \pi/2 \tag{3}$$

where B is the x coordinate of the arm base in the platform coordinate system (B = 0.574 meters). In the home position for the arm, $(x_b, y_b, \theta_b) = $ (-0.880, -0.356, $\pi/2$) and $(x_h, y_h, \theta_h) = $ (1.454, 0.365, 0.0).

To reach the surface of the object in the platform coordinate system (x_d, y_d), we travel in the θ_h direction by the sum of the length of the distance sensor ($L = 0.062$ meters) and the measured distance (D):

$$x_d = x_h + (D + L)\cos\theta_h \tag{4}$$

$$y_d = y_h + (D + L)\sin\theta_h \tag{5}$$

To calculate a point (x_s, y_s) on the surface of the object in the world coordinate system, we transform the coordinates from the platform reference frame to the world reference frame as follows:

$$x_s = x_p + x_d \cos\phi_p - y_d \sin\phi_p \tag{6}$$

$$y_s = y_p + x_d \sin \phi_p + y_d \cos \phi_p \qquad (7)$$

Note that we have used all of the measured data $[D, (x_b, y_b, \theta_p)$, and $(x_p, y_p, \phi_p]$ to calculate the point location on the surface (x_s, y_s).

2.2. Surface Normal Calculation

The location of the points on the surface are calculated at 100 Hz (100 Hz is the nominal rate. The measured rate is about 75 Hz). At 10 Hz, we would like to calculate the surface normal and the arm goal. To calculate the surface normal, we fit a polynomial to the surface points and calculate the slope of the curve. There are tradeoffs in choosing the order of the polynomial and the number of data points to use to estimate the parameters. A higher order polynomial (cubic or quadratic) has more parameters and requires more data points to estimate the parameters. Furthermore, a high order polynomial might not provide a good fit to a surface with discontinuous surface normals (e.g., a box). Our goal is to obtain the best estimate of the surface normal in a small neighborhood of the currently measured point on the surface. Thus, we would like to fit a curve with a small number of parameters using a small number of data points. We decided to fit a line using M points, where M is a user-defined parameter that can be adjusted to improve performance.

We have performed two series of experiments: the first in 1992 and the second in 1993. During the first series of experiments, we performed ten distance measurements each cycle (at 100 Hz) and averaged them. Using the average distance, we calculated a point on the surface of the object. Every tenth cycle (at 10 Hz), we fitted a line to the 10 data points. Thus, $M = 10$ for the first series of experiments.

During the second series of experiments, we performed four distance measurements each cycle. For each of the four measurements, we calculated a point on the surface of the object. Every tenth cycle (at 10 Hz), we fitted a line to N sets of 40 data points (for the data in this paper, $N = 5$). Thus, $M = 40N = 200$ for the second series of experiments.

Given M points on the surface, we determine the maximum and minimum values for each of their coordinates, x and y. If the spread in x is greater than the spread in y, we assume that $y = f(x)$. Otherwise, we assume that $x = f(y)$. The maximum speed for the platform is 0.45 meters/second. Thus, the maximum distance traveled in 0.1 seconds is 0.045 meters. If the maximum spread is less than 0.003 N meters, we do not calculate the surface normal. Otherwise, we use least squares to fit a line to the data points.

Let ψ be the normal to the line and let θ_s be the surface normal in the platform coordinate system, then:

$$\theta_s = \psi - \phi_p \qquad (8)$$

Typical experimental values for the surface normal (θ_s) for the first series of experiments are displayed in Fig. 3.

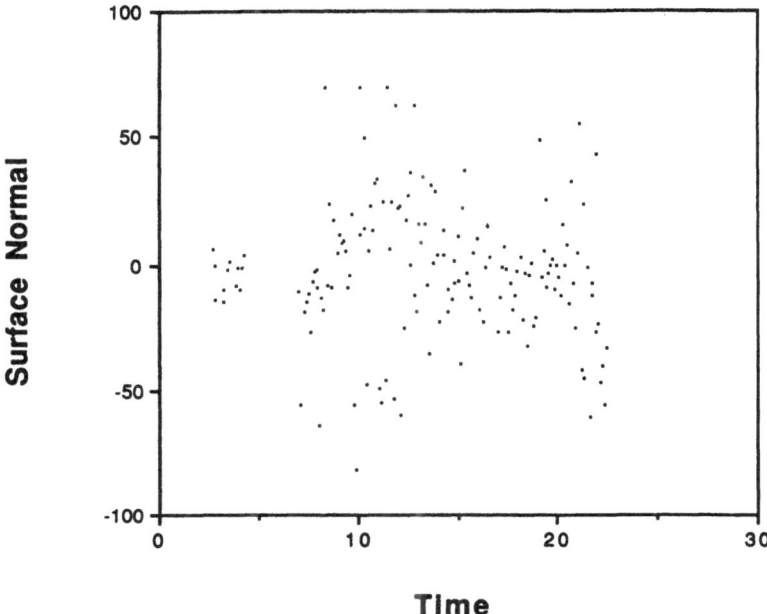

Time

Fig. 3. Calculated values for the surface normal for the first experiment. The units are degrees.

It is clear that the values displayed in Fig. 3 are much too variable or noisy to be fed as input target to the manipulator. When the input to the arm has high frequency noise, the arm will oscillate. To provide a smooth input to the arm, we filtered the calculated values of the surface normal (θ_s). Let θ_g be a running average of the calculated values for the surface normal:

$$\theta_g(t+1) = (1 - \mu)\theta_g(t) + \mu\theta_s(t) \tag{9}$$

When the parameter μ is 1.0, the average value is equal to the input ($\theta_g = \theta_s$). When the parameter μ is 0.0, the average is a constant that does not depend on the input. For the first series of experiments, $\mu = 0.1$. For the second series of experiments, $\mu = 0.5$.

Let θ_c be the target value for the orientation of the arm. We let the target track the average value of the surface normal. Let ϵ be the difference between the target and the average value: $\epsilon = \theta_g - \theta_c$. We limited the allowable rate of change ϵ to a given value δ. Thus, if $\epsilon > \delta$, then $\epsilon = \delta$ and if $\epsilon < -\delta$, then $\epsilon = -\delta$. Finally:

$$\theta_c(t+1) = \theta_c(t) + \epsilon \tag{10}$$

For the first series of experiments, $\delta = 0.01$. For the second series of experiments, $\delta = 0.03$. Thus, the four parameters (M, N, μ, and δ) can be used to smooth the time varying input to the arm.

2.3. Arm Goal Position Calculation

The calculation of a goal for the arm is illustrated in Fig. 4. The current orientation of the arm is θ. In the current position, the schematic follows the arm from the wrist (w), to the hand (h), and past the surface detector (d)

to the surface (s). The figure also displays the goal configuration of the arm (at orientation θ_c). Given the orientation, the goal for the hand (x_c, y_c) is calculated as:

$$x_c = x_d - (D_g + L)\cos\theta_c \tag{11}$$

$$y_c = y_d - (D_g + L)\sin\theta_c \tag{12}$$

where the desired value for the distance is D_g = 0.102 meters in the experiments described here.

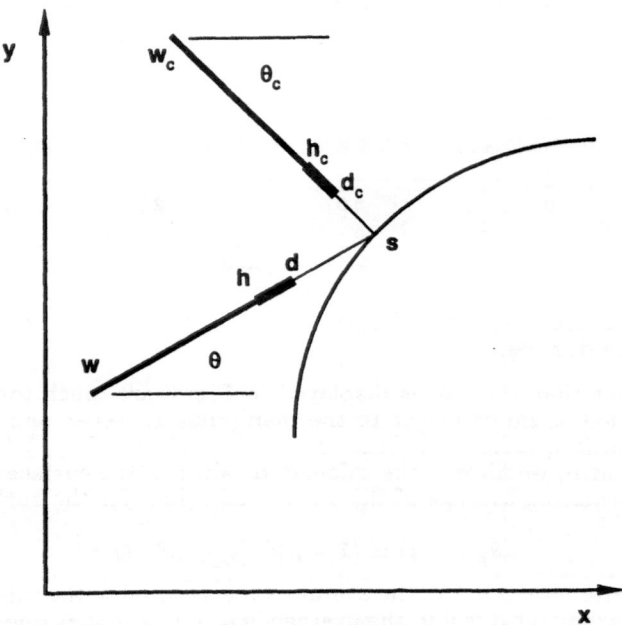

Fig. 4. The calculation of the next goal for the arm.

3. Experimental Results and Discussion

The data displayed in this section (and in Fig. 3) were collected during two experiments (the first during the 1992 series and the second during the 1993 series). In the first experiment, the arm began in the center of a curved surface. During the first experiment, the platform made a linear motion to the right, made a linear motion to the left past the starting point, and made a linear motion to the right to the starting point. In the second experiment, the platform moved in an arc about a barrel and returned to its initial location.

Figure 5 displays the distance measurements from the range finder during the first experiment. The goal is to maintain a distance of 0.102 meters. Clearly, the measured values show that the end-effector can be more than two centimeters from the goal. The low frequency of the curve would seem to indicate that an error accumulation takes place over time in the calculational system resulting in a gradual drift of some of the components from their target values.

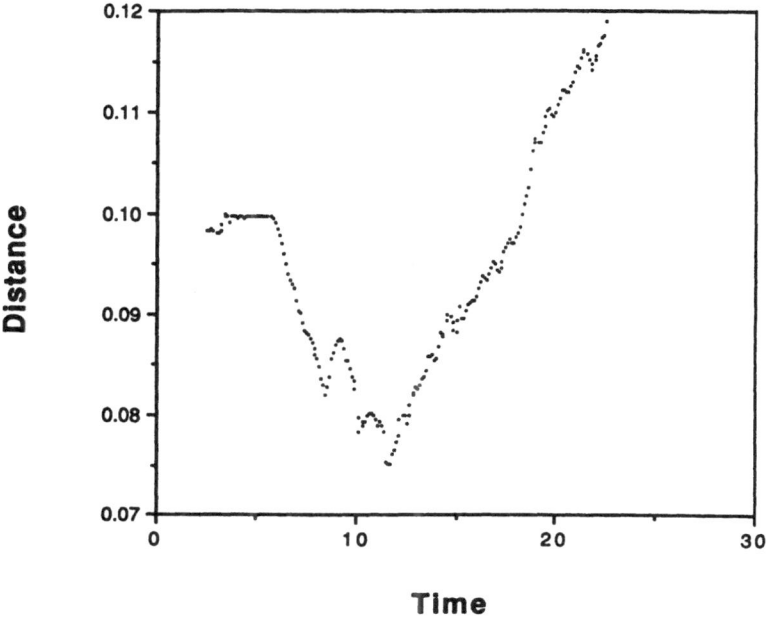

Fig. 5. The distance from the range finder sensor to the unknown surface for the first experiment. The units are meters.

To explore the sources of the low frequency errors in Fig. 5, we will examine each of the potential errors in the calculation. The range finder produces valid readings within 4 centimeters of the desired distance. Since all of the distance measurements in Fig. 5 are within 0.04 meters of the desired distance, the distance measurements are valid.

Figure 6 is a plot of the points (x_s, y_s) on the surface of the barrel calculated in the world reference frame from integration of the measured data over the entire platform-arm-sensor chain. There is significant noise in the data and both the low and high frequency errors can be observed. The very large scattering in the surface normal results which were observed in Fig. 3 correlate with the high frequency variations observed on Fig. 6.

A large part of the low frequency errors, however, cannot be corrected through any type of filtering. Although the reckoner for the platform can produce slight integration errors, the major sources of the large observed errors in Fig. 6 are thought to reside in the rolling contact of the platform's wheel on the floor. When moving along a circle, with the two driving wheels steered at different angles, the effective point of contact of the wide rubber-coated wheels varies slightly under the wheels with slight irregularities on the floor. This generates errors in the wheel velocity targets that are calculated to fulfill the rigid body constraint (which exists between the wheels in this configuration). From these errors, slight slippage of the wheels on the floor results, progressively accumulating to large errors, *undetectable with the odometry sensors.*

In a first step toward remedying this problem, a composite control architecture which accommodates for violations of the interwheel rigid body kinematic constraints in addition to controlling the position of the platform wheel system has recently been developed [7] and has shown dramatic improvements on the platform's control accuracy when used on HERMIES-III.

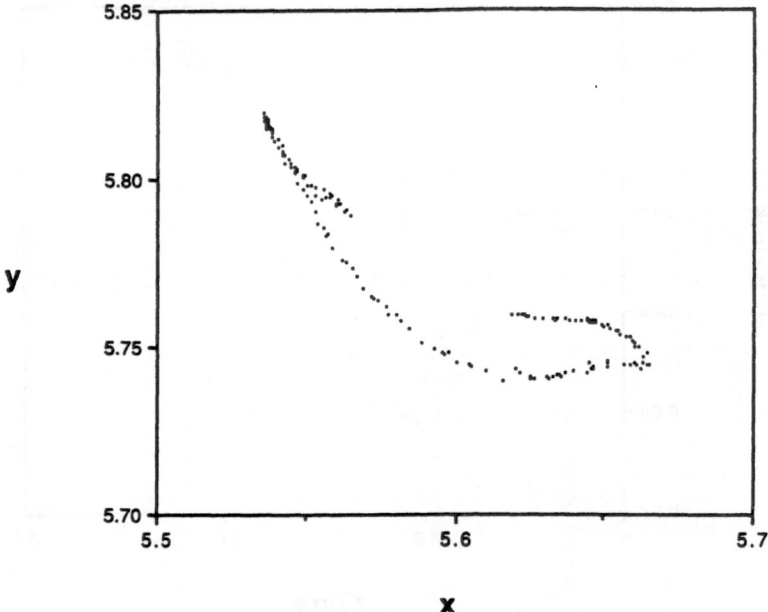

Fig. 6. The calculated position of the unknown surface in the world coordinate system for the first experiment. The units of x and y are meters.

The manipulator arm was also investigated as the possible source of some of the error; however, the manipulator system was consistently following the target values for both position and orientation as calculated from Eqs. (1) to (12). As an example of this, Fig. 7 displays a time log of the target $(\theta_c + \pi/2)$ and the measured values (θ_b) for the manipulator yaw angle. The targets move smoothly and the measured values follow the targets accurately, with an expected lag in execution of the order of one to two cycles of the 10 Hz calculational scheme. The smooth motion of the yaw targets demonstrates that the filtering parameter (μ) and the velocity bound (δ) produce a smooth signal from the noisy input data in Fig. 3. The time delay between the targets (set points) and the execution, therefore, cannot explain either the low frequency divergences exhibited in Fig. 5 or the large scattering displayed in Fig. 3.

To explore the sources of the low frequency errors in Fig. 5, we have examined each of the potential errors in the calculation. Calculation of points on the surface in world coordinates depend on the distance sensor, the arm configuration, and the platform configuration. While the points in Fig. 6 have low frequency errors in world coordinates, they have continuous values in the arm coordinate system that is required for the calculation in Fig. 4. For any value of the target yaw angle, the measured distance should be equal to the target distance. We conjecture that we have errors in our values for the Denavit-Hartenburg parameters.

As mentioned, we have performed two series of experiments: the first in 1992 and the second in 1993. In the first experimental series, conservative values were chosen for the parameters with the objective to produce a robust system.

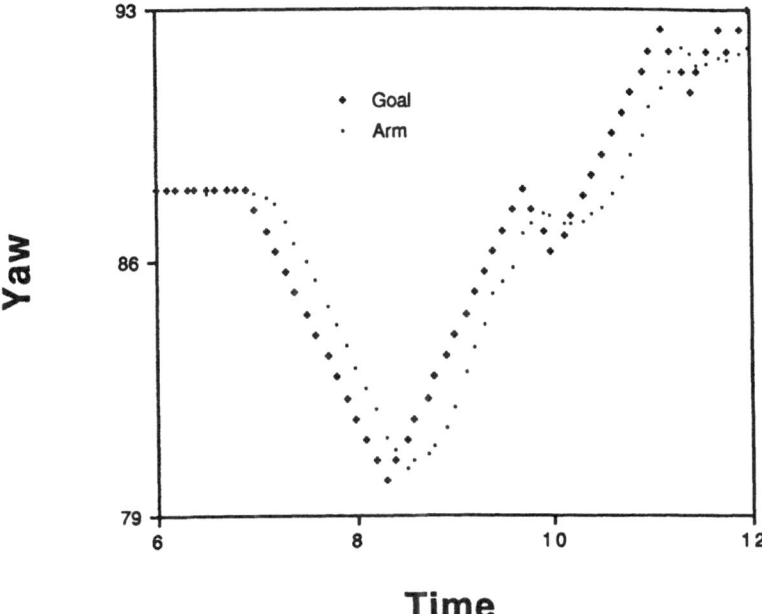

Fig. 7. Target and measured values for the yaw angle for the first experiment. The units are degrees.

The motivation for the second series was provided by a quest for the possible sources of errors encountered during the first series, and experimental investigations of the data provided by the range finder. For another project, we used the range finder to measure the distance from a platform to the floor and observed unexpected discontinuities in the data. We performed a series of experiments to investigate the behavior of the Analog to Digital board. We increased the precision of the output from the board and concluded that we should not average the output from the board. With new confidence in the data from the range finder, we decided to initiate the second series of experiments.

To improve our understanding of the sources of the scatter in the surface normal data in Fig. 3, we collected all of the points on the surface of the barrel for 100 cycles. Typical results are displayed in Fig. 8. Each cycle, we performed four distance measurements and calculated a point on the surface of the barrel. Thus, 400 data points are displayed in the figure. The nominal time for 100 cycles is one second. The actual time to collect the data in Fig. 8 was 1.34 seconds (about 75 Hz). The circles in Fig. 8 are the data that was collected every tenth cycle.

For the first series of experiments, we determined the best line for the data between each pair of circles and obtained the scattered results in Fig. 3. For the second series of experiments, we determined the best line for the data between six circles and obtained the results in Fig. 9. Each line depends on 200 data points. For the next calculation of surface normal, we drop 40 data points and add 40 data points. This method produced much more continuous results than our previous method. It is clear from Fig. 9 that the surface normal calculation does not exhibit the very large scattering observed in Fig. 3.

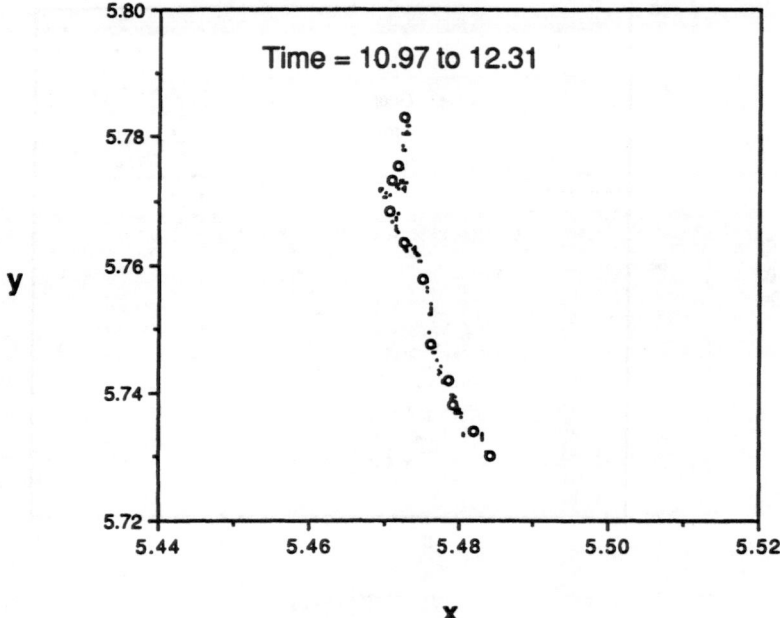

Fig. 8. All of the calculated points on the surface of the barrel in the world coordinate system for 100 cycles. The data identified by a circle were collected every tenth cycle. The units of *x* and *y* are meters.

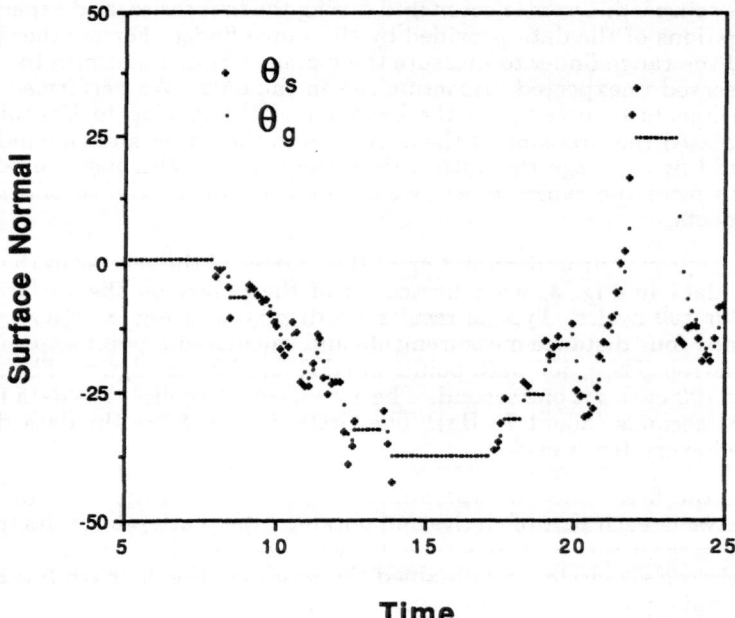

Fig. 9. Calculated values for the surface normal and their average values for the second experiment. The units are degrees.

As mentioned previously, for each set of 200 data points, we determine the spread in x and y. If the maximum spread is less than 15 mm, we do not calculate the surface normal. For the first seven seconds of the data in Fig. 9, the platform is not moving enough to generate sufficient spread in the data and we do not calculate the surface normal. The platform circles the barrel, stops, and returns. Similarly, during the period when the platform stops and reverses direction (from 14 seconds to 17 seconds on Fig. 9), the platform is not moving enough to generate sufficient spread in the data.

Figure 10 is a plot of the points on the surface of the barrel. The data in Fig. 10 are much more consistent than the data in Fig. 6. The surface measurements begin at the lower right, follow the barrel to the upper left, and return. The last thirty points (out of 200) are displayed with a new plotting symbol (a small plus) and show the greatest deviation from the initial curve. For the last 30 points, the platform has stopped moving. Thus, the deviations may be due to faulty values for the Denavit-Hartenberg parameters.

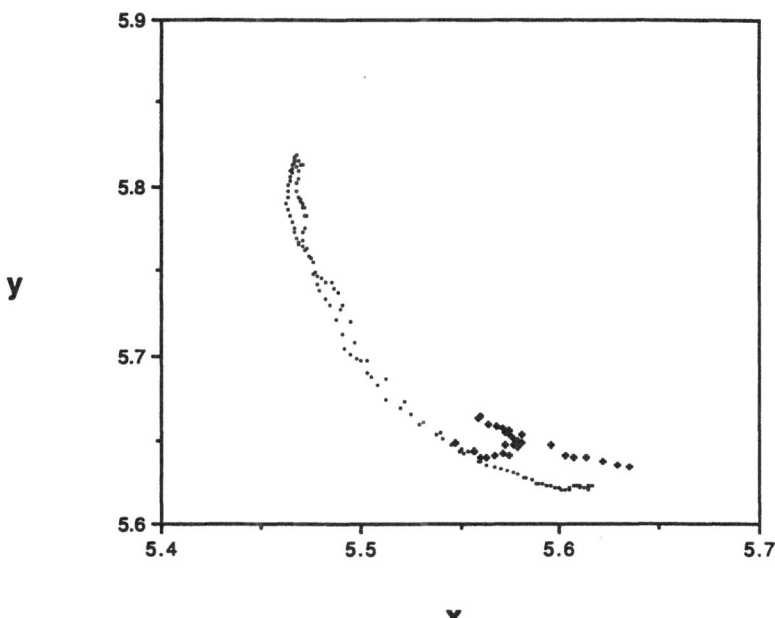

Fig. 10. The calculated position of the unknown surface in the world coordinate system for the second experiment. The units of x and y are meters.

Figure 11 displays a plot of the target and measured values for the yaw angle. The changes in yaw angle are much larger than in Fig. 7. In Fig. 7, the values of the yaw angle are within 10 degrees of the initial value (90 degrees). In Fig. 11, the values decrease from 90 degrees to about 53 degrees before increasing to 102 degrees. For the second series of experiments, the velocity bound (δ) was increased by a factor of 3 (from $\delta = 0.01$ to $\delta = 0.03$). The improved response is visible during the decrease in yaw angle in Fig. 11. Several times, the target catches the running average and stops decreasing.

We have also used two approaches to reduce the low frequency errors observed in Fig. 5: feedback from the distance sensor to adjust the parameter L in Eqs. (11) and (12), and feedback from the distance sensor to directly adjust the position goal for the hand. Neither approach has made large reductions in the errors. We are continuing our efforts to reduce the errors.

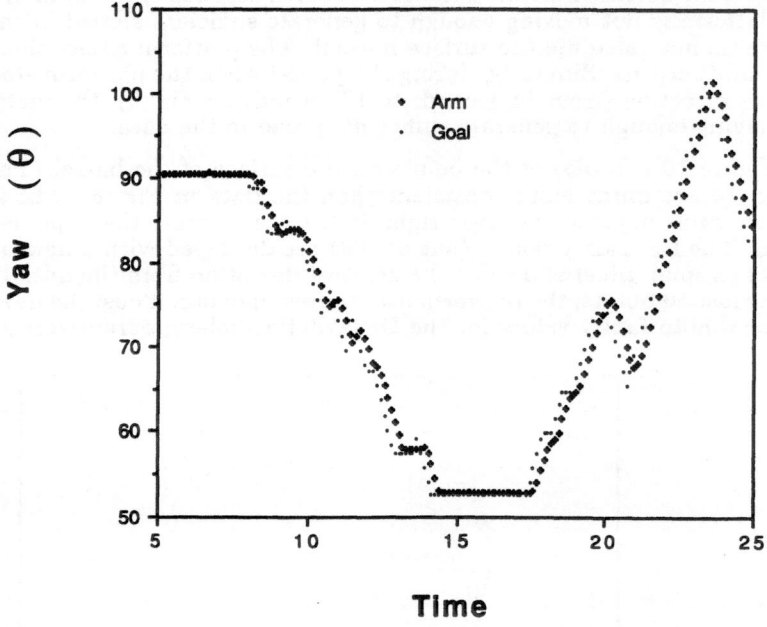

Fig. 11. Target and measured values for the yaw angle for the second experiment. The units are degrees.

4. Concluding Remarks

We have performed a series of surface-following experiments with a mobile manipulator. Each successive experiment has included improvements to reduce some unexpected errors which were observed in the initial tests. We have implemented several methods which have helped to improve the results of Fig. 3. However, much more improvement still seems to be needed on which we are focussing our efforts. Our results demonstrate the importance of experimental robotics. All of our calculations worked in simulation. To calculate points on the surface of an object, we require data from a range detector, a manipulator, and the platform. The errors and uncertainties that exist in each of these subsystems can propagate and accumulate in such ways as to produce results like those displayed in Fig. 3, while also being undetectable by the internal sensors and encoders. No simulator could display the effects of these errors and provide insight into their cause and potential remedies. We are using our continuing experiments to further our understanding of these error problems.

5. Acknowledgment

This research was performed under sponsorship of the Engineering Research Program, Office of Basic Energy Sciences, of the U.S. Department of Energy, under contract DE-AC05-84OR21400 with Martin Marietta Energy Systems, Inc.

References

[1] F. G. Pin et al., "Autonomous Mobile Robot Research Using the HERMIES-III Robot," Proceedings of the IEEE RSJ International Workshop on Intelligent Robots and Systems, IROS '89, September 4–6, 1989, Tsukuba, Japan, 251–256.

[2] D. B. Reister, "A New Wheel Control System for the Omnidirectional HERMIES-III Robot," *Robotica* 10, 351–360 (1992).

[3] J. P. Jones et al., "Hetero Helix: Synchronous and Asynchronous Control Systems in Heterogeneous Distributed Networks," *Robotics and Autonomous Systems* 10(2–3), 85–99 (1992).

[4] M. A. Unseren, "Input Relegation Control for Gross Motion of a Kinematically Redundant Manipulator," Oak Ridge National Laboratory Technical Report No. ORNL/TM-12165, October 1992.

[5] J. E. Baker, "Empirical Characterization of a High Intensity Proximity Sensor," *Sensors* 10(7), 29–32 (1993).

[6] J. E. Baker, "Terrain Following and Mapping of Arbitrary Surfaces Using High Precision Proximity Data," *Optical Engineering* 32(5), 1117–1124 (1993).

[7] D. B. Reister and M. A. Unseren, "Position and Constraint Force Control of a Vehicle with Two or More Steerable Drive Wheels," accepted for publication in *IEEE Transactions on Robotics and Automation*.

Section 9
Space Robotics and Flexible Manipulators

The major issues discussed in this section consist of a control aspect of flexible manipulators and key technologies for space robots. The control of robotic manipulators with flexible structural links as their arms is one of the most important problems to be considered especially for space robotic applications. A number of other key technologies must also be integrated and evaluated experimentally to build dexterous space robots.

Konno, Uchiyama, Kito, and Murakami discuss the controllability of the structural vibration of flexible manipulators. It is shown that a spatial flexible manipulator may have some vibration uncontrollable configurations and the 'modal accessibility' concept helps us to understand their physical interpretation.

The paper by Lucibello, Panzieri, and Ulivi addresses a state steering problem, that is, to move the robot between two given equilibrium points defined by assigned positions of its end-effector. It proposes a learning algorithm that belongs to the category of parameters learning and is applicable to flexible manipulators. The robustness of the learning algorithm with respect to unmodelled high-frequency dynamics is also discussed.

The paper by Yoshida presents the experimental research activities on space robotics with the EFFORTS simulators that can mechanically simulate the planar floating dynamics. It describes the experimental results on target capture operations, collision experiments to confirm the idea for impact dynamics representation with the Extended Inversed Inertia Tensor, and force control with soft compliance.

The paper by Hirzinger describes the key technologies developed for the space robot technology experiment ROTEX that flew with shuttle flight STS 55 end of April 93. It is shown that multisensory gripper technology, local (shared autonomy) sensory feedback control concepts, and the powerful delay-compensating 3D-graphics simulation (predictive simulation) in the telerobotic ground station have led to the success of ROTEX.

Configuration-Dependent Controllability of Flexible Manipulators

A. Konno, M. Uchiyama, Y. Kito and M. Murakami

Department of Aeronautics and Space Engineering
Tohoku University
Aramaki-aza-Aoba, Aoba-ku, Sendai 980, Japan

Abstract— In this paper, we discuss the controllability of the structural vibration of flexible manipulators. In the spatial flexible manipulator, some parameters change depending upon the arm's configuration. Ease of vibration suppression is configuration-dependent. We have defined the controllability of the structural vibration of flexible manipulators in our previous study. In order to understand the physical interpretation of vibration uncontrollable configurations, we propose the "modal accessibility" concept which indicates how well the actuators can affect the mode. The configuration in which all actuators are unable to affect at least one of the vibration modes of the manipulator is vibration uncontrollable. To clarify this point, we performed experiments. Experimental results show that all actuators can't suppress at least one of the structural vibration modes in the vibration uncontrollable configurations.

1. Introduction

A great deal of effort has been put in the control of robotic manipulators with flexible (elastic) structural members as their arms. Light-weight manipulators are expected to realize quick motion in industrial applications. For space robotic applications, such manipulators have an advantage of low transportation cost when sent the space from the earth. However, such light-weight manipulators have some problems because of their structural flexibility. One of the problems of such light-weight manipulators is the vibrations that occur due to its structural flexibility. Such light-weight manipulators which have structural flexibility are called as "flexible manipulators."

The research on the vibration control strategy has been enthusiastically pursued for a long time. What seems to be lacking, however, is a discussion on structural vibration controllability. Some researchers have analyzed the controllability of structural vibration of light-weight manipulators [1, 2]. However, little is known about physical interpretation of vibration uncontrollable configurations.

In the spatial flexible manipulator, some parameters change depending upon the arm's configuration. Controllability of the structural vibration is also configuration-dependent. When we calculated the optimal gains for vibration suppressing control on various configuration of the 2-link 3-joint type experimental spatial flexible manipulator in our laboratory using the optimal regulator technique, we got discontinuous gains in some of the configurations.

On further research, we found that some vibrations were uncontrollable in the configurations corresponding to the discontinuous optimal gains. Since such configurations were found to lie in the manipulator's work space, at which, on first impression, they seemed to be controllable, we were very much interested in exploring them further.

In this paper, we discuss the controllability of the structural vibration of flexible manipulators. In order to understand the physical interpretation of vibration uncontrollable configurations we applied the modal analysis technique to the flexible manipulator system and found that at the vibration uncontrollable configurations, actuators cannot affect some modes. We will use the term "modal accessibility" to refer to the concept of how well the actuators can affect the mode. We consider a 2-link 3-joint type flexible manipulator as an example, since this type of flexible manipulator has some interesting vibration uncontrollable configurations that we mentioned above. To study these configuration, we performed experiments with the 2-link 3-joint type experimental flexible manipulator. Experimental results show that all actuators are unable to affect at least one of the vibration modes of the manipulator in the vibration uncontrollable configurations. This help us to understand the "modal accessibility" concept.

2. Dynamics of flexible manipulator

Regardless of modeling approach, equations of motion of flexible manipulators can be written as follows:

$$\tau = M_{11}\ddot{\theta} + M_{12}\ddot{e} + h_1(\theta, \dot{\theta}, e, \dot{e}) + g_1(\theta, e) \tag{1}$$

$$0 = M_{21}\ddot{\theta} + M_{22}\ddot{e} + Ke + h_2(\theta, \dot{\theta}, e, \dot{e}) + g_2(\theta, e) \tag{2}$$

where, $\theta \in \mathbf{R}^{n \times 1}$ is the joint angle vector, $e \in \mathbf{R}^{m \times 1}$ is the deflection vector, $M_{11} \in \mathbf{R}^{n \times n}$, $M_{12} \in \mathbf{R}^{n \times m}$, $M_{21} \in \mathbf{R}^{m \times n}$, $M_{22} \in \mathbf{R}^{m \times m}$ are inertia matrices $h_1 \in \mathbf{R}^{n \times 1}$, $h_2 \in \mathbf{R}^{m \times 1}$ are centrifugal and Corioli's forces, $g_1 \in \mathbf{R}^{n \times 1}$, $g_2 \in \mathbf{R}^{m \times 1}$ are gravity matrices, $K \in \mathbf{R}^{m \times m}$ is the stiffness matrix and τ is the joint torque vector.

In the equation of motion, two distinct parts can be recognized as Eq. (1) and (2). Eq. (1) is related to the overall motion of the system while Eq. (2) is related to the elastic motion.

3. Controllability of vibrations

3.1. Controllability matrix

In the previous section, we derived dynamic equations of flexible manipulators as Eq. (1) and (2). Let us analyze Eq. (2) which is related to the elastic motion of the flexible manipulator to study the controllability of structural vibrations.

In case $\ddot{\theta} = \dot{\theta} = 0$ and $\ddot{e} = \dot{e} = 0$ in Eq. (2), we get

$$Ke_o + g_2(\theta, e_o) = 0, \tag{3}$$

where e_o stands for the static deflection of the links by gravity force. To delete the gravity term in Eq. (2), we consider the elastic deflection from e_o expressed by

$$\Delta e = e - e_o. \tag{4}$$

In this discussion, we assume that the centrifugal and Corioli's term can be ignored and $g_2(\theta, e)$ can be approximated as $g_2(\theta, e_o)$. Incorporating this assumption in Eqs. (2) and (4) we get

$$M_{22}\Delta\ddot{e} + K\Delta e = -M_{21}\ddot{\theta}. \tag{5}$$

Eq. (5) is the equation of elastic motion around e_o. In Eq. (5) $-M_{21}\ddot{\theta}$ can be regarded as the input vector to the system, therefore, the question is whether we can suppress vibrations (elastic motions) by appropriate $\ddot{\theta}$. Since M_{21} depends upon θ, e is not always controllable by $\ddot{\theta}$.

Although joint torques are regarded as inputs to the manipulator, as joint torques produce joint accelerations, we can consider joint accelerations as inputs to the elastic motion subsystem. Next we discuss the controllability of structural vibrations defined in [1] regarding $\ddot{\theta}$ as the control input to the elastic motion subsystem.

From Eq. (5), we obtain the state equation by

$$\begin{bmatrix} \Delta\dot{e} \\ \Delta\ddot{e} \end{bmatrix} = \begin{bmatrix} 0 & I \\ -M_{22}^{-1}K & 0 \end{bmatrix} \begin{bmatrix} \Delta e \\ \Delta\dot{e} \end{bmatrix} + \begin{bmatrix} 0 \\ -M_{22}^{-1}M_{21} \end{bmatrix} \ddot{\theta}, \tag{6}$$

which is rewritten as

$$\dot{x} = A(\theta, e)x + B(\theta, e)\ddot{\theta}, \tag{7}$$

where $x = [\Delta e^T \quad \Delta\dot{e}^T]^T$. $A \in \mathbf{R}^{2m \times 2m}$, $B \in \mathbf{R}^{2m \times n}$ and $x \in \mathbf{R}^{2m \times 1}$. M_{21} and M_{22} are functions of θ and e, usually A and B depend upon θ and e. Controllability matrix also depends upon θ and e, and is given by

$$G_c(\theta, e) = \begin{bmatrix} B & AB & \cdots & A^{2m-1}B \end{bmatrix}. \tag{8}$$

If

$$\text{rank } G_c < 2m, \tag{9}$$

some structural vibrations are uncontrollable.

As mentioned above, the controllability of structural vibrations of spatial flexible manipulators are configuration-dependent. Therefore, it is indispensable to calculate configuration matrix at all the configurations of the manipulator in order to investigate the controllability of vibration.

We would like to find the configuration of the manipulator at which the vibration become uncontrollable. From controllability matrix described in Eq. (8), there is no way to tell what the physical interpretation of the vibration uncontrollable configurations is. Thus to clearly understand these configurations, we apply the modal analyses technique to the elastic motion equation.

3.2. Modal accessibility

In order to introduce the modal accessibility concept, we transform Eq. (5) into a set of modal equations. To this purpose we apply the eigenvalue problem. Eq. (5) can be rewritten as follows:

$$\Delta \ddot{e} + M_{22}^{-1} K \Delta e = -M_{22}^{-1} M_{21} \ddot{\theta}. \tag{10}$$

Then we consider the diagonalization of $M_{22}^{-1} K$. The eigenvalue problem can be written as follows:

$$M_{22}^{-1} K \Phi_r = \Omega_r \Phi_r \quad (r = 1, \cdots, m). \tag{11}$$

For simplicity, we consider Ω_r to be distinct i.e. $\Omega_r \neq \Omega_s$ for $r \neq s$. The eigenvectors Φ_r given by Eq. (11) can be arranged in the $m \times m$ square matrix

$$\Phi = [\Phi_1, \cdots, \Phi_m]. \tag{12}$$

Using the square matrix Φ, we consider

$$\Delta e^* = \Phi^{-1} \Delta e. \tag{13}$$

Matrix Φ depends on θ and e, but now we are considering the small motions around constant values of θ and e, so we can write

$$\Delta \ddot{e}^* = \Phi^{-1} \Delta \ddot{e}. \tag{14}$$

Substituting Eqs. (13) and (14) into Eq. (10), we get a set of modal equations

$$\ddot{e}^* = \Omega e^* + \Gamma \ddot{\theta}, \tag{15}$$

where

$$\Omega = -\Phi^{-1} M_{22}^{-1} K \Phi = -\mathrm{diag}[\Omega_1, \cdots, \Omega_m],$$

$$\Gamma = -\Phi^{-1} M_{22}^{-1} M_{21}.$$

It is obvious from Eq. (15) that if Γ has some rows which consist of only 0s, the vibration modes corresponding to these rows are uncontrollable.

We use the term "modal accessibility" to refer to the concept of how well the actuators can affect the vibration mode. This concept is similar to the mode controllability concept for the control of flexible structures whose controllability doesn't depend on their configuration [3]. Modal accessibility concept is also similar to the "accessibility" concept of how well the actuators can affect the structural vibrations of flexible manipulators [2], but vibration modes are not taken into account in the computation of "accessibility." Our concept is the extension of the "accessibility" concept by taking vibration modes also into account. Thus, we use the term "modal accessibility" [1]. Let us define "modal accessibility" by $\sqrt{\gamma_r \gamma_r^T}$ where γ_r is the rth row vector of Γ.

---

[1] "modal affectibility" may be better to explain this concept.

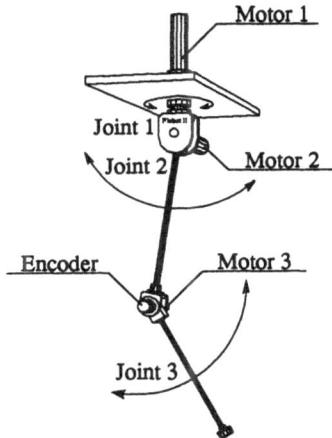

Figure 1. Overview of the experimental flexible manipulator FLEBOT II.

Next we discuss the relationship between modal accessibility and controllability of vibration. Eq. (15) can be cast in the state form as

$$\left[\begin{array}{c} \Delta \dot{e}^* \\ \Delta \ddot{e}^* \end{array} \right] = \left[\begin{array}{cc} 0 & I \\ \Omega & 0 \end{array} \right] \left[\begin{array}{c} \Delta e^* \\ \Delta \dot{e}^* \end{array} \right] + \left[\begin{array}{c} 0 \\ \Gamma \end{array} \right] \ddot{\theta}. \tag{16}$$

Since Eqs. (6) and (16) describe the same physical system, we can discuss the vibration controllability of the system by Eq. (16). Hence, the vibration controllability matrix can be rewritten as follows:

$$G_c^* = \left[\begin{array}{ccccccc} 0 & \Gamma & 0 & \Omega \Gamma & \cdots & \Omega^{2m-1} \Gamma \\ \Gamma & 0 & \Omega \Gamma & 0 & \cdots & 0 \end{array} \right] \tag{17}$$

Matrix Ω is a diagonal matrix, implying that Ω^j ($j = 2, \cdots, 2m-1$) is also a diagonal matrix. It is obvious that if Γ has rows which consists of only 0s, G_c^* cannot be of full rank. On the other hand, in case G_c^* is not of full rank, at least one row vector of Γ consists of only 0s.

In case $M_{22}^{-1} K$ has multiple eigenvalues, mode unaccessible configurations are not always the same as the vibration uncontrollable configuration.

4. An example with experiments

To simplify our discussion, we consider the horizontal movement of our experimental flexible manipulator called FLEBOT II with 2 elastic links and 3 actuated joints (Figure 1) as an example.

4.1. Experimental setup

FLEBOT II is driven by DC servo motors with velocity feedback. The tip deflections of each link of this manipulator are measured by strain gauges located

Table 1. Mechanical parameters.

Parameter	Notation	Value
Length of links	l_2 [m]	0.50
	l_3 [m]	0.44
Stiffness of links	E_2I_2 [Nm^2]	41.81
	E_3I_3 [Nm^2]	13.23
Mass of joints	m_2 [kg]	1.5
	m_3 [kg]	0.3

at the root of each link assuming that the links are massless slender beams. Mechanical parameters of this manipulator are presented in Table 1.

Since this manipulator is driven by velocity commands, we develop the vibration suppressing control scheme for the velocity commanded flexible manipulators [1] and apply this control scheme to FLEBOT II.

4.2. Modeling of FLEBOT II

In the modeling of flexible structure, there are some approximate method like Rayleigh's method, Ritz's method, Galerkin's method, assumed-mode method, Holzer's method, Myklestad's method, finite element method etc. [4]. In Holzer's method, flexible strictures are modeled by lumped-mass and massless elastic bodies. The lumped-mass and the massless elastic body are called as "station" and "field", respectively. We apply a technique which is similar to the Holzer's model to construct a lumped-mass and spring model for spatial flexible manipulators [5]. Joints, stations and fields can possibly be located at the same point. For example, assuming that a field is located at the same point as that of the joint, we can construct a model of the manipulator which has elasticity only at the reduction gears of the joint.

FLEBOT II is modeled as shown in Figure 2. Joint 1, station 1 and joint 2 are located at the same point while station 2 and joint 3 are at the same point. Horizontal deflections of field 2 and field 3 are denoted as δ_{z2} and δ_{z3} respectively (Figure 2). The elastic angular deflection ϕ_{y2} around y_2 axis can also be considered as a member of the deflection vector, but if the moments of inertia of the stations are small enough to be ignored, ϕ_{y2} depends upon δ_{z2} and δ_{z3}. For the clarity, we ignore the moments of inertia of the stations. δ_{z2} and δ_{z3} are controlled by joint 1.

Assuming that the elastic deflections of the links are small, we can separate the equations of horizontal motion from the equations of the whole motion of this manipulator as follows:

$$\tau_h = M_{11h}\ddot{\theta}_h + M_{12h}\ddot{e}_h \tag{18}$$

$$0_{2\times1} = M_{21h}\ddot{\theta}_h + M_{22h}\ddot{e}_h + K_h e_h. \tag{19}$$

where

$$\tau_h = \begin{bmatrix} \tau_1 \end{bmatrix}, \quad \theta_h = \begin{bmatrix} \theta_1 \end{bmatrix}, \quad e_h = \begin{bmatrix} \delta_{z2} & \delta_{z3} \end{bmatrix}^T$$

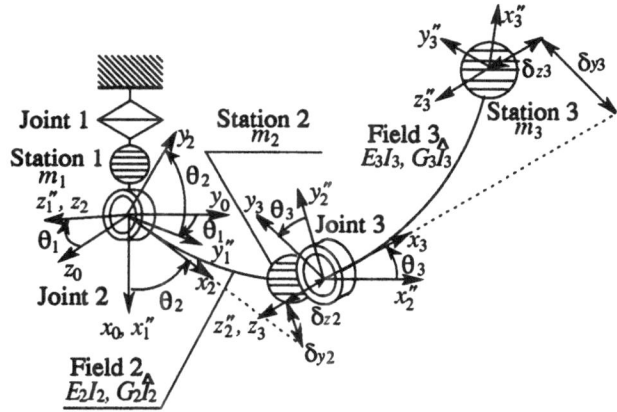

Figure 2. Lumped-mass and spring model of FLEBOT II.

4.3. Vibration controllability

Eq. (19) can be cast in the state form as

$$\left[\begin{array}{c} \Delta \dot{e}_h \\ \Delta \ddot{e}_h \end{array} \right] = \left[\begin{array}{cc} 0 & I \\ -M_{22h}^{-1} K_h & 0 \end{array} \right] \left[\begin{array}{c} \Delta e_h \\ \Delta \dot{e}_h \end{array} \right] + \left[\begin{array}{c} 0 \\ -M_{22h}^{-1} M_{21h} \end{array} \right] \ddot{\theta}_h. \qquad (20)$$

which is rewritten as

$$\dot{x}_h = A_h(\theta, e)x_h + B_h(\theta, e)\ddot{\theta}_h. \qquad (21)$$

Using Eq. (21) we evaluate the controllability matrix at various configurations and the minimum singular values of the controllability matrix are plotted in Figure 3. The configurations in which minimum singular value of the controllability matrix is 0 are vibration uncontrollable configurations.

4.4. Modal accessibility

In Figure 3, we see many vibration uncontrollable configurations lying in the manipulator's work space. To understand the physical interpretation of these configurations we study modal accessibility of this manipulator in this section. Since $\Phi^{-1} M_{22h}^{-1} M_{21h}$ is a 2×1 matrix, the absolute value of the 1st row and the 2nd row are the modal accessibility of the 1st and the 2nd mode of vibrations respectively.

Figure 4 shows vibration mode shapes of horizontal elastic motion of FLE-BOT II. Modal accessibility of these vibration modes at various configurations are plotted in Figures 5 and 6. Comparing Figures 5 and 6 with Figure 3, we can see the relationship between the mode unaccessible configurations and the vibration uncontrollable configurations. In these vibration uncontrollable configurations all actuators are unable to affect 1st vibration mode or 2nd vibration mode. Of course real flexible manipulators have infinite vibration modes, and

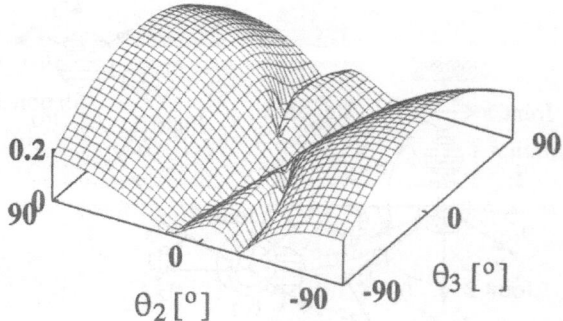

Figure 3. Minimum singular value of controllability matrix.

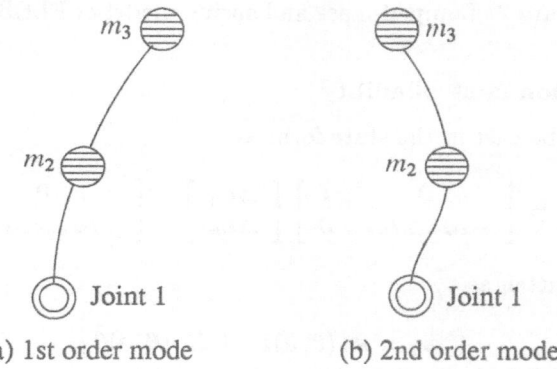

(a) 1st order mode   (b) 2nd order mode

Figure 4. Vibration modes.

may have other higher mode unaccessible configurations. It depends upon the model.

4.5. Vibration control strategy of FLEBOT II

In this section, we describe our control strategy as proposed in [1]. As we have mentioned above, we regard $\ddot{\boldsymbol{\theta}}$ as the input to the elastic motion subsystem. So using the equations of elastic motion, we calculate the appropriate joint acceleration $\ddot{\boldsymbol{\theta}}_e$ which suppresses the structural vibration of the manipulator.

To calculate $\ddot{\boldsymbol{\theta}}_e$ for vibration suppression, we use an optimal regulator technique assuming a performance index as follows:

$$J = \frac{1}{2} \int_0^t (\boldsymbol{x}_h^T \boldsymbol{Q} \boldsymbol{x}_h + \ddot{\boldsymbol{\theta}}_h^T \boldsymbol{R} \ddot{\boldsymbol{\theta}}_h) dt. \tag{22}$$

$\ddot{\boldsymbol{\theta}}_e$ to optimize J is given by

$$\ddot{\boldsymbol{\theta}}_e = -\boldsymbol{R}^{-1} \boldsymbol{B}^T \boldsymbol{P}(t) \boldsymbol{x} \tag{23}$$

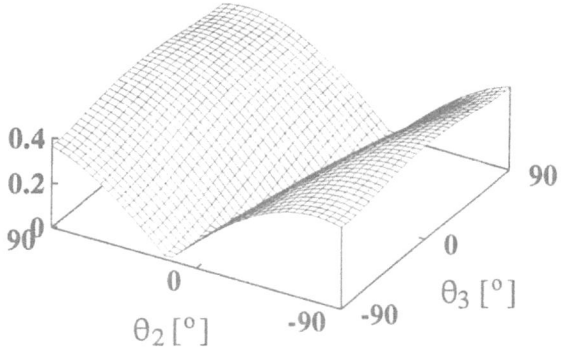

Figure 5. Modal accessibility of 1st vibration mode.

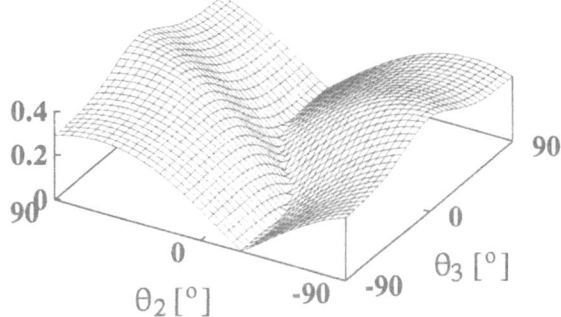

Figure 6. Modal accessibility of 2nd vibration mode.

where $P(t)$ is a solution of the Riccati equation:

$$\frac{d}{dt}P(t) = -A^T P(t) - P(t)A + P(t)BR^{-1}B^T P(t) - Q. \qquad (24)$$

We have calculated the optimal steady state gains for $t = \infty$ at various configurations with $Q = \text{diag}[800\ 800\ 800\ 800]$ and $R = 1$ as shown in Figure 7. Since the matrices A_h and B_h in Eq. (21) are functions of θ and e, the optimal steady state gains $R^{-1}B^T P$ in Eq. (23) are also functions of θ and e. As we see in Figure 7(a)~(d), we get discontinuous gains at vibration uncontrollable configurations shown in Figure 3.

Each joint of this manipulator is driven by a high-gain velocity servo. Therefore, it is difficult to command the servo to follow a specified torque. Instead, the servo follows the velocity command accurately. Therefore, we assume that

$$V_{ref} = K_{sv}G_r\dot{\theta} \qquad (25)$$

where V_{ref} is the reference voltage vector to the velocity servo, K_{sv} is a diagonal matrix whose diagonal elements are the voltage/velocity conversion

540

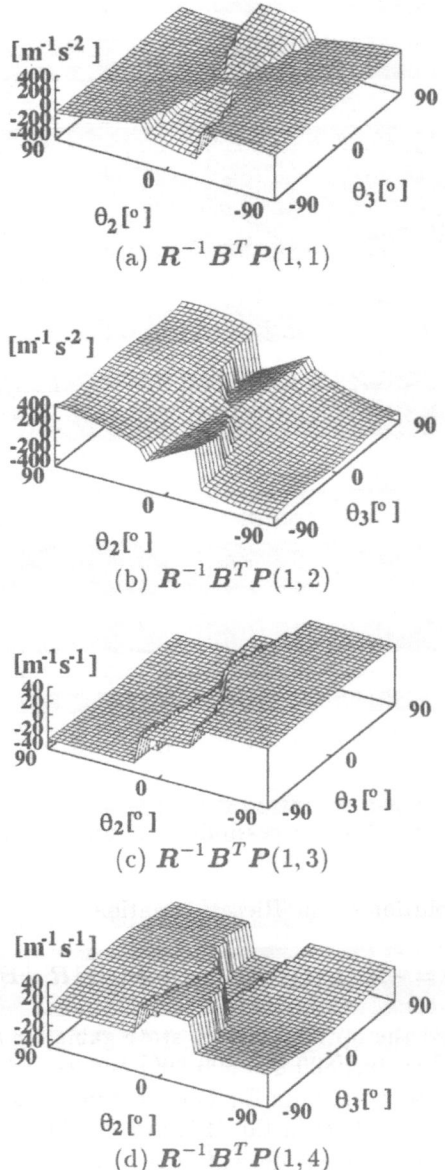

(a) $R^{-1}B^{T}P(1,1)$

(b) $R^{-1}B^{T}P(1,2)$

(c) $R^{-1}B^{T}P(1,3)$

(d) $R^{-1}B^{T}P(1,4)$

Figure 7. Optimal gain table for vibration control.

coefficients of the sensors for velocity feedback in the servo, and G_r is a diagonal matrix whose diagonal elements are the gear ratios of the servo. The joint acceleration $\ddot{\boldsymbol{\theta}}_e$ for vibration suppression is realized by the voltage references

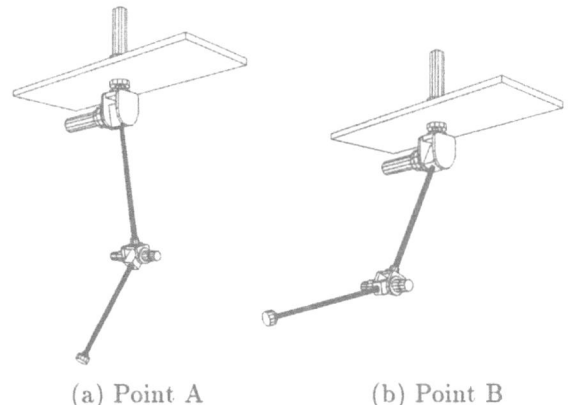

(a) Point A   (b) Point B

Figure 8. Singular configurations.

given by

$$V_{ref} = G_r K_{sv} \int \ddot{\theta}_e \, dt \tag{26}$$

To achieve harmony between the vibration suppression and the joint motion control, we combine the two velocity commands for vibration suppression and joint motion control into one. Finally, we calculate the velocity command V_{ref} by

$$V_{ref} = G_r K_{sv} \left\{ \dot{\theta}_{hd} + K_{vt}(\theta_{hd} - \theta_h) + \int \ddot{\theta}_e \, dt \right\} \tag{27}$$

where suffix d stands for the desired value.

4.6. Experimental results

We consider point A $(\theta_2, \theta_3) = (10°, -47°)$ (Figure 8(a)) and point B (θ_2, θ_3) $= (-25°, -52°)$ (Figure 8(b)) picked up from the 1st vibration mode unaccessible configurations and the 2nd vibration mode unaccessible configurations, respectively. As can be seen in Figure 7, we can't determine the optimal gains at the vibration uncontrollable configurations, thus, the points A and B are the neighborhoods of vibration uncontrollable configurations. At point A and point B, we command joint 1 to move in a continuous path as described by

$$\theta_{hd}(t) = \left(\frac{t}{T} - \frac{1}{2\pi} \sin \left(2\pi \frac{t}{T} \right) \right) (\theta_h(T) - \theta_h(0)) + \theta_h(0) \tag{28}$$

where t $(0 \leq t \leq T)$ is the time synchronized to zero at the beginning of the movement, T is the time to be spent from $\theta_h(0)$ to $\theta_h(T)$.

In the experiment we set $T = 1/3$ (s), $\theta_h(0) = 0°$, $K_{vt} = 4$ (s^{-1}) and sampling time 10 ms. $\theta_h(T)$ is 10° at point A and 15° at point B.

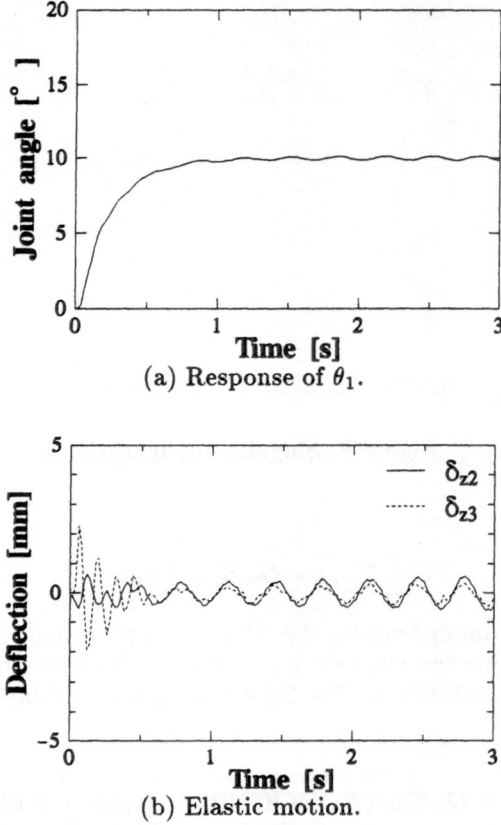

(a) Response of θ_1.

(b) Elastic motion.

Figure 9. Responses in a 1st mode unaccessible configuration (Point A).

Responses of θ_1 and elastic motion at point A and point B are plotted in Figures 9 and 10, respectively. At the point A, $\ddot{\theta}_1$ can neither produce nor affect the 1st vibration mode. Thus only 2nd vibration mode can be produced in the beginning (from 0 (s) to 0.5 (s) in Figure 9). Due to non-linearity and perhaps some other reasons, 1st vibration mode also appears later (from about 0.5 (s)). Since joint 1 can't affect the 1st vibration mode, the 1st vibration mode remains while the 2nd vibration mode is suppressed. On the contrary, at point B, we can see only the 1st vibration mode during the transient response and the 2nd mode vibration remains while the 1st vibration mode is suppressed.

5. Conclusion

We discussed the controllability of structural vibrations of a spatial flexible manipulator. We showed that the vibration controllability may depend upon the arm's configuration. Thus a spatial flexible manipulator may have some vibration uncontrollable configurations. We also propose the "modal acces-

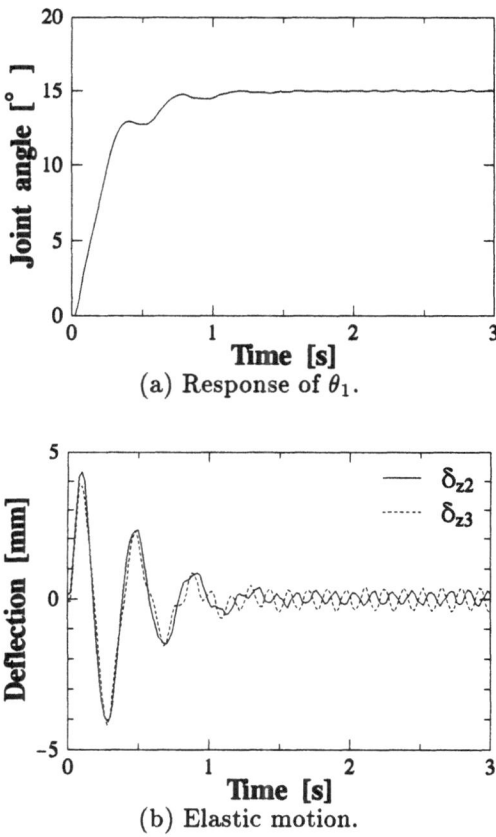

(a) Response of θ_1.

(b) Elastic motion.

Figure 10. Responses in a 2nd mode unaccessible configuration (Point B).

sibility" concept which helps us in understanding the physical interpretation of the vibration uncontrollable configurations. In the vibration uncontrollable configuration, actuators can't affect at least one of the vibration modes. Along with the physical interpretation of the vibration uncontrollable configurations the modal accessibility concept can also assign the order of priority for the vibration uncontrollable configurations. For example, if we assume structural damping to exist, first few modes of unaccessible configurations are much important than higher order mode unaccessible configurations and the modal accessibility map can clearly gives us the region corresponding to the first few modes.

We also considered an example and analyzed the vibration controllability and modal accessibility of the flexible manipulator. Experimental results helped us to understand the mode unaccessible configurations.

The authors express their gratitude to Dr. Praveen Bhatia for his kind help in writing this paper.

References

[1] M. Uchiyama and A. Konno. Computed Acceleration Control for Vibration Suppression of Flexible Robot Manipulators. *Proc. 5th Int. Conf. on Advanced Robotics*, pp. 126–131, 1991.

[2] S. Tosunoglu, S. H. Lin, and D. Tesar. Accessibility and Controllability of Flexible Robotic Manipulators. *Trans. of ASME, J. of Dynamic Systems, Measurement and Control*, 114:50–58, 1992.

[3] L. Meirovitch. *DYNAMICS AND CONTROL OF STRUCTURES*. John Wiley & Sons, Inc., 1990.

[4] L. Meirovitch. *Analytical Methods in Vibrations*. Macmillan Publishing Co., Inc., 1967.

[5] M. Uchiyama and A. Konno. Modeling of the Flexible Arm Dynamics Based on the Holzer's Method. *Proc. RSJ 1st Robotics Symposium*, pp. 247–252, 1991. (In Japanese).

Point to point learning control
of a two-link flexible arm

P. Lucibello   S. Panzieri   G. Ulivi

Dipartimento di Informatica e Sistemistica
Università degli Studi di Roma "La Sapienza"
Via Eudossiana, 18 – 00184 ROMA, ITALY

Abstract— *Repositioning by learning of a two-link flexible arm is dealt with. A finite dimensional algorithm is presented which trial by trial search for a control which moves the arm between two given configurations in a finite time interval. Effectiveness of these algorithms is proven both by theoretical arguments and experimental results. Robustness with respect to unmodeled dynamics is also addressed.*

1. Introduction

Output tracking by learning has received and is receiving a large attention in the literature. Interesting results have been achieved both for rigid manipulators [1, 2, 3, 4] and for robots having elasticity in the joints [5, 6, 7] or in the links [8, 9]. In most cases the theoretical studies have been validated by experimental results.

Other robotized tasks exist, where output tracking is not required. Typically, an elementary robot operation consists in moving the robot between two given equilibrium points, defined by assigned positions of its end effector. This kind of motion cannot be efficiently implemented by means of output tracking which puts unnecessary requirements. Indeed, suppose that a trajectory connecting the given equilibrium points is selected and that a learning algorithm is set up to track this trajectory. It may happen that during a trial the robot reaches the desired equilibrium point even if the selected trajectory is not exactly tracked. This would cause the update of the control for the next trial and the robot could no longer reach the desired equilibrium point at the end of the new trial.

The task described is to be classified as a state steering problem, for which ad hoc learning algorithms are needed, different from output tracking ones. On the basis of these motivations, the problem of steering the state of a control system by learning has been theoretically [10, 11, 12] investigated. These algorithms are based upon the selection of a finite dimensional subspace of the linear space of the control functions defined over a finite time interval. The search for the steering control is restricted to this subspace, thus building up a finite dimensional algorithm.

In this paper learning control of a nonlinear flexible arm is considered. An investigation on a two-link arm, with the second link flexible, is developed, showing that by using state feedback the system can be transformed in a perturbed linear one, for which a general theory was presented in [13]. We also prove that the learning algorithm proposed is robust not only with respect to sufficiently small nonlinear plant perturbation, but also with respect to unmodeled high frequency dynamics. This problem, known as spillover, was pointed out in [14] in case of output regulation of linear flexible structure for $t \to \infty$. In [3, 12] it was shown that a similar problem occurs also in case of learning control and that it can be overcome using suitable filters designed in a linear setting.

The implementation of the learning algorithm is detailed next for an experimental flexible arm built in the authors' laboratory. The experimental results, presented at the end of the paper, confirm the validity of the approach.

2. State steering by learning

Consider a linear control system

$$\dot{x}(t) = Ax(t) + Bu(t), \quad x(0) = x^\circ, \tag{1}$$

where $x(t) \in \mathbb{R}^n$ is the state and the control function $u(\cdot)$ belongs to the linear space U of piece-wise continuous function mapping the time interval $[0, t_f]$ onto $\mathbb{R}^m$, and a linear invertible map

$$C : \mathbb{R}^{n+m} \to \mathbb{R}^{n+m}. \tag{2}$$

Problem 1 Given the control system (1), the initial condition x° and the map C, find a piece-wise continuous control $u(t)$, $t \in [0, t_f]$, such that

$$C \begin{bmatrix} x(t_f) \\ u(t_f) \end{bmatrix} = \psi_d \in \mathbb{R}^{n+m}, \tag{3}$$

with ψ_d assigned. ‡

Definition 1 We refer to

$$\psi = C \begin{bmatrix} x(t_f) \\ u(t_f) \end{bmatrix} \tag{4}$$

as the system positioning at time t_f.

Proposition 1 Suppose that the pair (A, B) is controllable, then Problem 1 is solvable.

Proof

Dynamically extend (1) by

$$\dot{u}(t) = v(t), \quad u(0) = u^\circ, \tag{5}$$

with $v(t)$ a new control and u° arbitrary. Since integrators extension does not destroy controllability, the extended system is controllable. Hence, there exists a control $v(t)$ which steers the state of the extended system from the initial condition to the point

$$\begin{bmatrix} x_f \\ u_f \end{bmatrix} = C^{-1} \psi_d. \tag{6}$$

Integration of (5) finally provides a continuous solution to Problem 1.‡

If the plant is not exactly known in a direct way, the computation of a steering control is not possible. A learning strategy can in this case be applied in order to improve task performance trial by trial.

To this end, restrict the search for the steering control to the following type:

$$u(t) = \sum_{k=1}^{n+m} \alpha_k \phi_k(t), \quad \alpha_k \in I\!R, \tag{7}$$

where $\phi_k : [0, t_f] \in I\!R^m$ are piece-wise continuous functions.

At time t_f, one has

$$x(t_f) = e^{At_f} x^\circ + \sum_{k=1}^{n+m} \alpha_k \int_0^{t_f} e^{A(t_f - t)} B\phi_k(t)dt,$$

$$u(t_f) = \sum_{k=1}^{n+m} \alpha_k \phi_k(t_f) \tag{8}$$

These equations are rewritten in matrix form as:

$$\begin{bmatrix} x(t_f) \\ u(t_f) \end{bmatrix} = \begin{bmatrix} e^{At_f} x^\circ \\ 0 \end{bmatrix} + W\chi, \quad \chi = \begin{bmatrix} \alpha_1 \\ \vdots \\ \alpha_{n+m} \end{bmatrix}. \tag{9}$$

Clearly, if W is invertible a steering control of the type (7), which solves Problem 1, exists. The reader may refer to [15] for examples of linear functions ϕ_k such that W is invertible under the hypothesis of plant controllability.

Equation (9) is the basis on which a learning algorithm can be developed. Suppose that the system is exactly initialized at each trial and that, at the $j - th$ trial, the control function is given by

$$u^{[j]}(t) = \sum_{k=1}^{n+m} \alpha_k^{[j]} \phi_k(t), \tag{10}$$

where $\alpha_k^{[j]}$ denotes the value of the parameter α_k used in the $j - th$ trial. In general, a superscript $[j]$ identifies the function or the parameter used in the $j - th$ trial. At time t_f of the $j - th$ trial, the system positioning error is expressed by

$$e^{[j]} = \psi^{[j]} - \psi_d = C \begin{bmatrix} e^{At_f} x^\circ \\ 0 \end{bmatrix} + CW\chi^{[j]} - \psi_d. \tag{11}$$

Let the control function be updated by correcting the vector χ with the equation

$$\chi^{[j+1]} = \chi^{[j]} + Ge^{[j]}. \tag{12}$$

Premultiply (12) by CW to obtain

$$CW\chi^{[j+1]} = CW\chi^{[j]} + CWGe^{[j]} \tag{13}$$

and use (11) to get

$$e^{[j+1]} - C \begin{bmatrix} e^{At_f} x^\circ \\ 0 \end{bmatrix} + \psi_d =$$

$$e^{[j]} - C \begin{bmatrix} e^{At_f} x^\circ \\ 0 \end{bmatrix} + \psi_d + CWGe^{[j]}. \tag{14}$$

The linear and finite dimensional dynamics of the error, defined over the countable set of the trials, are then given by

$$e^{[j+1]} = (I - CWG) e^{[j]}. \tag{15}$$

where I is the identity map on $\mathbb{R}^{n+m}$. If the map

$$E = I - CWG \tag{16}$$

has all the eigenvalues in the unit open disk, $e^{[j]} \to 0$ uniformly as $j \to \infty$.

The existence of a map G such that the eigenvalues of E are in the unit open disk is proven by the fact that CW is invertible. In fact, by setting

$$G = (CW)^{-1} \tag{17}$$

E turns out to be equal to the null map. Choosing

$$G = \lambda(CW)^{-1}, \quad 0 < \lambda < 2 \tag{18}$$

may result in a more robust algorithm.

The above discussion justifies the following definition.

Definition 2 The linear space of all vectors ψ of dimension $n + m$ is the state space of the learning algorithm and ψ its state. ‡

Owing to the uniform convergence, all perturbations which are not persistent and sufficiently small in norm are rejected, including nonlinear perturbations, whereas sufficiently small persistent perturbation will correspond small persistent errors (Total Stability Theorem [16]). In particular, we consider plant perturbations of the type

$$\dot{x}(t) = Ax(t) + Bu(t) + \varepsilon\eta(x(t), u(t), t), \tag{19}$$

where ε is a small parameter. The linear part of this system obtained by setting $\varepsilon = 0$ is referred to as the unperturbed linear system. Without loss of generality, suppose that $x_f = 0$.

Let $u^\circ(t) \in \mathbb{R}^m, t \in [0, t_f]$, be a control function which steers to zero the state of the unperturbed linear system at time $t_f > 0$.

Define $B_u = \{w \in \mathbb{R}^m \mid |w - u^\circ(t)| < \nu_1, t \in [0, t_f]\}$ and let $\xi(t; u)$ be the solution of the unperturbed linear system corresponding to a control $u(t) \in B_u, t \in [0, t_f]$. Define the bounded set $B_x = \{z \in \mathbb{R}^n \mid z = \xi(t; u), t \in [0, t_f], u(t) \in B_u, t \in [0, t_f], \}$.

Suppose that $\eta(\cdot)$ is continuous on $B_x \times B_u \times [0, t_f]$. Then, in [13] is proven that for sufficiently small ε, this type of nonlinear perturbation is nonpersistent for the linear learning algorithm designed on the basis of the unperturbed linear system and therefore it is rejected.

In particular, robustness with respect to plant parameters uncertainties is ensured when the plant dynamics continuously depend on them.

It is emphasized that all the perturbations considered must be sufficiently small in order to preserve algorithm convergence. However the experimental results later presented show that, for the laboratory arm used, "significant" perturbations are rejected by the implemented algorithm.

3. The experimental arm

The arm used in the experiments is a direct drive planar chain with two revolute joints and two links, the second of which –the forearm– is very flexible [17,19]. A sketch of the arm is given in figure 1.

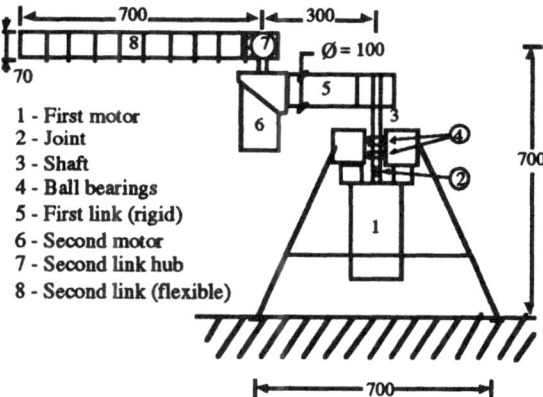

Figure 1. sketch of the arm

The first link (the rigid one) is 300 mm long; it has been obtained from an aluminum pipe of 100 mm diameter and 5 mm thickness. The second link has been designed to be very flexible in a plane orthogonal to the two motors axes (the horizontal plane) and stiff in the other directions and torsionally. It is composed, as depicted in fig. 2, of two parallel sheets of harmonic steel coupled by nine equispaced aluminum frames, which prevent torsional deformations. Each sheet is 700 × 70 × 1mm.

The frames are square, 70 × 70 mm, with a circular hole drilled at the center to reduce their weight. They have a sandwich structure: two external layers of 2 mm thick aluminum enclose an internal layer of 1 mm neoprene. The external layers continuity is interrupted on the opposite sides by four parallel 1 mm cuts, which preserve the neoprene continuity. Being the neoprene very flexible, two parallel hinges are obtained so that the steel sheets are free to bend only in the horizontal plane. The total mass of the flexible link is about 2 kg.

The inertia of the first and second links are respectively $J_1 = 0.447$ Kg m^2 and $J_2 = 0.303$ Kg m^2.

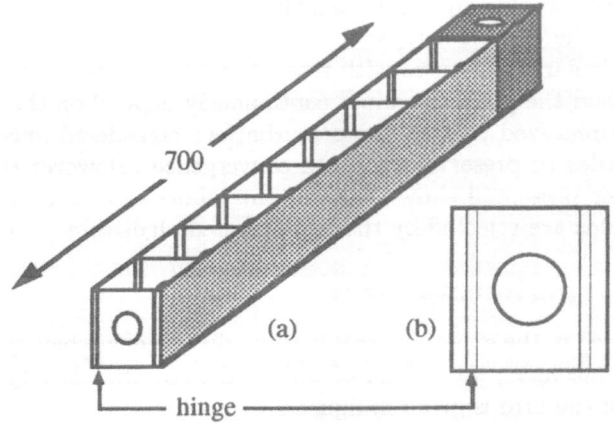

700

(a)   (b)

hinge

Figure 2. flexible link structure

The motors driving the two links can provide a maximum torque respectively of 3.5 Nm and 1.8 Nm and are supplied by two 20 KHz PWM amplifiers. Each motor is equipped with a 5000 pulses/turn encoder and a D.C. tachometer.

The second link deformations are measured by an optical sensor with a precision of about 0.1 degrees [19]. The light of three lamps positioned along the link is reflected by three rotating mirrors on three detectors. When a detector is illuminated, the position of the corresponding mirror is measured by an encoder. In this way it is possible to know the lamps displacement during the experiments and to reconstruct the link shape with a good approximation.

The electronics used to control the arm is outlined in fig. 3. It has been purposely built and provides the functional interface between the arm sensors and actuators and the control computer, a 16 Mhz HP Vectra 386SX equipped by a Burr-Brown PCI-20001 parallel communication board. To address a particular device (actuator or sensor) the computer outputs his address on 5 of the available 32 lines and then sends or receives the required values on other 16 lines. Part of the remaining bits are used for the housekeeping of the electronics (reset, status, etc.).

The blocks concerning the motors, the tachometers and the encoders are rather conventional. They mainly consists in the necessary converters and counters and the related electronics. The displacement sensor logic contains a speed regulator for the motor driving the mirrors and a phase locked loop which increases by a factor of 16 the number of the pulses provided by the encoder (a 500 pulses/turn unit).

The deflections measured by this sensor are rather noisy and require some processing to improve their quality. In the experiments, a state observer has been used [18], which allows also a clean estimation of the deflection velocities.

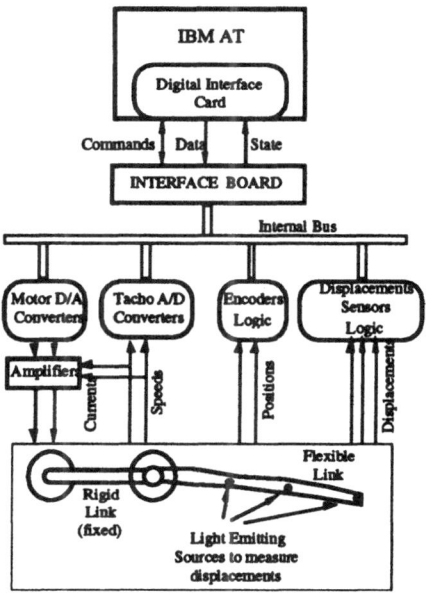

Figure 3. Block diagram of the interfacing system

4. Flexible arm repositioning by learning

The theory so far presented is now applied to the repositioning control of the two-link flexible arm, under the hypothesis that its material points move on a horizontal plane. We suppose that the arm experiences a motion which is the sum of a rigid motion plus a small elastic deviation. We further suppose, supported by the experience, that the elastic displacement field can be described, with respect to a floating frame of reference, by the linear combination of a finite number of low frequency modes of deformation. Moreover, under the hypothesis of small elastic displacements, the dependency of the material points velocities on the elastic displacements is neglected.

Let $\theta_1(t)$ and $\theta_2(t)$ be the first and second joint rotations, respectively, and $\xi(t) \in I\!\!R^s$ the vector of the amplitudes of the modes of deformation. Set

$$\zeta(t) = \begin{bmatrix} \theta_2(t) \\ \xi(t) \end{bmatrix}, q(t) = \begin{bmatrix} \theta_1(t) \\ \zeta(t) \end{bmatrix}. \tag{20}$$

Let the virtual work of the torque applied at the second revolute joint be given by $\tau_2(t)\gamma^T \delta\zeta$. Then, the Lagrange equations of motion are of the type:

$$\begin{bmatrix} b_{11}(\theta_2(t)) & b_{12}(\theta_2(t)) \\ b_{12}^T(\theta_2(t)) & M \end{bmatrix} \begin{bmatrix} \ddot{\theta}_1(t) \\ \ddot{\zeta}(t) \end{bmatrix} + n(q(t), \dot{q}(t)) +$$

$$+ \begin{bmatrix} 0 & 0 \\ 0 & D \end{bmatrix} \begin{bmatrix} \dot{\theta}_1(t) \\ \dot{\zeta}(t) \end{bmatrix} + \begin{bmatrix} 0 & 0 \\ 0 & K \end{bmatrix} \begin{bmatrix} \theta_1(t) \\ \zeta(t) \end{bmatrix} = \begin{bmatrix} \tau_1(t) \\ \gamma\tau_2(t) \end{bmatrix},$$

$$\tag{21}$$

where $n(\cdot)$ is the vector of Coriolis and centripetal terms and M, D and K are the mass, damping and stiffness matrices of the second flexible link, respectively.

The equations are detailed in [19], where model based algorithms have been tested. Using a learning approach to the control, less information is required. The necessary data shall be referred in the implementation description.

To add some damping and at the same time to decouple the state variables $\theta_1(t)$ and $\dot{\theta}_1(t)$ from the others, a proportional and velocity loop is closed around both joints

$$\tau_i(t) = -k_{di}\dot{\theta}_i(t) - k_{pi}(\theta_i(t) - v_i(t)), \quad i = 1, 2, \tag{22}$$

where $v_i(t)$ are the new input variables.

With this position, and setting $k_{p1} = \rho k_{d1}$, the system (21) can be expressed as

$$\dot{\theta}_1 = -\rho\theta_1 + \rho v_1 - \frac{1}{k_{d1}}(b_{11}(\theta_2)\ddot{\theta}_1 + b_{12}(\theta_2)\ddot{\zeta} - n_1(q, \dot{q})),$$

$$M\ddot{\zeta} + \bar{D}\dot{\zeta} + \bar{K}\zeta = \gamma k_{p2}v_2 - b_{12}^T(\theta_2)\ddot{\theta}_1 - n_2(q, \dot{q}). \tag{23}$$

where $\bar{D}$ and $\bar{K}$ are the damping and elastic matrix D and K modified by k_{d2} and k_{p2}.

Decoupling is obtained using an high gain strategy on the first link increasing the value of k_{d1}.

If we suppose that the value k_{d1} is big enough, and the values of $\dot{\theta}_1(t)$ and $\ddot{\theta}_1(t)$ are small along the trajectory, the nonlinear system (23) can be approximated by the linear one:

$$\dot{\theta}_1(t) = -\rho\theta_1(t) + \rho v_1(t),$$
$$M\ddot{\zeta}(t) + D\dot{\zeta}(t) + K\zeta(t) = \gamma k_{p2}v_2(t), \tag{24}$$

and (23) can be seen as a perturbed linear system of the type (19).

Note that the fast dynamics obtained increasing the value of k_{d1} can be expressed in term of

$$\frac{dz(\sigma)}{d\sigma} = -d_{11}(\theta_2(t))z(\sigma) \tag{25}$$

where $z(\sigma)$ describes the fast transient, σ is the fast time, and $d_{11}(\theta_1(t))$ is the first entry of the inverse of the inertial matrix in (21). Due to the fact that $d_{11}(\theta_1(t))$ is always positive we have convergence of the state of the whole system on the slow manifold defined by the first equation of (24). If the initial state is not on the slow manifold, this may result in a lower bound for t_f, related to the speed of convergence of the fast dynamics.

A further question which needs to be addressed is algorithm robustness with respect to unmodeled high frequency dynamics.

In open loop, the number of vibration modes used in flexible arm modeling influences the accuracy of arm motion prediction, whereas, in closed loop, neglected high frequency modes of vibration may be detrimental for the stability of learning algorithms.

In order to clarify this point, consider in modal coordinates the second equation — the one relative to the flexible link — of system (24)

$$\ddot{\nu}_i(t) + 2\omega_i\bar{\omega}_i\dot{\nu}_i(t) + \omega_i^2\nu_i(t) = \beta_i v(t),$$

$$\nu_i(t) \in \mathbf{R}, \nu_i(0) = \nu_i^0, \dot{\nu}_i(0) = \dot{\nu}_i^0, \quad i = 1, 2, ..., s \tag{26}$$

where ω_i is the $i - th$ natural circular frequency, $\bar{\omega}_i$ the $i - th$ damping factor and β_i a constant. To investigate the spillover effect, suppose that an arm model with the lowest $r < s$ modes of vibration is used to design the learning algorithm. Consistently, also suppose that $(2r + 1)$ measures, ψ_h, are used in arm repositioning specification.

When using the reduced model in algorithm design, the system positioning $\psi_h, h = 1, \ldots, 2r + 1$, are interpreted as functions of the control input $v_2(t_f)$ and of the first r modes of vibration. However, actually their are functions of all s modes of vibrations, that is

$$\psi_h = \sum_{j=1}^{s} (X_{hj}\nu_j(t_f) + Y_{hj}\dot{\nu}_j(t_f)) + \mu_h v_2(t_f),$$

$$h = 1, ..., 2r + 1, \tag{27}$$

with X_{hj}, Y_{hj} and μ_h proper constants. For algorithm design purposes, however, j runs from 1 to r. When implementing the learning algorithm, (27) tell us that the contributes of the neglected high frequency modes are fed back. This mechanism may instabilize the learning algorithm which has been designed on the basis of the reduced model.

We now investigate the stability of the learning algorithm under the hypothesis that the neglected modes of vibrations are in a high frequency range. Redefine ω_h as $\omega_h/\sqrt{\varepsilon}, h = r + 1, \ldots, s$, and, in order to preserve the modal damping level, $\bar{\omega}_h$ as $\bar{\omega}_h/\sqrt{\varepsilon}$. Next singularly perturb the system by letting $\varepsilon \to 0$. The fast system

$$\frac{d\tilde{\nu}_h(\sigma)}{d\sigma} = -2\omega_h\bar{\omega}_h\tilde{\nu}_h(\sigma), \quad h = r + 1, ..., s, \tag{28}$$

where σ is the fast time and $\tilde{\nu}_h(\sigma)$ describes the fast transient of $d\nu_h(t)/dt$, is globally exponentially stable and Tikhonov Theorem applies [20]. The slow system is given by

$$\ddot{\nu}_i(t) + 2\omega_i\bar{\omega}_i\dot{\nu}_i(t) + \omega_i^2\nu_i(t) = \beta_i v_2(t),$$

$$\nu_i(0) = \nu_i^0, \dot{\nu}_i(0) = \dot{\nu}_i^0, \quad i = 1, 2, ..., r,$$

$$2\bar{\omega}_h\dot{\nu}_h(t) + \omega_h\nu_h(t) = 0,$$

$$\nu_h(0) = \nu_h^0, \quad h = r + 1, ..., s. \tag{29}$$

From these equations, we see that the slow solution relative to the neglected modes is not affected by the control input. Then even if $v_2(t)$ depends on

$$\dot{\nu}_h(t_f), \nu_h(t_f), \quad h = r + 1, ..., s. \tag{30}$$

no instability occurs at the slow level.

For $\varepsilon \neq 0$, by Tikhonov Theorem we have that for sufficiently large t the actual solution is uniformly approximated by the solution of the slow system. Then, provided that t_f is sufficiently large, the difference between the learning dynamics corresponding to the slow system and the actual learning dynamics is a continuous function of ε and vanishes for $\varepsilon = 0$. Hence stability is preserved for sufficiently small ε.

Last, the disturbance due to the neglected dynamics is not persistent. Indeed, if at the end of a trial $\psi = \psi_d$, the control is no longer updated and at the next trial the same point of the learning state space is reached. Hence the state of the learning system converges on an equilibrium point in the subspace Ψ of the learning state space $I\!\!R^{2s}$ defined by the constraint $\psi = \psi_d$.

Proposition 4 The disturbance due to neglected sufficiently high frequency modes of vibration is small and non persistent for sufficiently large t_f.‡

We stress the fact that we cannot select the equilibrium point of Ψ to be reached. To fully appreciate this point, consider the case of moving the end point of the arm between two given mechanical equilibrium points. If the final point is a mechanical equilibrium point, the state derivative is zero at time t_f. Let $x(t) \in I\!\!R^{2s+1}$ denote the full state and $x_r(t) \in I\!\!R^{2r}$ denote the state of the reduced system. The repositioning equation in terms of the state of the reduced system is expressed by

$$\dot{x}_r(t_f) = A_r x_r(t_f) + b_r u(t_f) = 0,$$
$$\Gamma x_r(t_f) = y_d, \tag{31}$$

where we have used the general notation of (1), $\Gamma x_r(t_f)$ is the position of the end point at time t_f and y_d is the desired one. The equilibrium point that the state $[x^T(t_f), u(t_f)]^T$ of the learning algorithm reaches belongs to the subspace defined by (31).

Hence, even if the end point reaches the desired position, the arm may not be in equilibrium at time t_f since there is no guarantee that at time t_f the derivative of the full state of the arm, is zero. This means that for $t > t_f$ the arm could start to move again, if the neglected dynamics are not sufficiently fast and damped.

5. Implementation of the learning controller.

For the implementation of the learning controller only the first mode of deformation has been retained and equations (21) have been linearized assuming:

$$b_{11}(\theta_2(t)) = J_1 + J_2, \qquad b_{12}(\theta_2(t)) = 0,$$
$$n(q(t), \dot{q}(t)) = 0. \tag{32}$$

The model parameters values are reported in [19]. The control torque are given by equation (22) with

$$k_p = \begin{bmatrix} 100.18 \\ 1.16 \end{bmatrix}, \qquad k_d = \begin{bmatrix} 2 \\ 0.8 \end{bmatrix}. \tag{33}$$

The joint-clamped frequency of the deflection mode is $f_1 = 1.25$ Hz; the first of neglected ones is $f_2 = 7.25$ Hz.

As stated by *Problem 1*, the learning control formulation requires the values of the input at $t = t_f$. The approach of *Proposition 1* has been applied. Two integrators have been added (one for each control channel) and the extended system state has been considered. Both initial and final positions of the arm are equilibrium configurations. This means that the control torques at time $t = t_f$ must be zero. Then the state of the two integrators at time $t = t_f$ are set equal to the final desired position of the joints.

The implementation of the learning controller can be subdivided in four distinct phases.

Choice of the functions $\phi(t)$. ; Simple time polynomials t, t^2, $\cdots$ have been selected. Taking into account the numbers of state variables of the extended system concerning the two links, the following functions have been used:

$$\phi_i(t) = \begin{bmatrix} t^i \\ 0 \end{bmatrix}, \qquad for\ i = 1, 2, 3$$

$$\phi_i(t) = \begin{bmatrix} 0 \\ t^{i-3} \end{bmatrix}, \qquad for\ i = 4, \cdots, 8. \tag{34}$$

Computation of the learning matrix G. The main problem consists in computing matrix W, which, in the case of the extended system, is given by the integral appearing in the first of equations (8). Each column of W has been computed applying a single ϕ_i at the input of the system model and simulating it in the interval $[0, t_f]$.

This and the previous step are to be carried out only once, on the nominal data.

Implementation of the learning iteration (12). The value of parameter λ in (18) has been set equal to 0.7, a good compromise between robustness and rate of convergence. The implementation of this step is quite straightforward by means of a digital computer. It must be carried out off-line at the end of each trial and does not need a particular computational efficiency.

Implementation of the low-level joint controllers. This is a conventional issue. It should be noted that in the experimental implementation the derivative part of the controller has been realized in continuous time, by means of analog loops. In this way a smoother behavior has been obtained and no implementation delay affects the stability of the system.

6. Experimental results

Several experiments have been worked out with the described set-up and have confirmed the validity of the proposed approach. The tests have revealed other non-linearities besides those explicitly addressed by the model; in particular dry friction and saturation of the motor drives. In any case the learning controller worked properly and demonstrated its characteristics of robustness.

Here, a typical experiment is described, during which the links move at the maximum speed allowed for no saturation. The arm has to move from the

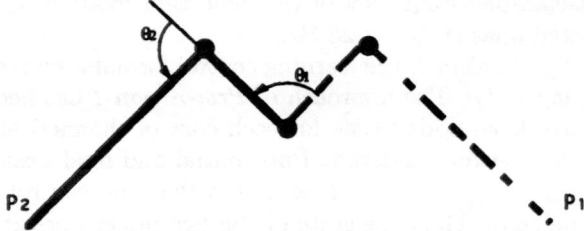

Figure 4. Starting and ending arm configurations

equilibrium point P_1 ($\theta_1 = 0$ rad, $\theta_2 = -\pi/2$ rad) to the equilibrium point P_2 ($\theta_1 = \pi/2$ rad, $\theta_2 = \pi/2$ rad), drawn in fig. 4, in 2.5 seconds. In both points ξ, the tip deflection, is null. Each trial starts from P_1; at the end the arm is repositioned in the same point.

The graphs show the values of the different variables at $t = t_f$ as a function of the iteration number.

In particular figures 5 and 6 refer the error positions of the two joints, which converge to zero in a smooth way.

The deflection of the tip and of the middle point of the flexible link can be seen in figures 7 and 8 respectively. The effect of the controller on the final deflection is evident, showing the ability of this algorithm to control also flexible mechanical structures.

As for the velocities, they are not referred, because they are under the sensitivity threshold of the measuring chain from the first iteration.

Finally, figures 9 10, 11, and 12 show the joint trajectories, the middle point deflection and the tip deflection wrt time during the different trials. Note that, even if the trajectories are not a priori specified, they result to be quite reasonable.

7. Conclusion

We have presented a learning strategy for the repositioning control of a two-link flexible arm. By restricting the search for the steering control to some finite dimensional subspace of the input space, finite dimensional learning algorithms have been developed, which need a finite memory for their implementation. Robustness with respect to a very large class of perturbations, including unmodeled high frequency dynamics, has been proven by means of theoretical arguments and experimentally verified on a laboratory arm.

References

[1] S. Arimoto, S. Kawamura, F. Miyazaki. Bettering operations of robots by learning. J. Robotic Systems, 1, 1984.
[2] G. Casalino, G. Bartolini. A learning procedure for the control of movements of robotic manipulators. Proc. IASTED Symp. on Robotics and Autom., Amsterdam, The Netherlands, 1984.

[3] A. De Luca, G. Paesano, G. Ulivi, A frequency domain approach to learning control: implementation for a robot manipulator, IEEE Trans. on Industrial Electronics, vol. 39, Num. 1, 1992.

[4] Arimoto S. (1990). Learning control theory for robotic motion. Int. J. of Adaptive Control and signal Processing, 4, pp. 543-564.

[5] P. Lucibello, On learning control for a class of flexible joint arms, Proc. of Int. Symp. on Intelligent Robotics, ISIR, Bangalore, 1993.

[6] A. De Luca, G. Ulivi, Iterative Learning Control of Robots with Elastic Joints. IEEE Conf. on Robotics and Automation, Nizza, FR, May 1992.

[7] F. Miyazaki et al., Learning control scheme for a class of robots with elasticity. 25th IEEE Conf. on Decision and Control, Athens (GR), 1986.

[8] M. Poloni, G. Ulivi, Iterative learning control of a one link flexible manipulator. IFAC Symp. on Robot Control, SYROCO '91, Vienna, Austria, 1991.

[9] P. Lucibello, Output tracking of flexible mechanism by learning. Proc. of IEEE Int. Symp. on Intelligent Control, Glasgow (U.K.), 1992.

[10] P. Lucibello, Learning control of linear systems, Proc. American Control Conf., 1992.

[11] P. Lucibello, Point to point polynomial control of linear systems by learning, Proc. IEEE Conf. on Decision and Control , 1992.

[12] P. Lucibello, Output regulation of flexible mechanisms by learning, Proc. IEEE Conf. on Decision and Control , 1992.

[13] P. Lucibello, Steering the state of nonlinearly perturbed linear systems by learning. IFAC Workshop on motion Control for Intelligent Automation, Perugia (I), 1992.

[14] M. J. Balas, Active control of flexible systems, JOTA, vol.25, no.3, pp. 415-436, July 1978.

[15] W. M. Wonham, Linear Multivariable Control: a Geometric Approach, Springer-Werlag, 1979.

[16] W. Hahn. Stability of motion. Springer-Verlag, 1967.

[17] P. Lucibello, G. Ulivi, Design and realization of a two-link direct drive robot with a very flexible forearm. To appear on Int. J. of Robotics and Automation.

[18] S. Panzieri, G. Ulivi, Design and Implementation of State-Space Observers for Flexible Robots. IEEE Conf. on Robotics and Automation, Atlanta, GE, 1993.

[19] A. De Luca, L. Lanari, P. Lucibello, S. Panzieri, G. Ulivi, Control experiments on a two link robot with a flexible forearm, 29th. IEEE Conf. on Decision and Control, Honolulu, USA, 1990.

[20] P. Kokotovic, H.K. Khalil, J. O'Reilly. Singular perturbation methods in control: analysis and design. Academic Press, 1986.

This paper has been partly supported by *ASI* founds under contract RS15791.

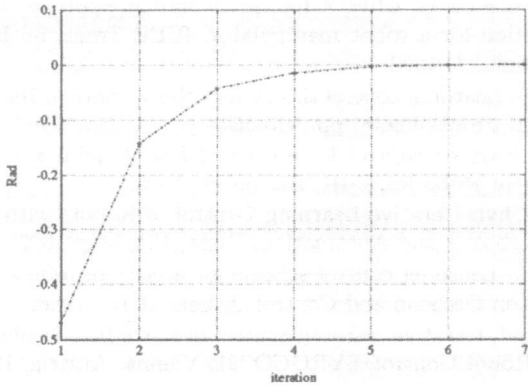

Figure 5. Position errors of joint 1 wrt iterations

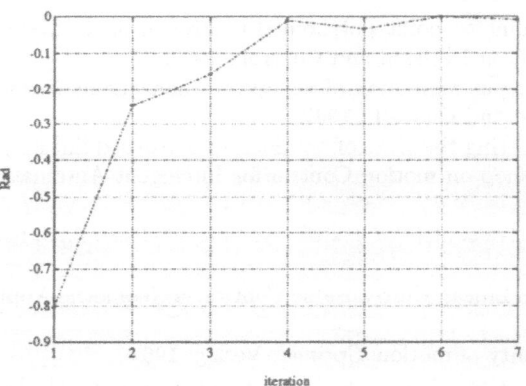

Figure 6. Position errors of joint 2 wrt iterations

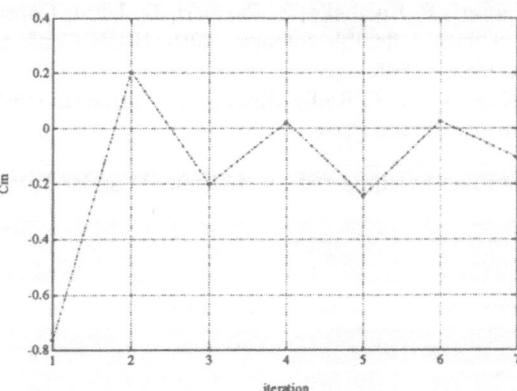

Figure 7. Tip position error wrt iterations

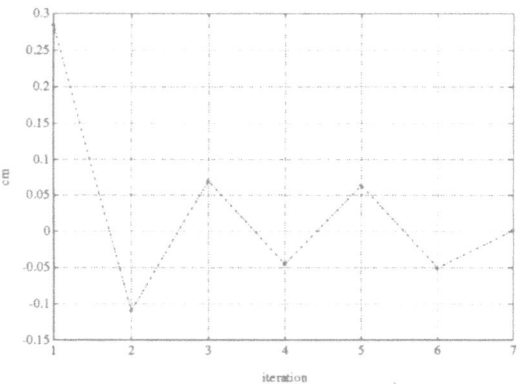

Figure 8. Position error of the middle point wrt iterations

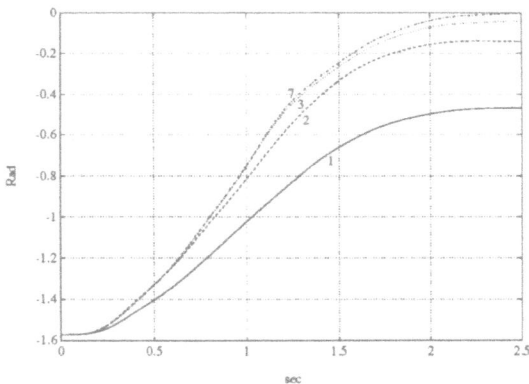

Figure 9. Time evolution of joint 1 trajectory for different iterations

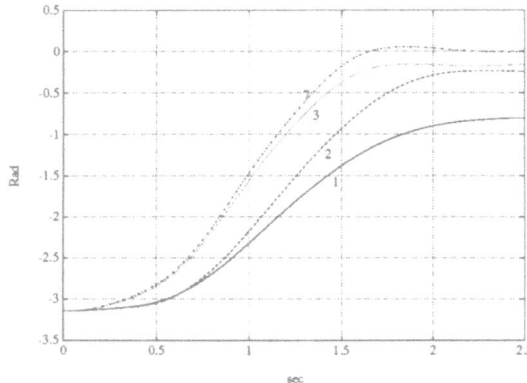

Figure 10. Time evolution of joint 2 trajectory for different iterations

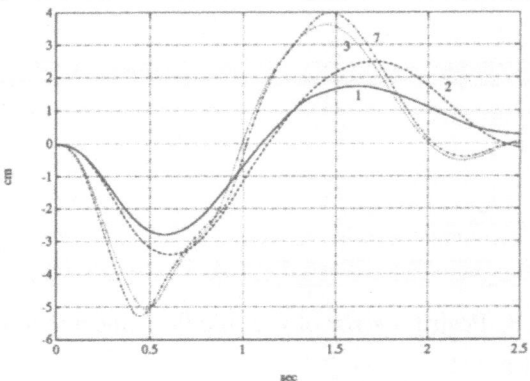

Figure 11. Time evolution of the middle point deflection for different iterations

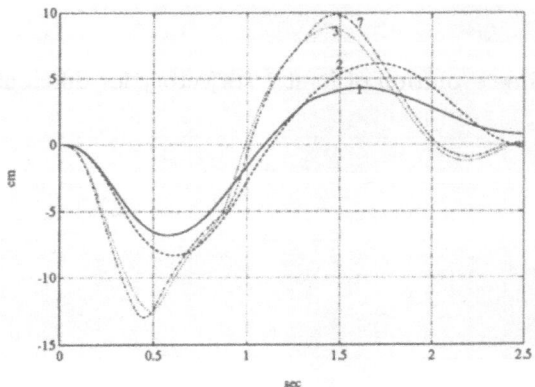

Figure 12. Time evolution of the tip deflection for different iterations

Space Robotics Research Activity with Experimental Free-Floating Robot Satellite (EFFORTS) Simulators

Kazuya YOSHIDA

Department of Mechanical and Aerospace Engineering
Tokyo Institute of Technology
O-okayama, Meguro, Tokyo 152, Japan

Abstract— In order to study and validate practical availability of dynamics models and control schemes for future in-orbit space robots, a research group at the Tokyo Institute of Technology has developed the Experimental Free-FlOating RoboT Satellite (EFFORTS-I and II) simulators, which enable us to examine 2-dimensional psuedo-micro gravity motion dynamics by air lift system. The robot models comprising a satellite base body and articulated manipulator arm(s), make horizontal motion without mechanical disturbances or external forces. This paper presents the discussion of the following topics relevant to space robotics frontier: (1) hardware design and development of the EFFORTS simulators, (2) dynamics modeling and basic cotrol concepts for space free-floating manipulators, (3) target chasing and capturing experiments, (4) modeling of impact dynamics at the target capture or collison, and its experimental validation, and finally, (5) practical force control for dexterous manipulation.

1. Introduction

For successful promotion of space projects, robotics and automation should be a key technology. Autonomous and dexterous space robots could reduce the workload of astronauts and increase operational efficiency. However a drawback in controling space robots is the lack of a fixed footbase in orbital environment, different from ground-based manipulators. Any motion of space arms will induce reaction forces and moments in the base, which disturb its position and attitude: the behaviour is governed by the nature of floating link dynamics. Another drawback in space robotics research and development is difficulty of the simulation of micro-gravity environment in on-earth laboratories. We need to qualify the developed control concepts, schemes, and softwares for the practical application, using actual hardwares with simulating space environment.

In order to cope with these unique problems for space robotics, the author has been developing control concepts and hardware simulators, since 1985. The *"Generalized Jacobian Matrix (GJM)"* or dynamic Jacobian is a significant and useful idea for Resolved Motion Rate Control, as well as Resolved Acceleration Control, of free-floating space manipulators [1][2][3]. On the other hand, the *"Extended-Inversed Inertia Tensor"* is a concept to represent the dynamic characteristics of floating links in presence of external force input or collision [4][5]. Based on this idea, we can predict the impulse caused by collision, as well as the motion after capture or rebound.

For the experimental study, the author has been developing laboratory models of robot satellite supported on air bearings. The first model, *Experimental Free-FlOating RoboT Satellite Simulator (EFFORTS-I)*, was developed in 1987 and an on-line control scheme with the GJM was successfully demonstrated with target capture experiments during 1987-1990 [1][6][7]. Improving the first model, the author designed the second one, EFFORTS-II, which comprises a satellite base with inertial motion sensors and a couple of arms with wrist force sensors, and carried out a series of basic experiments which provide a basis for such dexterous manipulation as structure assembly and deployment.

This paper presents both the theoretical concepts and the corresponding experiment results. It is organized in the following way: Section 2 outlines the EFFORTS simulation system and Section 3 introduces the basic modeling for space floating manipulators. Afterward the author discusses the topics on specific problems. Target capture operation appears in Section 4, the collision dynamics issue in Section 5, and a primitive compliant force control topic in Section 6.

2. Design and Development of EFFORTS simulators

In order to study real mechanical dynamics and demonstrate the practical validity and effectiveness of control methods using actual sensors, computers and mechanical assemblies, we need experiments with a laboratory model. However the micro-gravity simulation is not so easy, critical matter, because we cannot obtain natural 3D zero-gravity or perpetual free-falling environment on the earth. In general, the following methods could be available for emurating psuedo-zero-gravity:

1. Do experiments in an airplane flying along a parabolic trajectory or a free-falling capsule. In this case, we can observe pure nature of micro-gravity, but such experiment costs extremely high and is inconvenient.

2. Do experiments in a water pool with the support of neutral buoyancy. This is especially good for training of astronauts' activities, but from a micro-gravity dynamics point of view, water current and drag disturbes the dynamical motion, e.g.[8].

3. Suspend an experimental model by tethers to cancel the vertical gravitational motion. Whether a passive or an active counter-balancing is employed, the design of quick response, vibration free and simple suspension system is a key point, e.g.[9][10].

4. Support an experimental model by air-cushions or air-bearings. This is a simplest method, however, the motion is restricted on a horizontal plane, e.g.[1][11].

5. Calculate the motion which should appear in zero-gravity environment based on a mathematical model, then force the corresponding mechanical model to move according to the calculation. This method is called as a *'hybrid'* simulation, a combination of mechanical and mathematical models, e.g.[12]. However, the author personally feels that the method is good only for kinematical simulations, not for dynamical ones, due to the computation and servo-control bandwidth limitation.

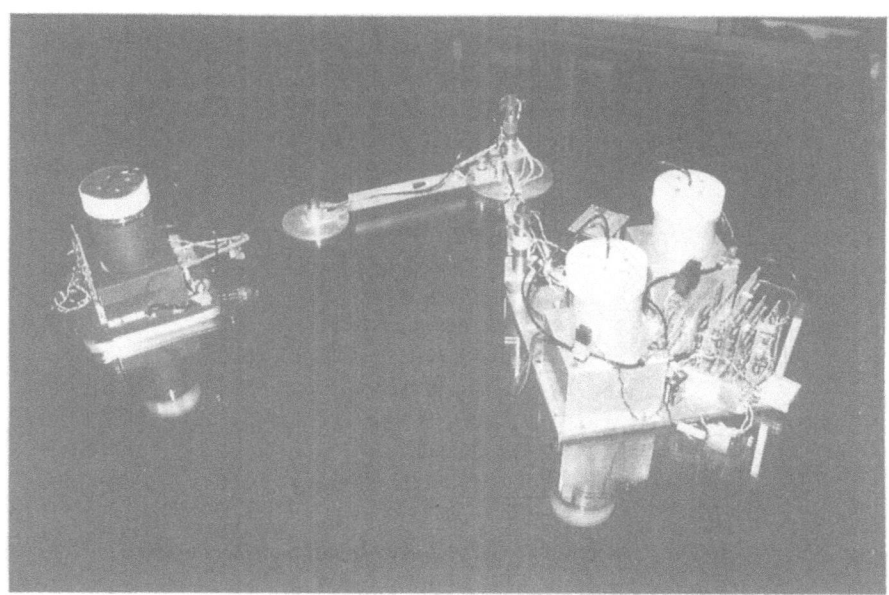

Fig.1 The EFFORTS-I (1987)

Fig.2 The EFFORTS-II (1992)

Each method has advantages and disadvantages, and we should carefully chose the method to satisfy the purpose of the experiment. The previous papers [8]-[12] offer good examples, each method satisfying its particular requirment. There is no perfect

564

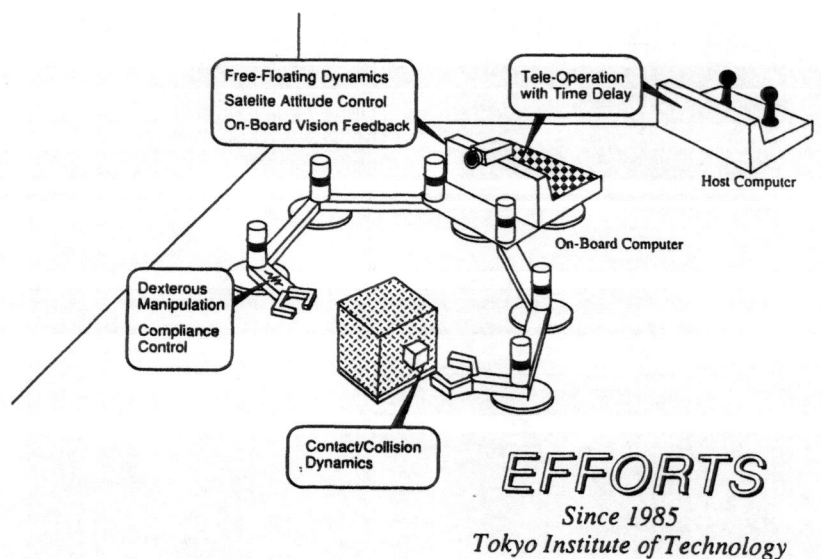

Host Computer

On-Board Computer

EFFORTS

Since 1985
Tokyo Institute of Technology

Fig.3 The Experimental Free-Floating Robot Satellite Simulator system

method in all aspects, so that we should combine the results of a serious of different experiments. The hybrid simulator (5.) is good for kinematic motion understanding of the total system and qualification of specific flight components or evaluation of integrated sensor-controller-actuator system performance, but it cannot simulate pure mechanical dynamics, especially, in the presence of contact forces or collisions. For such cases, the air-floating simulator (4.) is particularly essential although the observation is limited in a plane. As is discussed in Section 5 for collision dynamics, we can examine the validity of a dynamics model and estimate unknown parameters through the hardware floating experiments. The knowledge then could be available in the mathematical model of the hybrid simulation.

Here the present auther adopted the hardware floating experiment (4.), because we want to observe the behavior of mechanical links under the law of nature with smallest disturbance, by the simplest apparatus, and examine basic concepts and principles.

Fig.1 shows a photograph of the developed Experimental Free-Floating Robot Satellite Simulator-I with a single 2 DOF manipulator arm. The detailed specification and experimentally identified physical dimensions are published in [1] and [6]. The first model equips a gas-bomb type air supply and a wire-less communication system on the robot body, so that it is completely free from any umbilical cords thereby freely floats without any mechanical disturbances. However due to the limitation of air supply capacity, flight time is limited to one or two minutes. An acceleration environment of the simulator was evaluated to approximately 10^{-4} to $10^{-3}G$, in other words, the friction coefficient of the developed air-bearing system was 0.0001-0.001.

The second model EFFORTS-II is shown in **Fig.2**. This one equips a couple of arms and on-board inertial sensors: accelerometers for x and y direction, and a rate gyro around the vertical axis. To remedy the flight time limitation, we removed gas-bombs but employed an umbilical code, a flexible air-hose, to supply pressurized air from a compressor. In addition, employed wired connections for measurement and control between the robot and a ground-based computer, but these umbilicals will be removed except the air-hose, by installing an on-board camera, a computer, and attitude control devices on the base body.

Mechanical design of the model-II is almost same as the previous model: each link of arms has 0.25 [m] and each joint has a DC motor with a planetary gear train and an optical encoder, driven with a local velocity servo controller. At the tip of arms, installed is a simple gripper which can grasp a hand-rail type fixture. A 2-axes x-y force sensor is built in the wrist joints. Springs of the sensor provide both mechanical compliance and force data by their deformation. To support the robot model, aluminium air-pads of 0.1 [m] diameter are used: three at the base body and one at each elbow or wrist joints. 2.0 [kg/cm^2] air supplied, each pad can ideally support approximately 75 [kg].

Key research issues the author's research group can examine using the EFFORTS simulator are illustrated in **Fig.3**, and listed as follows:

- free-floating space robot dynamics and target capture control,

- impact dynamics at contact/collision with a floating target,

- dexterous manipulation based on force/compliant control,

- satellite attitude control coordinating with manipulator motion,

- tele-operation/autonomous control based on the on-board vision feedback, and

- tele-operation from a distance with large time delay

First three topics are emphasized in this paper.

3. Modeling and Control of Floating Manipulator Systems

3.1. dynamics equations

To discuss floating manipuator dynamics, consider a general model that a robot satellite has plural arms including solar paddles, reaction wheels or other appendages. This is regarded as a chain of links in a tree configuration consist of $n + 1$ rigid bodies, connected with n articulated joints, as shown in **Fig.4**. Assume that l pieces of arms are mounted on the satellite main body, and the arm k has n_k pieces of links, then $n = \sum_{k=1}^{l} n_k$.

Flexible arms like solar paddles can be treated as segmented virtual rigid links connected with elastic hinges. To avoid confusion and complexity, this paper will not explicitly discuss the flexibility.

Assume that the system freely floats in the inertial space, no external force is applied and no orbital motion is considered, hence the system Lagrangian is composed of purely kinetic energy of each links, and can be written in the following form:

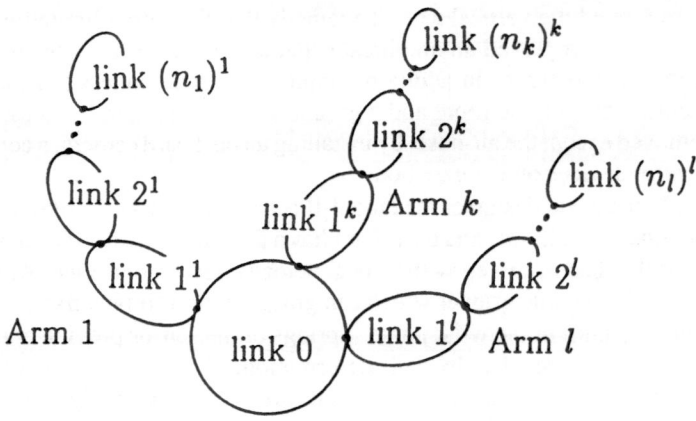

Fig.4 A general model for space robots

$$\mathcal{L} = \frac{1}{2}\{(\sum_{k=1}^{l}\sum_{i=1}^{n_k} m_i^k v_i^{kT} v_i^k + \omega_i^{kT} I_i^k \omega_i^k)$$
$$+ m_0 v_0^T v_0 + \omega_0^T I_0 \omega_0\} \tag{1}$$

where m_i^k is mass, I_i^k inertial tensor, v_i^k linear velocity, and ω_i^k angular velocity of link i of arm k. The suffix 0 stands for the base body.

Assumption of no external forces or moments leads to momentum conservation: that gives us an important information on the motion restriction among velocity variables. The relationship cannot be expressed by angle variables directly, but only by an unintegrable differential equation; say "***non-holonomic***" restriction [1]. Let linear and angular momenta with respect to the mass center of the satellite base body (link 0) be P_0, L_0, respectively, we obtain:

$$\begin{bmatrix} P_0 \\ L_0 \end{bmatrix} = \begin{bmatrix} wE & w\tilde{r}_{0g}^T \\ w\tilde{r}_{0g} & H_\omega \end{bmatrix} \begin{bmatrix} v_0 \\ \omega_0 \end{bmatrix} + \begin{bmatrix} J_{Tw} \\ H_{\omega\phi} \end{bmatrix} \dot{\phi}$$
$$= constant. \tag{2}$$

where $\dot{\phi}$ is a joint velocity vector.

Using above two equations, we finally obtain the equation of motion in the following form:

$$H^* \ddot{\phi} + \dot{H}^* \dot{\phi} - \frac{\partial}{\partial\phi}\{\frac{1}{2}\dot{\phi}^T H^* \dot{\phi}\} = \tau \tag{3}$$

where

$$H^* \equiv H_\phi - [J_{Tw}^T, H_{\omega\phi}^T] \begin{bmatrix} wE & w\tilde{r}_{0g}^T \\ w\tilde{r}_{0g} & H_\omega \end{bmatrix}^{-1} \begin{bmatrix} J_{Tw} \\ H_{\omega\phi} \end{bmatrix} \tag{4}$$

$$H_\omega \equiv \sum_{k=1}^{l} \sum_{i=1}^{n_k} (I_i^k + m_i^k \tilde{r}_{0i}^{kT} \tilde{r}_{0i}^k) + I_0 \tag{5}$$

$$H_{\omega\phi} \equiv \sum_{k=1}^{l} \sum_{i=1}^{n_k} (I_i^k J_{Ri}^k + m_i^k \tilde{r}_{0i}^k J_{Ti}^k) \tag{6}$$

$$H_\phi \equiv \sum_{k=1}^{l} \sum_{i=1}^{n_k} (J_{Ri}^{kT} I_i^k J_{Ri}^k + m_i^k J_{Ti}^{kT} J_{Ti}^k) \tag{7}$$

$$J_{Tw} \equiv \sum_{k=1}^{l} \sum_{i=1}^{n_k} m_i^k J_{Ti}^k \tag{8}$$

$$J_{Ti}^k \equiv [k_1^k \times (r_i^k - p_1^k), k_2^k \times (r_i^k - p_2^k), \dots,$$
$$k_i^k \times (r_i^k - p_i^K), 0, \dots, 0] \tag{9}$$

$$J_{Ri}^k \equiv [k_1^k, k_2^k, \dots, k_i^k, 0, \dots, 0] \tag{10}$$

$$r_{0g} \equiv r_g - r_0 \tag{11}$$

$$r_{0i}^k \equiv r_i^k - r_0 \tag{12}$$

r_i^k : position vector of mass center of link i of arm k
p_i^k : position vector of joint i of arm k
k_i^k : unit vector indicating joint axis direction of link i of arm k
r_0 : position vector of mass center of satellite base body
r_g : position vector of total mass center of the system
τ : internal torque vector applied at joints
and the tilde operator stands for a cross product such that $\tilde{r}a \equiv r \times a$.
All position and velocity vectors are defined with respect to the inertial reference frame.

3.2. basic manipulator control law

For a basic manipulator control, the author discusses velocity-level Cartesian control methods for a manipulator hand, commonly known as "*Resolved Motion Rate Control.*"

There are two different operational reference frames for space manipulators, which distinguish the basis of control strategy: they are *inertial-space frame* and *satellite on-board frame*. In the *on-board* frame operation, the relationship between the manipulator basis and a handling object, both being fixed on the satellite frame, is invariant, even though the satellite translates, rotates or vibrates. Therefore, well-known developed control schemes for ground-based manipulators can be straightforwardly applied to it. However in the *inertial* frame operation, for example the space robot attempts to catch a floating target, both of the robot and target motion must be described in only common back-ground: the *inertial frame*. Therefore, to achieve the floating target capture, we have to consider the hand control with respect to the *inertial* frame: such operation is a space-borne, unique issue that we've never experienced in on-earth robotics.

Define the inertial operation space at velocity for the manipulator hand k as $\nu^k = [\nu_h^{kT}, \omega_h^{kT}]^T$, then it is kinematically expressed as a function of ν_0, ω_0 and $\dot{\phi}$,

$$\nu^k = J_s^k \begin{bmatrix} \nu_0 \\ \omega_0 \end{bmatrix} + J_m^k \dot{\phi} \tag{13}$$

where

$$J_s^k \equiv \begin{bmatrix} E & -\tilde{p}_{0h}^k \\ 0 & E \end{bmatrix} \tag{14}$$

$$p_{0h}^k \equiv p_h^k - r_0 \tag{15}$$

p_h^k : position vector of hand k

$$J_m^k \equiv \begin{bmatrix} 0 \ldots 0 & k_1^k \times (p_h^k - p_1^k) & k_2^k \times (p_h^k - p_2^k) \\ 0 \ldots 0 & k_1^k & k_2^k \end{bmatrix}$$

$$\begin{bmatrix} \ldots & k_n^k \times (p_h^k - p_n^k) & 0 \ldots 0 \\ \ldots & k_n^k & 0 \ldots 0 \end{bmatrix} \tag{16}$$

J_s^k and J_m^k are Jacobian matrices for a satellite motion dependent part and a manipulator dependant part, respectively. The latter is identical to the Jacobian for ground-based manipulators, however, described with respect to the inertial frame. Eq.(13) is obviously different from ground-based manipulators, because the variables ν_0 and ω_0, which represent motion of a footing base of the manipulator, are involved in the basic kinematics.

Again there are alternative operational strategies concerning the satellite motion. One is to control ν_0 and ω_0 arbitrary using gas jet thrusters which yeild external forces and moments. The other is no use of external force/torque devices. In the latter case, the system momentum is throughly conserved, hence the manipulator dynamics and the satellite dynamics are coupled by the momentum conservation law.

Note that attitude control devices such as reaction wheels or momentum gyros do not appear explicitly in the formulation, however those devices generating internal moments are treated as a version of manipulator arms.

Rearranging eq.(2) using the momenta with respect to the inertial space P and L, we obtain:

$$\begin{bmatrix} P \\ L \end{bmatrix} = \begin{bmatrix} wE & -w\tilde{r}_{0g} \\ w\tilde{r}_g & \hat{H}_\omega \end{bmatrix} \begin{bmatrix} \nu_0 \\ \omega_0 \end{bmatrix} + \begin{bmatrix} J_{Tw} \\ \hat{H}_\phi \end{bmatrix} \dot{\phi}$$

$$\equiv I_s \begin{bmatrix} \nu_0 \\ \omega_0 \end{bmatrix} + I_m \dot{\phi} \tag{17}$$

where

$$\hat{H}_\omega \equiv \sum_{k=1}^{l} \sum_{i=1}^{n_k} (I_i^k - m_i^k \tilde{r}_i^k \tilde{r}_{0i}^k) + I_0 \tag{18}$$

$$\hat{H}_\phi \equiv \sum_{k=1}^{l} \sum_{i=1}^{n} (I_i^k J_{Ri}^k + m_i^k \tilde{r}_i^k J_{Ti}^k) \tag{19}$$

Providing the total momentum constant: $(P^T, L^T)^T = const.$, we can solve eq.(17) for v_0 and ω_0, and substitute them into the basic kinematics (13), then obtain the following significant expression.

$$
\begin{aligned}
\nu^k &= J_s^k \begin{bmatrix} v_0 \\ \omega_0 \end{bmatrix} + J_m^k \dot{\phi} \\
&= (-J_s^k I_s^{-1} I_m + J_m^k) \dot{\phi} \\
&\equiv J^{*k} \dot{\phi} + \nu_c^k
\end{aligned}
\tag{20}
$$

where the suffix c stands for the initial value.

This is a relationship to connect the inertial hand-operation space and joint-actuation space directly, implying the satellite motion dynamics (17) or (2). The matrix J^{*k} was firstly derived by the author in 1986 and named as the Generalized Jacobian Matrix (GJM) because it covers all options including even on-ground applications [1]-[3], or we can say *Dynamic Jacobian* because it's a dynamic expression unlike the conventional kinematic Jacobain.

Resolved motion rate control is an inverse-kinematics based control method for manipulators, in which method, the hand operational command at velocity in Cartesian space is resolved into the joint motion command to execute the given task. The control law is defined using the inversion of Jacobian matrix. Then we obtain the control laws for both *on-board* and *inertial* operations:

$$\textit{For satellite on-board operation} \quad : \quad \dot{\phi} = [J_m^k]^{-1} \nu^k \tag{21}$$

$$\textit{For inertial-space operation} \quad : \quad \dot{\phi} = [J^{*k}]^{-1} \nu^k \tag{22}$$

In the equatons, the Jacobians are significantly different: the above is kinematic, whereas the below dynamic. However both have exactly the same form, that brings us an important benefit that the same control frameworks developed for the ground-based robotics can be applied to the space robotics only by replacing the Jacobian matrices.

4. Target Capture Operation

Floating target capturing experiments were carried out by EFFORTS-I simulator during 1987-1990, and the on-line Resolved Motion Rate Control (RMRC) using the Generalized Jacobian Matrix (GJM) with a vision feedback was successfully validated in the real hardware system.

Fig.5 dipicts a mathematical representation of the experimental setup. The following values are measured at each sampling interval Δt: $\Omega_0 \in R$ satellite attitude, $\phi = [\phi_1, \phi_2]^T \in R^2$ manipulator joint angles, $p_r \in R^2$ position of manipulator endpoint and $p_t \in R^2$ position of the target.

As a 2 DOF horizontal motion system, we hence consider endpoint position and velocity, no orientation of a gripper. The desired endpoint velocity v_d is determined with a position error vector $p_t - p_r$ as

$$v_d = \frac{p_t - p_r}{\Delta t} \tag{23}$$

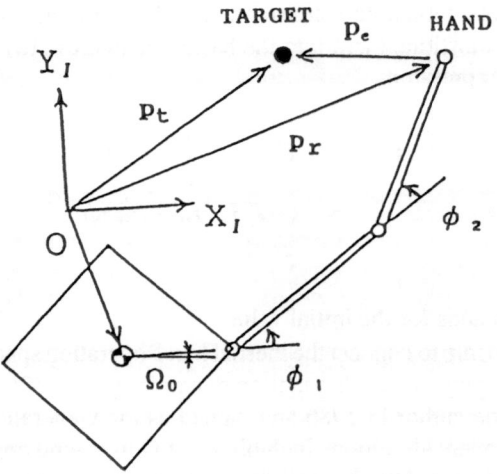

Fig.5 Notations for target capture operation

The joint motion command $\dot{\phi}_d$ is obtained with the inversion of the GJM.

$$\dot{\phi}_d = [J^*(\Omega_0, \phi_1, \phi_2)]^{-1} v_d \tag{24}$$

If the computed joint command exceeds a hardware limitation $|\dot{\phi}_d| > \dot{\phi}_{max}$, then it is modified into

$$\dot{\phi}_d \leftarrow \frac{\dot{\phi}_{max}}{|\dot{\phi}_d|} \cdot \dot{\phi}_d \tag{25}$$

The control loop is described as follows: the hand and joints of the manipulator arm, a reference point of the base satellite, and the target point are marked with light emitting diodes (LED), then their motion are monitored by an overhead CCD camera. Video signals of the LED markers are analysed into the position data p_t, p_r and the satellite attitude data Ω_0 by the Video Tracker (VT), a commercially available video analysis system, and put into a personal computer (PC) via a GPIB communication line. The control command $\dot{\phi}_d$ computed in the PC is transmitted to the robot model through a wire-less communication line. Manipulator joints are controlled with a local velocity feedback to precisely execute the velocity command.

Conditions for the control is specified as:

maximum joint velocity	$\dot{\phi}_{max}$	: 15.0 [deg/sec]
control sampling interval	Δt	: 100 [msec]

Computation is executed by i80386+80387 processors using C language. Position sensing by the VT requires 1/30 seconds and approximately 25 milli-seconds are needed for the calculation of the GJM and its inversion.

Note that the developed control scheme is based on the vision from the ***ground-fixed*** camera, however the examined scheme is equivalent to the one using a ***satellite***

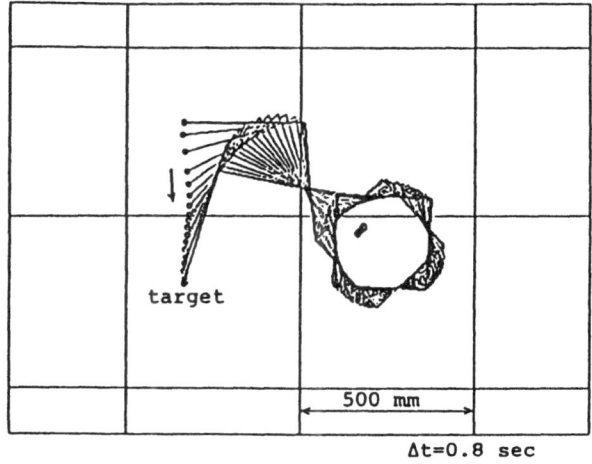

Fig.6 Capture experiment with a stational target

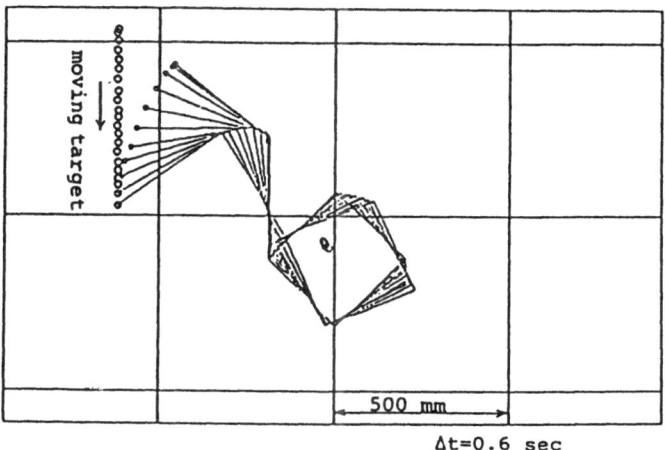

Fig.7 Capture experiment with a moving target

mounted (on-board) vision, because what we significantly need is the position error vector between the hand and target: $p_t - p_r$, which can be observed in any coordinate frame, i.e. the presented scheme can be totally mounted on the satellite.

Fig.6 shows a typical result of capture operation of a stationary target. At initial state, the manipulator hand was located about 0.5 meter from the target, and commanded to go straightly to the target according to eqs.(23)-(25). The result shows that, from the initial point to the target, the hand traveled smoothly in spite that the base satellite rotated as much as 40 degrees by the satellite/manipulator interaction. The maximun hand velocity during the operation was 0.07 [m/s]: the presented control method was stable against relatively slow control interval ($1/\Delta t$=10 [Hz]).

To achieve the chase and capture, or *rendez-vous* with a moving target, the endpoint velocity command (23) was modified with the information of the target velocity $\dot{p}_t$.

$$v_d = \frac{p_t - p_r}{\Delta t} + \dot{p}_t \tag{26}$$

With this small modification, the manipulator came to work remarkably smoother for the capture of both a stationary and a slowly moving target. **Fig.7** shows one of results of capturing a moving target at $\dot{p}_t = 0.05$ [m/sec].

The rendez-vous with a much faster moving target was studied in ref [7], considering a workspace of the manipulator mounted on the floating base.

5. Collision Dynamics Issue

For a theoretical representation of floating-linkage dynamics in presence of external forces, like collision impact forces, the auther has been developing the *Extended-Inversed Inertia Tensor (Ex-IIT)* concept [4][5]. This paper skips the detailed derivation of the Ex-IIT, but the tensor is basically driven with considering a time derivative of eq.(17) caused by the external force input $\mathcal{F}_h = (f_h^T . n_h^T)$ acting on the hand (see **Fig.8**).

$$\frac{d}{dt}\begin{bmatrix} P \\ L \end{bmatrix} = R_h^T \begin{bmatrix} f_h \\ n_h \end{bmatrix} \tag{27}$$

Finally, we obtain the following dynamics equation with respect to the hand acceleration in the inertial frame.

$$\alpha_h = G^* \mathcal{F}_h + d^* \tag{28}$$

where

$$G^* \equiv J^*(H^* + \lambda)^{-1}J^{*T} + R_h M^{-1} R_h^T \tag{29}$$

is the Ex-IIT and d^* is a velocity dependent term [5].

In the above equation, the parameter λ is a matrix named *"Virtual Roter Inertia"* to represent joint condition during the force input.

$$\lambda \equiv \text{diag}[\lambda_1, .., \lambda_n] \quad \in R^{n \times n} \tag{30}$$

$$0 \leq \lambda_i \leq \infty$$

For example, $\forall \lambda_i = \infty$ stands for *'all joints fixed'* and $\forall \lambda_i = 0$ for *'all joints free'*, otherwise *'back-drivable to some extent'*.

Using the Ex-IIT, we can define the virtual mass or effective mass of floating links, which concept expresses the inertia we can feel when we push the link system towards a certain direction. The virtual mass is defined by

$$m_p^* \equiv \frac{|f_h|}{|\alpha_h|} = \frac{1}{n^T G_{11}^* n} \tag{31}$$

where G_{11}^* is the left-above 3×3 part of the Ex-IIT associate with the translational motion, n is a unit vector indicating the pushing direction.

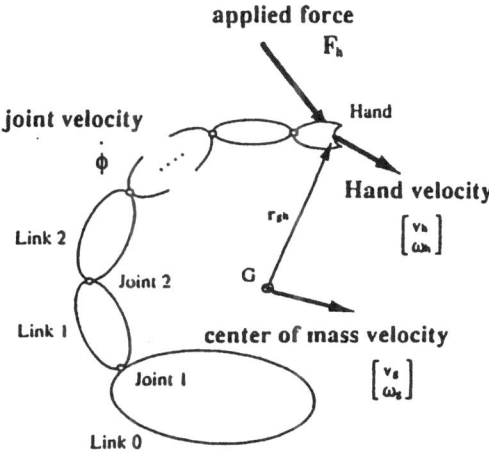

applied force
F_h

joint velocity
$\dot{\phi}$

Hand

Hand velocity
$\begin{bmatrix} v_h \\ \omega_h \end{bmatrix}$

r_{gh}

Link 2

Joint 2

G

Link 1

center of mass velocity
$\begin{bmatrix} v_g \\ \omega_g \end{bmatrix}$

Joint 1

Link 0

Fig.8 General model in presence of external force input

With the virtual mass, we can get a simple expression for the collision of a target and the floating manipulator. Let v_t, v_h be the target and manipulator velocities projected on the collision approaching line n, we get

$$m_p^*(v_h' - v_h) = m_t(v_t - v_t') \qquad (32)$$

where m_t is target mass and $\{'\}$ indicates the rebounding velocity after collision. If well-known restitution (elastic) coefficient e is introduced,

$$v_h' - v_t' = e(v_h - v_t). \qquad (33)$$

Using these two equations, we can describe the velocity relationship among two floating systems across the collision.

Supposing inertial parameters are well known for both systems, still there are a couple of unknown parameters: the restitution coefficient e and the virtual rotor inertia at each manipulator joint λ_i involved in the Ex-IIT. It is considerably defficult to theoretically predict these values (if try, you need a perfectly precise model of the inertia and elasticity for all particles), but relatively easy to estimate them by experiment. Observation data of the velocities for both systems are provided, we can solve (32)(33) for m_p^* and e.

The collision experiment was carried out using the EFFORTS-I with several joint conditions. **Fig.9** shows the experimental setup. The robot model was floating and keeping still with $\phi_2 = 90$ degrees. A floating target was coming from various directions with θ and hit the manipulator, then rebounded. Observing the velocity relationship, the author has got plots on the magnitude of the virtual mass for each collision directions, as shown in **Fig.10**. There are three groups of plots: circular ones are for the case of joint *'free'*, square ones for *'locked'*, and black triangles for *'back-drivable'* with an actual gear trains and a rotor of the DC motor.

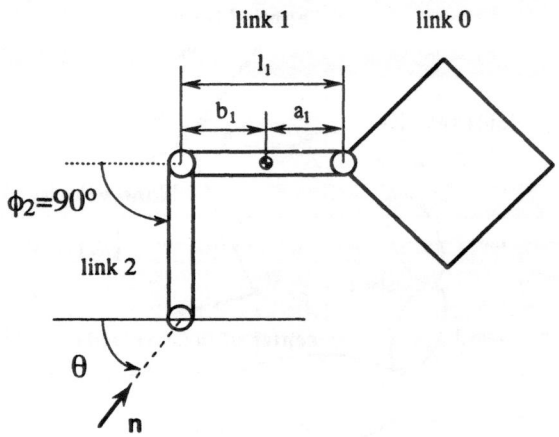

Fig.9 Collision experiment setup

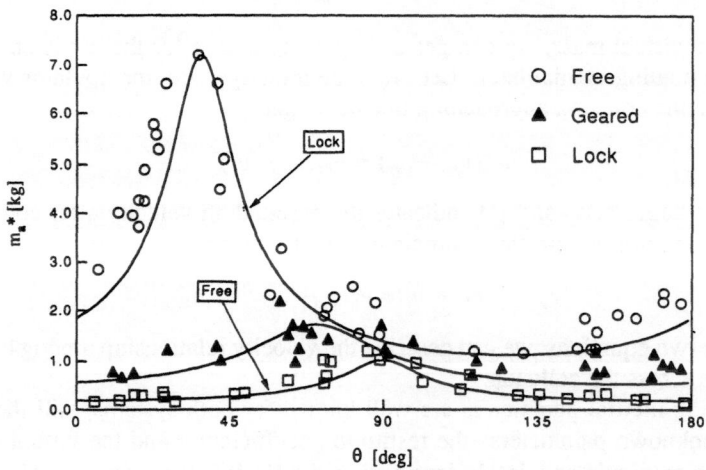

Fig.10 Virtual mass for various collision directions

The data tell us that we can feel the virtual mass of the system always heavier in locked joints case, middle in back-drivable, and lighter in free case, for each impact direction. Especially at $\theta = 37$ degrees with the locked joints, we feel heaviest of all, and the maximum value is exactly same as the total mass of the floating robot. Because at this special condition, the force input line shoots the system mass center, the robot makes any rotation after the collision but just translation. The rotation of the links makes the virtual mass small.

In the figure, the author drew theoretical lines calculated by the definition (31), using $\lambda_1 = \lambda_2 = 0$ for the free case and $\lambda_1 = \lambda_2 = \infty$ for the locked case. And for the back-drivable case, $\lambda_1 = \lambda_2 = 0.07[kgm^2]$ seems to give the best approximation. Theoretically estimating the λ from the catalogue-data of motor-rotor and gear-ratio, we've got $\lambda = 0.013[kgm^2]$, which value is much smaller than the experimental results: this implies that the catalogue-data does not include inertia of each gear particles, and other unknown motion resistance.

As for the restitution coefficient, e was identified approximately 0.214, in average. Although the data have large dispersion, there is no statistically significant difference among three joint conditions, as far as the presented experiments.

In general, it is almost impossible and non-sense to establish a precise mathematical model of the gears, actuators or elastic factors, therefore the experimental identification as presented above is extremely essential to this problem.

6. Force Control Primitives

Force feedback control is indispensable for fine manipulation, like space structures assembling with insertion, pushing and twisting actions. In space missions, such tasks will be commanded as satellite on-board operation, i.e. the manipulators should work with respect to the *on-board* coordinate frame, and the manipulator system, including the satellite base and handled object, comprise a mechanical closed-loop chain when a contact occurs. In this sence, there is no significant difference between force controls in space and on ground (see **Fig.11**). However, the drawbacks will be found in practical implementation, for example, due to limitation of computational power available in orbital environment, or a thin and extremely long communication line from a ground control station. The latter is particularly serious for *tele-operation* because it causes harmful time-delay.

This paper discusses a simple compliant force control to cope with the low computational capability, or low control frequency, and employs a classic, linear force accomodation control or *damping control* [13].

The basic idea of the force accomodation control is just to modify the hand operational velocity in Cartesian space with a force signal and a linear accomodation matrix α. If ν_j is a Cartesian velocity command given by an operator's joystick, and $\mathcal{F}$ the contact force between the hand and an obstacle, the modified motion command ν_d is determined by

$$\nu_d = \nu_j - \alpha\mathcal{F}, \tag{34}$$

which will be resolved into the joint space command using eq.(21).

A control blockdiagram for this simple scheme is dipicted in **Fig.12**.

Whiteny [13] analyzed the stability of the scheme and obtained the following result (notations correspond with **Fig.12**):

$$(Ts)^2 K_c \alpha K_e < 1 \tag{35}$$

where Ts is a digital control sampling interval, K_c a total gain of the servo controller, and K_e a stiffness of the environment.

This tells us an important knowledge that for a harder (stiffer) environment we must use a smaller gain α or a higher-speed (providing smaller Ts) computer, on the

576

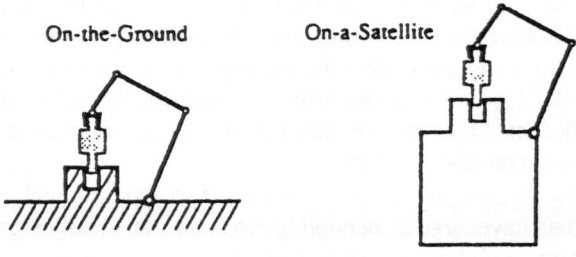

On-the-Ground   On-a-Satellite

Fig.11 Force control by a terrestrial arm and a space arm

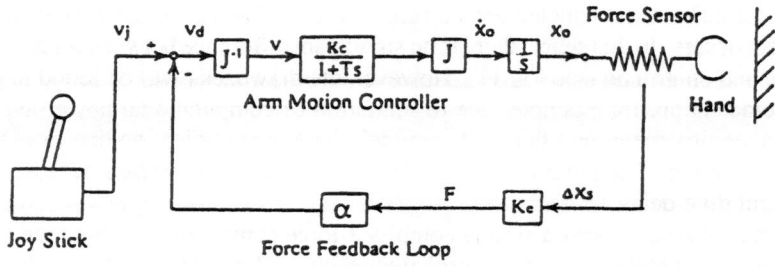

Fig.12 A diagram for force accomodation control

other aspect, for a slower-speed computer use a smaller gain or operate in a softer environment. In the experimental study, the author adopts a soft, easily deformable force sensor to provide a lower K_c, against a kind of common sense that the force sensor must be as stiff as possible, and obtains a good result using a relatively slower-speed computer.

This primitive force control was examined with the EFFORTS-II simulator. We employed a *notebook-style* personal computer with the i80386SX processor, for motion control for two reasons: one is that computational power of such personal com-

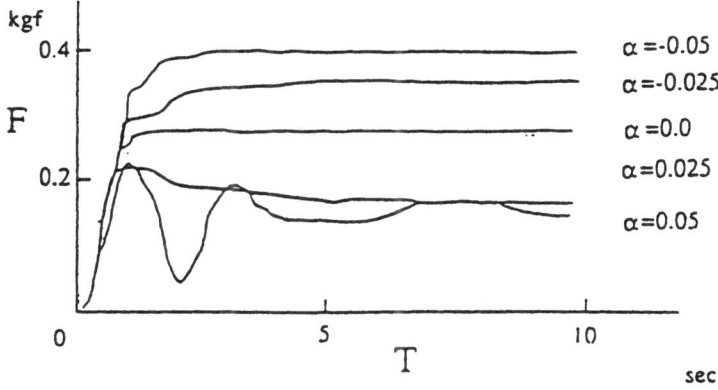

Fig.13 Step response of the implemented force control for various gains

puters is regarded to be equivalent to the real flight model which will be available in practical space systems in late 1990s or at the beginning of the 21st century [14], the other is that this type of PC can be easily installed on the experimental floating robot model.

Not a multi-task processing having been employed, the processor must serially care for all sensor inputs and servo-controller outputs, the computational sampling interval is thereby incredible slow for the force control: 500-1000 [msec], about 1000 times slower than the advanced terrestrial manipulators for research-use. However it must be realistic of space-use. To cope with this low computational power, based on the stability criteria of eq.(35), the author adopts low stiffness springs of 1.0 [kgf/cm] in force sensors, less than one-1000th of commonly used stiffness . The compliance of the springs works to reduce the environmental stiffness at contact, hence offers stabler operation.

Fig.13 shows experimented step-responses when the hand hits a target fixed on the base satellite with several force feedback gains, where $\alpha=0$ shows natural compliance. The positive gain makes the arm compliant: the larger α becomes, the lighter and the more vibrationally the hand reacts, to the contrary, the negative α stiffens the system and increases stability.

7. Conclusion

This paper presented the experimental research activity on the basic of space robotics with the Experimental Free-Floating Robot Satellite (EFFORTS) simulators, which can mechanically simulate the planar floating dynamics. The author exhibited (a) the experimental results on target capture operations to verify the inertial space RMRC method with the Generalized Jacobian Matrix concept, (b) collision experiments to

578

confirm the idea for impact dynamics representation with the Extended-Inversed Inertia Tensor, and (c) force controls with soft compliance to provide stable manipulation under the limitation of computational power. As discussed in Section 2, these experimental data of the planar floating systems could convince us the basis of the theoretical ideas and provide the significant knowledge for future more practical 3D simulations.

References

[1] *Space Robotics: Dynamics and Cotrol,* ed. by Xu and Kanade, Kluwer Academic Publishers, 1993.

[2] Y.Umetani and K.Yoshida: "Continuous Path Control of Space Manipulators Mounted on OMV," *Acta Astronautica,* vol.15, No.12, pp.981-986, 1987. (Presented at the 37th IAF Conf, Oct. 1986)

[3] Y.Umetani and K.Yoshida: "Resolved Motion Rate Control of Space Manipulators with Generalized Jacobian Matrix," *IEEE Trans. on Robotics and Automation,* vol.5, No.3, pp.303-314, 1989.

[4] K.Yoshida et al.: "Modeling of Collision Dynamics for Space Free-Floating Links with Extended Generalized Inertai Tensor," *Proc. of 1992 IEEE Int. Conf. on Robotics and Automation,* Nice, France, pp.899-904, 1992.

[5] K.Yoshida and N.Sashida: "Modeling of Impact Dynamics and Impulse Minimization for Space Robots," *Proc. of 1993 IEEE/RSJ Int. Conf. on Intelligent Robots and Systems,* Tokyo, pp.2064-2069, 1993.

[6] Y.Umetani and K.Yoshida: "Experimental Study on Two Dimensional Free-Flying Robot Satellite Model," *Proc. of NASA Conf. on Space Telerobotics,* vol.5, pp.215-224, Pasadena, CA, 1989.

[7] K.Yoshida and Y.Umetani: "Control of Space Free-Flying Robot," *Proc. of 29th IEEE Conf. on Decision and Control,* pp.97-102, 1990.

[8] D.Akin and R.Howard: "Neutral Buoyancy Simulation of Space Telerobotics Operations," *SPIE Vol.1612 Coorperative Intelligent Robotics in Space II,* pp.414-420, 1991.

[9] Y.Xu et al: "Control System of Self-Mobile Space Manipulator," *Proc. of 1992 IEEE Int. Conf. on Robotics and Automation,* pp.866-871, 1992.

[10] H.Fujii, H.Yoneyama and K.Uchiyama: "Experiments on Cooperative Motion of a Space Robot," *Proc. of 1993 IEEE/RSJ Int. Conf. on Intelligent Robots and Systems,* Tokyo, pp.2155-2162, 1993.

[11] R.Koningstein, M.Ullman and R.H Cannon,Jr: "Computed Torque Control of a Free-Flying Coorperating-Arm Robot," *Proc. of NASA Conf. on Space Telerobotics,* vol.5, pp.235-243, 1989.

[12] H.Shimoji, M.Inoue, K.Tsuchiya, et al.: "Simulation System for a Space Robot Using 6 Axis Servos," *Proc. of 11th IFAC Symp. on Automatic Control in Aerospace,* pp.131-136, 1989.

[13] D.Whiteny: "Force Feedback Control of Manipulator Fine Motions," *ASME J. of Dyn. Sys. Meas. Cntl,* pp.91-97, June 1977.

[14] M.Oda et al: "The ETS-VII, the World First Telerobotic Satellite," *Proc. Artificial Intelligence, Robotics and Automation, in Space '92,* pp.307-318, 1992.

ROTEX -
The First Space Robot Technology Experiment

G. Hirzinger

D L R

(German Aerospace Research Establishment)
Oberpfaffenhofen, D-82234 Wessling

Abstract— The paper describes the key technologies developed for the space robot technology experiment ROTEX that flew with shuttle flight STS 55 end of April 93. During this "spacelab-D2"-mission for the first time in the history of space flight a small, multisensory robot (i.e. provided with modest local intelligence) has performed a number of prototype tasks on board a spacecraft in the most different operational modes that are feasible today, namely preprogrammed (and reprogrammed from ground), remotely controlled (teleoperated) by the astronauts using a control ball and a stereo-TV-monitor, but also remotely controlled from ground via the human operator as well as via machine intelligence. In these operational modes the robot successfully closed and opened connector plugs (bajonet closure), assembled structures from single parts and captured a free-floating object.

Key technologies for the success of ROTEX have been its multisensory gripper technology, local (*shared autonomy*) sensory feedback control concepts, and the powerful delay-compensating 3D-graphics simulation (predictive simulation) in the telerobotic ground station.

1. Introduction

Automation and robotics (A&R) will become one of the most attractive areas in space technology, it will allow for experiment-handling, material processing, assembly and servicing with a very limited amount of highly expensive manned missions (especially reducing dangerous extravehicular activities). The expectation of an extensive technology transfer from space to earth seems to be much more justified than in many other areas of space technology.

This is one of the main reasons why several activities towards space robotics have started in a number of countries, one of the largest projects being the space station's mobile servicing center (MSC) with its two-arm special dexterous manipulator system to be built by the Canadian space agency. NASA's own big robot project, the flight telerobotic servicer (FTS), was cut down some time ago apparently due to excessive development costs, but we are sure that for a space station reduced in size, and may be only temporarily habited by astronauts, the use of robots will become a major issue again. Other remarkable mid-term projects are the Japanese space station "remote manipulator system" (JEM-RMS) or the Japanese ETS-VII project, a experimental flight telerobotic servicer for maintenance and repair of space systems. While the shuttle manipulator arm, which has been flown with the space-shuttle a number of times in the past (including spectacular actions like the solar-max satellite rescue in 1984), may be seen as a predecessor system for the Canadian MSC, there has been no space robot experience in Europe in the past. The European Space Agency ESA is going to prepare the use of robots in the European part COLUMBUS of the space station. Apparently those nations having less background and history in manned spaceflight have a strong interest in space robotics. On the other side we have a strong belief that for complex, partly autonomous robots with extensive ground control capabilities it would be too risky to leap from zero experience to a fully operational system; therefore we have proposed in 1986 the space robot technology experiment ROTEX, which has meanwhile successfully flown in space (spacelab mission D2 on shuttle flight STS 55 from April 26 to May 6, 93). ROTEX contained as much sensor-based on-board autonomy as possible from the present state of technology, but on the other side assumed that for many years cooperation between man and machine via powerful telerobotic structures will form the basis for high-performance space robot systems operable especially from a ground station, too. Thus ROTEX tried to prepare all operational modes which we can foresee in the coming years (not including the perfectly intelligent robot that would not need any human supervisor), and it also tried to prepare the most different applications by not restricting its prototype tasks to internal servicing operations, but also aiming at assembly and external servicing (e.g. grasping a floating satellite). The key features of this space robot project are explained in the sequel.

2. The basic features and goals of ROTEX

The experiment's main features were as follows (fig. 1):

- A small, six-axis robot (working space $\sim$ 1 m) flew inside a space-lab rack (fig. 2). Its gripper was provided with a number of sensors, especially two 6-axis force-torque wrist sensors, tactile arrays, grasping force control, an array of 9 laser-range finders and a tiny pair of stereo cameras to provide a stereo image out of the gripper; in addition a fixed pair of cameras provided a stereo image of the robot's working area.

- In order to demonstrate servicing prototype capabilities three basic tasks were performed:

  a) assembling a mechanical truss structure from three identical cube-link parts

  b) connecting/disconnecting an electrical plug (orbit-replaceable-unit-ORU-exchange using a "bajonet closure")

  c) grasping a floating object

- A variety of operational modes was verified,

- teleoperation on board (astronauts using stereo-TV-monitor)

- teleoperation from ground (using predictive computer graphics) via human operators and machine intelligence as well

- sensor-based off-line programming
(teaching by showing in a virtual graphics environment on ground including sensory perception, with sensorbased automatic execution later on-board).

The operational modes were based on a unified control approach for which we have coined the term *tele-sensor-programming*.

• Typical goals of the experiment were:

- To verify joint control (including friction models) under zero gravity as well as μg-motion planning concepts based on the requirement that the robot's accelerations while moving must not disturb any μg-experiments nearby.

- To demonstrate and verify the use of DLR's sensorbased 6 dof-handcontrollers ("control balls") under zero gravity.

- To demonstrate the performance of a complex, multisensory robot system with powerful man-machine-interfaces (as are 3D-stereo-computergraphics, 6 dof control ball, stereo imaging), in a variety of operational modes that especially include on-line teleoperation and off-line programming from ground.

During the mission in addition to executable programs a number of different joint controllers including friction observers have been sent up to the robot and tested by our colleagues from the University of Paderborn. However in this paper we will focus on the telerobotic aspects of ROTEX.

Local sensory feedback from the multisensory gripper was a key issue in the telerobotic concepts used, so we will address the gripper's components in more detail.

3. Sensory and mechatronic aspects

The ROTEX end-effector was a complex multisensory two-finger gripper. The gripper sensors belong to the new generation of DLR robot sensors based on a real multisensory concept with all analog preprocessing and digital computations performed inside each sensor or at least in the wrist in a completely modular way (fig. 3). Using a high speed serial bus with 375 kBaud only two signal wires are coming out of the gripper (carrying all sensory information), augmented by two 20 kHz-power supply wires, from which the sensors derive their DC-power supply voltages via tiny transformers themselves.

In the gripper 15 sensors are provided, in particular (fig. 3 and 4):

a) an array of 9 laser range finders based on triangulation, one "big" sensor (i.e. half the size of a match box) for the wider range of 3-35 cm, and 4 smaller ones in each finger for lower ranges of 0-3 cm.

b) A tactile array of 4 x 8 sensing elements (conductive rubber "taxels") in each finger. The dimensions of the tactile area is 32 x 16 mm. The binary state of each taxel is serially transmitted through analog multiplexers without additional wiring.

c) A "stiff" 6 axis force–torque sensor based on strain–gauge measurements and a compliant optical one. Originally it seemed necessary to make a final decision between these two principles, but as indicated in fig. 3 and 4 they finally were combined into a ring–shaped system around the gripper drive, the instrumented compliance being lockable and unlockable electrically. Shaping these sensors as rings around the gripper drive shows up different advantages:

- it does not prolong the axial wrist length

- it brings the measurement system nearer to the origin of forces–torques and yields a better ratio of torque range to force range than achievable with a compact form.

The optically measuring instrumented compliance was e.g. described in /10/. For more drawings and pictures see e.g. /16/.

The stiff, strain–gauged force–torque sensor is a new design no longer based on spokes or bars but membranes. It performs automatic temperature compensation based on the temperature characteristics as stored during the calibration process and continues operating reliably with reduced accuracy if one of the strain–gauges is damaged.

d) A pair of tiny stereo CCD-cameras, the CCD's plus optics plus minimum electronics taking a volume smaller than a match–box, too.

e) An electrical gripper drive, the motor of which is treated like a sensor with respect to the data bus and the 20 kHz power supply connections. An ESCAP stepping motor has been converted into a brushless dc motor by integrating hall sensors. The basic problem was to transform the motor's high-speed rotational motion into a fairly slow axial motion to move the fingers. What we gained with this motor–gearing combination is a small prismatic drive (applicable also in a robot joint), which used as gripper drive allows to exert grasp forces up to 300 N with a gripper weight of 5 N and a grasp speed of about 15 cm/sec; without measuring and feeding back grasp forces we arrived at a feedforward grasp force control resolution of ≈ 1 N (< 1 % of max force) with high repeatability. Reduction rate referring to the finger rotation is $\approx 1 : 1000$.

With more than 1000 tiny SMD electronic (fig. 5) and several hundred mechanical components the ROTEX gripper is probably the most complex multisensory endeffector that has been built so far. The gripper was not space qualifiable on component level especially because SMD technology is not yet generally permitted in space; so it had to undergo vibration, temperature, off-gasing and EMC-tests as a whole; we finally got a so-called "waiver", i.e. a special permission, from NASA. As a positive experience during the flight all gripper sensory and drive systems (which were permanently sending their temperatures down in addition to the measuring values) remained below 40°C, while being operable up to 80°C. All components worked absolutely reliable.

The flight version of the 6 axis ROTEX arm (without end-effector) was built by the German Space Company DORNIER using conventional space qualifiable technology (fig. 6a).

With a weight of nearly 400 N it however was not able to sustain itself on ground. Thus we built a light weight arm for the astronaut training with new joints integrated into carbon fibre grid structures /23/, see fig. 6b. It shows up a 1:1 ratio between own weight and load capability (100 N both). Presently a new version of this arm is provided with torque-controlled joints based on inductive torque sensing; for us it leads into a new generation of multisensory, torque-controlled (instead of position-controlled) robots.

4. Telerobotic Control: The Tele-Sensor-Programming Approach

The telerobotic control in ROTEX was in a unified way based on our *tele-sensorprogramming* approach; it comprises on-line teleoperation on board and on ground as well as sensorbased-off-line programming on ground and subsequent on-board execution (following kind of a "learning by showing" philosophy). Basically this approach has two main features

- a *shared control concept* (see e.g. /24/) based on local sensory feedback at the robot's site (i.e. on-board or in a predictive ground simulation), by which gross commands were refined autonomously providing the robot with a modest kind of sensory intelligence (fig. 7 and 8a). However due to processor limitations, on-board sensory feedback in ROTEX was restricted to force-torque and range finder signals (see below). Active compliance as well as hybrid (pseudo-)force control using nominal sensory patterns in the sensor-controlled subspace, based on MASON's C-frame-concept /4/ was realized locally. Gross commands in this framework may originate from a human operator handling the control or sensor ball (a 6 dof non-force-reflecting hand controller) or alternatively from an automatic path planner. The techniques used for projecting gross commands into the position and sensor-controlled subspaces have been discussed in a number of previous papers (e.g. /24/). Feedback to the human operator - in case of on-line teleoperation - was provided only via the visual system, i.g. for the astronaut via stereo TV images, for the ground operateur mainly via predictive stereo graphics (the stereo

TV images being an add-on for verification). This allowed us to realize a unified control structure despite of the fairly large round-trip signal delays of up to 7 seconds. Indeed predictive 3D-Stereographic simulation was a key issue for the big success of this space robot experiment, and in our opinion is the only efficient way to cope with large signal delays. Of course for these kind of ideas to work the same control structures and path planners had to be realized on-board as well as in the predictive graphics ground station. And this in turn meant that not only the robot's free motion but also its sensory perception and feedback behaviour had to be realized in the "virtual environment" on ground.

- an *elemental move* concept, i.e. any complex task like the dismounting and remounting of the bajonet closure is assumed to be composed of elemental moves, for which a certain constraint-frame- and sensor-type-configuration holds (to be selected by the operator e.g. using a set of predefined alternatives), so that automatic sensorbased path refinement is clearly defined during these motion primitives. As an example the bajonet closure screwing operation, but also the approach phase before grasping the closure, were typical sensor-based elemental moves, the screwing operation being a nice example for shared control as well, as the hand rotation was position- (or better orientation-) controlled using gross command projections, while the motion along the rotation axis was locally force controlled so that forces arising due to the screwing operation werde immediately compensated.

Basically the elemental move concept as realized in the ROTEX system requests various definitions and prodecures, in particular

- defining (or graphically demonstrating) the initial and goal situations (positions augmented by nominal sensory patterns); of course in case of on-line teleoperation the gross-path in between is also given by the operator, else it is generated later (i.e. on-board) by the path planner.

- providing the a-priori knowledge on the C-space configuration and the type of shared control (active compliance or using nominal sensory patterns).

- procedures for mapping sensory errors into positional/rotational errors (e.g. using a neural net training that allows to realize sensor fusion, too).

- procedures for mapping positional/rotational errors into motion commands.

- procedures for recognizing actual and goal states, thus determining e.g. the end of an elemental move. It seems worth mentioning that of course the robot in its real world in general is not able to reach both precisely, the nominal position as well as the nominal sensory pattern; this conflict was resolved easily by using projections of these nominal data into the position and sensorcontrolled subspaces. If in the goal state all 6 degrees of freedom are sensor-controlled, then in correspondence with a "relative reference philosophy" explained below of course the sensor information has absolute priority.

To us it seems important to emphasize again that the ROTEX tele-sensor-programming concept with its elemental move features comprises sensorbased on-line teleoperation by hand of predictive graphics simulation (including e.g. remote active compliance) as well as a corresponding off-line-programming version, which may be characterized as "sensorbased teaching by showing". Hereby the robot is graphically guided through the task (off-line on ground), storing not only the relevant hand frames but also the corresponding nominal sensory patterns (graphically simulated) for later (e.g. associative) recall and reference (fig. 7) in the on-board execution phase, after these data packages have been sent up. Thus we might talk of storing "situations" sent to the on-board path planner. Indeed this mode of tele-sensor-programming is a form of off-line-programming which tries to overcome the well-known problems of conventional approaches, especially the fact that simulated and real world are not identical. But instead of e.g. calibrating the robot (which is only half the story) **tele-sensor-programming** provides the real robot with simulated sensory data that refer to **relative** positions between gripper and environment, thus compensating for any kind of inaccuracies in the absolute positions of robot and real world.

An impressive verification of these concepts was given during the mission when the ground operator in on-line teleoperation stepped through the ORU-exchange task **without** waiting at the end of the elemental moves until the robot in space had confirmed reaching the goal situation of the corresponding elemental move.

Realistic simulation of the robot's environment and especially the sensory information (fig. 9) presumably perceived by the real robot was of crucial importance for this approach and its explained in more detail in /22/.

Nevertheless there are errors to be expected in any graphics simulation compared with the real world and therefore not only e.g. the gross commands of the TM command device (control ball) were fed into the simulation system, but also the real sensor data coming back from space (including the real robot position) were displayed, recorded and compared with the presimulated ones. In a future stage not yet realized in the D2-flight these comparisons might lead to an automatic update of e.g. world model, sensor models etc.. All sensory systems of ROTEX worked perfectly and the deviations between presimulated and real sensory data were minimal (fig. 10). This was one of the many positive surprises of ROTEX.

In the telerobotic ground station (fig. 12) a number of computers were connected via a VME-bus shared memory concept, especially powerful SGI (Silicon Graphics) "power vision" systems that allowed to display (in stereographic technology) the different artifical workcell views in parallel, simulating the workcell cameras, the hand cameras and an optional observer view which was varied by a control ball. During the ROTEX mission we did not overlay real and simulated images, instead the real endeffector's position was indicated by the hand frame and the real gripper's position by two patches in the graphics scene. In addition the graphics system permanently displayed real and simulated sensory data in form of overlayed bars (fig. 9) or dots (in case of the tactile arrays, see lower left part of fig. 9), while an additional SGI system displayed the time history of simulated and real sensory signals shifted by the actual delays, thus correlated in time (fig. 10).

5. Catching the floating object

There was only one exception from the local sensory feedback concept in RO-TEX. It refers to (stereo-) image processing. In the definition phase of ROTEX (around 1986) no space qualifiable image processing hardware was available; nevertheless we took this as a real challenge for the experiment "catching a free-floating object from ground" (fig. 8b). In contrast to contact operations as necessary in case of assembly we may deal here with a nearly perfect world model, as the dynamics of an object floating in zero g are well known. Hand-camera information on the free-flyer's pose (relative to the gripper) was provided on ground using alternative schemes; one was based on the "dynamic vision approach" as given in /20/, using only one of the two tiny hand-cameras, the other one was a full stereo approach realized in a multitransputer system. In both cases the thus "measured" object poses were compared with estimates as issued by an extended Kalman filter that simulates the up- and down-link delays as well as robot and free-flyer models; this Kalman filter /9, 21/ predicts (and graphically displays) the situation that will occur in the spacecraft after the up-link delay has elapsed and thus allows to close the "grasp loop" either purely operator controlled, or via shared control, or purely autonomously (i.e. solving an automatic rendezvous and docking problem). Fig. 11 shows photos of the TV-scene out of one of the hand cameras immediately before successful, fully automatic grasping from ground despite of 6,5 sec round-trip delay, following the image processing approach in /20/. This automatic, ground-controlled capture of the free-flyer was one of the many spectacular actions of ROTEX.

6. Conclusion

For the first time in the history of space flight a small, multisensory robot (i.e. provided with modest local intelligence) has performed a number of prototype tasks on board a spacecraft in the most different operational modes that are feasible today.

Key technologies for the success of ROTEX have been

- the multisensory gripper technology; with 16 sensors (e.g. a stiff and a more compliant optical force-torque-sensor, 9 laser range finders, a stereo camera and tactile arrays) and more than 1000 electronic components the ROTEX gripper presumably is the most complicated robot gripper built so far; nevertheless it worked perfectly during the mission. The stereo images out of the hand camera as well as those from the workcell camera were impressive.

- local (*shared autonomy*) sensory feedback control concepts, refining gross commands autonomously by intelligent sensory processing

- powerful delay-compensating 3D-stereo-graphic simulation (predictive simulation), which included the robot's sensory behaviour.

In addition to the overall performance observations the initialization phase showed interesting effects when different adaptive joint control parameters including friction observers were uploaded. Due to missing gravity the joints had no preloading and thus the controllers had to get along with backlash effects etc..

In-flight-calibration of the robot using the finger-tip laser range finders improved its positioning performance.

The experiment also clearly showed that the information and control structures in mission control centres for future space robot applications should be improved, allowing the robot operator on ground direct access to the different types of uplinks and providing him with a continuous TV-transmission link.

Close cooperation between man (astronaut or ground operator) and machine comprising different levels of robot autonomy was the basis of the success of ROTEX. It was clearly proven that a robot system configured in this flexible way of arbitrary and fast switching between the most different operational modes will be a powerful tool in assisting man in future space flight projects and it was impressively shown that even large delays can be compensated by appropriate estimation and pre-simulation concepts.

A number of terrestrial spin-off effects of DLR's (German Aerospace Research Establishment) space robotics work are visible even today. Just to mention one example: The 6 dof handcontrollers (control balls) used during the mission by astronauts and ground-teleoperators have been redeveloped meanwhile into a *space control mouse* for 3D-graphics, which will be distributed and manufactured in license now also by the computer-mouse company LOGITECH.

7. References

/1/ S.Lee, G. Bekey, A.K. Bejczy, "Computer control of space-borne teleoperators with sensory feedback", Proceedings IEEE Conference on Robotics and Automation, S. 205-214, St. Louis, Missouri, 25-28 March 1985.

/2/ J. Heindl, G. Hirzinger, "Device for programing movements of a robot", US-Patent: No. 4,589,810, May 20, 1986.

/3/ G. Hirzinger, J. Dietrich, "Multisensory robots and sensorbased path generation". Proceedings IEEE Conference on Robotics and Automation, S. 1992-2001, San Francisco, April 7-10, 1986.

588

/4/ M.T. Mason, "Compliance and force control for computer controlled manipulators", IEEE Trans. on Systems, Man and Cybernetics, Vol SMC-11, No. 6 (1981, 418-432).

/5/ G. Hirzinger, K. Landzettel, "Sensory feedback structures for robots with supervised learning". Proceedings IEEE Conference, Int. Conference on Robotics and Automation, S. 627-635, St. Louis, Missouri, March 1985.

/6/ G. Hirzinger, J. Heindl, " Sensor programming, a new way for teaching a robot paths and forces torques simultaneously". 3rd Int. Conference on Robot Vision and Sensory Controls, Cambridge, Massachusetts/USA, Nov. 7-10. 1983.

/7/ T.B. Sheridan, "Human supervisory control of robot systems". Proceedings IEEE Conference, Int. Conference on Robotics and Automation, San Francisco, April 7-10, 1986.

/8/ B.C. Vemuri, G. Skofteland, "Motion estimation from multi-sensor data for tele-robotics", IEEE Workshop on Intelligent Motion Control, Istanbul, August 20-22, 1990.

/9/ G. Hirzinger, J. Heindl, K. Landzettel, "Predictive and knowledge-based telerobotic control concepts". IEEE Conference on Robotics and Automation, Scottsdale, Arizona, May 14-19, 1989.

/10/ J. Dietrich, G. Hirzinger, B. Gombert, J. Schott, "On a Unified Concept for a New Generation of Light-Weight-Robots", Proceedings of the Conference ISER, Int. Symposium on Experimental Robotics, Montreal, Canada, June 1989.

/11/ F. Lange, "A Learning Concept for Improving Robot Force Control", IFAC Symposium on Robot Control, Karlsruhe, Oct. 1988.

/12/ J. S. Albus, "A New Approach to Manipulator Control: The Cerebellar Model Articulation Controller (CMAC)", Transactions of the ASME, Journal of Dynamic Systems, Measurement and Control, pp. 221-227, Sept. 1975.

/13/ S. Hayati, S.T. Venkataraman, "Design and Implementation of a Robot Control System with Traded and Shared Control Capability", Proceedings IEEE Conference Robotics and Automation, Scottsdale, 1989.

/14/ L. Conway, R. Volz, M. Walker, "Tele-Autonomous Systems: Methods and Architectures for Intermingling Autonomous and Telerobotic Technology", Proceedings IEEE Conference Robotics and Automation, Raleigh, 1987.

/15/ G. Saridis, "Machine-Intelligent Robots: A Hierarchical Control Approach", in Machine Intelligence and Knowledge Engineering for Robotic Applications, NATO ASI Series F. Vol. 33, Springer Verlag 1987.

/16/ G. Hirzinger, J. Dietrich, B. Gombert, J. Heindl, K. Landzettel, J. Schott, "The sensory and telerobotic aspects of the space robot technology experiment ROTEX", Proc. i-SAIRAS 2th Int. Symposium Artificial Intelligence, Robotics and Automation, in Space, Toulouse, France, Sept.30-Oct.2, 1992.

/17/ G. Hirzinger, J. Heindl, K. Landzettel, B. Brunner, "Multisensory shared autonomy - a key issue in the space robot technology experiment ROTEX", IEEE Conf. on Intelligent Robots and Systems (IROS), Raleigh, USA, July 7-10, 1992.

/18/ J. Funda, R.P. Paul, "Efficient control of a robotic system for time-delayed environments". Proceedings of the Fifth International Conference on Robotics and Automation, pages 133-137, 1989.

/19/ P. Simkens, "Graphical simulation of sensor controlled robots". PhD. Thesis, 1990, KU Leuven.

/20/ D. Dickmanns, "4D-dynamic scene analysis with integral spatio-temporal models", Fourth Int. Symposium on Robotics Research, Santa Cruz, Aug. 1987.

/21/ Christian Fagerer and Gerhard Hirzinger, "Predictive Telerobotic Concept for Grasping a Floating Object", Proc. IFAC Workshop on Spacecraft Automation and On-Board Autonomous Mission Control, Darmstadt, Sept. 1992

/22/ B. Brunner, G. Hirzinger, K. Landzettel, J. Heindl, "Multisensory shared autonomy and tele-sensor-programming - key issues in the space robot technology experiment ROTEX", IROS'93 International Conference on Intelligent Robots and Systems, Yokohama, Japan, July 26-30, 1993.

/23/ G. Hirzinger, A. Baader, R. Koeppe, M. Schedl, "Towards a new generation of multisensory light-weight robots with learning capabilities", IFAC'93 World Congress, Sydney, Australia, July 18-23, 1993.

/24/ G. Hirzinger, J. Heindl, K. Landzettel, B. Brunner, "Multisensory shared autonomy - a key issue in the space robot technology experiment ROTEX", IEEE Conference on Intelligent Robots and Systems (IROS), Raleigh, USA, July 7-10, 1992

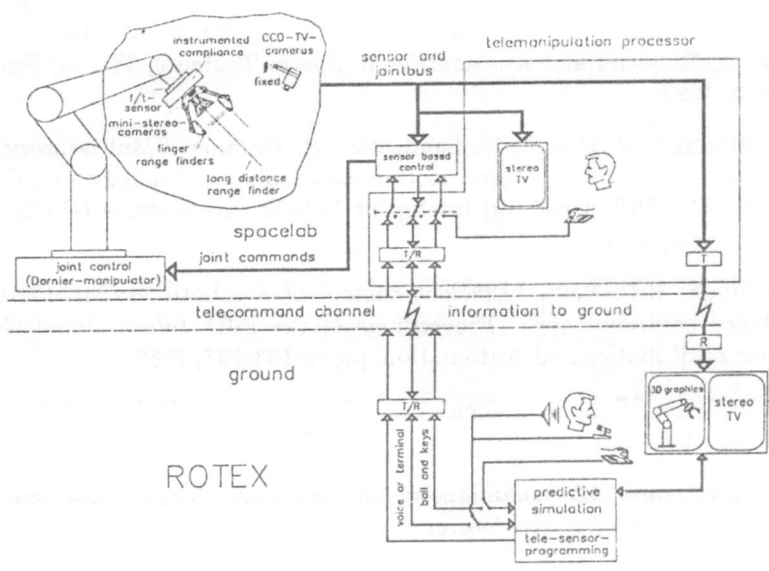

ROTEX

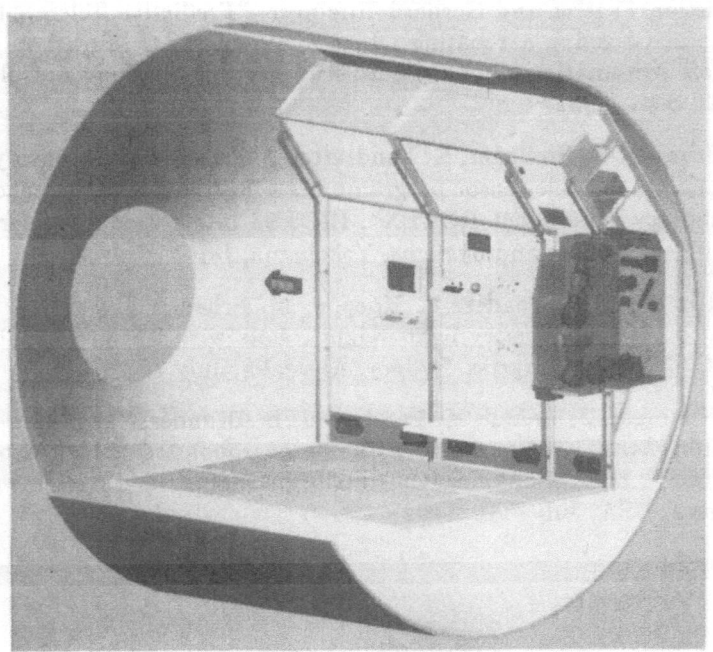

Fig. 1   Schematic representation of ROTEX (upper part)
and integration in spacelab (picture, courtesy of DORNIER)

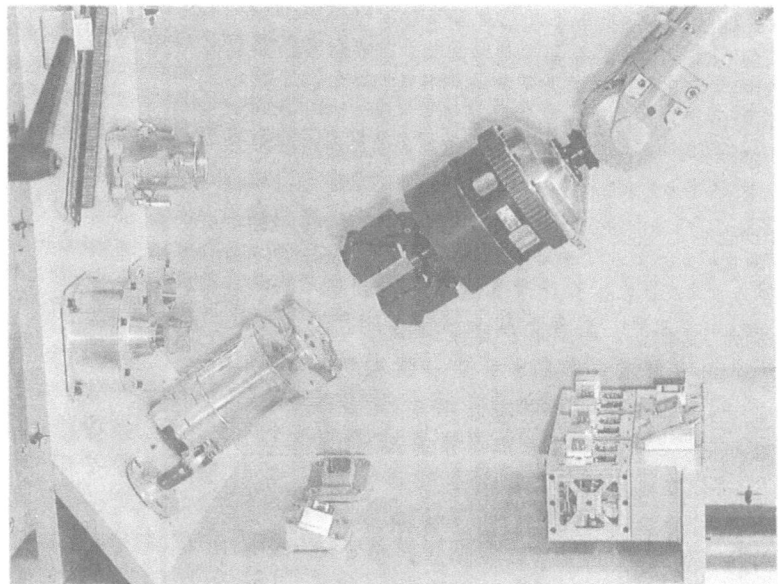

Fig. 2   ROTEX robot and experiment set-up in DLR laboratory,
where the multisensory gripper is below the three-part truss
structure just in front of the bajonet closure (the "ORU")

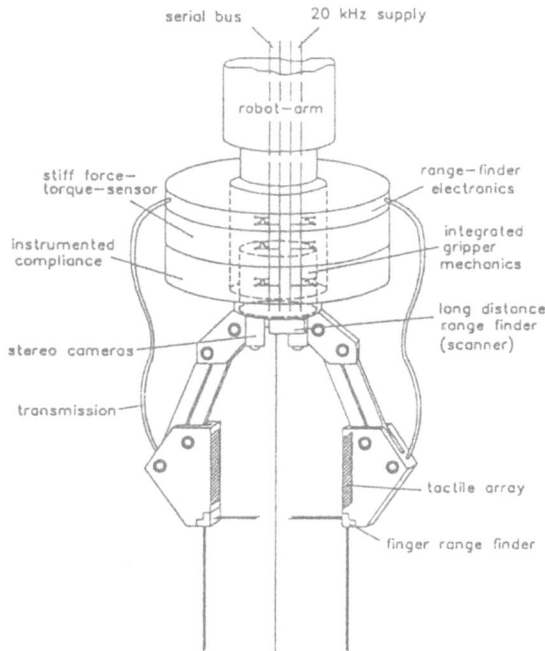

Fig. 3   Schematic drawing of the multisensory endeffector (gripper)

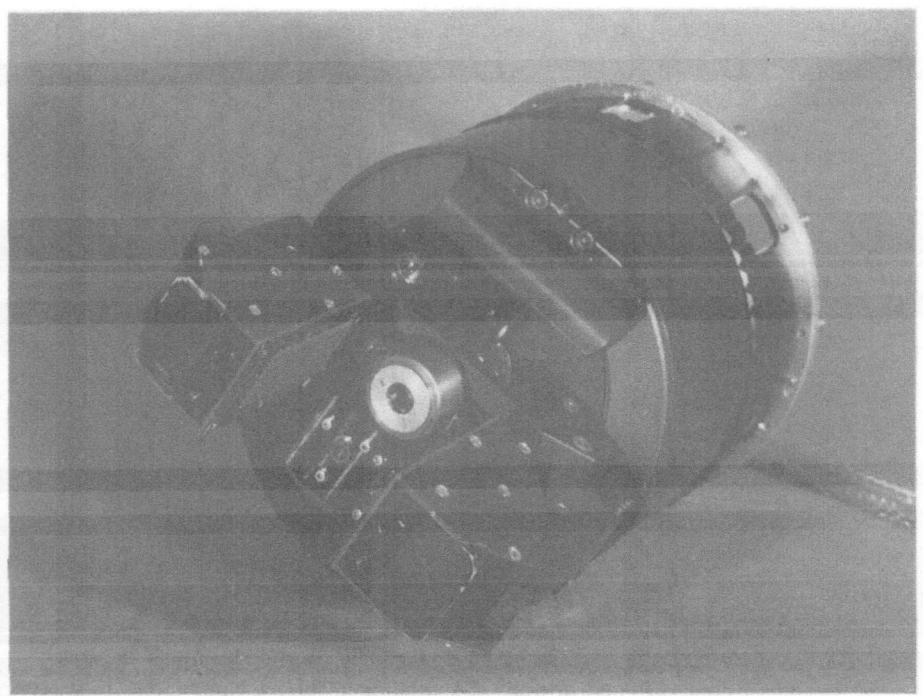

Fig. 4   The multisensory ROTEX gripper integrating 16 sensory systems

Fig. 5   The multisensory ROTEX gripper integrates
more than 1000 tiny SMD-components

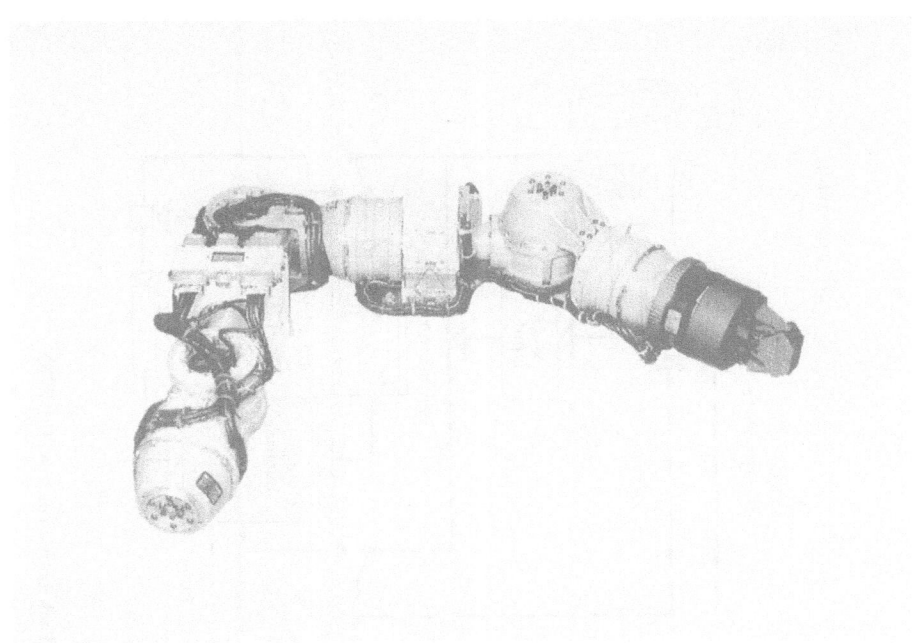

Fig. 6a   ROTEX flight robot built by DORNIER (courtesy of DORNIER)

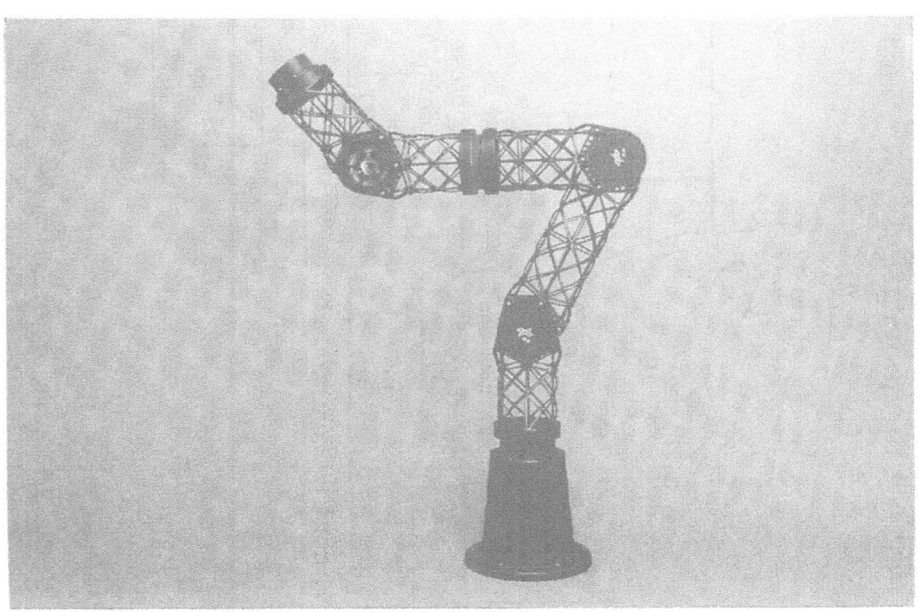

Fig. 6b   DLR-light-weight robot for astronaut training

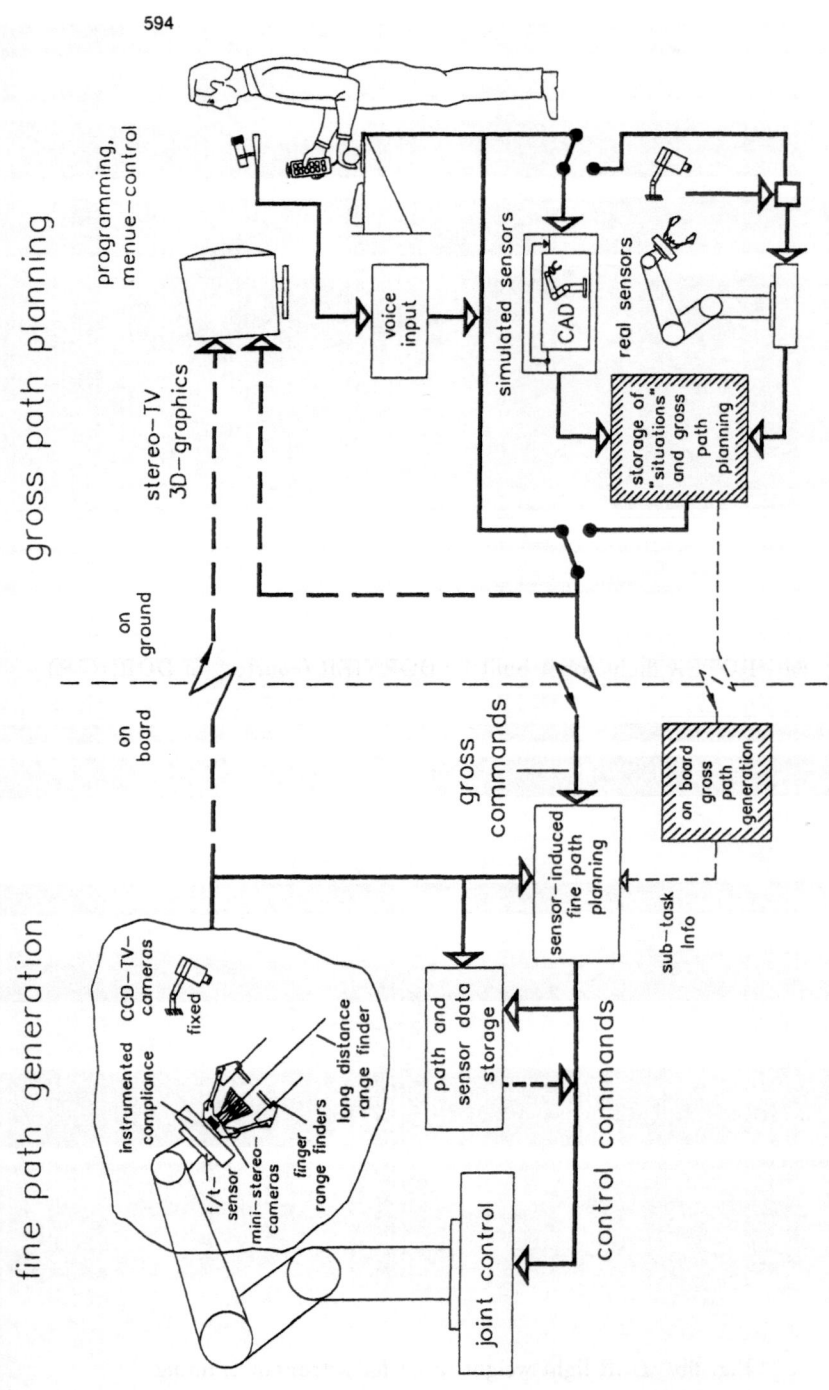

594

Fig. 7   Overall loop structures for the sensor-based telerobotic concept of ROTEX

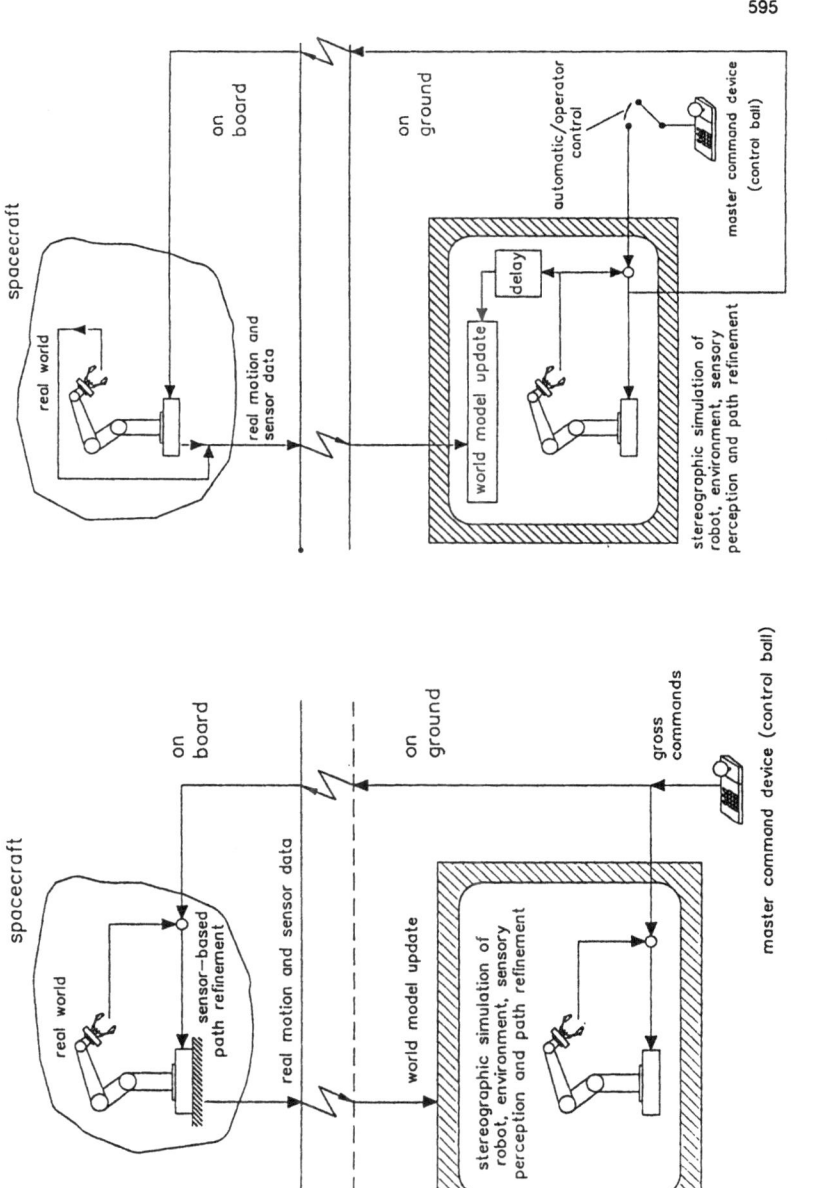

Fig. 8   Presimulation of sensory perception and path refinement in case of teleoperation from ground

a) local on-board sensory feedback
   (e.g. tactile contact)

b) sensory feedback via groundstation
   (grasping a free-flyer)

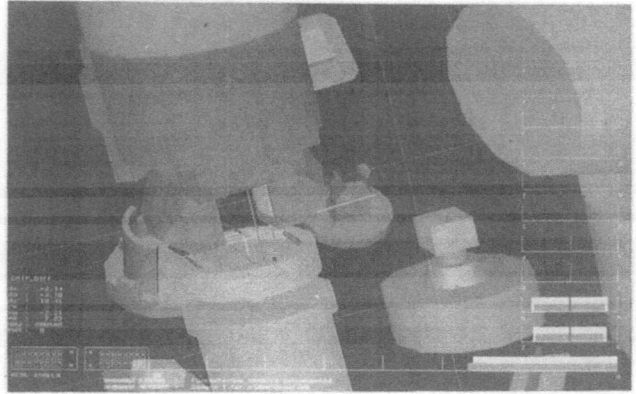

Fig. 9
Sensorsimulation: Range finder simulation in the "virtual" workcell environment. In addition to the 5 simulated rays out of the gripper (fig. 3) the bars in the right lower corner indicate the same simulated (bright) and the corresponding real (dark) range values as registrated by the real robot.

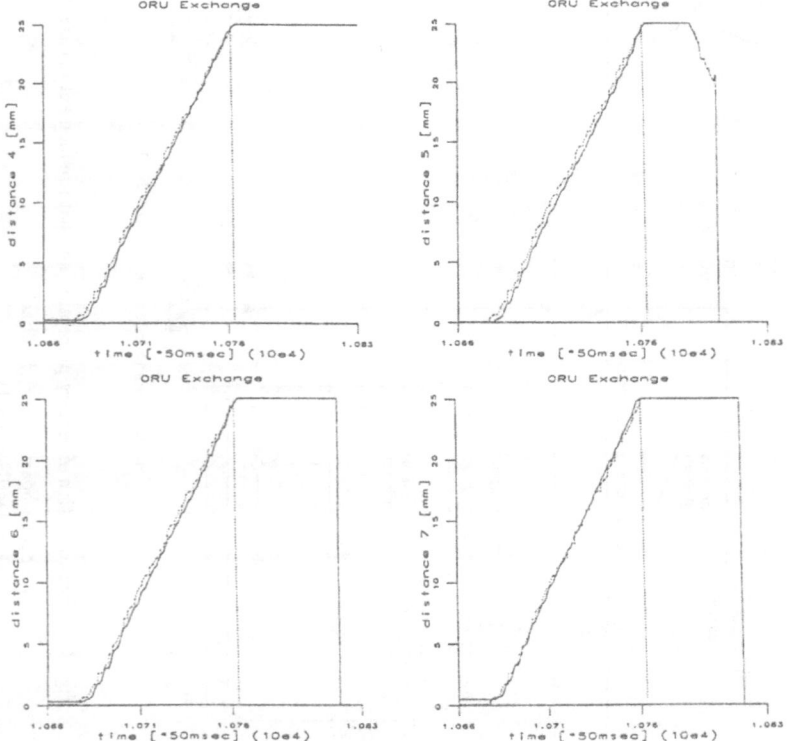

Fig. 10

Correlation between presimulated (for comparison delayed) and real sensory data (in closed loop each) was nearly perfect in ROTEX. These recordings of the four finger-range-finders pointing "downwards" were made during sensorbased teleoperation when removing from the ORU bajonet closure (fig. 9).

Fig. 11

Two subsequent TV-images out of one of the hand cameras shortly before grasping the free flyer automatically from ground. The dark areas at the left and right lower part are the gripper jaws.

Fig. 12   A part of the telerobotic ground station which was
used for training by the astronauts, too.

Fig. 13   Payload specialist Hans Schlegel tele-operating the robot
via stereo TV display and the 6 dof control ball (courtesy NASA)

Lecture Notes in Control and Information Sciences

Edited by M. Thoma